Airport Planning and Development Handbook

Other McGraw-Hill Books of Interest

Airport Planning and Development Handbook

A Global Survey

Dr. Paul Stephen Dempsey

Professor of Law
Director, Transportation Law Program
University of Denver

McGraw-Hill

New York   San Francisco   Washington, D.C.   Auckland   Bogotá
Caracas   Lisbon   London   Madrid   Mexico City   Milan
Montreal   New Delhi   San Juan   Singapore
Sydney   Tokyo   Toronto

Library of Congress Cataloging-in-Publication Data

Dempsey, Paul Stephen.
　　Airport planning and development handbook : a global survey /
Paul Stephen Dempsey.
　　　　　p.　　cm.
　　Includes index.
　　ISBN 0-07-134316-4
　　1. Airports—Planning.　I. Title.
TL725.3.P5D5　　　1999
387.7'36—dc21

99-30202
CIP

McGraw-Hill

A Division of The McGraw·Hill Companies

　2 3 4 5 6 7 8 9 0　AGM/AGM　9 0 4 3 2 1

ISBN 0-07-134316-4

The sponsoring editor for this book was Shelley Ingram Carr, the editing supervisor was Stephen M. Smith, and the production supervisor was Pamela A. Pelton. It was set in Garamond by Joanne Morbit of McGraw-Hill's Hightstown, N.J., Professional Book Group composition unit.

Printed and bound by Quebecor/Martinsburg.

McGraw-Hill books are available at special quantity discounts to use as premiums and sales promotions, or for use in corporate training programs. For more information, please write to the Director of Special Sales, McGraw-Hill, 11 West 19th Street, New York, NY 10011. Or contact your local bookstore.

This book is printed on recycled, acid-free paper containing a minimum of 50% recycled, de-inked fiber.

To my sons

Stephen Patrick
(born August 3, 1993)

Connor Josef
(born February 1, 1995)

and

Dylan Thomas
(born January 15, 1998)

That they might explore the world by passing through
the gateways described herein

Contents

Contents

6 Site Selection and Environmental Planning   *235*

7 The Air Side: Airfield Design and Construction   *271*

8 The Land Side: Terminal Design   *319*

Acknowledgments

The planners and designers of the facilities described in this book, as well as their airport authorities, deserve thanks for the time they took to answer this author's many questions and to provide invaluable insights as to how they went about their work. Their generosity in granting permission to reproduce the various graphic material sprinkled throughout this book has contributed to better understanding of the corresponding prose and is also appreciated.

Several individuals were particularly helpful in opening doors to enable this author to have access to key airport designers and planners. Professor Masao Sekiguchi of Komazawa University did this in Japan. Reinhard Zeiler of the Munich Airport Authority did this in Malaysia. Peter Trautman of the Munich Airport Authority did this in Germany. Jon Cross of the Federal Aviation Administration provided information and insight invaluable to portions of this book. Security experts Billie Vincent and Jalal Haidar also provided crucial insights. George Brewer, formerly of the Federal Aviation Administration; Gil Carmichael, former Federal Railroad Administrator; and Professors Andrew Goetz and Joseph Szyliowicz of the University of Denver Intermodal Transportation Institute reviewed portions of the manuscript and offered constructive criticism. To these fine men, this author is indebted.

He is also indebted to his tireless research assistants, John Ridgley, Stephanie St. John, and Sheri Straily, all Juris Doctor candidates at the University of Denver at the time they helped out with this book, and to his tireless secretary, Gregoria Frangas. Diane Burkhardt of the library at the University of Denver College of Law was extremely helpful in securing books, articles, and documents that became essential source materials for this book. Compiling material on such a global scale was a massive undertaking and could not have been accomplished without them.

This author would also like to thank the Hughes Research Foundation for its support in enabling him to visit many of the airports described in this book. Such support has manifested itself in two prior books (*Denver International Airport: Lessons Learned* and *Law and Foreign Policy in International Aviation*) and several scholarly articles.

The world is buzzing with airport activity. Airlines, their passengers, freight, and mail are demanding more capacity from land-based infrastructure. World economic activity and leisure travel are dependent on there being enough runways, gates, terminals, and baggage facilities to satiate demand. It is this author's hope that this book will make a small contribution to the development of the transportation infrastructure so essential to a global economy, and to making the world a global village.

Dr. Paul Stephen Dempsey

1

A Global Survey of New Airports and Expansion Projects

"Ninety percent of aviation is on the ground. Only 10% is in the air."—GLEN CURTIS, AIRCRAFT DESIGNER AND ENTREPRENEUR

"Transportation is a basic part of the economic/social/ cultural infrastructure, which affects the efficiency of all other business activities in a community and the quality of life of its residents. The ability of a city to retain existing industries, and attract new ones, is uniquely dependent upon the adequacy, convenience, and reasonable pricing of its airline service."[1]—MELVIN BRENNER, VICE PRESIDENT, TRANS WORLD AIRLINES

In Chap. 1, we examine:

- Major new airport and airport expansion projects around the world either planned, recently completed, or under way
- Projects totaling US$200 billion, including US$98 billion in airport activity in Asia, Oceania, and the Middle East; US$50 billion in North America; US$46 billion in Europe; US$4 billion in South America; and US$2 billion in Africa
- The projected growth rates of regional air transport markets
- Governmental efforts to provide the necessary infrastructure to lure economic growth and prosperity

Introduction

Airport development projects are some of the largest infrastructure projects a city, region, or nation may undertake. They are also

1

enormously capital-intensive ventures, often among a nation's most resource- and time-consuming public works projects. From initial conception to ultimate completion, the process of planning, financing, designing, and constructing a new airport, or expanding an existing one, can consume more than a decade of time and hundreds of millions of dollars.

Airports are an integral part of the commercial aviation system—the places where airlines and their customers converge. Airports are the portals through which airlines pass (from the air side) and passengers or freight pass (from the land side) to join in union in a movement of people and commodities from one corner of this increasingly shrinking planet to another.

In an accelerated attempt to keep pace with rapidly growing passenger and cargo demand for air transportation, governments the world over are devoting enormous economic resources to building new, or expanding existing, airports. New airport expansion and development is a by-product of the soaring growth in demand for air transportation, on the one hand, and the desire of communities to inspire economic prosperity, on the other. In a growing number of the world's villages, demand exceeds capacity. Airport congestion creates delay, inefficiency, and increased costs for the airlines which must land at and take off from these bottleneck facilities, as well as inconvenience, delay, and opportunity costs for their passengers. Congestion may manifest itself in terms of inadequate airport runway, apron, terminal, or surface transport capacity. An airport that needs more capacity can try to expand incrementally—adding another runway or two, paving more aircraft apron parking area, expanding the terminal, concourses, or number of gates, or adding new highway or rail lines, if space and financial resources permit. But expansion is often constrained by land, cost, and environmental considerations, coupled with political opposition, usually inspired by the NIMBY ("not in my backyard") syndrome.

The capital requirements of new airport infrastructure development are enormous. Both the airline industry's International Air Transport Association (IATA) and the U.N. International Civil Aviation Organization (ICAO) concur in their projections that, worldwide, US$250 billion will be spent for airports by the year 2010, of which US$100 billion will be required for the Asia-Pacific region alone.[2] The World

Bank projects that US$300 billion will have to be spent over the next decade to cope with an anticipated doubling of air traffic.[3] Thus, airport expansion and, more problematic, new airport development, face numerous obstacles.

In this chapter, we survey US$200 billion in major new airport and airport expansion projects around the world either recently completed, under way, or planned at the dawn of the new millennium. *But note: The present chapter is only a representative sample of airport projects, not an exhaustive compilation. More projects are under way than are summarized here.*

As we shall see in the next chapter, with the growth in passenger and freight demand, several major new "green field" airports are being built around the world (Table 2.1). Actually, some are not being built on green fields at all, but are built on brown landfill in bays and estuaries. While a few "green field" sites are available for wholly new airports, in most instances, existing airports must be marginally expanded to accommodate more flights, more aircraft, more passengers, and more freight. Yet most existing airports are hemmed in by urban development, which is expensive and politically difficult to raze in order to accommodate airport expansion.

In subsequent chapters, we examine the two principal motivations for airport development or expansion—growing demand outstripping inadequate capacity, and the desire of governments to build infrastructure to lure additional economic growth. Table 1.1 reveals the world's fastest-growing airports, as measured by total passengers.

The population demographics of developed nations are favorably disposed to fueling air traffic growth, given the maturing character of the population, the increased wealth of seniors, the time afforded by retirement, and the attractiveness of travel as a form of recreation. The increasing wealth of workers in developing nations may also fuel air travel, as disposable income grows. Globalization of trade also serves as a catalyst for business travel.

In the chapters that follow, we examine the indispensable role of governments and politicians in getting infrastructure projects under way. That is followed by an examination of the capital needs of airport development, considerations of finance and revenue. Another

Table 1.1. The World's Fastest-Growing Airports (by Total Passengers)

Airport	Growth (1996 vs. 1995), %	Rank
Cincinnati (CVG)	23.9	41
Salt Lake City (SLC)	14.2	36
Sydney (SYD)	14.0	39
Osaka (KIX)	13.9	42
Paris (CDG)	12.3	12
Seoul (SEL)	12.2	9
Atlanta (ATL)	9.7	2
Amsterdam (AMS)	9.7	20
Rome (FCO)	9.7	33
Bangkok (BKK)	9.5	32

important political issue to which a chapter is devoted is environmental concerns, particularly a potential stumbling block in developed nations. All that—the politics, need, capital, and environmental hurdles—must be jumped before ground can be broken on a new project. This book also examines the planning, design, and construction of new and expanded airports. It ends with the land-side needs of an airport, defined in terms of intermodal transportation access, and a review of the economic impact of new airport development and the symbiotic relationship between airlines and airports.

We begin our journey with a survey of major airport infrastructure development in Asia, for throughout much of the 1990s, it has been by far the fastest-growing air transportation region of the world.

Asia

If the nineteenth century was Europe's, and the twentieth America's, then the twenty-first century likely shall belong to Asia. Despite a plunge in Asian financial markets and currency valuation in 1997, much of the world's new airport infrastructure investment probably will be in Asia over the next several decades. Asia has half the world's people, an underdeveloped air transport system, and, throughout much of the last two decades, some of the most stunning economic growth rates on the planet. Some estimate that Asian governments will spend US$150 billion on new airports and airport expansion projects between 1997 and 2002.[4] One source notes:

The Asia-Pacific region is forecast to handle 400 million passengers yearly or over half the world's total by 2010, from 122 million passengers or one-third of global traffic in 1993. Spending on 16 current Asian airport projects [in 1996 exceeded] 50 billion US dollars, according to the Geneva-based Air Transport Action Group, but even this may not be enough to meet the anticipated travel and trade boom.[5]

In the past decade, the Asia-Pacific region has been the world's fastest-growing air transport market.[6] For much of the 1990s, many of the world's most profitable airlines were Asian. Traffic in this region grew from 30 million cross-border visitors in 1984, to 80 million in 1995; it is expected to double by 2005. Travel and tourism, the world's largest industry, by some estimates will employ 338 million people by 2005, one in five of whom will reside in the Asia-Pacific region.[7]

Projections of passenger growth in the Asia-Pacific market have been astounding. In the early 1990s, the Orient Airline Association predicted 7.5% annual growth for the Asia-Pacific region through the year 2000. The International Air Transport Association predicted between 7 and 8.6% through 2010.[8] The Organization for Economic Co-operation and Development (OECD) predicted inter-Asian traffic growth of between 8 and 9% over the next two decades. The U.N. International Civil Aviation Organization predicted between 9.3 and 10.8% between 1993 and 1995. McDonnell-Douglas predicted 9.7% through the year 2010. And the People's Republic of China (PRC) was expected to enjoy traffic growth in the range of 13.6% and 14.7%.[9]

More recent estimates anticipate an average 9.5% growth per year in the Asia-Pacific region through the year 2010.[10] Its share of total global scheduled passenger traffic is expected to grow from 25% in 1985 to 51% in 2010, thereby displacing North America as the world's busiest commercial aviation market.[11] By 2010, it is anticipated that the Asia-Pacific region will account for 398 million passengers a year, or about half the world's total.[12] In 1995, Asia-Pacific had only 112 million passengers, or 35% of the world's total.[13] One source has summarized the explosive growth in air traffic predicted for this region during the next several decades:

IATA has forecast that the total scheduled traffic to, from and within Asia-Pacific (and this does not include the expanding charter market) will grow from a 1990 figure of 87.3 million passengers to 132 million by 1995 and to 189 million by the end of the century. In other words, it will more than double within the normal gestation period of a new airport which is about 10 to 12 years. Which in turn means that either airport capacity has to double in that time, generating more departure slots, parking bays, and airport terminal space or traffic growth will be severely stunted, thousands of dollars are going to be wasted in holding patterns and passengers are going to find travel in the region a very frustrating experience.[14]

In 1998, IATA revised its 5-year Asia traffic growth projections down from 7.7% to 3.7%.[15] IATA's new projections are for 30 million fewer passengers than earlier predicted for the year 2001. On the other hand, the U.S. Federal Aviation Administration anticipates that the number of passengers flying between the United States and Asia could double by 2009.[16]

Total travel and tourism in the Asia-Pacific region increased 167% between 1985 and 1995. The number of passengers flown across the Pacific increased at an 8.6% annual rate between 1982 and 1992, compared with 5.2% on all other routes.[17] Asian airlines will place between 40 and 50% of an estimated US$150 billion in new aircraft orders between 1997 and 2001 to cover passenger demand which is estimated to grow at a rate of 7% a year for the next two decades.[18]

IATA's John Meredith notes, "Without drastic improvements, the forecast growth will soon overwhelm the capacity of Asia-Pacific's aviation infrastructure."[19] We now review many of the major new infrastructure developments in Asia, on a nation-by-nation basis, alphabetically:

Brunei

Brunei International Airport is spending US$120 million on airport expansion.[20] The projects include an aircraft maintenance center, re-fueling center, cargo village, and an "airport city" consisting of shops, hotels, restaurants and such.[21]

Cambodia

Beginning in the 1970s, political instability severely dampened passenger and cargo demand for travel to Cambodia, thereby stunting its economic growth. But with Pol Pot's Khmer Rouge at bay, traffic at Phnom Penh Pochentong Airport grew from 404,000 in 1993 to 650,000 in 1995, to 800,000 in 1996, and a projected 1,000,000 in 1998. Nonetheless, political instability may continue to impede economic growth and, therefore, air passenger and cargo traffic.

The Societé Concessionaire de l'Aeroport (SCA), a joint venture between the French construction company GTM-Enterprise and a Malaysian firm, Muhibbah Masteron Cambodia, was granted a 20-year concession to manage and develop the airport. SCA subcontracted airport administration to Cambodia Airport Management Services, whose shareholders are Aeroports de Paris (60%) and Malaysia Airports Berhad (40%). SCA added a terminal for domestic and international traffic, and opened an expanded departure lounge with capacity for 1 million passengers per year. It is investing US$120 million in a new 110,000-ft international terminal building scheduled for completion in 1999, and a 3600-m runway to be completed in 2002. The terminal will have telescopic bridges, state-of-the-art technology, lounges, restaurants, and a large shopping area.[22] The airport should be able to accommodate 1.7 million passengers by 2001. At Sianoukville, an upgrade project is under way to construct a 2.5-km runway to accommodate Boeing 737 size aircraft, and build a terminal, communications tower, and parking area.[23]

China

In 1991, China's passenger and cargo volume grew by 28%, in 1992, 33%, and in 1993, 20%. This placed an enormous strain on the capital requirements of the commercial aviation sector, and caused serious safety and operational problems. The People's Republic of China (PRC) concluded that its airlines and airports need capital and operational expertise, and in 1994, China opened its commercial aviation market (airlines and airports) to foreign investors.[24] The Civil Aviation Administration of China (CAAC) designated two of its airlines for foreign investment/operations, allowing foreign investment up to 35% and foreign voting rights up to 25%.[25] Among the most intriguing opportunities that appear to be on the table is the possibility of setting up a joint venture to build an airport in China. Construction costs in mainland China are a fraction of what they are anywhere else in the world.

China has undertaken a massive civil aviation infrastructure effort to build eight new airports and expand 12 existing facilities—including the nation's capital airport at Beijing.[26] Major airport projects in China also are planned or under way in Shanghai, Wuhan, Liangjiang, Haikou, Hangzhou, and Nanchang. By 1995, 43 airports were completed, under construction, or in the planning stages.[27] Between 1990 and 1998, 70 new airport terminal and expansion projects have been approved by the Chinese government, 11 new airports have opened, and 35 have completed renovation projects.[28] New airports recently opened include Fuzhou Changle, Guilin Liangjiang, Nanjing Lukou, Yichang Sanxia, and Zhengzhou Xinzheng. Other airport projects planned or under way include a new Mianyang Nanjiao Airport, a new Shanghai Pudong International Airport, a new Guangzhou Hua Du International Airport, and expansions at Beijing Capital, Xiamen Gaogi, Shenzhen, Shenyang Taoxian, Ghangsha Huanghua, Changchun Dafanshan, Chengdu, Haikou, Nanchang, Lanzhou, Kunming, Urumqi, Changbei, Meilan, Kaikou, Zhongchuan, Harbin, and Dalian Zhoushuizi.[29]

China has but 140 airports with commercial service, with only 12 capable of landing a 747; in contrast the United States has 500 commercial airports.[30] Seventy percent of China's air traffic is in the nation's eastern area which stretches from Harbin in the north to Kunming in the south.[31] By 2000, eastern China will have 22 new airports, while 10 more will have been expanded and upgraded.[32] China reportedly expects to build more than 40 new airports in the 1997–2000 period.[33] It has been estimated that China will need 400 new airports over the next decade.[34]

In China, civil and military airspace is separated, with approximately 85% of China's airspace being controlled by the military.[35] Between 1991 and 1995, the Chinese government spent nearly US$400 million to upgrade its air traffic management and control system.[36]. The Chinese government could spend as much as US$400–500 million on air traffic control equipment between 1997 and 2001, and possibly up to US$1.25 billion by 2005.[37] Total investment in airport projects in China is approximately US$12 billion.[38]

The following details the major airport projects in the PRC, on a province-by-province basis.

Beijing

Capital International Airport is undergoing an expansion that will include a new 268,000-m^2 H-shaped terminal building, aprons covering 480,000 m^2, and a five-story car park capable of accom-

1.1 *An air-side view of Beijing Capital International Airport, which is undergoing expansion, as are may airports in the People's Republic of China.* Air China.

modating 5000 vehicles.[39] The flagship project of a civil aviation infrastructure push across the country, the Capital expansion will cost a total of US$1.1 billion and be jointly financed by Japanese government and State Development Bank loans.[40] In 1996, the airport, built to handle 3 million passengers annually, handled more than 16 million passengers, with landings and take-offs continuing to increase by more than 20% annually.[41] Upon completion, scheduled for October 1999, the airport will be capable of handling 35 million passengers and over 780,000 tons of freight annually[42] (see Fig. 1.1).

Fujian

The Chinese government plans to allocate about US$24 billion over the next 5 years to develop infrastructure in the eastern and southern sectors of coastal Fujian. Among the major projects are the construction of the Changle International Airport in Fuzhou and the renovation of Gaoqi Airport in Xiamen City.[43]

Fuzhou Changle International Airport, in the capital city of Fuzhou, became operational in July 1997.[44] The airport, completed at a cost of US$390 million,[45] will handle all types of aircraft and accommodate 6.5 million passengers and 200,000 tons of freight per year.[46] A joint venture to fund the project was set up between Hume Industries (Malayasia) Bhd, Guoco Investment (China), and the Fujian Province Government. The agreement provides that the Fujian Province Government retain 50% ownership in the joint venture, with the balance being split among Hume and Guoco Investment. The joint venture firm will operate and fund the airport for 50 years.[47] Air traffic volume in Fuzhou has been growing at an annual rate of greater than 30%, and the new airport is designed to ease the burden on the city's other airport, Yixa.[48]

Xiamen   Xiamen Airport is under expansion.[49] Built in 1983, and capable of handling 350,000 passengers annually, the Gaoqi Airport saw more than 2.5 million passengers in 1993.[50] An expansion program to handle the increase included the extension of the runway to 3200 m, construction of a 70,000-m^2 terminal building, expansion of the aircraft parking apron, and construction of a 2700-m taxiway.[51] Hong Kong's Cathay Pacific Airways will take a one-third equity interest in the US$142.5 million expansion at Xiamen, which ranks as China's fourth largest airport in terms of passengers handled, size of facilities, and number of international services.[52] Renovation efforts were completed in early 1997.[53]

Guangdong

Guangzhou   A 20 billion yuan (US$2.4 billion) project is under way to relocate the existing Baiyan Airport 28 km from central Guangzhou in suburban Huadu city. The project is scheduled to begin in late 1999 with completion date set in 2002.[54] The airport project will include three runways and a 280,000-m^2 terminal,[55] which will make it the China's largest and airport.[56]

Shenzhen   Shenzhen, near Hong Kong, has doubled its airport's capacity.[57] A US$460 million project is under way to upgrade Shenzhen Airport into an international air traffic hub with parallel runways, including a 3900-m runway to accommodate 600-ton aircraft and 43 aprons to accommodate cargo aircraft.[58] Set for completion in 2000, the plan also includes two large passenger terminals, a cargo terminal facility, and a modern air traffic navigation system, as well as fuel and express mail facilities.[59]

Zhuhai   A new airport is being built at Zhuhai.[60]

Zhejiang

Fuyang Fuyang Airport is under expansion.[61] The central government has invested more than 20 billion RMB yuan (US$ 2.4 billion) in major infrastructure development, which includes the expansion of the existing airport at Fuyang.[62]

Hangzhou

The planned Hangzhou International Airport will be located on land reclaimed from the sea in Xiaoshan City,[63] across the Qiantang River from Hangzhou. The new airport, designed to ease the congestion at Jianqiao Airport in Hangzhou, will have two 3600-m runways and be able handle 8.6 million passengers annually. The project cost of 3 billion yuan (US$362 million) will be financed by the municipal and provincial governments.[64]

Jiangsu

Nanjing At Nanjing, 2.4 billion yuan (US$290 million) was spent building the new Lukou International Airport,[65] with 3600-m runways capable of handling jumbo Boeing 747-400s. The airport opened in June 1997.[66] Lukou International Airport, located 35 km from downtown Nanjing, includes a 3600-m runway which can handle fully loaded Boeing 747-400s, and is expected to become one of the major freight handling airports in the region.[67] The airport can handle 69,600 flights, 8.64 million passengers, and 120,000 tons of freight annually. A second phase of development will bring annual capacity to 40 million passengers and 1 million tons of freight.[68] Funding for the project was provided by the Civil Aviation Administration of China (CAAC), Jiangsu Provincial government, Nanjing City government, and some loans.[69] The airport will also be used as a back-up for the Hongqiao Airport in Shanghai.[70]

Shanghai At a cost of 13.3 billion yuan (about US$1.6 billion), Pudong International Airport is anticipated to be China's flagship airport.[71] When completed, the airport will have sufficient capacity to accommodate 20 million passengers and 500,000 tons of cargo a year.[72] In 1998, Shanghai completed construction of China's longest runway.[73]

Xuzhou Guanyin Airport, located 40 km from Xuzhou, opened in October 1997.[74] The new airport contains a 20,000-m² terminal building with five boarding bridges and a 3400-m runway.[75] The Xuzhou municipal government provided most of the US$105 million construction costs.[76]

Yangzhou   Shiqiao Airport, the city's first, became operational in late 1993.[77] The airport, which can accommodate light planes, was funded by moneys raised collectively by the farmers of Yangzhou.[78]

Hong Kong

Hong Kong's Chek Lap Kok Airport, built on an island landfill (near Lantau Island), opened in 1998.[79] The cost, including road and rail, and the longest road and rail suspension bridge in the world, is estimated at 156,364 billion Hong Kong dollars (more than US$20.1 billion).[80] The airport is planned to handle 35 million passengers a year.[81] Chek Lap Kok replaced Hong Kong's congested Kai Tak Airport.[82] Meanwhile, airlines are fearful that landing charges will be more than twice what they must pay at Kai Tak.[83]

India

By 1994, India had "opened up its aviation industry to new domestic carriers."[84] Major airport projects in India are planned or under way in Bombay, Calcutta, New Delhi, and Madras.

Karnataka

Bangalore   Plans to build a US$400 million international airport near Bangalore have been stalled by the Indian government's reluctance to approve a build-own-operate (BOO) scheme that would allow private ownership of the facility.[85] A memorandum of understanding was signed in December 1995 between the Karnataka state government and a consortium led by Tata Industries calling for the construction of an international airport on a BOO scheme.[86] The government has since decided not to grant private ownership saying that it goes against the existing policy of the Civil Aviation Ministry.[87] Another issue, the closure of a nearby domestic airport upon completion of the new facility, also remains unresolved. A committee has been formed to examine the feasibility of continuing the project on a build-operate-transfer (BOT) basis.[88]

Maharashtra

Bombay (Mumbai)   Sahar International Airport is undergoing a project that will expand its terminals, increasing annual capacity to 7.5 million passengers. The lack of available land hampers any further expansion of the facility.[89] A second international airport servicing Greater Mumbai is being planned for the Rewa-Mandwa region in the Raigad district.[90]

Navi Mumbai A domestic airport has been proposed by the Maharashtra government on a site near Kopar village in Navi Mumbai. The proposed airport will have a final runway length of 2700 m and accommodate Airbus 320 and Boeing 737 type aircraft.[91] The project will include a terminal building, five aprons, 16 hangars, a fire station, a fuel farm, a cargo terminal, an airport colony, hotels, and a shopping plaza.[92] The cost of the project is estimated around $490 million.[93]

Sinnar An international commercial airport is in the development stages near Nashik in northern Maharashtra. The proposed cargo facility is expected to reduce the freight load on Sahar International in Mumbai and ensure development of the North Maharashtra region.[94]

Indonesia

Among Asia-Pacific nations, Indonesia is second only to Papua New Guinea in its dependence on air transportation.[95] Indonesia is expected to have the fastest annual growth in international traffic during the 1990s, followed by Malaysia and Australia.[96] The government is planning to construct new airports at Medan and Padang in Sumatra (at a cost of US$358 million and US$174 million, respectively), and in central Lombok (at a cost of US$115 million).[97] The new Medan airport will be situated on a 1350-ha site east of Medan. Batam's Nang Nadim International Airport is expanding to two 4000-m runways, the same as Singapore's Changi.[98] Indonesia has also launched a US$276 million Eastern Islands Air Transport Project to upgrade airport infrastructure at Pontianak (capital of the West Kalimantan province) and Sorong (a major economic center of Irian Jaya), and to build a new airport at Samarinda (capital of the East Kalimantan province).[99] At Solo, on Sumatra, US$342 million will be spent upgrading the existing airport, building a new passenger terminal, extending the runway, building a parallel taxiway, and expanding the cargo facilities.[100] The government will also spend US$115 million on a new airport at Sengkol, on the island of Lombok.[101] According to one source:

> *Jakarta has 13 international airports. Three—Jakarta, Medan, and Ujang Padang—will become main hubs. Twenty-two secondary airports have been earmarked for development. Joint ventures between Indonesian and western companies are springing up to build new airports and expand runways of existing ones to accommodate the N-250, an Indonesian-built turbo plane, and larger aircraft.[102]*

Batam Island

Located 20 km offshore from Singapore, Batam Island opened a new international airport in 1995.[103]

Java

Jakarta   New airports are planned in Kuala Namu and Lombok Tengah.[104]

Kalimantan Island

PT Kiani Kertas, a paper and pulp plant, announced that it would build a US$20 million Makajang Airport with a 2700-m runway by the end of 1997.[105]

Japan

Japan spent US$24 billion between 1992 and 1997 on airport expansion.[106]

Aichi

Construction of the US$7 billion Chubu New International Airport regional airport is scheduled to begin in late 1998 on a man-made island located near Tokoname, and approximately 30 km from the site of the 2005 World Exposition in Japan.[107] The airport is to be ready for use in time for the Expo.[108]

Kitakyushu

A new state-of-the-art airport, built on a landfill site, is to open in 2005.[109] It will be "one of just four airports in Japan able to handle around-the-clock landings by jumbo jets when it opens."[110]

Kobe

A new US$2.3 billion airport is planned to be built by 2005 on reclaimed land on the shore of Osaka Bay. The airport will have a single 2500-m runway, capable of handling 60 take-offs and landings daily. The city projects 3.4 million passengers will use the airport in its first year. With roads and other facilities, the total cost of the project likely will exceed US$7 billion.[111]

Okayama

Okayama Airport was recently expanded.[112] Development of a freight terminal with facilities for storage and disposal of goods was undertaken, and offices for distribution companies will be built adjacent to the airport. This comes after the government declared the city of Okayama a Foreign Access Zone (FAZ). Okayama Airport is expected to cope with increased imports from other Asian countries.[113] It also plans to expand its runway from 2500 to 3000 m by March 2001.[114]

Osaka

Osaka's Kansai International Airport (KIA), which opened in 1994, achieved several firsts for Japan—it is the first Japanese airport to operate 24 hours a day, it is the first to be constructed and run by a private company, and it is the first to have been built on a man-made island in the ocean. Kansai International Airport withdrew a pledge to stay open 24 hours a day in an effort to reduce the financial burden of the new facility.[115] Built offshore from the city of Osaka, at a cost of 1.44 trillion yen (approximately US$15 billion), the new airport will be connected to the city by a bridge 2 mi (3.2 km) long.[116] Though KIA was expensive, a third airport serving Tokyo could cost US$23–35 billion.[117]

Tokyo

Tokyo's Narita Airport has historically been northeast Asia's dominant air hub. At 21.5 million passengers in 1995, it is near maximum capacity, for the airport has only a single runway.[118] A second runway is scheduled for completion at Narita Airport around 2001.[119] Tokyo's Narita and Haneda airports are critically short of space. Tokyo's Narita Airport, built in the 1970s, is near capacity, and hemmed in because of a decades-long dispute with nearby farmers. A 1966 plan called for Narita to have three runways; three decades later, because of a dispute with local farmers, it still had but one.[120] Until 1997, a farmer who owned a rice paddy in the middle of Narita's Terminal 2 apron vetoed airport expansion. Some predict Japan will be compelled to build a new airport by filling in part of Tokyo Bay.[121] The government is studying a report recommending construction of offshore airport islands in reclamation projects near Tokyo and Nagoya.[122]

Malaysia

Kedah (region)

A new international airport at Kedah was delayed due to budget cuts.[123]

Sarawak (region)

Bintulu   Bintulu New Airport is estimated to cost RM260 million (US$68 million), with a 2-year construction plan once the tender offers are made sometime in late 1997.[124]

Kuching   A new international airport is planned.[125]

Selangor (region)

Sepang   The Malaysian government is building an enormous airport 50 km south of Kuala Lumpur in Sepang.[126] Kuala Lumpur International Airport (KLIA) was "designed to be the most sophisticated in the region, [using an] airport-in-the-forest and forest-in-the-airport [concept] to be a monument of Malaysia's commitment to sustainable development."[127] KLIA is the largest airport in Southeast Asia, by acreage, with a control tower said to be "one of the tallest in the world."[128] Built to handle 25 million passengers per year,[129] the estimated cost is between RM8 and RM9 billion (US$2.1 to 2.3 billion) for the first phase.[130] The new airport features a 4-km (2.5-mi) runway, an aboveground train to shuttle international passengers from check-in counters to their gates at a midfield concourse, and an underground, state-of-the-art baggage-handling system. The terminal corridor containing duty-free shops spans almost a kilometer.[131] The new airport was scheduled for completion in time for the Commonwealth Games, hosted by Malaysia in 1998.[132] Unlike many other large projects that have been deferred due to the recent currency weakness, this project will "proceed as scheduled."[133] Malaysia is also contemplating building a Northern Regional International Airport at Penang, at a projected cost of between US$150 million and US$800 million.[134] The airport would have two runways on a 1200-ha airport site. It would serve the northern states of Perak, Penang, Kedah, and Perlis, which are responsible for 65% of Malaysia's semiconductor production.[135]

Langkawi (island of northwest coast of Malaysia)

A runway extension has been completed to accommodate Boeing 747s.[136]

Myanmar (Burma)

The government of Burma has incorporated construction of five international airports into the present 5-year economic plan. The airports are planned for Rangoon, Mandalay, Bago, Pagan, and either Taunggyi in southern Shan state or Than-dwe in western Burma.[137]

Mandalay

Mandalay International Airport is being built by Italian-Thai Public Development Co. Ltd.[138] The new airport,[139] built at a cost of US$357 million, will accommodate wide-body jets and have a peak handling capacity of 2000 passengers per hour.[140] The project, financed in part by a $150 million loan from the Thailand government, is expected to be completed in 1998.[141]

Rangoon

Work on extending the runway to handle jumbo jets was resumed at Yangon International Airport in 1997. Initially begun in 1987, work was halted after the Japanese government withdrew loan assistance because of the ruling junta's suppression of prodemocracy demonstrations in Burma.[142] In 1998, despite protests by the prodemocracy campaigners from the former Burma, Japan went ahead in lending US$11.6 million for the project. Completion is expected sometime in late 1998.[143]

Pakistan

At Lahore Airport, a new terminal and apron, a taxiway, and related infrastructure are planned to be completed by 1999 with this US$250 million expansion.[144] Karachi Airport was expanded in 1994.[145]

The Philippines

The Philippines has 207 airports, of which 87 are owned and operated by the national Air Transport Office, and four (Ninoy Aquino, Mactan-Cebu, Clark, and Subic Bay) are run by independent airport authorities.[146] In the Philippines, six domestic airports were being converted to international airports, the most significant of which is Mactan International Airport on Cebu Island.[147] The former Clark U.S. Air Force Base outside Manila also is undergoing a US$3 billion refurbishment to become a civilian passenger and cargo airport.[148]

Several other airport expansions are under way in the Philippines, including a US$293 million expansion and improvement of Manila's Ninoy Aquino International Airport, and US$297 million in projects to improve four regional airports (Zamboanga, Puerto Princesa, Laoag, and Cotabato) and six local airports (Jolo, Sanga-Sanga, Dipolog, Butuan, Tuguegarao, and Pagadian).[149] The airports of six municipalities (Puerto Princesa, Cotabato, Sanga-Sanga, Dipolog, Butuan, and Pagadian) in the southern Philippines will be upgraded in a US$167 million project.[150] The project includes "upgrading and extending runways, improving air navigation and safety facilities, expanding cargo facilities, constructing new passenger terminals, rehabilitating existing facilities and providing new equipment."[151]

Cebu Island

The Mactan (Cebu) International Airport was opened in the mid-1960s as a replacement for the old Lahuq Airport in Cebu City.[152] Financed in large part by a loan from the Japanese Overseas Economic Cooperation Fund, a second terminal building was added and the runway was expanded to handle Boeing 747-400s in 1996.[153] The second busiest gateway in the country after Manila, Mactan achieved its annual capacity of 3 million in 1995.[154] But traffic turned south in 1998 with a labor strike at Philippine Airlines and a deterioration in tourist traffic as Japanese and other Asian currencies weakened.[155] (See Fig. 1.2.)

General Santos City

General Santos Airport is a newly completed international airport.[156]

Manila

A second terminal at Ninoy Aquino International Airport costing US$160 million, completed in 1997, will accommodate 9 million passengers per year.[157] An international terminal (terminal 3),[158] with

1.2 *Mactan International Airport at Cebu is the Philippines' second busiest airport.* Photo by P. S. Dempsey.

planned capacity of 15 million passengers, is to be completed by 2001 at a cost of US$520 million.[159] In September 1995, DHL leased a vacant hangar from the Philippine Air Force at Villamor Airbase adjacent to the airport. DHL's cargo hub at Manila is equipped with more than US$3.5 million in state-of-the-art sorting equipment. It is part of an overall US$500 million investment program in Asia, which includes 5600 m^2 of warehouse space, 605 m^2 of office space, and 17,700 m^2 of aircraft parking space at Manila.[160]

Subic Bay

At Subic International Airport, a new US$12.6 million terminal was constructed in 9 months.[161] The departure of U.S. armed forces from the Philippines left that nation with facilities having attractive development opportunities. One was the former Clark Air Force Base, a 60,000-ha facility northwest of Manila. The government created the Clark Civil Aviation Complex (CCAC) on 4400 ha, which is designed to be the Philippines' premiere gateway international airport and the region's leading cargo center. The civil aviation support complex will occupy half of CCAC's 4400 ha, while a multimodal transportation center and commercial and industrial development will occupy the rest. Projected to cost between US$900 million and US$1.2 billion, Phase I of the CCAC project envisions a single runway capable of handling between 15 and 20 million passengers annually. Phase II is projected to be operational in 2006; it will consist of two parallel runways capable of handling 25 million passengers. Phase III, operating after 2016, will handle 50 million passengers per year.[162] FedEx opened an intra-Asia cargo hub in September 1995.

Singapore

In 1996, Singapore's Changi Airport handled 24.5 million passengers, up 5.7% from the preceding year, and 1.19 million tons of air cargo, up 7.7% from the preceding year. Traffic at Changi is expected to double and air freight volume triple over the next dozen years.[163] Long the most highly regarded airport in southeast Asia, Changi Airport is spending US$430 million to expand Terminal 1 by adding 14 new gates by 1999. At a cost of US$330 million, expansion of Terminal 2 raised airport capacity to 23 million passengers per year. A new Terminal 3, to be built on reclaimed land and completed in 2004, will have sufficient capacity for an additional 20 million passengers.[164,165] When completed, total passenger capacity at Changi will be 64 million passengers.[166] A fifth US$54 million cargo building is under construction, which will add 35,000 m^2 (376,600 ft^2), or 50% more capacity by 1998.[167] Also being added is a new airport hotel.[168] Singapore is hub to Singapore Airlines.

Changi Airport was voted one of the world's top two airports by 55,000 long-haul passengers polled by IATA.[169] (Britain's Manchester Airport shared the honor.) According to Communications Minister Mah Bow Tan, "To maintain Changi's position as a premier hub airport, and to enable us to continue to offer service to our airport users, the government has decided to start work on Terminal 3 before full capacity is reached in Terminals 1 and 2."[170] Construction is scheduled to run between 1999 and 2003 at a cost of 1.5 billion Singapore dollars (US$860 million). The terminal is designed to handle "600-seat 'super jumbos' in the 21st century."[171] Adding 20 million passengers a year to Changi's capacity, the new terminal will increase total capacity to 64 million."[172] While passenger growth is expected to double, air freight volume is expected to treble.[173] Work is also under way on a fifth cargo building, expected to cost 75 million Singapore dollars (US$43.4 million); it will add 35,000 m^2 (376,600 ft^2) or 50 percent more space by early 1998.[174] (See Fig. 1.3.)

1.3 *Singapore's Changi International Airport is one of the world's most modern and highly regarded. More than 60 airlines link it to more than 130 cities in 50 countries around the world.* Civil Aviation Authority of Singapore.

South Korea

Korea's major airports are Kimpo International, 17 km from Seoul, Kimhae International, 28 km from Pusan, and Cheju International, on Cheju Island.[175] Scheduled for opening in 2000, Seoul's Inchon International Airport will compete with Japan's Narita and Kansai as a regional hub for northeast Asia. Korea is spending some US$5 billion on the new Inchon International Airport near Seoul, which will have sufficient space for four runways and capacity for 100 million passengers per year at full build-out. Korea is also spending US$800 million on Muan County Airport, at the southern tip of the Korean peninsula, and US$820 million on airport improvement projects around the nation.[176] These include construction of a new international passenger and cargo terminal at Kimhae International Airport at Pusan, a new cargo terminal at Cheju Airport, an expanded passenger terminal at Ulsan Airport, a new terminal and runway at Yeosu Airport, the conversion of air force bases at Changju and Wonju to civilian airports, and upgrading facilities at Uljin and Yangyang airports.[177]

Sri Lanka

Opened in 1957, Colombo's Bandaranaike International Airport replaced an older airport which had opened in 1938. It has been expanded incrementally over the years, with a new runway opened in 1965 (the existing runway was transformed into a taxiway) and a terminal building opened in 1968. The terminal was expanded in 1976, and again in 1988. Annual passenger volume increased from 152,000 in 1972, to 2.3 million in 1997.

A US$15 million Phase II development plan to meet anticipated demand by 2010 is under review. It would add aprons and a pier with eight gates with loading bridges. It would also expand the terminal to allow a relocation of the duty-free area so as to reclaim the area conserved from the passenger waiting area. Phase II would also add a third air cargo village, with 12 runways, and new instrument landing technology. After maximum capacity was reached in 2010, or thereabouts, the airport might be able to expand onto an adjacent air force base.[178]

Taiwan

Taipei has begun a major expansion of Chaing Kai-Shek International Airport.[179] A new terminal was begun in 1991, suspended for 2 years, then restarted, and completed in 1997, giving the airport a capacity of 27 million passengers.[180] A second terminal, capable of handling 14

1.4 *A new terminal was completed at Chaing Kai-Shek International Airport in 1997. A second terminal is scheduled for completion in 1999.* CSK International Airport Office.

million passengers a year, is scheduled for completion in 1999.[181] These projects will double annual capacity to 30 million passengers a year.[182] (See Fig. 1.4.)

Thailand

A new US$3.6 billion Second Bangkok International Airport has been on the drawing boards since about 1960. Construction of a second international airport near the eastern seaboard has been under discussion for nearly two decades, and in 1996–1997, there were rumblings that the site might be moved.[183] Meanwhile, traffic has been growing robustly at the existing Don Muang Airport. Because it became apparent that the new airport would not be completed until 2003, the government decided to spend US$478 million upgrading Don Muang. In May 1997 the Thai cabinet reduced the budget for the second site (Nong Ngu Hao[184]), located in eastern Bangkok, from 97 billion to 68 billion baht (from US$2.4 billion to US$1.7 billion) because of a "change in the scope of construction."[185] Nevertheless, the airport is scheduled to open in 2004.[186] Thai International Airways is building a US$100 million aeronautical maintenance facility at U-Tapao.[187]

Turkmenistan

At Ashgabat Airport, a US$81 million project is under way to complete the terminal building and concrete apron, and install specialist airport systems.[188]

Uzbekistan

At Tashkent International Airport, a US$50 million plan is under way to "modernize its international terminal and rehabilitate the apron and taxiways."[189]

Vietnam

Fifty-four of Vietnam's 313 airports are in need of repair or rebuilding.[190] Much of the infrastructure of North Vietnam was destroyed by American bombing during the war which ended in 1975, though the U.S. military built significant infrastructure in the South. The government of Vietnam has approved a comprehensive US$6 billion airport development plan.[191] Vietnam is improving its airports at Hanoi and Ho Chi Minh City, and building a new airport at Da Nang.[192] It also plans to add three new international airports—the Chu Lai Airport at Haiphong, the Chu Lai Airport in the Dung Quat area, and the Long Thanh Airport near Ho Chi Minh City (formerly Saigon). The latter (the former U.S. Air Force Long Binh Airfield) is a candidate for replacement of Tan Son Nhat Airport. Tan Son Nhat is also undergoing expansion from its current capacity of 3.5 million passengers per year on projections that demand will be between 10 and 12 million passengers by 2010. (See Fig. 1.5.)

1.5 *Tan Son Nhat Airport had not undergone construction since U.S. military forces left Saigon in the early 1970s. Vietnam is currently expanding and updating this facility.* Photo by P. S. Dempsey.

According to IATA, Vietnam will face the fastest growth (17% per year) of all Asian nations by that year.[193] By 2020, the government would like to have six international airports. The Da Nang and Chu Lai airport projects are projected to cost US$1.2 billion.[194] Expansion of Noi Bai International Airport at Hanoi is projected to cost US$500 million. Vietnam's expansion of multiple airports around the country is projected to cost between US$5 billion and US$6 billion.[195] This includes a new international airport planned for Long Thanh between 2000 and 2005.[196]

Haiphong

Cat Bi Airport has been upgraded to handle Boeing 737-300 aircraft.[197]

Hanoi

Noi Bai International Airport is undergoing a US$50 million, two-phase upgrade to be complete by 2000.[198] A new five-story terminal is included.[199]

Ho Chi Minh City (Saigon)

At Tan Son Nhat International Airport, the Ministry of Transport (MOT) and the Vietnamese Civil Aviation of Vietnam (CAAV) announced long-term plans to spend US$1.8 billion to upgrade this facility, while in the short term (3-year plan) a US$400 million new international terminal is planned.[200] In December of 1996 a US$7 million "international standard cargo terminal" was opened to handle a 25% increase in cargo volume.[201] The airport is projected to service 40 to 50 million passengers and one million tons of cargo annually by the year 2010.[202] Currently, the airport handles 2 million passengers and 30,000 tons of cargo annually, with an increase in air traffic approaching 30% per year.[203] Among the projects under way or recently completed are the upgrade and expansion of the current terminal building, addition of a second runway, construction of a new cargo station, and construction of a new international terminal.[204] The government allocated US$8.7 million for the upgrade of the current terminal. It was completed, along with a second 3050-m runway, in early 1997.[205] A new cargo terminal, designed to handle 100,000 tons annually, became operational in late 1996.[206] Located on a 10,000-m² campus in 1997 the 3000-m² cargo facility handled 100 tons per day, of which 65% was in import goods and 35% in export goods.[207] Construction of the new international terminal is to be handled solely by Vietnamese companies. After initially considering a joint venture to include foreign companies, "the government favored the proposal that gives Vietnam complete control and management of the project," citing concerns over, "important matters," that would best be handled by Vietnamese managers and contractors.[208] The $200 million project is expected to increase passenger capacity to 8 million by 2000.[209] The CAAV is also considering Long Thanh Airport (formerly the U.S. Army Long Binh Airfield) as a possible candidate to replace Tan Son Nhat as the southern international airport.[210]

Table 1.2 summarizes nearly US$94 billion in major airport projects recently completed, under way, or planned in Asia.

Middle East

Bahrain

A new US$100 million passenger terminal at Bahrain International Airport on the Island of Muharraq was completed in September 1994,[211] raising the airport's capacity to 10 million passengers annually.[212] Voted the "Number One Airport in the Middle East" in the International Air Transport Association's (IATA) Airport Monitor Survey,[213] Bahrain International Airport handles an average of 55,000 flights and 3.6 million passengers a year.[214] The master plan calls for a second runway and new terminal by the year 2010.

Gaza Strip (Palestine)

At the Gaza International Airport at Rafah, tensions arose as Israel refused to allow the Palestinians to fully open their new airport because of security concerns.[215] Permission was finally granted after a new peace agreement between the Israelis and Palestinians was signed. Named the Yasir Arafat International Airport, the US$68 million facility opened in 1998 with a 10,100-ft runway and a Moorish-style terminal.[216]

Iran

At Iman Khomeini International Airport, a new US$50 million terminal is planned.[217]

Table 1.2. Major Airport Development and Expansion Projects in Asia

Nation	Airport	Development	Cost, million US$
Brunei	Brunei International	Redevelopment and expansion	120
Cambodia	Phnom Penh Pochentong	New international terminal and runway	120
China	Beijing Capital International	New terminal, aprons, car park	1,100
	Chengdu Shuangliu	Expansion	200
	Fuzhou Changle International	New airport (1997)	390
	Fuyang	Expansion	n.a.
	Guangzhou	New airport (2005)	2,400
	Guilin Liangjiang International	New airport (1996)	223
	Haikou Meilan	New airport	280
	Hangzhou	New airport	362
	Hong Kong Chek Lap Kok	New airport (1998)	20,100
	Kunming International	Expansion	120
	Macau	New airport (1995)	1,100
	Nanjing Lukou International	New airport (1997)	290
	Shanghai Pudong International	New airport (1999)	1,600
	Shenzhen	New airport (1991), expansion	460
	Urumqi Diwopu	Expansion	210
	Xiamen Gaogi Airport	Expansion	253

	Xuzhou	New airport	105
	Zhuhai	New airport	1,000
	Zhengzhou Xinzheng	New airport (1997)	145
India	Bangalore	New airport	400
	Bombay	New airport	n.a.
	Calcutta	Expansion	n.a.
	Navi Mumbai	New airport	490
	New Delhi	Expansion	n.a.
	Sinnar	New airport	n.a.
	Madras	Expansion	n.a.
Indonesia	Hang Nadim International (Batam)	Expansion	n.a.
	Kalimantan Island	New airport	20
	Lombok	Development	115
	Medan	Development	358
	Padang	Development	174
	Pontianak	Development	71
	Samarinda	New airport	43
	Solo	Development	342
	Sorong	Development	47
Japan	Aichi (Chubu International)	New airport	7,000
	Kitakyushu	New airport (2005)	n.a.
	Kobe	New airport (2005)	2,300

Table 1.2. Major Airport Development and Expansion Projects in Asia (Continued)

Nation	Airport	Development	Cost, million US$
Japan (cont.)	Okayama	Expansion	n.a.
	Okinawa (Naha International)	Domestic terminal	362
	Osaka (Kansai International)	New (1994)	14,400
	Tokyo (Narita)	Redevelopment	1,500
Korea	Seoul (Inchon International)	New airport (2001)	5,800
Malaysia	Bitilu	New airport	68
	Kuching	New airport	n.a.
	Langkawi	Runway extension	n.a.
	New Kuala Lumpur International (Sepang)	New airport (1998)	3,200
	Northern International (Penang)	Airport development	150–800
	Sultan Ismail (Senai)	Expansion	150
Myanmar	Mandalay International	Upgrade	357
	Rangoon	Runway extension	12
Pakistan	New Lahore International	New terminal, apron	250
Philippines	Clark International	New int'l terminal, cargo facilities, industrial park	900–1,200
	Davao	Runway extension, international building	105
	Mactan International (Cebu Island)	Conversion to international airport	n.a.

	Ninoy Aquino International (Manila	New international cargo terminal, new international passenger terminal, access road, improvements	680
	Subic International	Expansion	1,200
	Sixteen airports	New airports and airport expansions	464
Singapore	Changi	New terminal and expansions	1,974
South Korea	Inchon International (Seoul)	New airport (2000)	5,000
	Muan County	Expansion	n.a.
Taiwan	Chaing Kai-Shek International	Development	800
Thailand	Don Muang (Bangkok)	Expansion	478
	Global Transpark	New regional cargo center	1,500
	Second Bangkok International Airport	New airport	3,640
	U-Tapao	New heavy aircraft maintenance facility	100
Turkmenistan	Ashgabat	Terminal and apron	81
Uzbekistan	Tashkent International	Modernization	50
Vietnam	Da Nang International/Chu Lai	Expansion	1,200

Table 1.2. Major Airport Development and Expansion Projects in Asia (*Continued*)

Nation	Airport	Development	Cost, million US$
Vietnam (*cont.*)	Ho Chi Minh International (Saigon) Tan Son Nhat	Upgrades	1,800
	Noi Bai International (Hanoi)	Redevelopment	50
	Various airports	New airports and expansions	5,000–6,000

Israel

At Ben Gurion International Airport at Tel Aviv, a US$850 million project is under way to add a new terminal and increase capacity to 16 million passengers per year.[218]

Jordan

Aqaba/Eilat International Airport is to be expanded as a joint project with Israel at a cost of US$36 million.[219] An American feasibility study projected the cost of the airport project to be US$125 million, but the Israelis and Jordanians are exploring ways to trim the cost."[220]

Lebanon

At Beirut International Airport, a total airport reconstruction project is under way to expand the existing terminal, add a second runway, improve security, and add a new control tower as part of a US$490 million upgrade.[221] The airport was severely damaged by Israeli military forces during their occupation of Beirut. A redevelopment program for Beirut International Airport includes two new runways, a new terminal building and cargo facilities, a car park, and a control tower and navigational equipment.[222] A 2500-m breakwater was built to protect the 1900-m runway projection into the sea. The new runway will be 3880 m long, and roughly parallel to the existing runway. The first phase of construction, completed in 1997, includes a new passenger terminal with 28 check-in counters, and will ultimately have 84.[223] A second nine-gate concourse and two new runways comprise the second phase, currently under construction.[224] The project is being financed in part by a US$66 million loan from the European Investment Bank (EIB).[225] Upon completion, the airport will be able to accommodate 6 million passengers per year.[226]

The Lebanese government also plans to restore Riyak Airport and Renee Mouwad Airport. Each project is estimated to cost approximately US$50 million and increase capacity to 1 million passengers and 200,000 tons of cargo per year.[227]

Oman

Opened in 1973,[228] Muscat/Seeb International Airport approved a US$21 million expansion project in 1993.[229] At Salalah International Airport, a new international terminal is planned.[230]

Qatar

Expansion of the airport at Doha includes a new terminal with 14 air bridges, at an estimated cost of US$105 million.[231]

Saudi Arabia

King Fahd International Airport is a new airport located 22 mi northwest of Dammam.[232] It will be further expanded to handle 14 million passengers (from 6 million) by the year 2000.[233] King Abdulaziz International Airport (KAAIA) is undergoing an expansion project that includes the addition of a third public terminal. The expansion will permit the airport to handle 7 million domestic and 7.5 million international passengers by the year 2004.[234]

United Arab Emirates
Abu Dhabi

At Abu Dhabi International Airport, plans for development include a US$100 million terminal and an US$88 million runway.[235] The Abu Dhabi Civil Aviation Department announced a two-part development program for Abu Dhabi Airport that will increase capacity to 6.5 million passengers annually. The first phase includes construction of a second 100-m-diameter satellite that will be capable of handling new large aircraft (NLA). A new terminal building is planned for the second development phase.[236] Also under consideration is the construction of a second runway. A 60-m-wide, 4-km runway with associated taxiways catering to larger aircraft is planned for Abu Dhabi. The project is estimated to cost US$100 million.[237] A $19 million contract has been awarded to Aeroports de Paris for the installation of new radar systems at Abu Dhabi and Al-Ain airports.[238]

Al-Ain International Airport plans a US$100 million expansion, the first phase of which will include both freight and terminal buildings.[239] Al-Ain International Airport officially opened in March 1994. Located only 160 km east of Abu Dhabi, Al-Ain is utilized primarily as a transit point and technical stop for long-haul services between Europe and East Asia. The cost of the project was US$270 million.[240] With considerable traffic growth expected, plans are already under way to expand the airport by 50%, including doubling the size of the departure and check-in areas and adding a cargo terminal and a flight catering facility.[241] (See Fig. 1.6.)

1.6 *Designed by Paul Andreu, the first phase of expansion of the Abu Dhabi International Airport involves construction of a round satellite terminal connected to the main terminal by a people mover system. The second phase involves building a main terminal building similar to the existing terminal.* Civil Aviation Department of the Emirate of Abu Dhabi.

Dubai

Dubai Airport, the second busiest in the Middle East behind Cairo, is planning a US$540 million expansion program that will expand its capacity from 5 to 12 million passengers annually.[242] The program includes a new terminal, a new concourse, terminal expansions, new runways, a free-trade zone, a shopping center, a technological park, expansion of the cargo village, and other facilities.[243] The new terminal was completed in early 1998, and is expected to handle 2.5 million passengers annually.[244] The new terminal, Terminal 2, will serve as an interim facility while Terminal 1 is refurbished and a new concourse is built.[245] Upon completion, scheduled for 1999, Terminal 2 will continue as a VIP and charter facility.[246]

Table 1.3 summarizes nearly US$3 billion in major airport projects in the Middle East.

Table 1.3. Major Airport Development and Expansion Projects in the Middle East

Nation	Airport	Development	Cost, million US$
Bahrain	Bahrain	Expansion	100
Cyprus	Larnaca International	Redevelopment	300
Iran	Khomeini International	New terminal	50
Israel	Ben Gurion International	New terminal	850
Jordan	Aqaba/Eilat International	Expansion	36
Lebanon	Beirut International	Reconstruction	490
	Renee Mouwad	Restoration	50
	Riyak	Restoration	50
Oman	Muscat/Seeb International	Expansion	21
	Salalah International	New terminal	n.a.
Palestine	Yasir Arafat International	New airport	65
Qatar	Doha	Expansion	105
Saudi Arabia	King Abdulaziz International	New terminal	n.a.
	King Fahd International	Expansion	n.a.
United Arab Emirates	Abu Dhabi International	New terminal and runway	188
	Al-Ain International	Expansion	100
	Dubai International	Expansion	540

Oceania

Australia

Brisbane

The Brisbane Airport Corporation (BAC), which recently purchased the Queensland airport from the Australian government, has outlined the first phase of a 20-year master plan for development. The initial phase is estimated to cost A$300 million (US$179.25 million) and will include construction of a new parallel runway.[247] These efforts are still in their preliminary stages. A detailed master plan will be presented to the airlines, the government, and the community before being finalized. Initial responses from community groups have been negative.[248] Brisbane Airport is expected to see a threefold increase in passengers, to 33 million, in the next 20 years.[249]

Melbourne

Melbourne Airport was voted as the world's third best airport by 55,000 long-haul passengers polled by IATA.[250]

Sydney

A new US$300 million runway has been built at Sydney's Kingsford Smith Airport,[251] and a second airport west of Sydney at Badgery's Creek is under consideration.[252] Kingsford Smith Airport will undergo a A$300 million (US$194 million) renovation to provide visitors for the 2000 Olympics a "dramatic sense of arrival." The plan includes terminal expansion and development to accommodate new large aircraft.[253] Qantas has also undertaken a redevelopment program that will enlarge its domestic terminal to 6 times the original size. The renovated terminal will include separate departure and arrival levels, a two-level roadway, and a new check-in area. The project, slated for completion by November of 1998, will cost A$212 million (US$137 million).[254] Additionally, Ansett Australia will redevelop its Sydney terminal at a cost of A$167 million (US$108 million). Work will expand the current two-level building into a three-level complex, providing more check-in counters, increasing baggage handling, and accommodating an elevated roadway and an underground rail. Work is scheduled to be completed by December 1999.[255]

New Zealand

Auckland

A 3-year redevelopment and expansion project to double the size of the Jean Batten International Terminal was completed in December

1997 at a cost of NZ$180 million (US$103 million).[256] The expansion includes a 255-m-long glass wall along the front of the terminal, enlarged customs area, and a 5-m-long high-tech video wall, featuring sports, weather, and news. Auckland serves as the entry and exit point for 80% of New Zealand's tourists.[257]

Christchurch

A NZ$72 million (US$37.17 million) expansion of the international terminal building was completed in September 1998.[258] The expansion doubles floor space to 28,000 m² and doubles the handling capacity of the terminal to 900 passengers per hour.[259] International traffic at Christchurch rose 12% in 1997, with a steady growth around 8% expected to continue.[260]

Wellington

Wellington International Airport is building a new terminal that will process both international and domestic departures. The new terminal, meant to enhance the airport's user-friendliness and meet increased demand, is scheduled for completion in mid-1999 at a cost of NZ$85 million (US$53 million). The airport has witnessed a 40% growth in passengers in the past 5 years.[261]

Vanuatu

Vanuatu is a small island nation west of Australia, formerly known as New Hebrides. Two main airports, in the capital city of Port Vila and on the island of Santo, are being upgraded with the purpose of enabling both airports to handle Boeing 767 aircraft.[262]

Table 1.4 summarizes more than a billion U.S. dollars of airport projects in Oceania.

Europe

Some estimate that European governments will spend between US$65 billion and US$80 billion on airports and related infrastructure over the next 10 to 15 years.[263]

Albania

One of Europe's poorest countries, Albania has only one runway in the airport of its capital city, Tirana.[264]

Table 1.4. Major Airport Development and Expansion Projects in Oceania

Nation	Airport	Development	Cost, million US$
Australia	Brisbane International	Expansion	179
	Sydney Kingsford Smith	Runway, renovation	494
New Zealand	Auckland	Terminal expansion	103
	Christchurch International	Terminal	200
	Wellington	New terminal	53
United States	Guam Agana	Terminal expansion	264
Vanuatu	Port Vila	Upgrade	n.a.

Armenia

At Yerevan Airport, a US$28 million contract was awarded in late 1996 to a U.S.-based company to build a new airport cargo terminal.[265]

Austria

In 1996, Vienna International Airport completed its West Pier Terminal, equipped with 12 gates. It anticipates expansion of its airport to 41 gates (from 30) by 2000.[266]

Azerbaijan

At Baku International Airport, the German government may finance a second DM15 million (US$8.4 million) runway under favorable credit terms. The European Bank has agreed to finance a US$14 million upgrade of the air navigation system.[267] The airport is also undergoing a US$18.5 million terminal renovation.[268]

Belgium
Brussels

Brussels Airport opened on the occasion of the World Exhibition of 1958. The original terminal had sufficient capacity to handle 6 million passengers annually. By 1993, the airport was handling more than 10 million. By 1996, the airport was handling 13.6 million passengers, and nearly half a million tons of freight. In 1989, the airport adopted a modernization program, Zavantem 2000, to modernize the airport and expand capacity to between 20 and 25 million passengers per year. The three existing runways were upgraded to handle 80 movements per hour, up from 45 per hour in 1993. By 1994, the airport had laid 1.2 million m² of taxiways, platforms, and high-speed exits, installed a separate drainage system and underground tunnels for cable and pipes, and straightened a tunnel under runway 25R.[269]

Liege

TNT announced it would move its European hub from Cologne, Germany, to Liege International Airport, investing US$40–66 million in what "is expected to be among the most advanced [hubs] in Europe when completed in 1998."[270] One attraction for TNT was that the airport guaranteed night flights until the year 2015.[271]

Bulgaria

A planned US$100 million development for Sofia Airport has been ratified by the Bulgarian Parliament.[272]

Croatia

Zagreb Airport is planning for construction of a new passenger terminal, which would require the relocation of military facilities at the airport and the construction of road access to the new site. Completion of the terminal is contemplated for 2002, with expansion of the existing terminal in the interim.[273]

Czech Republic

At Prague Ruzyne International Airport, a new check-in terminal, called JIH 2,[274] was completed in June 1997 at a cost of 7 billion koruna (US$230 million).[275] Under study is the possibility of construction of a second new terminal to handle traffic up to 18 million passengers a year."[276]

Denmark
Copenhagen

Copenhagen International Airport, Scandinavia's largest, has a 10-year, DKK7 billion (US$1.1 billion) investment plan to build a new terminal and baggage handling system and upgrade cargo handling facilities.[277] The new cargo terminal has sufficient capacity to handle 350,000 tons of cargo and mail per year. Also added is a central baggage terminal capable of handling more than 8000 bags per hour, and a new underground parking garage with 5200 parking spaces. A distinctive wing-shaped roof has been added to Terminal 3, which houses arrival and departure areas, a restaurant, cafes, and a shopping area. Less than 100 m from check-in is a rail station where four trains per hour transport passengers to Copenhagen Central Station in only 10 minutes. A new 375-room hotel is also planned. The airport projects passenger volume of 17 million in 1997 will grow to 25 million by 2005.[278] (See Fig. 1.7.)

Billund

Billund International Airport, the second largest airport in Denmark, which began as a private airfield for Europe's biggest toy maker,

1.7 *Copenhagen International Airport is Scandinavia's largest. This is the view from air side.* Copenhagen Airports.

Lego,[279] is also planning expansion of passenger and cargo handling facilities.[280]

Estonia

At Tallin International Airport, a DEM28 million project is to repair the terminal building and add 2220 m^2 to the terminal.[281]

Finland

Helsinki-Vantaa Airport has a two-phase project under way. Phase 1, a middle terminal to connect the international and domestic terminals, is complete.[282] During the second phase, scheduled for completion in 1999, the middle terminal will become an "independent passenger unit."[283] A third, parallel runway also is under construction. Begun in 1996, the 9900-ft runway is scheduled for completion in 2002 at a planned cost of US$80 million.[284] And, as part of a US$43 million Finnish air traffic management integration system, Helsinki-Vantaa Airport received a new tower in what has been described as the "most significant air traffic management project in Finnish history."[285]

France
Champagne
At Vatry Airport, a new US$1.08 billion dollar air cargo platform includes a 20-year concession to manage the facility.[286] The platform is expected to be operational by the end of 1998.[287]

Marseilles
Marseilles International Airport, ranked second for freight and third for passenger traffic among French airports, is undergoing an expansion to double its passenger capacity to 10 million per year.[288]

Paris
Paris Orly's south terminal opened in 1961, while Orly's west terminal was added in 1971. It has three runways. Orly Airport has adopted a 5-year plan to increase passenger capacity which begins in 1998.[289] Total capacity at Orly is 31 million passengers and 300,000 tons of freight annually.[290] The first terminal opened at Charles de Gaulle Airport in 1974, and the first unit of the second terminal opened in 1981. At Roissy-Charles de Gaulle Airport, beginning in 1998, two new runways will be built parallel to the existing ones at a cost of FFr1.5 billion (US$270 million) over 5 years. (See Fig. 1.8.)

1.8 *An aerial view of Paris Charles de Gaulle Airport, France's largest.* Photo by G. Halary, Aeroports de Paris.

A further FFr10 billion (US$1.77 billion) will be spent over the next 4 years on the construction of Roissy 2F and Roissy 2E, which will form two sides of an oval, or peninsula structures at right angles to the main terminal.[291] The expansion will be accompanied by noise regulations and distribution of the taxes generated by the airport to the local communities.[292] Paris' first airport, Le Bourget, today specializes in business aviation.

Germany

Berlin

The Berlin Brandenburg International Airport (BBI), the first privately financed airport in Germany, will be built on the existing Berlin-Schoenefeld Airport site.[293] The bid for financing, planning, construction and operation will be awarded through an international competition.[294] Due for completion by 2007, the new airport is projected to cost US$4.52 billion, and have an annual capacity of 18 million passengers.[295] Redevelopment of the airport will include a new terminal facility, a second runway, and "ancillary installations" capable of handling 20 million passengers annually by 2010.[296] A second expansion could raise capacity to 30 million passengers.[297] The project is billed as the largest privatization effort in continental Europe; the winning consortium will take a 74.9% stake in Berlin Brandenburg Flughafen (BBF), the holding company (consisting of the federal government, city of Berlin, and state of Brandenburg) that owns Berlin's three airports.[298] The remainder may be sold at a later date provided satisfactory operating results. The tender will be a "hybrid acquisition and project finance" type consisting of a concession of at least 50 years to plan, build, and operate the new airport; a contract to operate Schoenefeld, Tegel, and Tempelhof until the new installation is operational (Tegel is to be closed in 2002, Tempelhof in 2007); and the assumption of an existing BBF debt, estimated at nearly DM750 million.[299] The project is to be completed "as far as possible without public financing," but public moneys (DM1.5–1.8 billion) will be provided for related infrastructure including highway, subway, and high-speed rail connections.[300]

The two consortia vying for the project are Hochtief and IVG Holding AG.[301, 302] The group headed by Hochtief includes Frankfurt Airport, ABB, and the Bankgesellschaft Berlin. The IVG consortium includes Dorsch Consult, the Dresdner Bank, and the Commerzbank.[303]

1.9 *The new Terminal 2 at Cologne/Bonn International Airport, designed by Helmut Jahn. In the foreground is a multistory car park and the glass canopy of the future intercity rail and S-Bahn station.*
Flughafen Köln/Bahn.

Cologne

Cologne/Bonn Airport (Flughafen Köln/Bonn, FKB) is undergoing a total infrastructure expansion valued at DM1.1 billion (US$650 million) over 4 years to include a new Terminal 2 (DM575, US$338 million), new parking lot area (DM70 million, US$41 million), a noise abatement infrastructure (DM120 million, US$70.6 million), and an underground railway station (DM100 million, US$59 million).[304] FKB's runway is designed for intercontinental connections and, unlike many major European airports, is not subject to restraints on night flights.[305] (See Fig. 1.9.)

Dortmund

Dortmund Airport plans to expand its 1050-m runway by 400 m to accommodate small jets.[306]

Düsseldorf

Düsseldorf is the second largest airport in Germany.[307] At Düsseldorf, the first of three fire-damaged airport terminals has reopened after

17 people died in April 1996, many from toxic gas.[308] The terminal has transparent floors and ceilings to reassure passengers that no blaze will again spread undetected through the ceiling.[309] The extended terminal complex will be completed by 2003. Düsseldorf's airport is the first German airport to be partially privatized.[310]

Frankfurt

Frankfurt is one of the busiest airports in the world. In 1997, Frankfurt Airport handled more then 40 million passengers for the first time in its history. Current expansions aim to increase its capacity to 60 million by 2005.[311] Frankfurt also handled 1.4 million tons of cargo and 144,000 tons of mail in 1997. Frankfurt's 1985 master plan called for expansion of the aprons and parking areas, and expansion of Terminal 1. In 1994, it opened Terminal 2. Frankfurt Main Airport, Europe's leading cargo handling airport, is expecting annual growth of 4.5% through the year 2010. To accommodate the growth, Frankfurt is building a three-phase second cargo center, CargoCity South, scheduled for completion in 1996, 2001, and 2006.[312] This 218.9-acre facility will "feature a rail line which can be used by intermodal air/rail cargo services to ease the burden on surrounding road networks."[313] Lufthansa Cargo began work in 1995 on a DM125 million (US$85.3 million) expansion of its cargo terminal to accommodate growth anticipated by the year 2010.[314] (See Fig. 1.10.)

Hamburg

Flughafen Hamburg will begin construction on a new terminal in 1999. It will replace Terminal 2, adding 11 aircraft positions.[315]

Hannover

Hannover has built a new terminal that added eight new aircraft positions.[316]

Leipzig Halle

Flughafen Leipzig Halle is investing US$705 million to build a second 3600-m runway, expand its terminal facilities, and build a central terminal railway access and car park.[317]

Munich

At a cost of more than US$7 billion, Munich's Franz Josef Strauss Airport (sometimes referred to as Munich II) opened in 1992, some three decades after initial planning for a new airport began, and a decade after construction began. The new airport is the first com-

1.10 *Frankfurt is one of the world's busiest airports. It is everything an airport should be—a major passenger, cargo, and intermodal transport center.* Flughafen Frankfurt.

pletely new airport to have been built in Europe since Paris Charles de Gaulle opened in 1974.[318]

Greece

Work began in 1996 on the new Athens International Airport, located 30 km outside of Athens in Spata. The new airport is scheduled for completion in 2001 at a cost of Dr3.204 billion (US$2.7 billion).[319] It should be open in time for the 2004 Summer Olympic games, to be held at Athens.[320] The existing Athens airport was voted as the world's most improved of 1997 by 55,000 long-haul passengers polled by IATA.[321]

Hungary

Aeroports de Montréal (ADM), partnered with a group of Canadian companies, won a US$120 million dollar contract for the expansion and modernization of the air terminal at Budapest's Ferihegy Airport.[322] The agreement includes management of the terminal for a 15-year period, after which the Hungarian authorities will manage the terminal.[323] The mayor of Budapest, Gbor Demszky, has called for an expansion of Ferihegy II into a major international airport.[324]

Ireland

The Irish Aviation Authority announced "plans for a state of the art air traffic control centre in Shannon which will cost US$20 million to construct and another $20 million plus to equip," operational by 2001.[325] 1996 capital expenditures of US$5 million included a new radar station in Mayo.[326] As an "Objective One" region in the EU, Northern Ireland has used EU funds to build a new airport.[327] Derry Airport is newly renovated.[328]

Italy
Milan

With a catchment basin of 7.5 million people, Milan ranks third behind London (10.5 million) and Paris (8.7 million). It is therefore expanding one of its two airports under the Malpensa 2000 project. Such expansion includes a new main terminal building connected to two remote satellites with enclosed skywalks.[329] By 2000, the new Terminal West will have a capacity of 18 million passengers a year, which will be in addition to the 6 million annual capacity of the existing Malpensa North Terminal, which will operate as a remote satellite.[330] The two existing runways are also being upgraded. The total cost of the project is 1,990 billion lire (US$1.2 billion).[331] (See Fig. 1.11.)

1.11 *At Milano a new passenger terminal and two satellite terminals have been constructed.* Aeroporto Milano Linate.

Rome

A new national air terminal was completed at Fiumicino Airport on July 31, 1997. A new major air terminal is currently under construction and scheduled for completion by 1999.[332] A US$2.5 billion project to expand and modernize Fiumicino and Ciampino airports in anticipation of the Holy Jubilee in 2000 has also begun. The project will include improvement of taxiing strips, runways, ground support, electronic landing systems, airplane parking facilities, car parking lots, shopping centers, commercial space, and hotels. This long-term project is expected to continue until 2030, ultimately increasing capacity to 60 million passengers annually.[333] A two-stage development at Fiumicimo Airport is scheduled for completion in 2005 and 2030 at a cost of US$4.5 billion.

Latvia

At Riga's Latvia International Airport, a US$200 million extension of the passenger terminal is planned for the first stage of a major reconstruction program.[334] A loan from the European Bank for Reconstruction and Development financed a runway extension to accommodate large aircraft. Latvian aviation officials are considering a separate cargo facility to be built and operated by a private company.[335]

Lithuania

Vilnius International Airport has been modernized with funds provided by the EIB.[336]

The Netherlands

The first flight at what would become Amsterdam Schiphol airport landed in 1916. Regularly scheduled service by KLM between Amsterdam and London began 4 years later. With 30 million passengers per year and 1.1 million tons of cargo, Schiphol ranks fourth among European airports in passenger volume, and second in cargo. In 1995, the Dutch Parliament approved a government white paper decreeing Schiphol would be allowed to grow into a "mainport," an airport that is a home base and central European airport for one of the global airlines.[337] (See Fig. 1.12.) Voted the world's fourth best airport by 55,000 long-haul passengers polled by the IATA[338] (Britain's Manchester and Singapore's Changi Airport shared the top slot with Australia's Melbourne taking third), Schiphol is also undergoing significant expansion. Approval was given in 1996 to add a fifth run-

1.12 *Amsterdam's Schiphol Airport is among the world's busiest, a tribute to the marketing acumen of KLM Royal Dutch Airlines, which has made the airport a major international hub.* Amsterdam Airport Schiphol.

way, planned to open in 2003.[339] The 1993–1998 five-year plan called for renovation of the terminal center, expansion of two piers, and adding an office center and hotel. The 1998–2003 five-year plan calls for additional terminal extension, pier widening and extension, an automatic transport system, multistory parking garages, an elevated bus track, and another hotel.[340] Because capacity has been capped at 44 million passengers and 3.3 million tons of cargo annually, Schiphol is studying off-site expansion. The alternatives are construction of a remote runway airport on an artificial island to built in the North Sea, connected with a high-speed rail line, at a cost of between NLG29 billion and NLG32 billion (US$15.3 billion and US$16.8 billion), or reconfiguring Schiphol on the same location combined with an overflow airport for charter and cargo operations, at a cost of NLG14 billion (US$7.36 billion). Whichever alternative is adopted, it must be completed by 2012.[341]

Norway

Oslo completed a new airport at Gardermoen, 25 mi north of the central business district, in 1998.[342] The new airport is Norway's largest development project ever. Its capacity will be 12 million passengers annually by 2000.[343]

Poland

In Cracow, Poland's second busiest airport at Balice is building a US$4.3 million terminal to be completed in 1999. It also plans to build a new cargo terminal, road improvements, and car park.

Portugal
Lisbon

Airport renovation at Lisbon is under way to prepare for EXPO '98.[344] Meanwhile, Portugal's government is studying possible sites for a new Lisbon-area airport because Lisbon's main airport, in Portela, is expected to "exceed capacity by 2010."[345] Two possible locations are Ota and Rio Frio.[346]

Madeira Island (Funchal)

At Funchal Airport, an ECU269 million (US$314 million) project is under way to add a new control tower, maintenance facilities, fuel depots, and aircraft parking spaces.[347]

Romania
Bacua

A US$330 million project is under discussion to upgrade Bacua Airport and build a commercial center.[348]

Bucharest

At Bucharest-Otopeni International Airport, a US$160 million modernization project is under way to include a new passenger terminal, a finger connecting two terminals, and a new cargo terminal.[349]

Russian Federation
Moscow

At Sheremetevo International Airport, a new 3.5-km runway was completed in July 1997 at a cost of US$70 million. Reconstruction of the second runway and construction of a third runway is also planned.[350]

St. Petersburg (Leningrad)

Lufthansa Cargo plans to build a US$12 million cargo terminal at Pulkovo Airport to be completed in 1999.[351]

Samara Oblast

Samara International Airport has adopted a US$650 million three-phase plan to modernize the airport over a 20-year span.[352] Phase I (first 5 years) involves an extension of a runway, expansion of air navigation facilities, and reconstruction of the passenger terminal. Phase II (between 5 and 10 years) involves construction of a hotel and training center, the lengthening and widening of a second runway, and construction of additional hangars. Phase III (between 10 and 20 years) involves private commercial development, as well as construction of additional hangars.[353]

Urals (Ekaterinburg)

Koltsovo International Airport has begun a US$120 million project to reconstruct two runways, increase capacity of the existing international terminal, and construct two additional terminals to accommodate 8 million passengers a year.[354]

Spain

Bilbao

The state-owned Spanish airport authority, Aena, invested 5.836 billion pesetas (US$40 million) into improvements to the airport at Bilbao in 1998.[355] An airport expansion is planned at Bilbao.[356]

Balearic Islands (Mallorca Island: Palma de Mallorca)

At Palma de Mallorca Airport, a new terminal opened in April 1997. Sixteen of 64 planned operating fingers are operational. Scheduled completion is 2008 at a planned cost of US$320 million.[357]

Canary Islands

The European Investment Bank (EIB) has loaned the Spanish government 11 billion pesetas (ECU68 million, or US$79 million) to finance major improvement works at the airports in Lanzarote and Fuertaventura and both of Tenerife's airports. The project involves construction of new passenger terminals and access roads at Tenerife-North, Lanzarote, and Fuertaventura airports and a taxiway at Fuertaventura, and improvement and expansion of one of the passenger terminals at Tenerife-South Airport.[358]

Catalonia (Barcelona)

At Barcelona Airport, a third runway and terminal buildings are estimated to cost 50 billion pesetas (US$339 million).[359]

Madrid

In May 1997 a new airport for Madrid was ruled out by Spain's Development Minister in favor of increasing to four or five the number of runways at the existing Barajas Airport.[360] The third runway at Barajas is scheduled for completion in December of 1998.[361] At 4.4 km it will be "the largest in Europe," costing 25.4 billion pesetas (US$180 million).[362] The third runway is part of a larger 150 billion peseta (US$1.05 billion) expansion project that includes a "new terminal, [a] control tower, subways and internal trains."[363]

Sweden

At Stockholm's Arlanda Airport, the addition of a third runway is scheduled for completion in 2001.[364]

Switzerland
Geneva

An expansion project is under way at Geneva-Cointrin Airport to include the addition of a new terminal finger, new check-in facilities, maintenance hangar extension, and modernized waste water treatment. Scheduled for completion in 2005, the terminal finger alone is estimated to cost SFr100 million (US$72 million), while the total project is planned at SFr400 million (US$289 million).[365]

Zurich

Zurich Airport was voted the world's fifth best airport by 55,000 long-haul passengers polled by IATA.[366] (Britain's Manchester and Singapore's Changi Airport shared the top slot, with Australia's Melbourne taking third and the Netherlands' Schiphol taking fourth.)

Turkey
Bodrum

A new airport is being built at Bodrum.[367]

Istanbul

Istanbul's Ataturk Airport is undergoing a US$300 million expansion to add a new international terminal.[368] Construction is scheduled to begin in late 1997, and is to last 30 months under a build-operate-transfer model.[369]

United Kingdom
Birmingham
Birmingham International Airport is undergoing a new terminal expansion costing 260 million pounds.[370]

London
At London's Heathrow Airport, a new Terminal 5, scheduled for completion in 2015, will be as large as all the other Heathrow terminals combined.[371] Because of continued controversy, construction is not expected to begin until 2000, with the opening date for phase 1 scheduled in 2005, 3 years later than the original plan.[372] The cost has escalated to £1.5 billion (US$2.5 billion) for the 60-gate terminal.[373] In the interim, while awaiting the opening of Terminal 5, the British Airports Authority (BAA) plans to invest £250 million (US$423 million) to improve the baggage handling system, having already paid British Airways "token" compensation of £200,000 (US$338,060) toward the airline's cost of handling passengers' lost luggage.[374] Approved in 1978, Heathrow's Terminal 4 was to have been the last major expansion at the airport.[375]

Currently underutilized, Stansted Airport nonetheless recently began a US$1.6 million, 3000-m^2 extension of its cargo shed so as to expand it to a total of 18,000 m^2.[376] Plans also have been laid to build a new offshore international airport in the Thames Estuary.[377]

Manchester
Voted one of the world's top two airports by 55,000 long-haul passengers polled by IATA[378] (Singapore's Changi Airport shared the honor), Manchester Airport gained government approval to build a second runway.[379] Construction has been hampered by protesters who burrowed 50 ft below the surface of the proposed runway.[380]

Southampton
Southampton International Airport was redeveloped in 1994.[381]

Table 1.5 summarizes more than US$45 billion in airport projects in Europe.

North America

Over the next two decades, the world air transport market is projected to grow between 5 and 6% a year, although North America is anticipated to grow at only about 4% a year.[382] This developed market is, by

Table 1.5. Major Airport Development and Expansion Projects in Europe

Nation	Airport	Development	Cost, million US$
Armenia	Yerevan	Cargo terminal	28
Azerbaijan	Baku	Runway, air traffic control	22
Austria	Vienna	Terminal expansion	900
Belgium	Brussels	Modernization	n.a.
	Liege	Cargo facilities	66
Bulgaria	Sofia International	Redevelopment	100
Czech Republic	Prague Ruzyne	Terminal	230
Denmark	Billund	Expansion	n.a.
	Copenhagen	Second terminal	1100
Estonia	Tallin	Terminal expansion	n.a.
Finland	Helsinki-Vantaa	Terminal, runway	80
France	Vatry (Champagne)	Air cargo facility	1080
	Charles de Gaulle (Paris)	New terminal and runways	2070
	Marseilles	Expansion	n.a.
Germany	Berlin/Brandenburg International	New airport	4600
	Cologne/Bonn	Expansion	700
	Frankfurt	New terminal, cargo center	2200
	Franz Josef Strauss (Munich)	New airport (1992)	7100

Table 1.5. Major Airport Development and Expansion Projects in Europe (*Continued*)

Nation	Airport	Development	Cost, million US$
Germany (*cont.*)	Franz Josef Strauss	Airport center, second terminal	1000
	Hamburg	Terminal expansion	600
	Leipzig Halle	New runway; expansion	705
Greece	Spata (Athens)	New airport (1997)	2700
Hungary	Ferihegy (Budapest)	Expansion	120
Ireland	Shannon	Traffic control upgrade	40
Italy	Fiumicino (Rome)	Terminal and runway expansion	4500
	SEA (Milan)	Malpensa 2000, cargo city, people mover	1200
Latvia	Latvia International (Riga)	Terminal expansion	200
Netherlands	Schiphol (Amsterdam)	Expansion	4100
Norway	Gardermoen (Oslo)	New airport (1998)	1850
Portugal	Lisbon Portela	Renovation	n.a.
	Funchal	Upgrades	314
Romania	Bacua	Upgrade	330
	Bucharest-Otopeni International	Modernization	160
Russian Federation	Sheremetevo International	New runway	70

	Pulkovo (St. Petersburg)	Cargo terminal	12
	Koltsovo International (Ekaterinburg)	Runway reconstruction and terminal expansion	120
Spain	Barcelona	New runway and terminal	339
	Bilbao	Improvements	40
	Canary Islands	Terminals and roads	79
	Madrid	Terminal and runway, plus other airports	3000
	Palma de Mallorca	New terminal	320
Sweden	Arlanda (Stockholm)	New runway	400
Switzerland	Geneva-Cointrin	Expansion	289
Turkey	Ataturk (Istanbul)	New terminal	300
	Bodrum	New airport	n.a.
United Kingdom	Heathrow (London)	New terminal	2540
	Manchester	Runway	n.a.
	Stansted (London)	Cargo expansion	2

some accounts, approaching maturity. Nonetheless, the creation of the "hub-and-spoke" distribution system, the dominant megatrend on the deregulation landscape, has created enormous peak capacity demand, straining the ability of hub airport infrastructure to keep pace.[383]

Canada
Calgary

At Calgary Airport, a 20-year plan is under way to "improve runway, taxiway and road development as well as expand air terminal facilities and services."[384] A phased expansion project scheduled for completion in 2015 includes the eventual addition of 12 new gates, expansion of concourses, aprons, and taxiways, and a new common-use domestic concourse. Phase 1, projected to cost US$103 million, is scheduled for completion in 2000. Phase 2, at US$250 million, is planned to begin in 2001 with completion scheduled in 2015. Phase 3, the long-term plan, will require at least one new concourse and further terminal enhancements.[385] Total expenditure for all three stages of Calgary's airport expansion will be about US$650 million.[386] Long-term development contemplates the construction of an additional runway, with associated taxiways and navigational and air traffic control facilities. (See Fig. 1.13.)

Edmonton

At Edmonton International Airport, a US$200–250 million renovation and expansion to triple the size of the existing terminal is scheduled to begin in 1998 with completion planned in 4 to 5 years. Financing is planned via a separately collected departure tax.[387] Included improvements are upgrading access roads; building a parkade, warehouse, and hotel; and enlarging the terminal aprons.[388]

Montreal

At Montreal's Dorval Airport, Aeroports de Montréal (ADM) plans a US$200 million expansion to the airport terminal. To improve traffic flow, the ADM also supports Dorval's proposed US$36 million plan to build a series of land-side roadway improvements, including overpasses, circular ramps, and underpasses.[389] ADM had planned to transfer all international flights from Mirabel to Dorval; however, these plans have been delayed by a February 1997 Quebec Superior Court ruling that included a "stop-work order on any construction at Dorval that would further ADM's plan to turn it into an international hub."[390]

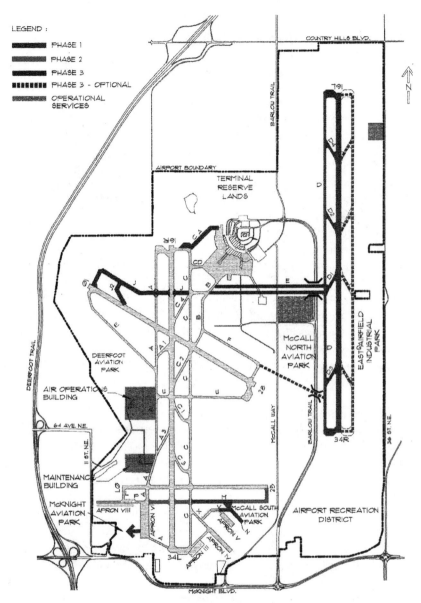

1.13 *Calgary International Airport contemplates the eventual construction of a new 14,000-ft parallel runway, with four high-speed exit taxiways, complete with full precision approach aids and airfield lighting systems, on the eastern side of the airfield.* Calgary International Airport.

Ottawa

Macdonald-Cartier International Airport plans to double the capacity of the main terminal within 10 years, which will be financed through passenger fees similar to those in effect in Toronto and Vancouver.[391]

Thunder Bay

At Thunder Bay, a new airport was completed in 1995 to prepare for the Nordic Championships.[392]

Toronto

Toronto's Lester B. Pearson Airport is undergoing a US$2 billion project to replace two of the existing three terminals with a single large horseshoe-shaped terminal with multiple piers radiating from it.[393] (See Fig. 1.14.)

Terminal 1, opened in 1964, has 23 gates; Terminal 2, opened between 1972 and 1979, has 32 gates; Terminal 3, opened in 1991, has 24 gates. The expansion, expected to be completed in 10 years, will increase the number of gates from 79 to 130.[394] The airport also has a phased air-side development project under way, which will add three runways over time to the airport's existing three runways. The first phase calls for construction of a new north-south runway by the year 2000.[395]

Vancouver

Vancouver is building a new international terminal building and runway. Vancouver's new 1.13 million square foot international terminal will have 15 gates, with sufficient capacity for 17 million passengers by the year 2005.[396] Vancouver International Airport underwent a C$350–500 million (US$231–300 million) expansion project, completed in 1996, which added a new international terminal, control tower, and runway.[397] Vancouver's new third parallel[398] all-weather runway, designed to accommodate the next generation of large passenger aircraft, was completed at a cost of US$100 million.[399] The new international terminal, built to be easily expandable, is already under construction to add seven more gates to the Canada-U.S. concourse; the gates are scheduled for completion in 2000.[400] In addition, Purolator Courier is constructing a US$10.2 million sorting facility at the airport.[401]

Yellowknife

At Yellowknife Arctic Airport, a new flight hangar will be added for national defense purposes as part of a US$9 million federal project.[402]

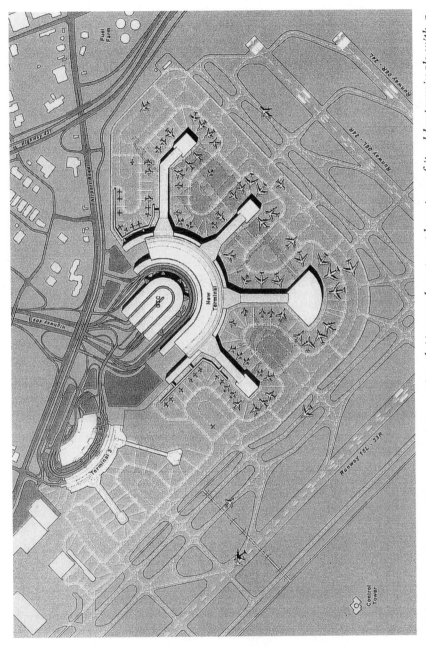

1.14 *Toronto's Lester B. Pearson International Airport plans to replace two of its older terminals with a single horseshoe-shaped terminal with piers radiating from it.* Greater Toronto Airports Authority.

Costa Rica

The Costa Rican government is negotiating for improvements to the Juan Santamaria International Airport. Improvements include the expansion of the terminal building from 168,000 m² to 309,000 m² at an estimated cost of $80 million.[527] Liberia International Airport opened in September 1996.[403]

Honduras

The Honduran government plans to build a new US$85 million international airport in Laguna del Pedregal just northwest of Tegucigalpa.[404]

Jamaica

At Montego Bay, construction of a new US$24 million airport terminal was scheduled for completion in December 1997.[405]

Mexico

The government of Mexico plans to spend US$27 million on modernizing its airports prior to their privatization."[406] Capacity is strained at Mexico City International Airport, leading to a proclamation by the Communications and Transport Secretariat (SCT) that studies are under way to consider "increasing air traffic capacity in the Valley of Mexico through the construction of at least one more airport," though no time frames were given.[407]

Montserrat

Construction of airport in the northern part of the island of Montserrat is to be financed by Britain.[408]

Panama

Tocumen International Airport was established in 1978.[409]

St. Vincent and the Grenadines

At Canouan Airport, a US$7.1 million project for airport development is planned.[410]

United States

U.S. domestic and international enplanements are anticipated to grow from 606 million to 920 million, or 52%, between 1996 and 2006.[412] Aircraft movements are expected to grow from 63 million operations in 1997 to 73 million in 2008.[413] It has been estimated that the United States will spend some $60 billion in airport upgrades within the next decade.[414]

The United States has some 567 commercial airports.[415] The United States' 50 busiest airports account for 81% of the nation's traffic. Twenty-three of these airports account for more than 20,000 hours of annual flight delay, costing the airlines more than $700 million a year.[416] In 1995, the Federal Aviation Administration designated 25 of the largest airports in the United States as "severely congested." Without additional capacity enhancements, this number is expected to climb to 29 by 2005.[417] The following are the airports with the longest gate-to-takeoff delays in the United States: (1) Newark International, (2) Honolulu International, (3) New York LaGuardia, (4) New York John F. Kennedy International, (5) St. Louis Lambert International, (6) San Francisco International, (7) San Diego International, (8) Salt Lake City International, (9) Miami International, and (10) Minneapolis/St. Paul International.[418] A ranking of America's best airports (based on speed of baggage delivery, ease of reaching gates, availability of ground transport, appropriate signage, cleanliness, attractiveness, quality of restaurants, and proximity of parking) found these the best of the bunch[419]:

1 Tampa

2 Pittsburgh

3 Charlotte

4 Nashville

5 Denver

6 Phoenix

7 Las Vegas

8 Salt Lake City

9 Atlanta

10 Baltimore

11 Honolulu

12 Seattle

13 Kansas City

14 Minneapolis/St. Paul

15 Oakland

Those ranking worst were Detroit, New York Kennedy, and Boston Logan. The following list summarizes several of the major U.S. airport projects recently completed, undertaken, or planned.

Albany, New York

Colonie Albany Airport, has opened a new $72 million 230,000-ft^2 terminal building.[420]

Atlanta

In 1980, Atlanta Hartsfield International Airport constructed wholly new terminal facilities on its existing airport site. The 1980 reconfiguration included three remote concourses from a new main terminal building; today, it has six concourses. (See Fig. 1.15.) Atlanta has a $2 billion, 5-year improvement program under way. A fifth runway for commuter aircraft, costing $418 million, was planned for opening in 2002.[421] Airport officials reevaluated the project and decided that a full 9000-ft runway should be added, at a cost of between $800 million and $900 million. A sixth runway may eventually be added, as well as another $200 million east-side terminal, to handle the projected 121

1.15 *Atlanta Hartsfield International Airport was reconfigured on its existing site in 1980. Atlanta introduced the parallel remote concourse concept, which is the most efficient layout for a connecting hub airport.* Denver International Airport.

million passengers who will flow through the Atlanta hub in 2015. After that, it will be time to consider building a second airport at Atlanta.

Austin

As a "peace dividend" resulting from the end of the Cold War, Bergstrom Air Force Base was decommissioned and transferred to the city of Austin in 1993. With the addition of a new parallel runway and a 500,000-ft^2 25-gate terminal, at a cost of $675 million, Bergstrom was converted to a civilian airport to replace Austin's Robert Mueller Airport, which cannot be expanded. Located only 8 mi from the Texas state capitol building Austin, Bergstrom was ready for cargo service in 1997 and passenger service in 1999.[422]

Baltimore

Baltimore/Washington International Airport completed a new international wing in its terminal in 1998.

Boston

Massachusetts is spending $4 billion on a tunnel to link downtown Boston to Logan Airport.[423] The airport is also adding new tiered roadways to link the Third Harbor Tunnel. Among land-side improvements at Boston Logan Airport are expansion and improvement of Terminal E, to add three new gates to accommodate wide-body aircraft and to complete a walkway connector between Terminals D and E; a 14,000-ft^2 shopping mall on Terminal C; a $100 million hotel; and a $120 million seven-story parking garage with spaces for 3150 automobiles linked to the four terminals by pedestrian walkways. Among air-side improvements are a new 5000-ft unidirectional runway, a centerfield taxiway, Category II navigational aids, and a $91 million airfield fueling system.[424]

Charlotte

Charlotte/Douglas International Airport plans to open a third parallel runway by 2000 to permit triple instrument flight rules (IFR) approaches, at a cost of $122 million.[425]

Chicago

Chicago Midway Airport began construction of a new terminal and parking garage in the summer of 1998. It will build two new concourses on the site where the original terminal was constructed. (See Fig. 1.16.) When completed, in the year 2002, the airport will have 3 times its prior terminal and concourse gate space. The $772 million overhaul will expand the number of gates from 29 to 41 and parking

1.16 *This model depicts Chicago Midway Airport's new terminal, piers, and parking garages, to be completed by 2002.* Photo by P. S. Dempsey.

spaces from 7311 to 12,862. Intermodal access to the Chicago downtown Loop will be via Chicago Transit Authority rail.[426] At Chicago O'Hare International Airport, the world's busiest, the parking garage is undergoing a $60 million face-lift.[427] Since 1993, the FAA has authorized $2.3 billion in passenger facility charge spending on airport improvements at O'Hare.[428] Beginning in 1999, in a $1 billion face-lift project, O'Hare will be rehabilitated to look more like Helmut Jahn's ultramodern design for United Airlines' remote midfield concourse.[429]

Cleveland

Cleveland Hopkins International Airport would like to build a new 10,950-ft^2 airport terminal by 2000 at a cost of $180 million, and expand two existing runways at a cost of $43 million.[430] Cleveland is also building a midfield terminal for Continental Airlines.

Dallas/Fort Worth

Dallas/Fort Worth International Airport (DFW) was built in 1974 as a "green field" airport. DFW extended an existing runway in 1993, and constructed a new runway in 1996, at a cost of $300 million, allowing DFW to accommodate triple simultaneous instrument landings. Another runway, projected to cost $100 million, could be open by 2001.[431] DFW plans to add an eighth runway, and to extend three existing runways by 2000 ft. DFW is planning two new terminals in the horseshoe shape of its original terminals. It is also upgrading its existing terminals, adding piers and gates, and upgrading its people-mover system. Specifically, it is expanding Terminal 2W, constructing a Terminal 3W, constructing a partial Terminal 4W, and adding a $650 million people-mover system to provide high-speed movement of passengers and baggage between terminals.[432] DFW has also added a five-story parking garage. DFW anticipates such additional infrastructure will be necessary to accommodate a doubling of its traffic anticipated by 2015.[433] At Dallas Love Field, upstart airline Legend Airlines is building a new terminal and four-level parking garage.[434]

Denver

Completed in 1995 at a cost of $5.3 billion, Denver International Airport is the first new airport to have been built in the United States since Dallas/Fort Worth International Airport was completed two decades earlier. Denver intends to add a sixth runway, 16,000 ft in length, by 2000, at a cost of $75 million.[435] It is also moving its remote employee parking lots closer to the main terminal.

Detroit

Detroit Metropolitan Wayne County Airport is planning a fourth runway, scheduled for completion in 2001, at a cost of $116.5 million.[436] It is also building a midfield terminal. A $1.6 billion expansion project is under way at Detroit Metropolitan Airport that includes a new 74-gate midfield terminal, a fourth parallel runway, and a new south access road that will cut passenger travel time to the airport.[437] The midfield terminal will have 64 domestic gates and 10 "swing" gates to be used for either international or domestic service. Included within the midfield terminal project is an overhead tram that will transport passengers between gates and an on-site parking garage with 5000 spaces. Plans include the option to add nine more domestic and two more international gates by 2010. Northwest Airlines will occupy the terminal upon completion, scheduled for December 2000.[438] Additionally, an interim project to improve the existing facilities and reduce delays was completed in early 1998. The project included construction on the Michael Berry International Terminal, doubling capacity to 14,000 passengers daily[439]; increased passenger shuttle services; and additional moving walkways.[440]

Fort Lauderdale

Fort Lauderdale–Hollywood International Airport is planning to extend an existing short parallel runway by 2002 at a cost of $270 million.[441]

Las Vegas

At a cost of $442 million, Las Vegas has launched a number of airport improvement and expansion projects. It developed a satellite terminal with four concourse wings connected to a central hub in the form of an X, which at full buildout will nearly double the gate capacity of McCarran International Airport. Architecturally, it incorporates the theme of flight and the glitz of Vegas.[442] It is linked to Terminal 1 by an above- and belowground monorail. That terminal's lobby and baggage claim areas have been expanded as well. Airport information systems have been integrated. A general aviation runway was upgraded to 9770 ft.[443]

Los Angeles

John Wayne Airport has built a new terminal. Burbank is planning a new terminal for opening in 2003. But community opposition to expansion projects at Los Angeles International (LAX), Burbank-Glendale-Pasadena Airport, and the conversion of El Toro Marine Corps Air Station to a commercial airport have left plans for dealing with future traffic in the southern California region up in the air. The proposed expansion at LAX calls for an additional runway, a new terminal, and a reconfiguration of the existing facilities that would nearly double its size.[444] The project is estimated to cost $8–12 billion, which would make it the most expensive public works infrastructure project in the United States.[445] Estimates call for 157.4 million air passengers annually in southern California by 2020, with 98 million using LAX.[446] Air industry experts favor the expansion of LAX, combined with the development of El Toro into an international airport, to alleviate congestion and handle future growth.[447] However, community opposition is strong and local politicians are urging the city to consider other suburban airports to help disperse the traffic. Alternatives include utilizing Ontario International, John Wayne Airport, and Palmdale Airport. Ontario recently completed a $270 million expansion project that included a new twin terminal complex with 26 additional gates and 4900 new parking spaces.[448] The expansion will enable Ontario to handle 10 to 12 million passengers annually; however, current restrictions limit the airport to 125,000 commercial flights a year.[449] Activity at John Wayne Airport is also restricted. A 1985 settlement of a lawsuit brought by local residents of

Newport Beach restricts capacity there to 8.5 million passengers per year.[450] Meanwhile, proponents of expansion at LAX do not consider Palmdale, located 60 mi from LAX, as a feasible alternative.[451]

Memphis

Memphis International Airport completed a new parallel runway in 1997 at a cost of $146 million. Reconstruction of an existing runway is scheduled for completion in 1999 at a cost of $95 million.[452]

Miami

Miami International Airport will spend more than $2.5 billion for terminal, concourse, and parking garage expansion or rehabilitation, a new roadway system, and 11 new cargo buildings.[453] A new runway is scheduled for completion in 2000 at a cost of $149 million.[454] It is anticipated that Miami will spend $4 billion on its airport by 2006, adding a fourth runway; increasing the terminal to 121 gates (up from 103), including a 1-mi linear terminal for American Airlines; and adding 14,000 parking spaces (up from 8000). A $6.2 billion intermodal center linking the airport with high-speed rail, light rail, car rental, roads, and boats approaching via canal is also contemplated.[455]

Minneapolis

Minneapolis/St. Paul International Airport extended an existing runway to 11,000 ft in 1996 and is planning a new runway for completion in 2003 at a cost of $120 million.[456] It is also building a new $2.9 million charter terminal.[457]

Myrtle Beach, South Carolina

Myrtle Beach International Airport has launched a $40 million, 5-year project to expand a runway to 10,500 ft and build a new international terminal.[458]

Nashville

Nashville International Airport opened a 750,000-ft^2, 46-gate terminal in 1987 and added a fourth runway in 1989. It added 25 commuter aircraft parking positions in 1992, and completed a new International Arrivals Building in 1994. (See Fig. 1.17.) Since 1984, it has issued $296 million in airport revenue bonds, and dedicated $55 million of passenger facility charges to various projects.[459]

New York

The Port Authority of New York and New Jersey plans to spend $4.3 billion to overhaul John F. Kennedy International Airport (JFK) and $1.5 billion on a rail line linking Manhattan with JFK and LaGuardia

1.17 *Nashville's new international arrivals building has separate departure and arrival land-side levels.* Metropolitan Nashville Airport Authority.

Airports.[460] Kennedy's air traffic control tower was demolished and replaced with a new 321-ft-high tower. The project includes an improvement of runways, turnoffs, and aprons, and a massive reconstruction of the airport roadway system. A new $435 million Terminal 1 opened in 1998. Terminals 2 and 3 are undergoing a $150 million modernization. A new $1.2 billion three-story 1.5 million ft² international Terminal 4 is scheduled to open in 2000. It will replace the existing international terminal, which is actually six buildings that have been consolidated since the first one was built in 1957.[461] Terminals 5 and 6 are undergoing renovation. Terminal 7 underwent a $120 million renovation in 1991. In 1997, ground was broken on a $143 million expansion and improvement. Terminals 6 and 7 are nearing completion of a $220 million project to consolidate and upgrade them. A new $46 million parking garage has been built adjacent to Terminals 1, 2, and 3.[462] Japan Airlines, Northwest Airlines, United Airlines, Korean Air, and AEI freight forwarders are building or have built cargo terminals at JFK.[463]

At LaGuardia Airport, the six-block-long Central Terminal Building, opened in 1964, is undergoing a $340 million expansion and modernization.[464] More than $800 million will be invested by the year 2000 in a redevelopment program that includes such projects as modernizing and expanding the Central Terminal Building; improving roads, runways, and taxiways; and modernizing gate and passenger service areas.[465]

Newark

Newark's two main runways are just 800 ft apart, well below the 4300-ft separation the FAA approves for instrument landings. As a consequence, Newark is one of the most congested airports in the United States. Newark's principal departure runway is being expanded to handle more departures by larger aircraft.[466]

Orlando

Orlando International Airport is among the most visually attractive in the United States. Lakes and palm trees adorn a gardenlike setting viewed from elevated trams which link the main terminal with the remote concourse pods. Recently, the airport spent $1.2 billion doubling the number of airport parking spaces, constructing a fourth airside terminal with 16 gates and connecting people-mover system, expanding the existing terminal, improving roadways, and building a new air traffic control tower and taxiway.

Philadelphia

Philadelphia International Airport is hemmed in by Interstate 95 to the north, a naval shipyard to the east, its terminal area to the west, and the required lateral separation of its east-west parallel runways coupled with the Delaware River on the south.[467] So as to relieve commercial aircraft capacity demand on its two runways, it will spend $1.3 billion to purchase 30 acres, rebuild aprons and taxiways, remodel the terminal building, and construct one commuter runway (the 5000-ft runway will cost $210 million).[468] It will also add new terminals, a people-mover system, and commuter facilities and make roadway improvements.[469] The airport recently completed a $45 million terminal rehabilitation program, and a $130 million consolidation of Terminals B and C.[470]

Phoenix

Phoenix Sky Harbor International Airport is undergoing a $128 million program of parking expansion, roadway relocation, general aviation improvements, and building a facilities and services complex.[471]

Pittsburgh

Pittsburgh completed a new terminal facility in 1988 for $690 million.[472] A new terminal complex, called Midfield because of its place-

1.18 *Pittsburgh reconfigured its airport in 1992. The design includes an X-shaped midfield remote terminal.* Denver International Airport.

ment between the two main runways, opened on October 1, 1992. The terminal replaced the antiquated 1950s vintage terminal, tripling the size of the old complex. The 2.2 million ft^2 facility features an X-shaped design containing 75 gates.[473] (See Fig. 1.18.) Funds for the project were secured from a variety of sources, including about $228.5 million from the county, state, and federal governments. The remaining $554.5 million was raised from general revenue bonds, which will be repaid through airline landing fees and lease agreements. USAirways accounts for more than 80% of the traffic at Pittsburgh International Airport.[474] In late 1996, Pittsburgh International received an additional $8.9 million in FAA grants to rehabilitate a runway and taxiways, repair apron and taxiway settlements, acquire snow removal equipment, and provide sound insulation to some of the residences of the nearby communities.[475]

Portland, Oregon
Portland International Airport is planning a $300 million terminal expansion, a new 13-gate concourse, and an expansion of international gates.[476]

Raleigh-Durham
Raleigh-Durham International Airport opened a new Terminal A and a five-story, 2700-space parking garage.[477]

Sacramento

Sacramento International Airport has opened a new $95 million, 325,000-ft^2, 12-gate Terminal A, which will double the size of the airport.[478]

St. Louis

Lambert International Airport has received FAA approval for a plan to build a new 9000-ft runway, add 300,000 ft^2 of space and four gates to the main terminal, and add a new Y-shaped concourse with 35 new gates.[479] The project is estimated to cost $2.6 billion.[480]

Saint Claire County, Illinois

The U.S. government and State of Illinois built the $313 million Mid-America Airport, near Scott Air Force Base, east of St. Louis. As of 1998, no airlines were serving it.[481]

Salt Lake City

Salt Lake City's first two runways were constructed in 1937. Salt Lake City International Airport initiated a major runway development program in 1992. Located west of the terminal building, the 12,000-ft runway was completed in 1995, as were Concourse E and an international arrivals building. Today, the airport has three north-south runways, and a northwest-southeast crosswind runway. The airport is planning a $1.7 billion expansion, which includes a dynamic new main terminal and remote midfield concourses to enhance the efficiency of the airport as a connecting hub.[482] The new terminal and concourses are scheduled for completion in 2003.[483]

San Diego

The single runway at San Diego International Lindbergh Field is the busiest in the United States. The airport is studying the possibility of building a second runway.[484]

San Francisco

San Francisco has a $2.4 billion improvement and expansion program under way, including a new 2 million ft^2 international terminal constructed over the access roadways, to be completed in 2000.[485] The new terminal will have between 24 and 26 international gates capable of accommodating wide-bodied aircraft. It will also have 140,000 ft^2 of retail and concession space. Runway inadequacy has long been a problem at San Francisco, both because the four runways cross one another and because the parallel runways are too close for instrument landings during inclement weather, such as the fog which often hangs over San

Francisco Bay. A fifth runway is being planned, with construction to begin around 2001.[486] It will be built on landfill in San Francisco Bay.[487] An airport rail transit system will be completed in 2001, serving nine stations at the airport, and connected to the new Bay Area Rapid Transit (BART) station being built to link the airport to downtown and the bay area suburbs. A new parking garage adjacent to the international terminal will add 2100 parking spaces to the 3200 spaces in the garage opened in 1996. New roadways, rental car facilities, and air cargo facilities are also being built. By the year 2006, San Francisco International Airport expects to handle about 50 million travelers a year.[488]

San Jose

San Jose International Airport plans nearly $1 billion in improvements by the year 2010, including airport extensions, taxiway improvements, and a third terminal, which will bring the terminal space at the airport to 775,000 ft^2 and 40 gates.[489]

Seattle

If it can overcome the environmental hurdles, Seattle-Tacoma International Airport will spend more than $400 million to build a new runway between 5200 and 8500 ft in length.[490] Seattle is also considering adding gates to Concourse A by 2000 and constructing a North Unit Terminal within the next 10 to 15 years.[491]

Washington, D.C.

Metropolitan Washington Airports Authority is spending $2 billion on construction of new and expanded terminals at Washington Ronald Reagan National Airport and Dulles Airport.[492] It spent nearly $1 billion in capital improvements at Ronald Reagan National Airport, including a new North terminal with 35 gates, and an expansion of the main terminal with an additional nine gates.[493] National also added two new parking structures, with a total of 3500 parking spaces.[494] Dulles is expanding its main terminal building by an additional 600,000 ft^2, to 1.1 million ft^2. At completion, it will be 1240 ft long, consistent with architect Eero Saarinen's original design. (See Fig. 1.19.) Dulles also has a 422,000-ft^2 midfield concourse. Originally built with 17 gates, it is designed ultimately to accommodate 44 gates.[495]

Windsor Locks, Connecticut

Bradley International Airport in Windsor Locks is undergoing a $135 million renovation and expansion, which will demolish the

(*Text continues on p. 82.*)

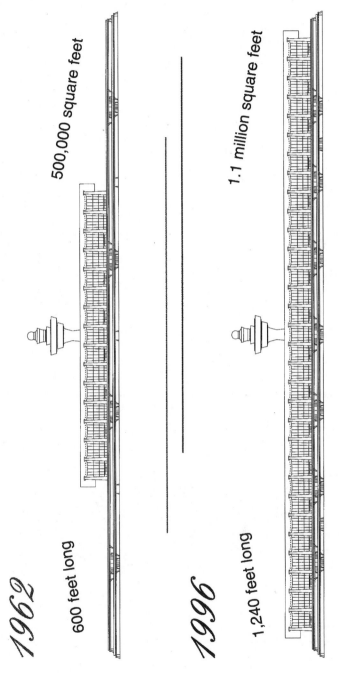

1962

600 feet long

500,000 square feet

1996

1,240 feet long

1.1 million square feet

1.19 *Completed in 1962, the main terminal at Dulles International Airport is undergoing an expansion that will more than double its capacity.* Metropolitan Washington Airports Authority.

Table 1.6. Major Airport Development and Expansion Projects in North America

Nation	Airport	Development	Cost, million US$
Canada	Calgary	Expansion and upgrade	650
	Edmonton International	Renovation and expansion	250
	Dorval (Montreal)	Expansion	236
	Lester Pearson (Toronto)	New terminal	2000
	Vancouver International	Expansion	330
	Yellowknife Arctic	New hangar	9
Costa Rica	Santamaria International (San Jose)	Expansion	105
El Salvador	El Salvador International	Expansion	n.a.
Guatemala	Guatemala City	New airport	n.a.
Honduras	Laguna del Pedregal	New airport	85
Jamaica	Montego Bay	New terminal	24
Mexico	Mexico City International	New airport	1000
	Monterrey International	Expansion	25
	Guadalupe	Expansion	15
St. Vincent and the Grenadines	Canouan	Airport development	7
United States	Albany (New York)	Runway extension; new runway	13

Airport	Project	
Austin	New terminal, improvements	335
Atlanta	New runway, terminal	2000
Baltimore/Washington	Terminal, runway	286
Bentonville	Expansion	125
Bergstrom International (Austin)	Conversion to civil airport; new runway and terminal	675
Birmingham	Runway, taxiway improvements	8
Boise	Runway extension	8
Boston Logan International	New tunnel	4000
Charlotte	New runway	122
Chicago Midway	Runway reconstruction	521
Chicago O'Hare	Runway extension, improvements	2274
Cincinnati	Reconstruction	375
Columbus	Terminal expansion, runway improvement	127
Cleveland Hopkins	Runway replacement, extension	223
Dallas/Fort Worth International	New runways, extensions, terminal improvements	496
Denver International	New airport opened in 1995; sixth runway	5375

Table 1.6. Major Airport Development and Expansion Projects in North America (*Continued*)

Nation	Airport	Development	Cost, million US$
United States (*cont.*)	Des Moines	Runway extension	22
	Detroit	New runway, terminal	1863
	El Paso	New runway; extension	38
	Ft. Lauderdale	Runway extension	270
	Ft. Myers	Terminal improvement; new runway	244
	Grand Rapids	Runway extension	58
	Greensboro	Runway extension	16
	Greer	New runway; extension	84
	Houston Intercontinental	New runways; extension	96
	Huntsville	Improvements	20
	Indianapolis	New runway, improvements	115
	Jacksonville	New runway	50
	Kahului	Runway extension	40
	Kansas City	Runway extension	12
	Las Vegas	Runway, terminal improvements	1585
	Little Rock	Runway extension	31
	Louisville	New runway	59

City	Project	
Lubbock	Runway extension	5
Memphis	Runway extension; new runway	241
Miami International	Expansion; new runway	2500
Midland	Runway expansion	5
Milwaukee	New runway, extension, realignment	93
Minneapolis	New runway; expansion	229
Myrtle Beach International	Runway expansion; new terminal	40
Nashville	New runway; expansion	351
New Orleans	New runway	400
Newark	Runway extension, monorail	630
New York Kennedy	Redevelopment	4300
New York LaGuardia	Terminal transit system, rail	479
Norfolk	New runway	75
Oakland Metropolitan	New runway; extension	83
Oklahoma City	Runway extensions; new runway	34
Omaha Eppley Field	Runway extension	n.a.
Orlando International	Redevelopment	1200
Palm Beach	Runway extensions	15

Table 1.6. Major Airport Development and Expansion Projects in North America (*Continued*)

Nation	Airport	Development	Cost, million US$
United States (*cont.*)	Philadelphia	Redevelopment	1300
	Phoenix	New runway; extension	300
	Pittsburgh	New terminal	690
	Providence	New terminal, apron	108
	Portland	Airfield improvements	131
	Raleigh-Durham	Runway extension; new runway	n.a.
	Richmond	Runway extension	45
	Reno/Tahoe	Runway extensions	n.a.
	Rochester	New runway; runway extensions	17
	Saint Claire County (Illinois)	New airport	313
	St. Louis	New runway, concourse, parking	2600
	Salt Lake City	Redevelopment	1700
	Sacramento	Improvements	190
	San Antonio	New runway; runway reconstruction	420
	San Diego	Improvements	124

San Francisco International	New terminal, transit system, parking, roadways	2400
San Jose	Runway extension	116
San Juan	Taxiway, runway improvements	133
Santa Ana	Runway extension	n.a.
Sarasota-Bradenton	New runway; extension	15
Savannah	New runway	20
Seattle	New runway; terminal expansion	1500
Spokane	New runway	11
Syracuse	New runway	55
Tampa	New runway, extension, reconstruction	119
Tucson	New runway, terminal improvements	30
Tulsa	New runway	115
Washington National and Dulles International Airports	New and expanded terminals	2000

Table 1.7. Major Airport Development and Expansion Projects in South America

Nation	Airport	Development	Cost, million US$
Argentina		ATC modernization	185
		Airport system restructuring	680
Bolivia	National system	ATC modernization	200
	La Paz, Cochabamba, Santa Cruz, Trinidad	Modernization	100
Brazil	Uberlandia	Upgrading	40
Chile	National system	ATC modernization	30
	Iquique	New terminal	5
	Tepaul/Puerto Montt	Replacement airport	5
	Concepcion, Punta Arenas, La Sarena, Calama	Improvements	17
	Santiago	Terminal development	150
Colombia	Barranquilla, San Andres	New ATC	38
	Bogota	Second runway	97
Ecuador	Quito	New airport	300
	Guayaquil	New airport	300

Peru	Lima, Arequipa, Cuzco, Iquitos	Modernization and expansion	506
Trinidad and Tobago	Trinidad	New airport	65
Uruguay	Montevideo	Expansion	180
Venezuela	Maiquetia International	Terminal expansion	60
	Falcon	New airport	n.a.

existing terminal, built in 1949, and replace it with a new 600,000-ft² terminal.[496]

Table 1.6 summarizes nearly US$50 billion in major airport projects in North America.

South America

More than US$4 billion in airport-related undertakings are on the drawing board in South America (see Table 1.7).[497]

Argentina

At Buenos Aires, a private group, Aeroisla, has expressed an interest in replacing both Ezeiza International Airport and Aeroparque Airport with a new airport built on an artificial island on the River Plate.[498] The Argentina 2000 consortium has proposed a plan to build a new, joint international-domestic airport 2.5 km offshore from Buenos Aires on the new island, with a rail connection between the airport and the city. The existing airport, Ezeiza, would be converted into an air cargo terminal.[499]

Bolivia

A modernization project for three airports (El Alto in La Paz, Viru Viru in Santa Cruz, and Cochabamba) is under way. These three airports account for 90% of domestic and international air traffic in Bolivia.[500] The project, undertaken by the U.S. consortium Airport Group International (AGI), will bring the airports up to a B international rating under International Air Transportation Association standards, at a cost of $200 million.[501] Increases in air traffic of 6% annually are projected, with Viru Viru International Airport at Santa Cruz becoming "a hub of regional and inter-American traffic driven by economic integration of the Mercosur."[502]

Brazil
Bahia

Salvador International, the top gateway into northeastern Brazil, is undergoing a 2-year expansion program at an estimated cost of $100 million.[503]

Brasilia

A five-stage modernization and expansion of Brasilia International Airport began in 1990. The first three stages were completed in

September 1996, costing US$200 million and bringing capacity to 4.5 million passengers. The fourth stage, estimated to cost US$55.9 million, includes a 23,595-m² addition to the airport complex, two boarding bridges, and a new 2700×45 m runway built to allow simultaneous operation with the existing runway. The fifth stage will include construction of the south satellite and seven more boarding bridges. Upon completion, the airport will be able to handle 8 million passengers per year.[504]

Curitiba

A new terminal was completed at Afonso Pena International.[505] The terminal is capable of handling up to 3.5 million passengers annually.[506]

Fortaleza

The passenger terminal at the new Pinto Martins International Airport opened in November 1997. This terminal will triple the airport's passenger capacity.[507]

Porto Alegre

A 3-year expansion and modernization project was started in January 1997 at Salgado Filho International. The project includes a new 35,000-m² three-story terminal with six boarding bridges, expansion of the apron, and work on the access road system. Salgado Filho will be equipped with an integrated information system allowing central control of all airport systems and services. The expansion, estimated at US$105.5 million, will bring capacity to 3 million passengers annually.[508]

Rio de Janeiro

Structural work on the second passenger terminal at Rio de Janeiro International Galeao will increase the degree of automation, making it a "smart facility." At an estimated cost of US$170 million, the expansion will increase the airport's handling capacity from 6 million to 13.5 million passengers.[509] A new cargo terminal has also been completed.[510]

Sao Paulo

Infraero, the government agency responsible for airport management and development, plans to begin construction on passenger terminal no. 3 at Guarulhos International Airport in 1999, with limited operations beginning in 2001 and completion scheduled for 2003. The total cost of the new terminal is $300 million.[511] Recent projects at the airport have included check-in and arrivals expansion, new air cargo terminals, and parking lots.[512] Construction of a new maintenance center and refurbishment of the passenger lounge and auxiliary runway at Sao Paulo Congonhas has also been completed.[513]

Other Airport Projects

Expansion and renovation of passenger terminals is planned for Pelotas, Maceio, Belo Horizonte, Pampulha, Uberaba, Uberlandia, Londrina, Campo Grande, Porto Velho, and Florianopolis.[514] Runway work is in progress at Goiania, Recife, Navegantes, and Florianopolis.[515] A new cargo terminal has been completed at Petrolina airport.[516] Infraero plans to expand the airports at Aracaju and Maceio.[517] Infraero also plans to begin construction of a new international airport at Belem.[518]

Chile
Santiago

Opened in 1994, Aeropuerto Arturo Merino Benitez has already exceeded its handling capacity of 3 million passengers.[519] The Chilean government plans to expand the terminal from 25,000 to 65,000 m², expand the apron, improve the parking lot, build a new control tower, and upgrade an electrical substation.[520] The contract for the expansion of the airport was awarded to SCL Terminal Internacional Santiago. The company plans to begin construction in May 1998, with completion of the project set for the year 2000.[521] Construction costs are estimated at US$150 million.[522] (See Fig. 1.20.)

Concepcion

Expansion and modernization is planned for Carriel Sur Airport. The project includes construction of a 6000-m² passenger terminal and expansion of the apron. The project, awarded under a 15-year concession, will bring handling capacity to 800,000 passengers annually.[523]

Colombia
Barranquilla

A 15-year concession for Barranquilla International Airport has been awarded to Spain's Aeropuertos Espanoles y Navegacion Aerea (AENA).[524]

Bogota

A second runway is being built at Eldorado International Airport under a build-transfer-maintain concession by Compañia de Desarrollo

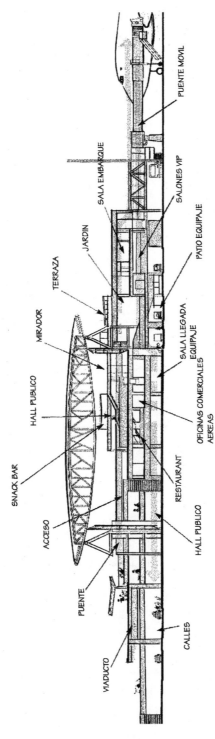

SNACK BAR
HALL PUBLICO
MIRADOR
TERRAZA
JARDIN
SALA EMBARQUE
PUENTE MOVIL
ACCESO
SALONES VIP
PATIO EQUIPAJE
SALA LLEGADA
EQUIPAJE
OFICINAS COMERCIALES
AEREAS
RESTAURANT
HALL PUBLICO
PUENTE
CALLES
VIADUCTO

1.20 *This artist's rendering depicts the link between land side and air side that a passenger traverses when passing through the international terminal at Santiago, Chile's, Aeropuerto Arturo Merino Benítez.* Santiago Aeropuerto.

85

Aeropuerto Eldorado S.A. (Codad) at a cost of US$145 million. Completion is expected by August 1998.[525]

Cartagena

A 15-year concession for Cartagena Airport was awarded to Schiphol Management of the Netherlands for US$2.5 million in October 1996.[526]

Ecuador

Ecuador's Civil Aviation Directorate (DAC) plans to replace the existing airports at Quito (Mariscal Sucre International) and Guayaquil (Guayaquil International) at an estimated cost of US$600 million.[528] The town of Puembo, about 10 mi east of Quito, appears to be the likely site for the new airport. Available at Puembo is approximately 3630 acres on which to build, compared with 257 acres at the current airport in Quito. The town of Daular, about 20 mi southwest of Guayaquil, is the leading site for replacement of Guayaquil International. Space there totals 4984 acres compared with 420 acres available in Guayaquil.[529] The concession, to be awarded in late 1998, also will include about US$50 million in road improvements leading to Quito and Guayaquil.[530]

Peru

The Peruvian government plans to expand and modernize Lima's Jorge Chavez International Airport. The airport, Peru's principal gateway, handled 3.9 million passengers and 80,000 tons of cargo in 1996.[531] A private concession, with bidding to begin in 1998, will include the total modernization of the airport and the construction of a second runway.[532]

Trinidad and Tobago

Tobago's Crown Point airport underwent a US$50 million renovation project that included a new terminal building, new baggage handling equipment, a first-class lounge, and a restaurant.[411,533] Construction, which will quadruple the number of passengers the airport can handle, was scheduled to be completed in December 1997.[534]

Uruguay

The Ministry of Defense is supervising a two-phase BOT privatization scheme for improvements to Montevideo's Carrasco International Airport.[535] The first phase, at an estimated cost of US$180 million, includes rehabilitation of paved surfaces, a drainage system, a new passenger terminal with parking, a new cargo terminal, new airport rescue and fire fighting facilities, and extension of the main runway to 3000 m and a secondary runway to 2250 m. The first phase of improvements is expected to handle the growth of the airport to 2010, with traffic estimated at 3 million passengers annually. A second phase of terminal expansion would handle growth beyond 2010.[536] Private operation is expected to begin by early 1999.[537]

Venezuela
Barcelona

The state of Anzoategui wants to improve its airport at Barcelona. A very early plan calls for expansion of the terminal and cargo storage facilities, and construction of a general aviation ramp. Economic feasibility is questionable.[538]

Caracas

Earlier attempts at privatization of Maiquetia International Airport Simon Bolivar in Caracas were met with violent demonstrations by the airport's unions of hugely overstaffed personnel and were abandoned. The government, however, wants to privatize its airports and plans to open bidding on renovation and expansion of Simon Bolivar in early 1998. The project calls for the redesign and expansion of the passenger terminals, the construction and operation of an airport hotel, and the feasibility study of a second runway. The airport has more than 80,000 operations and handles more than 4 million passengers yearly.[539]

Falcon

The state of Falcon has a tentative plan for construction of a "tourism development airport" at San Juan de los Cayos located near the rapidly developing resort area of Morrocoy. Again, economic feasibility is questionable.[540]

Sucre

The state of Sucre, on the northeastern Atlantic coast, has appointed the U.S. aviation consulting firm Aarotec to develop privitization strategies for three of the state's main airports. Antonio Jose de Sucre International at Cumana and the domestic facilities at Carupano and Guiria account for a combined 30,000 aircraft movements each year, with approximately one-third being commercial scheduled and charter traffic.[541]

Africa

Algeria

Air Algeria is building a US$111.7 million maintenance hangar at Houari-Boumediene International Airport, scheduled to be completed in 2000.[542]

Cameroon

The Cameroon government recently completed construction of an international airport in President Biya's province of origin.[543]

Egypt

Cairo

Cairo International Airport is expanding Terminal 2 and Terminal 3 to serve 5000 domestic and international flights during heavy traffic hours.[544] Scheduled for completion in 2000, the estimated cost of this project is US$380 million.[545]

Hurghada

US$1.8 million is being spent to upgrade to expand halls and parking areas at Hurghada Airport.[546]

Luxor

Expansion and upgrades at Luxor airport are budgeted at US$35 million, including a new terminal.[547]

Mersa Alam

A US$35 million build-operate airport scheme was recently awarded to a Kuwait contractor; the operating concession is to run for 40 years.[548] A 3-km-long runway, a terminal, and a control tower are included in the bid.[549]

Ethiopia

Ethiopia's Addis Ababa Airport is undergoing a US$240 million redevlopment. This includes a new runway with three exit taxiways and an airfield lighting system at a cost of US$68.5 million.

Gambia

Yundum Airport was upgraded by the U.S. National Aeronautics and Space Administration (NASA) in 1989 for U.S. space shuttle emergency landings.[550]

Kenya
Eldoret

The new Eldoret International Airport opened in September 1997.[551] Construction of the airport began in January 1995 amid much protest.[552] Some Kenyans accused the government of "diverting funds voted for projects in other parts of the country," such as the potholed Nairobi-Mombasa highway.[553] The airport is expected to be one of the region's largest, though it had to be downsized after the International Monetary Fund expressed concern that the airport would interfere with Kenya's ongoing economic reforms.[554] It was initially planned to cost US$50 million.[555]

Mombasa

At Moi International Airport, a US$82 million expansion and upgrade includes a new passenger terminal and runway capable of handling a Boeing 747; the project was completed in 1995.[556]

South Africa

In 1997, South African High Commissioner Carl Werth disclosed that Singapore companies were involved in studies to upgrade three of South Africa's international airports and to potentially develop two new airports.[557] South Africa's three main airports are located in Johannesburg in inland Gauteng province, coastal Durban in Kwazulu Natal, and Cape Town in the Western Cape province, on the southern tip of the African continent.[558] Commissioner Werth had cautioned that the projects were only in the exploratory phase but was encouraged by Singapore's involvement.[559] He also pointed to the lag in infrastructure development due to apartheid in the last 20 years and stated that, "when we talk about ports or airports, Singapore is the yardstick by which the world measures itself."[560]

Cape Town

Cape Town has seen a dramatic increase in demand since the world community ended its boycott of South Africa. In 1993, only three international airlines provided 10 flights a month. By 1998, 21 international carriers operated 160 flights per month. To meet the increasing demand, the new airport authority, Airports Company South Africa, has committed R1.5 billion to upgrade Cape Town International Airport, a program stimulated in part by the city's bid to host the 2004 Olympic Games.[561]

Johannesburg

U.S.-based Lea & Elliot Inc. has been retained to complete a feasibility study for a proposed US$60 million airport track transit system for the Johannesburg International Airport.[562] Lea & Elliott is a transportation consultant specializing in fare structures, fare collection systems, equipment specifications, test monitoring, revenue control procedures, and system audits.[563] The U.S. Department of Commerce estimates that the project could generate up to US$30 million in U.S. exports.[564]

Kruger National Park

By May 1997, the South African government had also approved a grand plan for a new international airport adjacent to the Kruger National Park.[565] The plan would consist of a 2-mile runway and airport on a green field site outside the Kruger border in the Mpumalanga province.[566] The plan was prompted by statistics indicating that there was a large untapped market for flights to Kruger Park.[567] The airport is planned to accommodate aircraft as large as Boeing 747s and is designed to handle up to 400,000 passengers a year in the first phase. The estimated cost is approximately US$32 million, with completion set for early 1999.[568]

La Mercy

This major airport project in South Africa includes upgrading the nearby facilities at the existing Durban International Airport, and eventual relocation of Durban's facilities to a new international airport at La Mercy.[569]

Swaziland

In Mbabane, Matsapha International Airport is undergoing a 3-year, US$4.5 million project to upgrade runways to Boeing 747 capability and build a new terminal.[570]

Zimbabwe

Harare

In Zimbabwe, the government has decided to build a more modern international airport in Harare to facilitate and increase tourism.[571] The US$363 million contract to build a new international airport went to a little-known Cypriot firm.[572] The selection of the Cypriot firm has been criticized because the firm is owned by Zimbabwe President Mugabe's nephew, Leo Mugabe.[573] At Harare Airport, the three-phase US$178 million upgrade includes a new international terminal, a control tower, and a supporting road network with related ground installations. The project was scheduled for completion in 1997.[574]

Zimbabwe Airports Improvement Project

Zimbabwe anticipates spending US$100 million upgrading 11 domestic and regional airports, including Bulawayo, which serves Zimbabwe's principal industrial district, and Victoria Falls, Kariba, and Hwange, which serve tourist destinations. This includes US$36 million for navigation and safety equipment, US$22 million for lighting and roads, US$20 million for buildings and runways, US$15 million for contingencies, and US$7 million for engineering services and training.

Table 1.8 summarizes nearly US$1.5 billion in major airport projects in Africa.

Summary and Conclusions

As this survey reveals, airports are being built or expanded in every corner of our shrinking planet. Enormous public resources are being devoted to providing adequate infrastructure to support transportation, the veins and arteries of the economic system. Airport bottlenecks can choke aviation with inefficiency; conversely, abundant airport capacity sometimes can be a catalyst for productivity, and thereby, broader economic expansion.

At the dawn of the twenty-first century, air transportation is poised for unprecedented growth. Airport infrastructure development is an essential ingredient in the efficient flow of passengers and commodities in a rapidly growing global economy, on a planet made smaller by air transportation. As portals through which people and high-value commodities pass, airports are gateways to the world. As catalysts for economic growth, airports assist the world's economy in growing richer and more diverse as each nation produces that for which it has a comparative advantage. The breadth of markets grows as transportation networks extend their reach. Enrichment cuts across social planes as well, as diverse cultures interact, cross-fertilize, and flourish.

In today's information age, where time is money, airport development and expansion are among the top priorities in a comprehen-

Table 1.8. Major Airport Development and Expansion Projects in Africa

Nation	Airport	Development	Cost, million US$
Algeria	Houari Boumediene International	New hangar	112
Cape Verde	Amilcar Cabral	Upgrading of terminal and communications	14
Egypt	Cairo International	Terminal expansions	380
	Hurghada	Upgrade	2
	Luxor	Expansion; upgrades	35
	Mersa Alam	New runway, terminal, and tower	35
Ethiopia	Addis Ababa	Redevelopment	240
Ivory Coast	Felix Houphonet-Boigny	Upgrade	9
Kenya	Eldoret International	New airport	50
	Mombasa	Expansion and upgrade	82
Senegal	Youff International (Dakar)	Upgrade of terminal, navigation, and communications	9
South Africa	Johannesburg	Transit system	60
Swaziland	Matsapha International	Runway upgrades	5
Zimbabwe	Harare	New airport	363
	11 airports	Upgrades	100

sive infrastructure design and development plan. Poor infrastructure, bad planning, and inadequate facilities yield constant problems of delay, inefficiency, and increased costs that restrain economic growth and prosperity. Airports are being pushed beyond their capacity to meet the needs of continued air traffic and economic growth.

By 2001, a third of the world's population will travel by air,[575] with passengers in the Asian-Pacific region accounting for about half of the world's total.[576] Governments worldwide are gearing up to meet this demand. The People's Republic of China, anticipating annual growth in air traffic of as much as 14%,[577] estimated that it would need to build 400 new airports over the next decade alone.[578] The United States, which by most standards already boasts a mature airport network, continues renovation and expansion of its facilities in an effort to increase efficiency and meet the rising demand. Whether renovating, expanding, or building "green fields," investments of hundreds of billions of dollars are being placed into the development of airports worldwide to help fuel local economies and link underdeveloped markets.

No longer saddled by antagonistic political ideologies, the global market can begin to realize its full potential. With the emergence of a global economy, the airport's role in this process becomes more clearly understood as the critical link between the capitalist and the consumer, its value increasing with the growing prosperity of the markets it serves.

2

Green Fields

"For all the growth that we see in the future...the single most important constraint is lack of airport capacity. We will have adequate airplanes, adequate air crews, adequate air carriers, adequate air traffic control. The science we have will accommodate the growth. The single, most potentially important constraint to this is airports."[1]
—DAVID HINSON, FEDERAL AVIATION ADMINISTRATION ADMINISTRATOR

"All travel is circular....The grand tour is just the inspired man's way of heading home."[2] —PAUL THEROUX, AMERICAN WRITER

In Chap. 2, we examine:

- Several of the world's new "green field" airports built, or being built, on wholly new sites
- The capacity and cost of these new airports
- What should be done with the existing airports that the new airports replace

Introduction

A chain is only as strong as its weakest link. The growth in flights and passengers can create congestion in the air or on the ground. At airports, bottleneck congestion may manifest itself on the land side (in terms of surface access), air side (in terms of runway, tarmac, and airspace), and in the terminal buildings sandwiched between the surface and air transportation modes. Some of that can be resolved with better utilization of scarce resources, such as technological advances in aircraft and aircraft navigation, multiple-carrier uses of ticket counters and gates, or peak period pricing. Ultimately, increased demand and congestion can cause airports to expand terminals and add runways.

Usually, the airport is hemmed in by unyielding urban development. Occasionally, with the right mix of need, political leadership, public support, financial ability, and available land, wholly new airports are built on "green fields." In this chapter we examine several of the new green field airports which have recently been or are being built around the world (Table 2.1). Subsequent chapters will elaborate on this introduction to the world's new green field airports.

This is by no means an exhaustive list. For example, the Mexican government wants to build an airport near Mexico City's congested Benito Juarez International Airport.[3] Earlier "green field" airports are also worthy of mention. In 1968, a regional airport authority was created and the first dirt turned on what was to become Dallas/Fort Worth International Airport.[4] Major reconfigurations of terminals on runways at existing airports also are worthy of mention, such as Atlanta's Hartsfield International Airport (reconfigured and rebuilt in 1980) and Pittsburgh International Airport (reconfigured and rebuilt in 1988).

In most instances, new airports are not built until existing airports are well beyond capacity. One source summarized the movement toward green field sites in Asia and the need therefore:

> *Inside [Hong Kong's] Kai Tak, the low-wattage functionality is redolent of a bygone era of austerity. Passengers are many, facilities few. When it rains, those passengers have to dodge the drips from the ceilings, as well as skirt the plastic buckets hastily produced by airport staff to catch the leaks. In short, the Kai Tak experience is one to be endured rather than enjoyed.*

> *Not that Kai Tak is uniquely grim. Kimpo in Seoul, Subang near Kuala Lumpur and Don Muang in Bangkok are international airports in the same ugly company. Victims of an unpredicted surge in air traffic over the past couple of decades, they have been stretched and enlarged and coaxed into the mid-1990s. Each will have to soldier on for at least a couple more years yet. But by 2002 all should have crumbled before the bulldozers; few who have passed through them will shed any tears.[5]*

Table 2.1. Major New Green Field Airports

Airport	Opening date	Projected cost, US$	Runways (ultimate capacity)		Passengers, millions (1997)
Munich Strauss	1992	7.1	2		18
Osaka Kansai	1994	14.4	1	(3)	17
Denver International	1995	5.3	5	(12)	35
Macau International	1995	1.1	1		2
Kuala Lumpur International	1998	3.2	1	(2)	16
Hong Kong International	1998	20.1	1	(2)	29
Oslo Gardermoen	1998	1.9	2		13
Shanghai Pudong International	1999	1.5	1		n.a.
Seoul Inchon International	2001	5.8	2	(4)	37
Athens Spata International	2001	2.7	2		n.a.
Guangzhou Hua Du International	2001	2.4	2	(3)	n.a.
Second Bangkok International	2003	3.6	1	(4)	25
Sydney Second Airport	2006	n.a.	n.a.		21

Designers of new airports have the benefit of hindsight. Given adequate economic and land resources, airport planners often can avoid the problems of the airports being replaced. They can build for capacity levels greater than immediate needs, and leave sufficient land for terminal, concourse, and runway expansion in the future.

Munich's Franz Josef Strauss Airport

Opened: May 17, 1992

Cost: US$7.1 billion

Runways: 2

Gates: 20 main, 14 auxiliary

Size: 5.4 mi^2

Passengers: 18 million (1997)

Franz Josef Strauss Airport replaced Munich-Riem Airport, which was opened in 1939 only 7 mi from central Munich. It was almost completely destroyed by Allied bombing in 1945.[6] With the opening of Strauss, Riem was converted to a trade fair and exhibition area. With one 2804-m (9200-ft) runway, Riem was designed to handle 7 million passengers a year. But by 1986, it was handling 8.4 million, by 1987, 9.3 million, and by 1990, 10.7 million. In 1989, 43% of all flights at Riem were delayed more than 15 minutes because of air traffic control and capacity problems.[7]

Strauss handled 13.5 million passengers in 1994, 26% more than Riem in its final year.[8] The new airport has sufficient capacity to handle 17 million passengers a year. The terminal can be expanded to accommodate 30 million passengers annually.[9]

Strauss' capacity will take some of the load off other German airports. Willi Hermsen, CEO of Munich's airport authority, notes, "Munich came just in time to take part of the overflow from Frankfurt." Already, Munich boasts that its airport is Germany's most punctual, with an 85% on-time rate for international flights and 92% for domestic flights. It guarantees 35-minute connections and 8.5 minutes for luggage to reach baggage claim.[10] (See Fig. 2.1.)

The new airport includes the Munich Air Cargo Center, large enough to accommodate seven 747 aircraft side by side. The airport also has a large Lufthansa hangar, which will be able to service six 747-400s

2.1 *Munich's Franz Josef Strauss Airport is growing both in traffic volume and capacity, as expansion plans are laid.* Flughafen München.

at once.[11] Given the inability of Frankfurt's airport to handle additional capacity, Lufthansa has begun to shift more and more of its operations to Munich.

In 1998, Flughafen München selected the German architectural firm of Koch und Partner to design a US$955 million second terminal, which will be connected to Terminal 1 by the München Airport Center, a six-story business center to open in 1999. The airport is also extending the southern end of Terminal 1 with a new walkway, airbridge, departure lounge, and three gates.[12]

Osaka's Kansai International Airport

Opened: September 4, 1994

Cost: US$14.4 billion

Runways: 1

Gates: 16

Size: 511 ha (1262 acres)

Passengers: 17 million (1997)

Although Kansai International Airport has but a single 3500-m (11,400-ft) runway, it is ultimately expected to be able to handle up to 454 arrivals and departures a day, accommodating 68,000 passengers and 3000 tons of cargo.[13] Ultimate capacity was projected at 160,000 takeoffs and landings, transporting 30.7 million passengers and 1.4 million tons of cargo, per year.[14] In 1998, these capacity projections were reduced to 130,000 flights per year.[15] Because aircraft approaches are over the ocean, Kansai is Japan's first 24-hour-a-day airport.[16] This greatly enhanced cargo utilization, for cargo landings could be cleared at night and transported through Japan when highway traffic is light, thereby saving warehousing expenses.[17]

Original projections of 600 flights per week, made when the Japanese economy was robust, have fallen. The airport opened at about two-thirds of its original expectations,[18] with 337 international weekly flights (compared with 630 when the airport achieves full operation)[19] and 469 domestic weekly flights.[20] That was still 40% above the capacity of the airport it replaced, Itami, which had serious capacity limitations because of noise slot controls. Itami will be left open for domestic operations,[21] although flights have been reduced from 340 a day to 121.[22] Nonetheless, in 1995, the year following its opening, Kansai was the world's fastest-growing airport, with a 226% increase in passengers. With 16.5 million passengers, Kansai was the world's 42d busiest airport in 1995.[23] By 1998, Kansai had 653 international departures per week, up from 338 when it opened in 1994, a 93% increase.[24]

Nevertheless, airport officials concluded they needed to begin airport expansion almost immediately after the airport opened.[25] The conventional approach would be to expand the island by 1730 acres to add two new runways (one parallel and one crosswind) and a second terminal building. The cost would be a staggering US$21.4 billion. A less conventional alternative would be to connect two-hundred 1000-

2.2 *At full build-out, Kansai International Airport could have three runways and two terminal buildings situated on a man-made island in Osaka Bay.* Kansai International Airport.

ft-long by 200-ft-wide steel boxes into a floating runway, an alternative which would cost an estimated US$17 billion.[26] (See Fig. 2.2.)

Denver International Airport

Opened: February 28, 1995

Cost: US$5.3 billion

Runways: 5 initially; 12 ultimately

Gates: 87

Size: 53 mi^2

Passengers: 35 million (1997)

America's self-proclaimed "Queen City of the Plains," Denver emerged as a hub for three airlines—United, Continental, and Frontier—in the first decade of airline deregulation. For a while, Denver's Stapleton International Airport was the world's fifth busiest. In 1980, the number of total passengers arriving, connecting, or departing in Denver had expanded to more than 20 million, and by 1986, it had jumped to over 34 million.[27] The FAA optimistically projected Denver would have 56 million passengers annually by 1995. Yet by 1995, the year Denver International Airport (DIA) opened, Denver had but a single hub airline, and served only 32 million passengers. Though initially projected to cost $1.5 billion, it actually cost $5.3 billion, a cost bloated by belated design changes insisted upon by its anchor tenant, United Airlines.[28]

Stapleton's runway configuration was poor. Its two parallel north-south and two parallel east-west runways were so close together that they could not accept instrument landings during periods of inclement weather, which has a propensity to occur at Denver's elevation of 5280 ft, at the western end of the Great Plains and eastern base of the Front Range of the Rocky Mountains. With weather and strong winds, Stapleton could sometimes be reduced to a one-runway airport. Because it was a hub for three major airlines—United, Continental, and Frontier—delay at Stapleton created national air transportation gridlock.

The design of the new airport attempts to rectify the airfield problems of Stapleton. It reserved sufficient land (53 mi^2) for 12 well-spaced runways, and opened with five. At its opening, DIA was the only airport in the world capable of three simultaneous Category III landings. DIA is Atlanta perfected, with three midfield parallel concourses, enabling efficient aircraft movements for expeditious hub operations. Sufficient room exists for adding two additional concourses, and expanding the terminal by one-third. (See Fig. 2.3.) At full build-out, sometime in the mid or late twenty-first century, DIA may have sufficient capacity to handle 200 million passengers per year.

2.3 *Opened in 1995, Denver International Airport was the first green field airport to have been built in the United States in two decades.* Denver International Airport.

Macau International Airport

Opened: November 9, 1995

Cost: US$1.1 billion

Runways: 1

Gates: 4

Size: 190 ha

Passengers: 2 million (1997)

The new Macau International Airport (MIA) has a single 3360-m runway with Category II ILS equipped access.[29] The terminal was built on a rock-cut platform at Ponta Cabrita on the eastern end of the Island of Taipa. (See Fig. 2.4.) The apron in front of the terminal is on reclaimed land. The runway is on an artificial island connected to the terminal apron by two taxiway bridges.[30]

The new airport was anticipated to handle 2.7 million passengers during its first year of operation, 4 million by the year 2000, and 6 million by the year 2010.[31] The Macau Airport Company, which built and operates the airport, anticipates that the MIA will gain up to 15% of Hong Kong's passenger traffic.[32] Macau offers airlines landing fees 10% lower than Hong Kong's Kai Tak Airport. A high-speed ferry could place passengers into downtown Hong Kong nearly as quickly as those arriving at Hong Kong's new Chek Lap Kok Airport.[33] Initial annual capacity of

2.4 *The new airport at Macau offers access to the western Pearl River Basin of China. The terminal was constructed on a rock platform cut from Taipa Island, while the runway, taxiways, and apron are on landfill.* Autoridade de Aviação Civil de Macau.

MIA is 4.5 million passengers and 120,000 tons of cargo. International flights are anticipated to account for 72% of operations.[34]

The second phase of the airport, opened in 1996, was focused on developing Macau's potential as a cargo hub and as an aircraft maintenance facility.[35] Site preparation on a 10,000-mi^2 cargo terminal began in 1995, while doubling the apron area for aircraft parking, which should allow capacity of 160,000 tons of cargo per year.[36] The anticipated tripling of cargo volume by the year 2011 will require more land reclamation, to be completed by 2001.[37] A second terminal is scheduled for the year 2010.[38]

Between 1996 and 1998, the new Macau International Airport handled 4.75 million passengers and 110,000 tons of freight. That is an impressive record indeed, considering the fact that before MIA was constructed, Macau was without an airport altogether. By 1998, Macau was served by 13 airlines, linking it to 25 cities, 13 of which were in mainland China.[39]

Kuala Lumpur International Airport

Opened: June 27, 1998

Cost: US$2.3 billion

Runways: 2 originally, 4 ultimately

Gates: 45

Size: 10,121 ha (24,998 acres)

Passengers: 16 million (1997)

Passenger capacity: 25 million

In 1996, Kuala Lumpur's Subang Airport handled 14 million passengers and 300,000 tons of cargo.[40] In 1997, the airport handled 16 million passengers. Such a growth rate indicated a strong need for a new airport.

Phase 1 of the new Kuala Lumpur International Airport (NKLIA) included two parallel 4000-m runways, capable of handling simultaneous wide-bodied aircraft takeoffs and landings, and main and satellite terminals, with a capacity of up to 25 million passengers a year. The main terminal building is an enormous 241,000 m^2 in size. Its contact pier is 95,000 m^2 in size, while the first satellite building covers 143,000 m^2. The terminals have 216 check-in counters, 69 immigration counters, 61 emigration counters, 16 transfer counters, and

26 customs counters. The initial phase of the airport has 80 contact, remote, or multiaircraft ramp stands. Cargo capacity will be 1 million tons in 2005, and will triple that at full build-out, matching the capacity of Asia's other major new airports.

Phase 2 of the airport anticipates sufficient capacity to handle 35 million passengers by 2008; the third phase anticipates capacity for 45 million by 2012.[41] By 2020, NKLIA's capacity will rise to 60 million.[42] With 25,000 acres, the airport has sufficient room for expansion to five runways so as to raise capacity ultimately to 100 million passengers annually.[43] Kuala Lumpur serves as a hub for Malaysian Airways, one of the world's fastest-growing airlines.

The Asian financial crisis did not delay the opening of the new Kuala Lumpur International Airport.[44] The new airport began commercial operations in 1998, taking over operations from the 33-year-old Subang Airport.[45] However, plans to build a new international airport in northern Malaysia off the coast of Kedah have been delayed. The new airport, estimated to cost 2.5 billion ringgit (US$660 million), was expected to replace Penang Airport as Malaysia's northern air hub for passengers and freight.[46] (See Fig. 2.5.)

2.5 *The main terminal building at Kuala Lumpur International Airport is 241,000 m² in size. In the foreground are parking garages, while in the background is the 143,000-m² remote terminal.* Malaysia Airports Berhad.

Hong Kong International Airport at Chek Lap Kok

Opened: July 6, 1998

Cost: US$20.1 billion (including intermodal facilities)

Runways: 1 originally, 2 ultimately

Gates: 38

Size: 1248 ha (3084 acres)

Passengers: 29 million (1997)

At a cost of US$20 billion, Hong Kong International Airport (HKIA), on the island of Chek Lap Kok, is the largest construction project ever undertaken in Hong Kong, and the world's most expensive new airport by a wide margin. HKIA replaced the crowded and slot-constrained Kai Tak Airport, one of the world's busiest,[47] with its single runway jutting out into the bay. Kai Tak was one of the world's most congested airports, with a land mass of only 310 ha serving nearly 30 million passengers a year.[48] It was built by the Japanese during World War II in the middle of Kowloon,[49] which is today the venue of some of the world's most valuable real estate.[50] More than 350,000 people were adversely impacted by the noise of Kai Tak.[51]

HKIA was built on 1248 ha, of which the 302-ha island of Chek Lap Kok and the 8-ha island of Lam Chau were excavated to a height of 6 m above sea level, while the remaining 938 ha was reclaimed from the sea. The airport opened with a single 3800-m-long runway, with a second to be commissioned within months thereafter.[52] Though HKIA opened initially with only one runway, its location allows 24-hour utilization. Kai Tak's maximum slot-controlled utilization rate was 28 flights an hour, although it sometimes reached 36. First-year volume at HKIA was expected to be 28 million passengers and 1.4 million tons of cargo.[53] HKIA is anticipated to take as many as 43 movements an hour. It also includes the world's largest passenger terminal, at 45 acres.[54] The 516,000-m^2 terminal opened with 38 gates, and will add another 10 gates and 34,000 m^2 when the remaining northwest pier is completed.[55] The airport also has 27 field aprons for passenger aircraft and 13 for cargo.[56]

The new airport is expected to handle 35 million passengers a year and 3 million tons of cargo, demand that is expected to materialize by the year 2002.[57] At full build-out, HKIA's ultimate capacity is anticipated to

2.6 *At full build-out, Hong Kong International Airport will have two terminals, two runways, and sufficient capacity to handle 87 million passengers a year.* Airport Authority Hong Kong.

be 87 million passengers and 9 million tons of cargo annually.[58] Hong Kong serves as a hub for both Cathay Pacific Airways and Dragonair. The airport also includes the world's largest air cargo facility, Superterminal 1, built on a 17-ha site and containing a six-story terminal building with a total floor area of 274,000 m^2.[59] (See Fig. 2.6.)

Oslo Airport Gardermoen

Opened: October 4, 1998

Cost: $1.9 billion

Runways: 2

Size: 270 ha

Passengers: 17 million (projected)

Construction began on a new airport at Gardermoen near Oslo on August 13, 1993, to replace the slot-constrained downtown airport at Fornebu. The airport opened in three phases—a runway was opened to charter flights in early 1998, a charter terminal was opened in mid-1998, and the airport became fully operational to scheduled traffic later that year. It has two staggered runways equipped with Category

2.7 *Like many modern airports, Oslo Airport Gardermoen placed its terminals between staggered parallel runways.* Illustration by A. Krogness. Oslo Lufthavn.

II instrumentation, 2950 and 3600 m long, capable of handling 80 all-weather aircraft movements per hour.[60] The airport will be able to accommodate 17 million passengers per year. It has 52 aircraft stands, 34 with passenger loading bridges. Its terminal building is 137,000 m², with 64 check-in counters, each equipped with SITA (a Geneva-based company) Common Use Terminal Equipment (CUTE) technology, and six baggage carousels. The SAS cargo facility will handle nearly 100,000 tons of cargo in its first year, but has sufficient capacity for twice that. The airport describes itself as "Europe's Safest and Most Efficient Airport."[61] Its marketing strategy encourages nonstop service, bypassing the Scandinavian hub of Copenhagen.[62]

The new airport is connected to downtown Oslo with an express train capable of carrying 170 passengers per hour at speeds of up to 210 km/h, reducing travel time to 19 minutes.[63] The US$3 billion spent building the new airport, upgrading the highways, and constructing the new rail line made this the largest land-based development project ever undertaken in Norway. (See Fig. 2.7.)

Shanghai Pudong International Airport

Scheduled opening: December 1999

Projected cost: US$1.6 billion

Runways: 1

Size: 32 km²

At a cost of 13.3 billion yuan (about US$1.6 billion), Pudong International Airport is anticipated to be China's flagship airport.[64] The airport will occupy 32 km², half of which is on reclaimed land. It was designed by Aeroports de Paris, in conjunction with Sodechanges.[65] The first phase, scheduled for completion in 1999, will provide capacity for 20 million passengers and 500,000 tons of cargo a year.[66] In 1998, Shanghai completed construction a 4000-m-long runway, China's longest, to accommodate future large aircraft.[67] The existing Hongqiao International Airport will continue operations, though it may be relegated to domestic operations.[68]

Seoul's Inchon International Airport

Scheduled opening: 2001

Projected cost: US$5.8 billion

Runways: 2 originally, 4 ultimately

Size: 5615 ha (13,869 acres)

Passengers: 37 million (1997)

The South Korean government is building the new Inchon International Airport to relieve congestion on the existing Kimpo International Airport. The new airport is expected to begin receiving flights by 2000.[69]

The new Inchon International Airport (IIA) will be built on reclaimed land between Yongjong and Yongyu Islands near Inchon, 52 km (about 30 mi) west of Seoul.[70] A new airport is needed at Seoul because the existing facility at Kimpo Airport (which opened in 1958) is saturated.[71] Kimpo handles 86% of Korea's international traffic, and 39% of its domestic traffic.[72] Kimpo serves as a hub for Korean Air Lines and Asiana Airlines, created in 1990. Asiana enjoyed meteoric growth, by 1997 flying 45 aircraft (at 3.7 years, the world's youngest fleet) to 46 destinations in 14 countries.[73] It has adopted an aggressive-growth strategy, which it expects to propel it to one of the 20 top airlines in the world by 2005.[74]

Seoul's Kimpo International Airport is already the eleventh busiest airport in the world, in 1995 handling 31 million passengers, 300,000 more than the design capacity of its domestic and international terminals, and 14.2% more than 1995.[75] During the 1990s, Kimpo experienced a 14% growth rate.[76] The Korean economy grew at a rate of more than 8.6% per year between 1962 and 1991.[77] Metropolitan Seoul has a population of 20 million passengers, who generate more than half of Korea's GNP. It has been estimated that a nation at Korea's stage of growth enjoys an air traffic rate of growth at about a 2-to-1 ratio of its annual growth in per capita income. That suggests that air traffic growth at Seoul may grow 12% or more per year.[78]

The site for the new Inchon International Airport is predominantly landfill of an area of Inchon Harbor between the islands of Yongjong and Yongyu, to create nearly 50 mi^2 of space.[79] The airport will occupy a land mass about 4 times the size of the new Hong Kong Chek Lap Kok Airport.[80] The new airport's original master plan included a single 3750-m (12,300-ft) runway equipped with Category IIIA navaids equipment. In Phase 1, it will be capable of handling 170,000 aircraft movements, 1.7 million tons of cargo, and 27 million passengers annually.[81] Two additional runways have since been added to the plan to handle the anticipated 10% annual growth in traffic.[82] Four or five 12,300-ft runways may be built at ultimate completion in the year 2020.[83] Airlines have urged that the new airport be opened with two runways, instead of the one runway designated in the airport master plan.[84]

The first-phase terminal building will be 357,000 m², measuring 3500 ft by 490 ft by 110 ft high, or 3.8 million ft², with 256 check-in counters, 44 gates, and 18 apron stands. The second-phase terminal will increase Inchon's total gates to 153 and apron capacity to 153 aircraft. (See Table 2.2 and Fig. 2.8.) Ultimately, the airport will have two four-

Table 2.2. Inchon International Airport

	First Phase	**Second Phase**
Airfield area	2900 acres	11,722 acres
Runways	Two—12,300 × 200 ft	Four—300 × 200 ft
Passenger terminal	3.8 million ft²	12 million ft²
Cargo terminal	1.9 million ft²	8.7 million ft²
Navaids	CAT-IIIa	Cat-IIIb
Expressway	6–8 lanes, 25 mi	8 lanes, 25 mi
Dedicated railway	None	36 mi double track
Passenger capacity	27 million	100 million
Cargo capacity	1.7 million tons	7 million tons
Movements	170,000 per year	530,000 per year

2.8 *At full build-out, Inchon International Airport will have a main terminal with two piers and several remote concourses, surrounded by four parallel runways.* Korea Airport Construction Authority.

story passenger terminals and four remote concourses, and accommodate 100 million passengers.[85] Two deluxe hotels and a large shopping center will also be built near the passenger terminals.[86]

The cargo terminal will be 175,000 m^2 in size, with expansion ultimately to 805,000 m^2, and will be capable of handling 7 million tons of freight.[87] When ultimately built out (in about the year 2020), the airport will have four runways with a capacity of 700,000 aircraft movements, 100 million passengers, and 7 million tons of cargo.[88] Already, cargo accounts for 23% of Korean Air Lines' revenue, and 17% of Asiana's revenue, among the highest such percentages in the world.[89]

Athens Spata International Airport

Scheduled opening: 2001

Projected cost: US$2.1 billion

Runways: 2

A new international airport, located 30 km southeast of Athens and just 12 km from the Athens Olympic Sports Complex,[90] is being constructed on a build-operate-transfer (BOT) basis by a consortium led by Germany's Hochtief.[91] The new airport will be operated by the Athens International Airport Company, a partnership between the Greek state and Hochtief. Hochtief holds a 40% equity stake and will manage the airport.[92] Cost of the project is estimated at 658 billion drachmas, or about US$2.1 billion.[93]

Upon completion of Phase I, scheduled for March 2001, the new airport will be capable of handling 16 million passengers annually, a 60% increase over the number of passengers using Athens in 1998.[94] Further expansion will allow for sufficient capacity to handle up to 50 million passengers a year.[95] The airport will have two runways, 4000×60 m, which will be able to handle up to 600 simultaneous takeoffs and landings a day.[96] The airport will be served by a new six-lane toll highway and rail line linking it to Athens.[97] Greece is preparing to host the Summer Olympic Games in 2004.[98]

Guangzhou New International Airport

Scheduled opening: 2001

Projected cost: US$2.3 billion

Runways: 2 originally, 3 ultimately

Size: 1453 ha (3589 acres)

Upstream from Hong Kong and Macau, several airports have been built, or are planned, for the Pearl River Basin. Zhuhai, on the west bank of the Pearl, opened a new airport in 1995, with a longer runway (at 4000 m, the longest in China outside Tibet) and double the terminal space of Macau,[99] as well as some of the world's most modern technology.[100] Shenzhen, on the eastern bank, opened a new airport (Huangtian) in 1991, which already is China's fifth busiest airport in terms of passenger volume, and fourth busiest in aircraft movements.[101] China is seeking foreign investment to build or expand more than 30 airports over the next several years.[102]

The third busiest airport in China (behind Beijing and Shanghai), Guangzhou's Baiyun Airport already handles 10 million passengers a year. Guangzhou's new airport will be built in stages on 10 mi^2, with the first phase to be completed in 2001, in time for the Ninth National Games, with capacity to handle 27 million passengers.[103] When fully built out, the airport will have sufficient capacity to handle 80 million passengers and 2.5 million tons of cargo annually.[104] More than 20,000 individuals in nine villages were displaced by the site.[105] Guangzhou is the provincial capital of Guangdong, which encompasses the Pearl River and its estuary, with a population of 62 million people.[106]

Second Bangkok International Airport at Nong Ngu Hao

Scheduled opening: 2004

Projected cost: US$3.9 billion

Runways: 1 originally, 4 ultimately

Gates: 46

Size: 3100 ha (7750 acres, or about 13 mi^2)

Passengers: 25 million (1997)

The Second Bangkok International Airport (SBIA) at Nong Ngu Hao is designed to relieve capacity constraints at the existing airport at Don Muang, which has ultimate capacity of 25 million annual passengers. The existing airport is already the tenth-fastest-growing airport in the world.[107] Bangkok's air passenger demand is anticipated to reach 35 million by the year 2000.[108] The new airport is being built 18.6 mi southeast of central Bangkok, near the Bangkok-Pattaya highway.[109]

The master plan for the Second Bangkok International Airport at Nong Ngu Hao calls for two 3700-m-long (12,210-ft) runways in Phase 1, with a second pair to be built later, along with a second terminal. The first phase of the new airport, originally scheduled to be completed by 2000, would have accommodated 30 million passengers. (See Fig. 2.9.) The second stage of the airport will probably include a concourse parallel to the main terminal, connected by an underground people-moving system. The airport was projected able to handle 38 million passengers in 2010, and 100 million at full build-out, in 2023.[110] The original plan called for the new airport to occupy nearly 13 mi^2 (3200 ha),[111] about 4 times the size of Don Muang.[112] With four runways (two runway pairs of 4000-m length each), SBIA could handle 112 aircraft movements per hour.[113]

This compares with the 25 million passenger per year capacity at Bangkok's existing airport, Don Muang, which was also expanded in the early 1990s. Don Muang's international terminal was doubled in size in 1995 (extended by 1000 ft), permitting 94 additional check-in counters and 64 immigration counters for arrivals and 36

2.9 *Though it has since been scaled down, the design for the Second Bangkok International Airport called for a 500,000-m^2 terminal with 56 gates—one of the largest in the world.* Murphy/Jahn.

for departures,[114] and its domestic terminal was tripled, raising capacity to 25 million by 1997.[115] The five-story international terminal cost US$100 million, and the domestic terminal cost US$478 million.[116]

In 1997, the Thai government decided to downsize SBIA, while expanding Don Muang. (Thailand's new ruling party, the Democrat Party, has since announced that it has reversed this decision.[117]) The government altered the master plan for construction at SBIA after markets failed throughout East Asia and the Thai baht was devalued. In an effort to meet budgetary constraints, the government decided to scale back the project at SBIA and slow development, moving back completion of the first phase to 2003, curtailing construction to one runway, and reducing projected capacity to 20 million passengers annually.[118] Funds were to be diverted into the expansion of the existing Bangkok International Airport at Don Muang to handle air traffic until completion of Phase 1 at SBIA in 2003.

A downsized Phase 1, to be completed in 2003, would include only one runway and two-thirds of the original terminal, giving the airport sufficient capacity to handle 20 million passengers annually. Completion of the second runway was moved back to 2005. Two runways would give the airport the ability to handle 112 flights per hour. The terminal would be expanded by 2007 so as to be able to handle another 10 million passengers, giving it capacity of 30 million passengers per year.[119] At full build-out (with four runways and a mirror-image terminal to the south), the new airport would be able to handle 100 million passengers and 6.4 million tons of freight per year.[120]

Since SBIA would not be open until 2004, at the earliest, additional capacity had to be found at Don Muang. Airports Authority of Thailand committed US$478 million to give Don Muang high-speed taxiways, parking aprons, expansion of both domestic and international terminals, the addition of a fifth airport pier with six gates, control tower construction, monorails, installation and improvement of automated baggage systems, and 12 additional parking bays (for a total of 100) all to be completed by 2000.[121] All this would theoretically give Don Muang sufficient capacity to handle 45 million (up from 33 million) passengers per year,[122] and 65 to 75 aircraft movements per hour (up from 45 per hour).[123] Bangkok is hub for Thai International Airways.

The latest reversal adopts the master plan as initially proposed, restoring SBIA's status as Thailand's primary international airport and regional aviation hub. The Thai government plans to accelerate the implementation of SBIA and seek private sector investment in airports throughout Thailand. A privatization study recommended that the Airports Authority of Thailand (AAT) set up a holding company to operate both SBIA and Don Muang, with AAT holding 70% of the new entity and the private sector sharing the rest.[124]

In addition, the expansion project at Don Muang carried over from the previous government has been revised. An earlier budget, approving 12 billion baht, has been cut back to 6 billion baht, with a further reduction recommended.[125] The expansion program will now consist of two phases to be completed in 2000 and 2003, respectively. Phase 1 includes construction of the east aircraft aprons, high-speed taxiways, a new aircraft pier with six contact gates, improvements to existing piers, expansion of the existing domestic passenger terminal, and construction of a new air traffic control tower. Phase 2 will include improvements to the automated baggage sorting systems and construction of a third passenger terminal.[126] In an effort to increase revenues, the Thai government doubled departure taxes on international flights from 250 to 500 baht (US$6 to US$12) per passenger.[127]

Sydney's Second Airport

The Australian federal government's plans to build a second airport near Sydney have been regarded as "an essential, long-term investment in the future of the Sydney and New South Wales economy."[128] It has been warned that the existing facility, Kingsford Smith Airport, will experience "severe capacity restraints within a few years."[129] Existing runway capacity can cope with demand until 2006 at which time an estimated A$2.4 billion (US$1.43 billion) a year would be lost to the Sydney economy unless a second airport exists to relieve capacity constraints.[130] However, a dispute between state and federal government over site location, the adequacy of an environmental impact study (EIS), and the recent approval of a very high speed train (VHST) connecting Sydney with Canberra have raised new questions concerning the future of a second airport for Sydney.

While the need for a second airport servicing the region is widely recognized, there is no consensus on what should be done. Badgerys Creek, west of Sydney, was originally chosen by the federal government as the site for the second airport. However, local and state opposition, as well as environmental questions concerning the adequacy of a draft EIS, have raised doubt about the feasibility of this site. Senior ministers, NSW Opposition, the West Sydney Alliance (a local citizens group), Finance Minister John Fahey, and the Carr government have reopened speculation that the airport could be built in Newcastle, Goulburn, or the recently privatized airport in Canberra.[131] Goulburn and Canberra became viable options after the government announced plans to build a $A3 billion (US$ 1.8 billion) VHST link between Canberra and Sydney.[132] The Australian government is postponing its decision on the second airport until after the next election, which must be held by March 1999.[133] This delay will mean that even if the project receives government approval, it will not become fully operational until at least 2006.[134]

The Role of the Existing Airport

Another issue to be addressed when building a new airport is what role will be given existing airports? Will they continue to serve as a reliever airport, as did Chicago Midway when O'Hare opened? Similarly, though Heathrow replaced Croydon after World War II as London's primary airport, Gatwick opened in the late 1950s to take on some of the demand. Will they be relegated to domestic or short-range air transport, as was Washington National when Dulles was built, as was New York LaGuardia when Idlewild (later renamed Kennedy) was built, as were Dallas Love and Fort Worth Meachum when Dallas/Fort Worth International Airport was built, and as was Laval when Montreal's Mirabel Airport was built? Perimeter rules prohibit long-range landings at Washington Reagan, New York La-Guardia, and Dallas Love airports, while intercontinental flights are required to land at Washington Dulles and Montreal Mirabel airports. Will they be condemned and put to an alternative use, as was Munich Riem when Strauss was opened, as was Denver Stapleton when Denver International Airport was opened, as was Fornebu when Oslo Gardermoen was opened, and as will be Berlin Tegel and Tempelhof Airports once Berlin Brandenburg International Airport is opened?[135] Or will they be transformed into general aviation facilities?

Keeping the existing airport open can jeopardize the economic well-being of the new airport. For example, Montreal's Mirabel has never realized its full potential because Dorval remained open. Houston's

George Bush Intercontinental experiences some traffic diversion to the older Hobby Airport. In contrast, though Denver International has some of the highest per passenger charges in the United States, passengers seeking to fly to Denver have little choice but to use DIA since the city closed Stapleton.

Summary and Conclusions

The majority of green field development, like most expansion and renovation, is occurring in the surging air transport markets of East Asia. Japan, Macau, Hong Kong, Singapore, Korea, China, and Thailand are all developing new international airports to deal with the record growth. Europe has recently completed or is beginning green field sites at Munich, Oslo, and Athens, and major reconfigurations at Berlin and Milan. The United States has built one, at Denver, though major reconfigurations have taken place at Atlanta and Pittsburgh, and a military conversion into a commercial airport was undertaken at Austin. Designed primarily to alleviate overcrowding and ease the burden on existing airports, these new facilities are built to accommodate the immediate needs of the region while allowing sufficient room for future development and expansion.

While these new airports increase the handling capacity of air traffic into the region, many also allow for larger aircraft, continuous operations, extended hours, and reduced noise problems. Munich's Franz Josef Strauss Airport can provide continuous operations during heavy winter storms by alternating between its two new runways.[136] Osaka's Kansai International Airport provides a runway approach over the ocean, which reduces noise pollution and allows for 24-hour operations.[137] Malaysia's Kuala Lumpur International Airport is capable of handling simultaneous takeoffs and has a two-phase expansion plan to raise airport capacity to 100 million passengers annually.[138]

These new airports represent thousands of hours of planning and billions of dollars of investment. They serve as models of air infrastructure planning and development at the close of this first century of air travel, and are expected to meet the growing demands of their regions well into the next. We shall examine each of these airports again in the ensuing chapters.

3

The Politics of
Airport Development

"Politics is the art of the next best."[1] —Otto von Bismarck-
Schoenhausen, Prussian Chancellor

"We have the best politicians money can buy." —Will Rogers,
American Humorist

In Chap. 3, we examine:

- The need for strong political leadership in the development
  and planning of new airports
- The dynamics of leadership in facilitating completion of
  major airports
- The delicate relationship between politics, economics, and
  necessity crucial to bring these massive projects to
  completion
- On a case-by-case basis, the role of politics in the
  development of several of the world's new airports
- The problem of corruption of political figures in large-
  infrastructure projects
- Legislation by the United States government to end
  corruption in public works projects abroad and at home
- Parallel legislation recently proposed by the Organization for
  Economic Cooperation and Development (OECD)
- Efforts by the International Monetary Fund (IMF) and the
  World Bank to adopt stricter standards and greater
  accountability when lending money to emerging market
  economies

Introduction

The first step in building a new airport or expanding an existing one is convincing the politicians or government transport ministers that such construction is desirable. The case for new construction or expansion can be made on grounds that demand has or soon will surpass the ability of existing capacity to satiate it, or that construction or expansion will enable the airport to attract more traffic so as to stimulate regional economic growth, or both. New "green field" airports often are promoted on grounds that noise at the existing facility is intolerable, that safety of residents in the approach path is jeopardized, and/or that the existing airfield has insufficient land for needed expansion.

New airports cannot be built, nor can existing airports be significantly expanded, without political support, and in most instances, without political leadership. Though some believe that large technological projects are developed on the basis of plans that have been developed in an analytically neutral way, both technical rationality and political rationality permeate the process.[2] In developed nations, political support will be essential to neutralize environmental, cost, and not in my backyard (NIMBY) opposition.

Vast economic resources must be dedicated to any massive infrastructure project like an airport. Typically, the business community will be mobilized to support the project, particularly where it perceives tax advantages and/or economic opportunity flowing from the project. In democratic nations, public support is essential, for it is the public that will pay for the infrastructure developments at issue. The U.S. Federal Aviation Administration has made the point in these terms:

> *Public acceptance is dependent upon whether the potentially affected public understands and accepts the need for the development; receives complete, truthful and unbiased information about the impacts; and recognizes that public concerns have been considered adequately and fairly....[T]he program for public involvement must be designed with careful consideration for not only providing accurate and unbiased information, but also for the perception of openness and completeness, along with a demonstrated commitment to the development of mitigation measures appropriate to the situation.[3]*

New airport projects often are the most significant public works project a nation will undertake—significant in terms of expense and employment. Politicians frequently want to be associated with massive infrastructure improvements the public will use so as to earn their confidence and gratitude. They want to be perceived as "visionaries," who build for the future and bring economic development to the citizenry. Some seek to put their names on large public works projects. Typically, politicians name airports after governmental leaders, popular figures, or public heroes. (See Table 3.1.)

Table 3.1. Airport Nomenclature

Airport	Named after
Atlanta Hartsfield	Atlanta Mayor William B. Hartsfield
Chicago Midway	World War II Pacific naval battle of Midway
Chicago O'Hare	World War II naval flier Lt. Edward O'Hare
Dallas Love	U.S. Cavalry officer Moses Love
Dayton Cox	Ohio Governor James M. Cox
Denver Stapleton	Denver Mayor Ben Stapleton
Fort Worth Meacham	Fort Worth Mayor A. C. Meacham
Houston George Bush Intercontinental	U.S. President George Bush
Las Vegas McCarran	U.S. Senator Patrick McCarran
Milwaukee Mitchell	Aviator Billy Mitchell
New York Kennedy	U.S. President John F. Kennedy
New York LaGuardia	New York Mayor Fiorello LaGuardia
Omaha Eppley Field	Hotel entrepreneur Eugene Eppley
Orange County John Wayne Field	Actor John Wayne
Paris Charles de Gaulle	French President Charles de Gaulle
San Diego Lindbergh Field	Aviator Charles Lindbergh
Sydney Kingsford Smith	Aviator Sir Charles Kingsford Smith
Washington Dulles	U.S. Secretary of State John Foster Dulles
Ronald Reagan Washington National	U.S. President Ronald Reagan

Politicians also see economic benefit in massive public works projects. Depending upon prevailing legal and ethical norms, or lack thereof, politicians may see in public works projects the potential for campaign contributions or direct bribes and kickbacks flowing from contractors. In either event, the public is disserved, for financial payments to politicians must be built into the construction price and ultimately passed on to the consumer.

But whether motivated by altruistic notions of enhancing national or regional economic growth, or the potential for personal economic or political gain, or both, the role of the politicians is both complex and indispensable for airport development. Though we will leave the significant political problems of environmental opposition to another chapter (Chap. 6), let us examine several case studies of the role of politics in airport development.

Red Shadow over Hong Kong

With a population of only 6 million people, Hong Kong is the world's eleventh largest economy.[4] With the expiration of the 99-year lease on the New Territories on the horizon, in 1984, Margaret Thatcher's British government and Deng Xiaoping's Chinese government concluded an agreement to return Hong Kong to China by June 30, 1997.

The sordid history of British annexation of Hong Kong began with the sale of opium by the British to the Chinese beginning in 1773. Widespread addiction of the Chinese people and the concomitant erosion of Chinese wealth led the Chinese emperor to issue an edict banning the trade of opium in 1800, which the British arrogantly ignored. Having had enough, in 1839, the Qing Dynasty sent Len Zexiu to Canton to seize and destroy 20,000 chests of British opium. The following year, the British assembled a naval flotilla near Macau, launching the Opium War. Defeated by the British, the Chinese emperor signed the Treaty of Nanjing of 1842, which ceded Victoria Island (Hong Kong) to the British in perpetuity, required the Chinese to pay the British 21 million Mexican dollars in reparation, and allowed full resumption of the opium trade. In 1860, during the second Opium War, British and French troops burned down the Emperor's Summer Palace, forcing cession of the Kowloon Peninsula near Victoria Island.[5] As Hong Kong grew, Britain signed a 100-year lease on the "New Territories" adjacent to Kowloon.

Though the opium trade had long since vanished, and millions of Chinese came to Hong Kong to escape political oppression and participate in the economic opportunity of a market economy, the Beijing government was determined to recover what had been taken by colonialists from China in disgrace. Beijing also recognized the marvelous economic resource that Hong Kong could be to a China united with it. Beijing refused to renew the lease, and the British concluded that Hong Kong was not really viable without the New Territories. Further, the Thatcher government was keen on improving relationships with China to give British commercial interests improved access to the growing Chinese economy.

Nonetheless, being a democratic government, the United Kingdom felt some obligation to the residents of Hong Kong (and domestic political pressure in England) not to cast them adrift in a politically and economically oppressive world of Maoist Marxism. Thus, the Thatcher government tried to exact some concessions on the political liberty front. Note that this too was done with some hypocrisy. Never before had a British colony been cut free without granting the colony's citizens a British passport and the opportunity to emigrate to the United Kingdom. Not so with Hong Kong. Neither British passports nor emigration to England were permitted Her Majesty's loyal subjects in Hong Kong.

Deng promised to grant economic autonomy to Hong Kong under a "one country, two systems" approach.[6] Both the joint British-Chinese declaration and the Basic Law (a miniconstitution to govern Hong Kong after 1997) promised that certain U.N. human rights principles would be observed after 1997, and that China would not legislate for Hong Kong.[7] As the transition approached, the British colonial governor allowed creation of a democratically elected parliament, which the Beijing government promptly replaced with an appointed parliament upon its acquisition. *The Wall Street Journal* summarized the political and economic difficulties of reunification:

> *Hong Kong's British-bred taste for democracy still threatens to clash with China's desire for political control. But China's fast-brewing, often-chaotic economy desperately needs Hong Kong's money and its technical and managerial skills. And Hong Kong's tycoons need business opportunities.*
>
> *Making it work won't be easy. Despite a similar language and culture, Hong Kong businessmen have frequently stumbled*

dealing with mainlanders. The businessmen who began crossing the border in the early 1980s were accustomed to British-run courts and feared Communist-run China's arbitrary ways. And the risks weren't just financial: Over the years, China jailed dozens of Hong Kong businessmen over commercial disputes.[8]

The inability of western businesses to enforce their commercial agreements in disputes with their communist joint-venture partners also chilled foreign direct investment in China. For example, McDonald's hamburger restaurant in Beijing was summarily dispossessed of its property in apparent violation of contract and property law.

The brutal and bloody crackdown on dissidents at Tiananmen Square in Beijing on June 4, 1989, sent shivers throughout Hong Kong, bringing a million people to the streets and creating an exodus of capital and talent abroad. Fear of Bejing's command economy also created anxiety. One must remember that 30 million Chinese starved to death in the famine that followed the Great Leap Forward of 1958–1962. But having secured a safe haven, many Hong Kong citizens returned home with a "wait and see" attitude. As a means of boosting confidence in the local economy, in late 1989 British Governor Sir David Wilson announced a new airport would be built for Hong Kong.[9] The airport was seen as "critical to Hong Kong's future as an international business center."[10] The master planning for a new airport was begun in 1990, and completed 2 years later.[11]

A dispute between the Beijing government and the British Hong Kong governor over financing of the new airport erupted shortly after the new airport plans were laid. China had objected to what it perceived to be excessive reliance on borrowed capital, which would cause a heavy burden of indebtedness after it assumed sovereignty over the British colony, in 1997.[12] The Chinese government insisted that three-fourths of the airport debt be retired by its opening date.[13] China also resisted on grounds that it had neither been consulted on the desirability of a new airport, nor been included in the decision-making process.[14] It objected to what it perceived to be an excessively costly gold-plated facility devised by the British, for whom China has had long-standing antipathy.[15] Moreover, Beijing was unhappy about Governor Chris Patten's proposals for enhanced democratic representation in the Hong Kong local government.[16]

Skeptics claimed the airport was devised as a scheme for "British colonials to make as much money as possible before 1997," when the British Hong Kong colony reverted to mainland China. Actually, Japanese firms won the largest construction contracts, although British firms were second.[17] Others note that the airport, its tunnels, expressways, bridges, and railways solve a number of transport problems, that they have been long planned, and that by packaging them together, they boost confidence in the economic future of Hong Kong even as the communist government comes to power, thereby dissuading a bailout of capital and skilled population.[18] A Memorandum of Understanding between the British and Chinese governments for a new airport was signed in 1991; the master plan was completed the following year; site reclamation began in 1993.[19]

The new airport is integrated into a comprehensive intermodal Port and Airport Development Strategy (PADS) that comprises 10 core projects, including new mass transit lines to serve the seaport and airport, a new city, new roads and bridges, new container terminals, and, of course, a new airport.[20] An agreement between the United Kingdom and China concluded in the summer of 1995 resolved the dispute on financial support for the airport and its railway.[21]

Did the US$20 billion transportation infrastructure package alleviate some of the anxiety suffered by the 6.3 million Hong Kong residents of being absorbed by its 1.2 billion Chinese neighbors? One cannot say what role it played in inspiring confidence in Hong Kong's economic future. But clearly, by the time of reunification, the anxiety level had lowered. By 1997, Hong Kong's investment in mainland China totaled US$100 billion (58% of China's foreign investment), making it the mainland's largest source of outside capital.[22]

The Swamp of the Cobras

In Thailand, they call it "the Night of the Barking Dogs." It is the night preceding the election, during which the political parties send out their minions to buy votes. The dogs bark as strangers walk up to houses with fistfuls of baht.

Originally built in 1914 as the Aviation Division of the Royal Thai Army, Bangkok's Don Muang Airport is the oldest international airport still occupying its original location.[23] Four generations after its founding, it had become among the world's most congested airports.

One source succinctly summarized the odyssey of the Cobra Swamp: "The 37-year saga of the Second Bangkok International Airport at Nong Ngu Hao reads more like a bad Hollywood script than the story of a pivotal infrastructure project. Aimed at securing Bangkok as the aviation hub of South-east Asia, the epic has faltered with every change in government since 1960."[24] The fundamental problem has been the absence of governmental continuity or strength to see the project through to completion, as well as conflicts over awarding contracts under successive governments.[25] One source bluntly stated, "the barracuda in successive governments have scrambled to divert some of the billions of dollars being spent their way."[26]

The idea for a second airport in Bangkok emerged in 1957. Two reasons were advanced: (1) the need to separate civil from military operations for purposes of safety and (2) the need for a facility capable of handling capacity needs when Don Muang reached saturation. In cooperation with the U.S. Federal Aviation Administration (FAA), in 1959 several sites were surveyed and three were recommended. In 1960, the Thai government commissioned a study for city planning for Bangkok. It concluded, "Bangkok should have a new commercial airport in order to separate civil aircraft from military ones as well as to accommodate the city growth." The proposed location was east of Bangkok, which was anticipated to be the venue for industrial growth.[27]

Thailand launched its first economic and social development plan in 1961.[28] That year, the Ministry of Transport and Communications surveyed alternative sites for a new airport and concluded the area of Nong Ngu Hao (English translation, Cobra Swamp) was most suitable.[29] (See Fig. 3.1.) An act was proclaimed in 1962 expropriating some 3200 ha (approximately 8000 acres) for the new airport site at Nong Ngu Hau 30 km east of Bangkok.[30] The site lay in a flood plain, flush with water which had been corralled by dikes into a series of small fish ponds. But the war in Vietnam brought the U.S. Air Force to Thailand, and Don Muang was expanded. In 1965, an FAA team reported that a fully developed Don Muang could be used until 1990, at which time the second airport at Nong Ngu Hao should be open.

In 1968, French and Japanese groups surveyed the site, and five proposals were presented for low-interest loans for new airport construction. In 1971, a group of Thai investors formed the Thai Airport Co., and formed a joint venture with the Northrop Airport Development

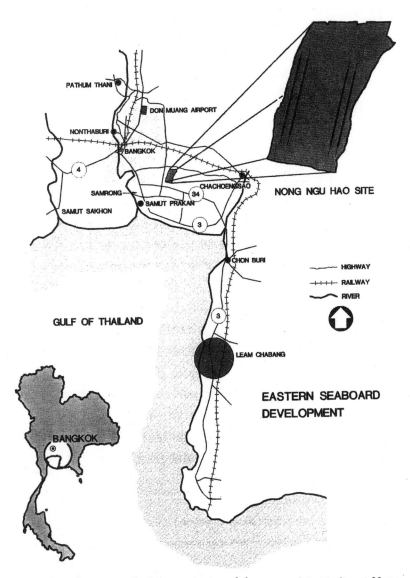

3.1 *This map reveals the proximity of the new airport site at Nong Ngu Hao to Bangkok and the existing Don Muang Airport.* New Bangkok International Airport.

Corp., which proposed that they jointly build and manage the new airport for 20 years, at which time they would transfer the airport free of charge to the Thai government. A contract was signed with the Ministry of Communications on February 1, 1973.[31] But with a student uprising against the military dictatorship on October 14, 1973, and replacement of the three-man junta (known by its critics as the "terrible trio") which had ruled Thailand since a 1963 coup,[32] the contract was rescinded.

Thailand adopted a new constitution in October 1974. In 1976, the military returned to power, and the Ministry employed TAMS (Tippetts-Abbott-McCarthy-Stratton) to study expansion at Don Muang and to find a site for the second Bangkok airport. The final study concluded Nong Ngu Hao was the most appropriate location.[33]

Taking over from the Directorate of Civil Aviation, the Airports Authority of Thailand (AAT) was created in 1979 as an airport planning and administrative body. In 1983, the Ministry of Communications signed a contract with a consortium led by Netherlands airport consultants NACO B.V. to develop a master plan; conceptual designs; preliminary designs for runways, buildings, utilities, and equipment; a soils study; and financial, economic, and environmental studies. In 1984, contracts were signed with the French government (Aeroports de Paris) for assistance for feasibility studies and master plans for the development of Don Muang Airport. NACO warned that Don Muang would reach saturation in 1994, when all activities should be transferred to the new airport at Nong Ngu Hao, with four runways. Aeroports de Paris predicted that Don Muang's two runways would reach saturation in 2016. TAMS predicted their saturation by 2000, and recommended a second airport at a site northwest of Don Muang.[34]

But the new airport was again suspended in 1984. A new domestic terminal was completed at Don Muang Airport in 1985, and an international terminal was scheduled for completion in 1987.[35] The earlier master plan to build the second airport was shelved in the 1980s because the 31,000-ha site at Nong Ngu Hao was believed too soft for construction. But the rapid growth at Don Muang, coupled with traffic projections, persuaded the government to approve the new airport project in 1991.[36] In 1991, Nong Ngu Hao was again declared the official site by the Anaud Panyarachun government. Deputy Prime Minister Samak Sundarevej alleged the airport plan was politically motivated, and that some politicians would benefit from the construction.[37] Design work began the following year.

In 1993, the first phase of the new airport was estimated to cost US$3.2 billion. This figure did not include land costs, because the Thai government has owned the land for 25 years, purchasing it for about US$40 a hectare.[38] But since then, more than 2000 families

3.2 *Though earlier governments had reserved land known as the "Swamp of the Cobras" for a future airport, inaction on the field led squatters to occupy the land and engage in fish farming.* Photo by P. S. Dempsey.

moved onto the swampy land (and some of the original families never moved off) and made a living farming and fishing there.[39] (See Fig. 3.2.) To resolve the political problems posed by the squatters, families were paid an average of 800,000 baht (US$32,000).[40] Removing the families delayed site preparation for 18 months.

In 1994, AAT short-listed three firms to bid on the terminal design under NACO's master plan—Murphy Jahn/TAMS (MJTA), Aeroports de Paris, and a consortium including Curt Fentress, designer of Denver International Airport. MJTA was low bidder, but the bid was still above AAT's budget, so AAT asked the qualified bidders to rebid. MJTA won the second round as well, though Aeroports de Paris protested. MJTA's detailed 500-page inception report was approved by AAT in January 1996.[41]

Murphy Jahn was concerned that the U.S. government might tilt in favor of the Fentress proposal, for Fentress/Bradburn was headquartered in Denver, the city of which then U.S. Secretary of Transportation Federico Peña earlier had been mayor. Murphy Jahn, a Chicago firm, enlisted the assistance of Illinois U.S. Senator Paul Simon to neutralize the U.S. government's preference for any U.S. firm. The Fentress proposal was ultimately rejected because it went beyond terminal design (for which the request for proposals had been limited) and attempted to integrate a reconfiguration of the airfield into a more rational design. This would require revamping the Master Plan, which by then had been cast in concrete. Thus, the Fentress design was rejected on a technicality.

Dedicated to going forward with the new airport project, the Chuan Leekpai government, in power from late 1992 until mid-1995, established a state company (officially formed in February 1996), the New Bangkok Airport Authority (NBAA), to take over the new airport project from AAT.[42] The Authority would be a joint venture whose shareholders were AAT and the Thai Finance Ministry, thereby weakening the influence of the military on the new airport project.[43] AAT would continue to supervise the other major airports in the country, including Don Muang, while NBAA would supervise the design and construction of the Second Bangkok International Airport.[44] NBAA has been characterized by some "by its commercial approach and by its more flexible management structure that is designed to exclude the air force from managing the future airport,"[45] and by others as having the "reputation of being ineffectual."[46] The new airport was to be completed by 2000 at a cost of 97.3 million baht (about US$3.9 billion).

As noted, under the previous government, AAT had issued the terminal contract to Murphy Jahn/TAMS. But in June 1996, NBAA put the design contract on hold, objecting to Helmut Jahn's futuristic design as not sufficiently reflecting "Thai character," as the contract required, and his use of glass and high-tech fabric in lieu of traditional wall and roofing material, as well as the cost estimates and technical specifications for various systems.[47] But the predecessor agency, AAT, had approved the design. Fearing their contract in jeopardy because of motivations external to the project (including continued efforts by Aeroports de Paris to unseat them), MJTA asked the U.S. embassy for help. The U.S. Ambassador to Thailand enlisted the aid of U.S. President Bill Clinton, who approached the Thai Prime Minister on the subject.[48]

In late 1996, on the eve of the inauguration of a new government, a US$452 million site preparation contract was signed with the Italian-Thai Development Company. Some claimed the contract was cost-inflated, and possibly politically influenced.[49] In November 1996, former Army General Chavalit Yongchaiyudh of the New Aspiration

Party was elected Prime Minister of Thailand. Two coalition partners, the Chart Pattana Party and the Social Action Party, fought with Chavalit's New Aspiration Party over control of the new airport project.[50] Members of the Social Action Party were alleged to have bought up large tracts of land near the new airport site.[51]

As the new coalition government assumed control, the economy had stagnated, exports had softened, loans were defaulting, the government was suffering budget deficits, and valuation of the baht was in jeopardy. Restoration of investor confidence and public belt tightening became top priorities of the new coalition government.[52] Citing "austerity measures" Chavalit shocked domestic and foreign investors by suspending work on Nong Ngu Hao, ordering the Transport and Communications Ministry to study expansion of Don Muang Airport. Chavalit also ordered NBAA to study whether Nong Ngu Hao should be built at all, how its budget might be reduced, and alternative building sites, including Bang Pu, 30 km south of Bangkok, a site which had never before been considered and would require partial sea reclamation.[53] The U.S. embassy in Thailand reported, "This set off a din of debate heard round the world, raising questions about the consistency and steadiness of Thai government decision making."[54] AAT was also displeased with the decision to consider mothballing the new airport. Ms. Spoar Rojnuckrin, AAT's deputy managing director, said, "I think Nong Ngu Hao should be developed immediately. A second international airport is needed."[55] Clearly, it was.

Domestically, charges were levied that partisan politics and alleged payoffs were driving these decisions.[56] Deputy Prime Minister Montri Pongpanich, head of the Social Action Party, charged that a "certain senior politician" had been bribed by a nearby government to sabotage the new airport project.[57] Democrat Chinawarn Boonyakiet charged the government had delayed the new airport project for "corrupt" reasons.[58] Others speculated that this was the former general's nod to his Air Force compatriots, who share the use of Don Muang with a military base directly north of its two runways and a military golf course between them. One source observed, "the suspicion is…that the former army chief is pandering to his chums in the air force who control Don Muang, via their control of the Airports Authority of Thailand….[T]he air force has been vigorously trying to derail Nong Ngu Hao so as to extend the life of Don Muang."[59] According to another:

The military plays an active role in commercial aviation in Thailand. Until recently, generals and marshals shared seats at Thai Airways' board of directors. The AAT, which runs the four largest airports in the country, is almost entirely in the hands of the military. Nong Ngu Hao is considered to be something of a thorn in the side of the air force as the new airport will take most of the international traffic from Don Muang and reduce its income.[60]

Yet another source noted, "The suspension of work on Nong Ngu Hao is seen as a victory for the air force, which now has control of US$1.9 billion in expansion funds for Don Muang...."[61]

Indecision and dissent among government officials halted work on Bangkok's second international airport. Some 5000 million baht have already been spent on land reclamation at the Nong Ngu Hao site. However, allegations that the Social Action Party has directly benefited from excessive expenditures at this location have some officials questioning the feasibility of the project. Finance Minister Amnuay Viravan questioned the cost of the sand used in the landfill at Nong Ngu Hao, which is 900 baht per cubic meter. SAP members claim that the true cost of the sand is only 520 baht per cubic meter, and the additional cost is from the equipment used to drain the construction area. Regardless, the landfill cost for only the runway will total 4650 million baht more than earlier projected, and this accounts for only 10% of the total landfill project. Opponents to the Nong Ngu Hao project also claim the new airport company, a state enterprise set up to run the project, is filled with SAP loyalists. Furthermore, opponents cited the severe account deficit as another reason to scrap the project and save funds by expanding Don Muang.[62]

SAP officials countered with mud slinging of their own. SAP leader Montri Pongpanich spoke of an unconfirmed report that accused an "influential Thai politician" of accepting a bribe by a neighboring country to scrap the Nong Ngu Hao project in a bid to eliminate Thailand as the region's main potential aviation hub.[63] Communication Deputy Minister Somsak Thepsuthin, a supporter of the site, claimed that Communications Minister Suwat Liptapallop opposed the project only because his father's company had failed to win the landfill tender at Nong Ngu Hao. The Engineering Institute of Thailand (EIT) came out in support of the project at Nong Ngu Hao. The institute claimed that land reclamation at Bang Pu would be even more costly, that the soil is of poorer quality, and that construction of an airport at that location would threaten the mangrove forests and raise environmental opposition.[64]

However, General Chavalit appeared determined to move the project to Bang Pu, where he planned to develop a modern city. This idea was supported by the Thai Olympic Committee, which wants to host the 2008 Summer Games. Conflicts of interest, interparty jealousies, and prime ministerial grand dreams seem certain to delay the project once again despite the enormous cost to the country and the people.[65]

In February 1997, the cabinet announced that Nong Ngu Hao would be downsized to one runway and two-thirds of the original terminal, while Don Muang would be expanded on a fast-track design/build approach with new high-speed taxiways, parking aprons, and expanded domestic and international passenger terminals, construction of a new wide-body aircraft pier (Pier 5 with six gates), a control tower, monorails, and improved automated baggage system.[66] The number of parking bays would be expanded from 101 to 125, enabling the airport to handle 75 aircraft an hour, while the international terminal would be expanded by nearly 1000 feet, permitting 94 additional check-in counters and 64 immigration counters for arrivals and 36 for departures.[67] The general engineering management contract was issued by AAT to the U.S.-led consortium T.A.P., consisting of TAMS Consultants (United States), Act Consultants (Thailand), and Provide Group Architects (Thailand). Construction would be divided into two contracts. Expansion of Don Muang was scheduled for completion in 2000.[68]

With downsizing, Nong Ngu Hao's cost would be reduced from US$3.9 billion to US$3.64 billion, while expansion of Don Muang would cost US$478 million.[69] Thus, the change would allow no real cost savings. Moreover, the capacity growth estimated by proponents of Don Muang's expansion was extraordinary—that Don Muang would be able to handle 45 million passengers per year, a figure beyond belief. By 1996, Don Muang was handling 24.5 million passengers, nearly piercing its full capacity of 25 million.[70]

When Chuan Leekpai regained power in 1997, his government restored the Nong Ngu Hao project, saying, "At this stage, the project has gone far beyond the stage where suspension or relocation can be considered. The government has already invested more than 2.5 billion baht (US$625 million). We cannot abandon it and restart elsewhere." Nonetheless, despite 30% completion of the costly site preparation, many predict the airport may not open as scheduled, a victim of the currency meltdown in Asia.[71]

Nong Ngu Hao is scheduled to open in 2004, and have an annual capacity of 20 million passengers (from its original 30 million). A second runway is scheduled for 2005.[72] By building out both airports, Thailand ends up with excessive combined capacity and no plans to link the two airports together with high-speed ground transport. If one is to be a domestic airport, and the other international, how are the passengers seeking to connect to make their way between airports? If both are to be domestic and international, then the airlines (particularly Thai International Airways) lose out on the economies and efficiencies of a single hub connecting point. If Don Muang is to be closed, then why is the government spending more than $400 million on a facility merely designed to be a Band-Aid until 2003? In any event, this may not be the final word on the subject. In recent decades, the average life span of a Thai coalition government is 2 years or less.

The fits and starts of the new airport project in Bangkok are a direct result of the type of parliamentary democracy that produces one short-lived coalition government after another, each with an interest in tinkering with the nation's largest public works project. Without continuity of focused leadership, the project has been floundering for 4 decades. As a consequence, Thailand denied itself an opportunity to capitalize on its superior geographic position relative to China, India, Indochina, and Indonesia, vis-à-vis either Singapore or Malaysia, both of which built twenty-first century airports capable of facilitating efficient regional business headquarters. We now visit a couple of governments south of Thailand which do not suffer from indecision, though strong autocratic governments have their costs as well.

The Multimedia Super Corridor and Second Industrial Revolution

South of Thailand lie two nations which are polar opposites to Thailand in terms of political longevity, leadership, and decisiveness—Malaysia and Singapore. Singapore was part of Malaysia until the latter, not wanting to be dominated economically or ethnically (Singapore is overwhelmingly Chinese), cast it adrift on August 9, 1965. Both are engaged in concerted economic rivalry. Both have dedicated themselves to achieving vigorous economic growth and prosperity, and have invested heavily in airport infrastructure as a means to that end.

Mahathir Mohammed has been Malaysia's Prime Minister since 1981. First elected in 1959, Lee Kwan Yew was Singapore's first Prime Minister. Serving for 31 years as Prime Minister, Lee is Singapore's current Senior Minister, and southeast Asia's elder statesman. Both are strong (some would say autocratic), decisive, and competitive leaders determined to achieve vigorous economic growth in a market economy. One source summarized Mahathir's persona:

> *He's blunt, he's abrasive and he's headstrong. His detractors regularly accuse him of autocratic tendencies. His supporters talk of a man of conviction. Over the past decade, Prime Minister Dr. Mahathir Mohammed has curtailed the power of the judiciary, clamped down on the press, jailed his critics and removed the privileges enjoyed by Malaysia's nine hereditary monarchs.*[73]

Of Lee, it has been said, "By his own hand, Lee turned tiny Singapore into a byword for excellence, efficiency and high-achievement. In his 3 decades in power, Lee built Singapore into a powerful economy, but his authoritarian style also turned Singaporeans—western commentators would say—into timid citizens."[74]

Prime Minister Mahathir calls it "Wasawan 2020" (English translation: "Vision 2020"), a plan to quadruple per capita income, double the size of the economy, and make Malaysia a fully developed country by the end of the second decade of the twenty-first century.[75] He launched a national automobile, the Proton Saga, which was quickly profitable. In 1991, the Malaysian government unveiled a plan to build the new Kuala Lumpur International Airport (NKLIA) 41 mi southeast of Kuala Lumpur, the nation's capital, at a projected cost of US$7.4 billion. It would replace the international operations of Subang Airport (renamed the Sultan Abdul Aziz Shah Airport, some 13.6 mi southwest of Kuala Lumpur), which has become increasingly congested despite terminal expansion and runway extension.[76] Its planners hope the new Kuala Lumpur airport will eclipse Singapore's Changi Airport as Southeast Asia's primary airline hub.[77] Singapore's Changi handled 24.5 million passengers in 1996, while Kuala Lumpur's Subang handled 14 million passengers. By 2008, passenger traffic is expected to double at Changi, while air freight will treble.[78]

Originally, the new Kuala Lumpur airport was being built by an Anglo-Japanese Consortium, AJAC.[79] But Prime Minister Mahathir severed the contract (and canceled US$6 billion of British contacts in

Malaysia) after London's *Sunday Times* ran a story alleging that a British company building a dam in Malaysia had given kickbacks to certain Malaysian politicians, and that senior Malaysian politicians were corrupt. Critics of the government have long accused it of allowing money politics and political patronage to flourish.[80]

The government denied the allegations, and in 1981 Mahathir adopted "Buy British Last" and "Look East" policies, banning British companies from getting new government contracts. Malaysia's "Buy British Last" policy also came on the heels of the Permodalan Nasional Berhad's (PNB) "dawn raid" on the London Stock Exchange (LSE) to gain control of British plantation conglomerate Guthrie.[81] The LSE immediately amended its rules to disallow such conduct and deny PNB control of Guthrie. Another problem occurred in 1994, when Malaysia suspended all government contracts with British firms following reports by the British press alleging that Malaysian leaders were involved in corrupt practices relating to major public projects and an alleged aid for arms deal.[82]

In June 1993, the government formed the Kuala Lumpur International Airport Berhad to take over development of the new airport from AJAC. Cost projections were reduced from AJAC's estimated $8 billion to $5 billion.[83] No one can fault the cost savings. But surely a government serving the best interests of the people would welcome press criticism of corruption, so as to root it out. (To his credit, in order to foster greater probity, Mahathir announced a new code of conduct for senior officials in early 1995.) Moreover, one cannot imagine that punishing British contractors would have any impact on the notoriously independent, sensationalist, and irreverent London newspapers.

The latest tiff between Malaysia and Britain involves reports carried by the British newspapers *The Observer* and *The Independent* accusing the Malaysian government of poisoning and killing Indonesian illegal immigrants who were detained pending deportation.[84] Prime Minister Mahathir claimed it was not worthwhile for the government to counter the allegations, saying "anything we do will be reported in an adverse way to make matters worse."[85] Chancellor of the Exchequer Gordon Brown said that the British government had no control over what the media reported, and that any complaint received by the government would be forwarded to the newspapers.[86]

Mahathir said that these allegations were just another attempt by the foreign media to frustrate Malaysia and make "money out of other people's miseries."[87] These reports will not affect ties between Britain and Malaysia, which are described as good. A visit by Queen Elizabeth corresponded with the Commonwealth Games which took place in Malaysia in September 1998.[88]

Nonetheless, Mahathir's growth plan is ambitious. It is dedicated to luring high-tech industries and higher-value-added production to Malaysia, while enhancing skills production and productivity of the work force.[89] In 1994, he announced plans for a Multimedia Super Corridor, a 14 by 50 km (9 by 30 mi, or about the size of Singapore) zone extending from Kuala Lumpur City Centre (a collection of modern office buildings, including the Petronas Twin Towers, the world's tallest buildings) at one end, to the new Kuala Lumpur International Airport on the other, with the future administrative capital of Putrajaya (the world's first paperless government complex) and the high-tech 7000-ha (17,300-acre) city of Cyberjaya (English translation: "Cyber Success") in between. A large number of high-tech firms committed to locate in the Corridor, including Microsoft, AT&T, Siemens, NCR, British Telecommunications, and Mitsubishi.[90] Buoyed by Malaysia's economic miracle, and nearly nonexistent unemployment, Mahathir's coalition government won an unprecedented 64% of the vote, and 161 seats in the 192-seat parliament, in the 1995 election.[91]

Mahathir is among Asia's most decisive leaders. During Asia's economic crisis, Mahathir blamed currency speculators and the free market for Malaysia's woes, then artificially raised the value of the currency and pegged it to the U.S. dollar. He sacked his finance minister, his chief political rival, then arrested him on (trumped-up, some say) charges of sodomy and treason.

Prime Minister Mahathir has also been heavily involved in airport design decisions, from choosing the color of the terminal roofs (green), to insisting that the jetways be walled with glass, the train connecting the remote terminal be aboveground, the runways be grooved to disperse water, and the airport be on a fast-track basis so as to be open in time for the 1998 Commonwealth Games. It is well that the nation's leading government official has taken an interest in the new airport. Whatever the political costs of autocracy, having a decisive and powerful leader in charge keeps the project moving forward expeditiously, though as we shall see in Chap. 8, haste has its costs as well. (See Fig. 3.3.)

3.3 *Under the strong leadership of Malaysia's Prime Minister Mahathir Mohammed, construction of the new Kuala Lumpur International Airport was put on a fast track to meet the opening of the Commonwealth Games in 1998.* Photo by P. S. Dempsey.

Much the same can be said of Lee's economic miracle at Singapore. In the late 1970s, Lee launched the "Second Industrial Revolution," which shifted the economy from a labor-intensive focus to capital-intensive and high-tech industries, and gave corporations tax incentives to use Singapore as a regional headquarters and international purchasing center. Though upon independence, in 1965, Singapore had a per capita income of $700 a year, by 1996, its per capita income was $24,000, on a par with France and higher than Britain.[92] Despite the relatively high cost of land and labor in Singapore, it continues to be an attractive venue for foreign investment because of its unrivaled infrastructure, including the highly efficient Changi Airport and surface transportation links thereto. For 7 years in a row, surveys ranked Changi first among Asian hub airports by aviation experts, academics, and employees, and Singapore Airlines among the world's finest air carriers. Among airports, Hong Kong's Chek Lap Kok ranks second, Korea's Inchon ranks third, and Japan's Kansai ranks fourth.[93]

The Island of the Swallows

South Korea's military dictatorship ended in 1987. The center-right New Korea Party lost its grip on the South Korean parliament in the April 11, 1996 elections. Nonetheless, parliamentary maneuvering gave its candidate, Kim Young-Sam, the presidency. Former Presidents Chun Doo-hwan and Roh Tae-woo were tried and convicted of sedition and corruption. A former four-star general, Roh had accumulated a secret US$650 million slush fund. Among the projects for which Roh was suspected of having received bribes was the Inchon International Airport at Yongjong Do (English translation: "Swallow's Island").[94] Initial construction for the Inchon International Airport began in November 1992.[95] Chun was sentenced to death. Roh was succeeded by Kim in February 1993. New elections were scheduled for 1997, and Kim was constitutionally barred from seeking reelection.[96]

Queen City of the Planes

With the oil bubble bursting in the early 1980s, and the minerals economy flat—both major Rocky Mountain industries—Colorado was faced with skyrocketing unemployment, a growing office vacancy and business bankruptcy rate, and an outmigration of population. As the east-west and north-south hub of the North American west, transportation was among the few of Colorado's industries with some sustaining vitality. Jobs and land proved to be the driving forces behind the effort of local political leaders in building Denver International Airport. The needs of the airlines that served Denver were only a secondary motivation, and their reluctance was viewed by the political leaders as a nuisance. A subsidiary consideration was alleviating the noise on Denver residents in the flight path of the existing Stapleton International Airport.

Like many cities, Denver faced a declining tax base as the white middle class abandoned the city to African-American and Hispanic populations and moved to the suburbs. After the federal courts ordered forced busing of children to achieve school desegregation,[97] the Colorado electorate passed a state constitutional amendment which stifled Denver's attempts to expand its city limits. A new airport became a means to another end—to radically expand the geographic boundaries of the city by persuading the county in which the airport would be built (Adams County) that it, too, could ride the engine of economic growth.

Whatever the merits of the concept of a new airport as a means of stimulating economic growth in a depressed economy, or to meet the pressing transportation needs of a constipated national air traffic system (Stapleton International Airport's runways were too close together to permit simultaneous instrument landings), DIA's implementation is another story indeed. With billions of dollars on the table, friends of the two mayors who built DIA feasted, some of that under the politically correct, but increasingly controversial, policy of affirmative action. They would show their appreciation with generous campaign contributions and business opportunities. It is often beyond the stamina of mortal men to refrain from temptation, particularly when billions of dollars are on the table. DIA is also a story of politics, power, lust, and greed, and a large number of pigs feeding at the trough. "Cronyism" would become a major issue, with the new airport in the center of the storm.

Without politics, public projects such as these are not built. With politics, we nearly always see need overstated, cost understated, and the friends of the politicians showered with dollars. DIA became the subject of a dozen investigations by Congressional committees, grand juries, and federal agencies as diverse as the Federal Bureau of Investigation, Securities and Exchange Commission, and General Accounting Office, with the unenviable task of sorting through what was done legally, and what was not.

In 1983, Denver Mayor Bill McNichols faced a challenge in the Denver mayoral election race from a former state legislator, Federico Peña. In February 1983, Peña strongly criticized the notion of building a new airport: "In terms of access, convenience and land use impacts, development of a new regional airport represents an inferior choice....At present, the commitment and financial resources required to build such a facility do not exist."[98] He believed expansion of Stapleton onto the Rocky Mountain Arsenal "represents the best long-term option available."[99] The Arsenal, a huge piece of undeveloped land directly north of Stapleton International Airport, had been used by the U.S. Army in wartime to manufacture nerve gas, and subsequently by chemical manufacturers to produce pesticides. It was some of the most contaminated land in the world west of Chernobyl.

In May 1983, Peña defeated incumbent mayor Bill McNichols.[100] Federico Peña, Denver's first Hispanic mayor, had campaigned on a theme of "Imagine a Great City." As mayor, Peña was a builder in the Keynesian tradition, with public works projects including not just a new airport, but a convention center, new library, bridges, highways, viaducts, a baseball stadium, and assorted civic improvements designed to put people to work and jump-start a stagnant economy.

Denver Mayor Peña, Colorado Governor Roy Romer, the Greater Denver Chamber of Commerce, and the city's two newspapers rolled out a bandwagon to convince the Federal Aviation Administration and the metropolitan Denver public to embrace the concept of a new airport. Traffic projections were wildly optimistic, while cost projections were grossly understated.

Adams County, which lay north of Denver, led the opposition to expand Stapleton onto the Arsenal. It retained a Washington law firm and threatened to fight expansion in the courts. It could mobilize two powerful arguments—violations of federal regulations concerning aircraft noise, and the additional environmental risks that would be associated with the construction and operation of airport facilities on severely contaminated land.

Negotiations were begun in February 1984 and, by January 1985, Denver and Adams County officials signed a Memorandum of Understanding (MOU) whereby Adams County would cede (pending voter approval) approximately 50 mi^2 of virtually uninhabited semi-desert land east of the Arsenal on which Denver would build a new airport, and upon its opening, Denver would close Stapleton Airport.

This agreement met with a mixed reception. The local business community supported it enthusiastically. Denver was experiencing a serious economic downturn in the wake of the oil bust, and an active Greater Denver Chamber of Commerce had developed a strategy to promote growth through a series of large projects.[101] A new airport was a welcome addition to the list. On the other hand, the airlines which would use the facility were unenthusiastic about a new airport because their costs of operation in Denver would undoubtedly rise. The city's and the FAA's projections were robustly challenged by the airlines which would bear the burden of an airport they believed was both too costly and, frankly, not needed.[102] But their protestations were rebuffed by a city steadfastly determined to spend its way out of poverty.

Annexing 53 mi^2 of Adams County on which to build a new airport, however, was possible under the state constitution only if Adams County voters approved the annexation. An election was scheduled for May 17, 1988. The campaign was a heated one because of the City of Denver's unpopularity with the Adams County electorate. The proponents, however, persuaded Denver Mayor Federico Peña to keep a low profile while popular Governor Roy Romer campaigned actively via the "oatmeal circuit" throughout the county, wearing a leather bomber jacket and promising economic growth over breakfast with the good citizens of Adams County. In a fit of arrogance, Romer promised to "roll over and crush" the new airport opponents.[103] The pro-airport lobby was amply funded, and won 56% of the vote.

Airport critics within Denver then demanded their own referendum on the issue. Airport proponents were confronted with a difficult decision: to hold an election, which was not legally required, and run the risk of seeing the project defeated, or push ahead with the new airport. Mayor Peña and his advisors perceived the risk of losing to be high and even though the vote would only be advisory, a negative one would make it difficult for the city to proceed with the project. Still, they decided that the Mayor, who had been elected on a populist platform, could not deny the citizenry the right to vote on the largest public works project in the city's history.[104]

The election was held on May 16, 1989. The supporters of the project included the political establishment, the business community, and the media. The opposition, on the other hand, was weak, divided, and possessed only limited resources. The result was an overwhelming victory (63%) for the new airport.[105]

After two terms as Denver mayor (from 1983 to 1991), Peña would leave DIA midstream, but not until a mayoral challenge by Republican Don Bain (who threatened to mothball the site), prompted Peña to take DIA well beyond the point of no return. Peña signed a sufficient number of eleventh hour construction contracts that the cost of mothballing the enormous project would likely subject the city to many billions of dollars in potential liability by bond holders who had invested in Federico's field of dreams.

After Federico Peña served a short term as a Denver investment consultant, he was appointed by the newly elected President, Bill Clinton, as Secretary of the Department of Transportation. Among the key reasons for this selection was the transportation experience Peña acquired as a result of starting the airport project, as well as the ethnic (Hispanic) diversity he brought to Clinton's cabinet. More than anyone else, Peña is responsible for the decision to build the new airport. As DOT Secretary, Peña would facilitate the transfer to DIA of the several hundred million dollars of federal money he had negotiated as Denver's mayor. Oddly, as DOT Secretary, Peña would not encourage other cities to do as Denver had done and build new airports.

The world's media converged on Denver in 1993 to experience the widely touted opening of the first new American airport in over two decades, and the grandest of them all. But lo, the airport was not to open, cursed with nefarious gremlins sabotaging the hardware and software of the world's largest and most sophisticated baggage handling system in the world's newest, largest, and most sophisticated airport. The news teams had to report something, so they filmed a baggage system that was every traveler's worst nightmare—one that ate bags. (See Fig. 3.4.)

The airport would not open on October 29, 1993, its initial opening date. Nor would it open on December 19, 1993, its second scheduled opening date. After the December 19 date fell through, Denver Mayor Wellington Webb promised, "We are going to open March 9 [1994] if I have to unload baggage myself." But again, another scheduled date would be missed by a baggage system that apparently couldn't promptly be fixed.

Opening was postponed again, this time to May 13, 1994. The world's media would trot out again, only to report that the baggage system had an insatiable appetite for the bags it was devouring. Comedians feasted on the morass, and Denver's new airport became a national joke.[106] Among the acronyms given DIA were

3.4 *The decision to add a highly sophisticated automated baggage system, on a scale never attempted before, and in an unrealistic time frame, well after construction of the new Denver International Airport had begun, led to costly and embarrassing delays.* Photo by P. S. Dempsey.

Denver Imaginary Airport, Doesn't Include Airplanes, Delayed It Again, Denver International Aggravation, and Dumbest Idea Anywhere.[107]

Delayed openings inspired the news media to probe more deeply, and they found a multitude of explanations for the delays and allegations of ineptitude, incompetence, corruption, cronyism, and fraud. The local newspapers had been generally supportive of the airport, but once the national media began to do the serious undercover journalism the local papers had not, they too felt compelled no longer to give DIA the light-handed treatment that characterized their reporting since the inception of the project.

From the city's perspective, the airport was becoming a local political liability and a national embarrassment, as well as a growing financial burden. After the May 13, 1994, opening date fell through, the city hired a consulting firm to evaluate whether the BAE system could be fixed, and if not, what options the city could explore. (BAE Automated Systems is a subsidiary of the British BIR conglomerate.) For its part, BAE, which designed and built DIA's infamous highly sophisticated automated baggage system, insisted that the time frame for completion was unrealistic from the outset, that the city had dragged its feet in assuring that other contractors were out of the way (for example, blocking the tunnels which link the distant concourses), and that the city had failed to order surge protectors for the electricity. In any event, city hall knew that it had to get DIA open at any cost, lest Webb's reelection in 1995 be jeopardized.

The recommendation was that a primitive "tug and cart" back-up baggage system be installed (the original plan before United Airlines asked for a high-tech bag system). United Airlines, DIA's principal tenant, wanted the BAE system finished in its Concourse B, and in order to do so, got the city's permission to take over the Concourse C loop, thereby depriving Concourse C carriers of an automated baggage system.

In the end, United agreed to pay a good portion of the $1 million per day delay costs, and the city put BAE under United's arm, though United refused to allow DIA to open until its rival, Continental Airlines, committed to abandon its Denver hub.[108] But, and this was crucial, the agreement provided that DIA must open not later than February 28, 1995, coincidentally, fully 2 months before the next Denver mayoral election. Webb could not face the voters with an airport still inoperable, and chewing up millions of dollars in costs.

Finally, the airport opened, on its fifth scheduled date, February 28, 1995. The 16 months of delay consumed $100 million in cost overruns and $361 million in financing costs.[109] What originally was a $1.5–$1.7 billion airport had turned into a $4.8–$5.3 billion airport, even though it had fewer gates than the airport it replaced.[110]

Webb's predecessor as Denver Mayor, Federico Peña, would implicitly point the finger at Webb, claiming, "When I left office, I thought the airport was in fairly good shape." On DIA's opening day, *The Philadelphia Inquirer* reported:

> Peña, on hand for the opening, spent much time defending the project derided as Federico's Folly.

> "This is not the time for Monday-morning quarterbacking," he said, before blaming others for the problems.[111]

But let he who is without sin cast the first stone. In selling his vision of a great city, Peña had overstated the need for the new airport, while understating its costs—56 million passengers were projected for 1995, but only 32 million showed up; costs were originally projected to be $1.5 billion, but actually were $5.3 billion.[112] So as to put DIA past the point of no return, Peña signed scores of contracts in his closing days as Mayor for which design specifications had not even been written, leading to costly contract modifications and cost overruns. Peña had also put into place the city's minority contracting program, ensuring many contracts would go to firms other than those with the lowest bid. Later, the U.S. Supreme Court ruled Colorado's minority preferences for state transportation contracting unconstitutional.[113]

In 1991, Wellington Webb had run for mayor as the man who could bring DIA in on time and on budget. In the 1995 mayoral campaign, mayoral candidate and city councilwoman Mary DeGroot insisted that "three-quarters of a billion dollars in delay costs and additional costs [are] associated directly with Wellington Webb's management. We went through five opening dates, 16 months of delay. I don't believe that is a sign of good management."[114] Her position was summarized as, "My opponent has given away city hall, and its lucrative contracts, to his friends. I can do better ethically, and in managing Denver."[115] Webb took the position that "I am the proven leader who got the new airport built and other projects done. My opponent is trying to trash my character and downplay my accomplishments."[116]

Accusations of ethical lapses and cronyism—giving jobs, contracts, and other city favors to friends, family, and campaign contributors[117]—dogged Webb after he became Denver mayor in 1991, and became key issues in his 1995 re-election battle.[118] Polls showed that Denver voters thought Webb's administration more corrupt than that of any previous Denver mayor, and that the public was particularly concerned over contracts surrounding DIA.[119] In the May 2 municipal election, DeGroot beat Webb by 97 votes, forcing a June runoff. DeGroot would push the cronyism issue, hard. After Webb insisted on documentation, DeGroot published a list of 64 alleged ethical lapses she culled from local newspapers.[120] Mud was slung in both directions. Webb would accuse DeGroot of using the "big lie" tactics of Nazi Germany.[121] DeGroot would respond by labeling Webb's campaign literature a "bald-faced lie."[122]

The correlation between political contributions and city contracts was profound. Two months after being elected Mayor of Denver, Webb assigned the key role of co-lead to the airport bond underwriting to an investment banking firm that donated $30,000 to his campaign for election, the largest single contributor, accounting for roughly 6% of Webb's campaign war chest (one-third was donated under a different corporate name, possibly to obfuscate its source).[123] Similarly, a Wall Street bond underwriting firm which had donated $11,000 to Webb's mayoral campaign in 1992 received $5.8 million in airport work.[124] In 1994, the U.S. Securities and Exchange Commission (SEC) responded by adopting Rule G-37, which attempts to prohibit large Wall Street investment houses from using campaign contributions as a means of securing lucrative municipal bond deals.[125] In reviewing the rule, the U.S. Court of Appeals observed, "underwriters' campaign contributions self-evidently create a conflict of interest in state and local officials who have power over municipal securities contracts and a risk that they will award the contracts on the basis of benefit to their campaign chest rather than to the governmental entity."[126]

In Wellington Webb's 1995 bid for reelection, 22% of his campaign contributions came from airport contractors, consultants, and concessionaires.[127] One mayoral candidate charged, "The city of Denver is for sale under Wellington Webb."[128] These types of accusations would prompt widespread investigations by federal and state agencies as to "whether city officials steered millions of dollars in airport work to brokerage houses, law firms and consultants because of political contributions."[129]

Four senior DIA officials—airport director George Doughty,[130] engineer Ginger Evans, public relations director Richard Boulware,[131] and aviation director Ed Trommeter[132]—pointed fingers at city hall and spoke of incompetence, ineptitude, political favoritism, and/or ethical lapses. In the mayoral election, the foul smell of cronyism was the central theme of Mary DeGroot's criticism of Wellington Webb.

Another issue of national import became the question of affirmative-action contracting, which while unquestioned as "politically correct" during the 1960s, 1970s, and 1980s, became the focus of widespread scrutiny in the 1990s. Federico Peña was Denver's first Hispanic mayor; Wellington Webb was Denver's first African-American mayor. During Federico Peña's and Wellington Webb's tenures as mayor, the city of Denver awarded between a fifth and quarter of its contracts to companies owned by minorities and women.[133] During the same period, only 1% to 2% of the companies in the Denver Metropolitan Statistical Area were minority-owned.[134] As construction of the new airport accelerated, the contracts awarded to firms designated as "disadvantaged business enterprises" grew robustly, reaching a peak of $75 million in

1991, Webb's first year in office, and totaling more than $300 million during Peña's and Webb's combined terms as mayors.[135] Many criti- cized the minority contacting program as a rationale to dole out con- tracts to politically connected contractors.[136] As *The Denver Post* reported:

> *Mayor Wellington Webb is a firm supporter of the city's effort to steer work to minority and women businesses, saying it helps level the field and gives disadvantaged companies a chance at winning city work.*
>
> *A review of the program, however, shows that many minority concerns winning city work have had close ties to the ad- ministrations of Webb and Federico Peña. Further, a few of those firms won city work despite lacking in-depth experience in areas covered by their contracts.[137]*

Some of Denver's multimillion dollar minority "disadvantaged busi- ness enterprise" firms were accused of "[b]loated payrolls, over- charging and political influence."[138] Others were alleged to be minority front companies ("beards"),[139] with sloppiness in the over- sight process resulting in 32 ineligible minority or female owned companies awarded $34 million in city contracts over 3 years.[140]

Four years earlier Webb had campaigned as the candidate with the experience to complete the new airport project on time and on bud- get. In the 1995 election, Webb took credit for the completion of the new Denver International Airport. But in fact, DIA was completed 16 months late, and conservatively estimated, was at least $750 million over budget.

Denver is one of the few cities in the United States still blessed with two independent newspapers. Both endorsed, supported, and cheered the construction of DIA early on. But both were appalled by its implementation by the Webb administration, and turned on Webb. The *Rocky Mountain News* argued, "there should be political consequences for lackadaisical management at crucial moments in the construction of Denver International Airport, and for excessive cronyism in Webb's first couple of years as mayor."[141]

Wrote *The Denver Post:*

> *[W]hen Webb ran for office in 1991, he promised that the hallmark of his administration would be the efficient man- agement of the DIA project. Instead, the project was repeat-*

edly delayed—and its cost was grossly inflated—by poor management. At critical times during the project, in fact, top leadership jobs on the project remained vacant for months and months.

While it's true that Webb inherited the project under unfavorable political terms, its most costly and most embarrassing dimension—the trouble-plagued automated baggage system—was a commitment made on his watch.

That aspect of the project, more then any other factor, put DIA late on schedule and way over budget....

In short, Webb's primary argument for being elected in 1991 was his pledge to manage the DIA project—and he faltered in that task.[142]

On another occasion, *The Denver Post* opined, "the mayor spent excessive time fretting over which retail and food companies would win lucrative concession contracts; on what public act projects would grace the terminal; and on a minority-contracting program. These trivialities distracted City Hall from the really important issues, such as whether all six of the promised runways were fully funded, and if the high-tech baggage system could be up to speed on opening day."[143] Among the allegations of mismanagement leveled at the Webb administration were:

- Webb was slow to fill critical management positions, leaving the airport's top slot vacant for months after the death of city engineer Bill Smith and resignation of aviation director George Doughty.

- Webb's administration spent money so fast that some staffers began to refer to DIA's coffers as the "candy store."[144]

- Webb caved in to United Airlines' belated demands to radically overhaul the airport's design, approving several "scope changes," including moving the planned location of Concourse C and building a full basement in the terminal after construction was well underway, and adding a state-of-the-art fully automated baggage handling system to the entire airport (although such had never before been attempted on such a scale) swelling construction costs by 50%, while not amending a manifestly unrealistic time schedule.[145]

- The belated airline construction design changes (estimated at half a billion dollars), coupled with a quarter of a billion dollars in capitalized interest caused by delays, left DIA with at least $750 million in cost overruns.[146]

- Webb agreed to build out Concourse A to 20 gates despite the fact that its principal tenant, Continental Airlines, was in Chapter 11 bankruptcy, and would sign leases with terms no longer than 5 years, vis-à-vis the industry norm of 25- to 30-year terms.

- Webb failed to advocate that Continental Airlines be allowed to keep its maintenance base at Stapleton Airport, despite the fact that it would have retained about 2000 good-paying jobs for Denver, given Continental incentive to maintain its strong presence in Denver, and resulted in no more than three flights a day on one of Stapleton's north-south runways, thus avoiding flying over the Park Hill community which had been such a vehement foe of Stapleton.

In the month preceding the opening of DIA, the Denver daily newspapers ran exposes harshly critical of the project. The Denver papers engaged in some measure of self-flagellation, admitting their role in selling the city's and chamber of commerce's spin on DIA to the public with overly optimistic projections of demand and overly conservative projections of cost. They had failed to engage in hard investigatory journalism early on, when the critical decisions were made, and seemed to feel some responsibility for a project clearly out of hand. In banner front-page headlines, *The Denver Post* described the project as "DIA: Dream to Disappointment," while the *Rocky Mountain News* published a multipart series "DIA: What Went Wrong?", which uncovered "new details about a monument to miscalculation."[147] The editorial page of *The Wall Street Journal* slammed DIA as "Peña's Folly," laying the blame on President Clinton's Secretary of Transportation, who in an earlier incarnation as Denver's mayor began construction of the new Denver airport.[148] In a cover story, the conservative *American Spectator* magazine also labeled DIA as "Peña's Plane Stupidity."

However, once DIA opened, on February 28, 1995, the political issue of DIA subsided considerably.[149] Webb insisted that the public put the past behind them and look forward to work toward the success of DIA. He admitted he had made mistakes, but assured the electorate he had learned from them.[150]

But the cost overruns and the virtual monopoly enjoyed by United Airlines by virtue of Continental's departure made it a continuing issue for each origin and destination passenger at DIA, costing $40 per round-trip ticket,[151] and costing passengers beginning or ending their trips in Denver an additional $500 million per year.[152]

DeGroot herself proved to be a poor challenger.[153] Her suggestions that affirmative action be scrapped and replaced with a contracting process based on economic disadvantage, rather than race and gender,[154] would have sold well in the Republican Congress, but went over like a lead balloon in liberal Denver, represented in Congress by feminist Pat Schroeder. DeGroot's frontal assault on affirmative action would enable Webb to tap out-of-state campaign donations to the tune of $100,000, particularly from Washington, D.C., and New York. That, coupled with her suggestion that, in her administration, the police would engage in "maximum harassment" of gang members, allowed Webb's supporters to falsely[155] tag DeGroot as a "racist."[156] This enabled Webb to turn out the vote in the north Denver African-American and west Denver Hispanic communities. Webb carried 90% of the black and Chicano voters, while DeGroot carried only 60% of the southeastern white districts.[157] While using the race card would ensure Webb a heavy turnout of minority voters, it would strain race relations in a city already characterized as suffering from "white flight" to the suburbs.[158] Race relations in Denver and across America would be further strained in the O. J. Simpson trials later that year.

It was an atmosphere ripe for incumbent reelection. The people of Denver were feeling good about their city. The airport was open, finally. The new baseball stadium was a gem. Californians were bailing out and moving to Colorado. The local economy was roaring. No one seemed concerned with the long-term implications of growth and sprawl—that twenty-first century metropolitan Denver might one day look like today's metropolitan Los Angeles.

Webb won the mayoral election, 54% to DeGroot's 46%.[159] As the state Republican chairman observed, "Sometimes the devil you know is preferable to the devil you don't."[160]

Moreover, only a quarter of Denver's metropolitan residents reside within Denver's city limits, and they are disproportionately minority and lower income vis-à-vis suburban residents. Given that a higher percentage of suburban residents would be regular users of DIA, and pay the inflated costs attributable to cost overruns, they might exact political retribution on a politician whom they perceived to be guilty of mismanagement of an essential part of regional transportation infrastructure.[161] Yet suburbanites were denied a vote on whether Webb should be retained or put out to pasture.

Where there's smoke, there's usually fire, and a dozen formal investigations began of DIA by federal and state agencies and grand juries. DIA had consumed half a billion dollars of federal money, and the federal government had a right to assess whether the money had been properly spent. Federico "Imagine a Great City" Peña, who as Denver mayor conceived and began DIA, was now President Clinton's Secretary of Transportation, so the Congressional Republicans too were anxious to find wrongdoing to embarrass a cabinet officer and his President.[162]

Several Congressional committees and federal agencies—from the Federal Bureau of Investigation, to the U.S. Department of Transportation, Federal Aviation Administration, Department of Justice, and Securities and Exchange Commission—would investigate various aspects of DIA, and the role of Mayors Peña and Webb in it.[163] Grand juries would be convened to assess allegations of construction fraud, bogus surety performance bonds, minority-contracting irregularities, altered documents, and spending airport revenue on a $10,000 trade junket to Gabon (which has, and will likely have, no direct air service to Denver).[164] Craig Silverman, a former attorney under Denver District Attorney Bill Ritter, whom Silverman attempted to unseat, claimed Ritter mismanaged the grand jury investigation of DIA by appointing an unenergetic lawyer to prosecute the case. A former U.S. DOT investigator, John Deans, was prepared to testify about "unlawful and illegal diversion of federal aviation funds at DIA," but was muzzled by a U.S. attorney with ties to former Denver mayor and DOT Secretary Federico Peña.[165] The DOT found Denver had unlawfully diverted $6.2 million in airport revenue,[166] and a lawsuit brought against the city and the bond houses by bond holders for providing false and misleading information was settled out of court for $4.5 million.[167] By 1998, lawsuits were pending against contractors who allegedly shorted the amount of cement that was put into concrete for the runways.[168] But as the dust settled, most of the investigations and lawsuits came up with insufficient evidence to convict.

The Windy City

Chicago O'Hare is the world's busiest airport. The older Chicago Midway Airport is bursting at the seams. Does Chicago need a third airport? The City of Chicago and the airline industry (United and American Airlines at O'Hare and Southwest Airlines at Midway) say no. The Illinois Department of Transportation (IDOT) and state politicians say that a new airport is essential to maintaining Chicago's presence as a premier air transport route.

The proposed site for the new airport is in rural Peotone, about 30 mi south of downtown Chicago. Two proposals have been offered by the IDOT, which would develop Peotone as a supplemental airport to O'Hare and Midway. The first proposal is for a small airport handling about 1 million passengers annually with room for future expansion. Preliminary cost estimates show that a small airport could be built for as little as $290 million.[169] A second proposal involves a larger, full-size airport that would include three runways and 37 gates, handling 6.3 million passengers and costing more than $1.6 billion (which could grow to as much as $2.7 billion with bond interest and finance charges if publicly funded).[170] The airport would be built and financed either by the state or in partnership with a private developer.

A feasibility study undertaken on behalf of the IDOT reported that a third airport servicing the Chicago area would save travelers hundreds of millions of dollars a year in lower airfares and create nearly 535,000 jobs in the region.[171] The study cited limited capacity and lack of competition at O'Hare as proof of the necessity of a third airport. It revealed that passengers flying out of O'Hare pay $626 million in higher fares and spend 19% more per mile than travelers using the 56 largest airports in the country.[172] The IDOT contends that these higher costs are hurting the state economy and threaten Chicago's "geographic air transportation advantage" as a stopover for connecting flights. IDOT further contends that the limited capacity and lack of competition at O'Hare threatens service to smaller communities within the state and the Midwest. State Transportation Secretary Kirk Brown said, "Downstate communities already have lost 11% of their commuter seats because of lack of capacity."[173] According to the feasibility study, by the year 2020 more than 44 airports could lose service to and from O'Hare, including Bloomington-Normal, Moline, Peoria, Champaign-Urbana, and Decatur.[174]

David Fuscus, a spokesman for the Air Transport Association (ATA), questions the findings of the study. "Chicago has one of the most highly developed aviation infrastructures in the country. The combination of airports…offers tremendous competition and tremendous service to the people of Illinois. We think it's [Peotone] a bad idea and shouldn't be built."[175] He contends that IDOT's comparison of fares was flawed, and that, "market, not capacity, determines costs."[176]

Though there was solid support for a third airport at the state level (then-Governor Jim Edgar, U.S. Representatives Jesse Jackson, Jr., and Henry Hyde, as well as all gubernatorial hopefuls), the plan still faced an uphill climb. United and American, which controlled 80% of all landing and takeoff slots at O'Hare,[177] and the City of Chicago (Democrat Mayor Daley) successfully lobbied the Clinton Administration to reject the proposal.[178] As a consequence, in March 1997, the FAA removed the Peotone site from its National Plan for Integrated Airport Systems, which essentially blocks eligibility for federal funding of the project.[179]

It was not always so. Mayor Daley supported the construction of a third Chicago airport in the early 1990s, when it was going to be within the city limits. At the time, a Daley aide said, "He's so enthusiastic about the new airport, he would like to open it tomorrow." But when Daley's preferred site failed to get state approval, he pronounced the third airport "dead, dead, dead." As the state legislature moved to seize control of O'Hare and Midway, Daley cut a secret deal with Gary, Indiana, which as an interstate arrangement, shielded the Chicago airports from control of the Illinois legislature.[180]

The Daley administration solicited proposals from public relations firms seeking a $350,000 contract to fight the Peotone airport. Funds for the consulting contract were initially to be paid from the $3 user tax (passenger facility charge) on airline customers who board at Chicago's airports. However, these funds are limited by the FAA to "airport planning and development," which must be related to safety, security, or flight capacity issues; airport noise mitigation; or enhancing air carrier competition.[181] Charitably, United and American are considering paying the cost of this consulting effort. A United Airlines spokesman said that such a decision by the airlines would "save the taxpayers twice: They wouldn't be paying for the public relations firm or for the construction of an airport that just isn't needed."[182]

Another airline, Southwest, operating out of Midway, threatened to pull operations out of Chicago if a third airport was built.[183] Midway has been earmarked by the city for renovation and refurbishment. Much of the financing for the project is to be handled by the tenant airlines (of which Southwest and American Trans Air are the largest) at Midway.[184] The proposal calls for a new terminal building with three concourses that would double the size of the current terminal, and a six-story parking building.[185] The agreement for renovation at Midway lets the airlines financing the project "off the hook if a third airport that has a 'direct impact' on Midway is built."[186] "If the new airport is controlled 'in whole or in part' by the city, airlines can take a hike with just 60 days' notice. If the city is not involved, the agreement can be terminated on Dec. 31, 2008."[187]

A recent decision by the U.S. Senate Committee on Commerce, Science, and Transportation approving the addition of 100 commercial flights a day at O'Hare is seen as another obstacle to efforts for a third airport. Against the objections of local groups opposed to any expansion of O'Hare, the Senate committee approved the conversion of unused slots being held for military operations, in an effort to increase competition and provide better service to smaller cities in the Midwest.[188] O'Hare already handles 2400 flights daily, and federal reports show it with more near-collisions than any other U.S. airport.[189]

The sides are clearly divided, but even within the parties, agendas differ considerably. The state would like to see economic development in the south suburbs and consumer options (especially with regard to service to and from Midwestern destinations) expanded. State and federal politicians with constituents in the area are doing what they do best, posturing. Residents, who make up the Suburban O'Hare Commission, support the airport in so far as it will relieve congestion at O'Hare; while local residents of Peotone oppose development "in their backyard." Mayor Daley, who was the biggest supporter of plans to build a third airport at Lake Calumet, opposes any airport outside the Chicago metro area that would divert taxes collected by the city from airlines currently operating at O'Hare and Midway. As explained in Chap. 11, United and American enjoy a "non-competitive advantage" at slot-constrained O'Hare Airport, the world's busiest. Southwest Airlines considers its fleet "movable assets," and threatens to discontinue service to Chicago if the state attempts to increase competition among low-cost carriers by building a third airport. Early reports show the odds are heavily stacked against an airport at Peotone. But these words were written by *The Economist:*

> *[Mayor Daley] would do well to study history. Chicago grew from a swampy marsh into a city when local leaders had the foresight to build a canal making it possible to pass from*

the Great Lakes to the Mississippi River. Later, the city boomed when it became the centre of the country's railway system. There's a lesson here.[190]

Legislative Efforts to Arrest Bribery and Corruption

With airports as projects worth hundreds of millions of dollars in contracts, and some local officials intent on exacting bribes, the potential for corruption is overwhelming. Corruption of government officials, of course, is inimical to the public interest in the efficient expenditure of public resources. The United States has long been alone in prohibiting its contractors from engaging in such unsavory practices. More recently, the Organization for Economic Cooperation and Development (OECD) has attempted to abate such conduct. Table 3.2 reveals the world's most corrupt nations, as ranked by Transparency International, a Berlin-based international good-government advocacy group. As Frank Vogl, vice chairman of the organization said, "We simply can't see how countries can sustain economic growth when they're perceived to have massive problems of corruption."[191]

Origins of the U.S. Foreign Corrupt Practices Act

In the United States, the Foreign Corrupt Practices Act (FCPA) emerged as a direct result of investigations undertaken by the U.S. Securities and Exchange Commission (SEC) in the mid-1970s that uncovered questionable foreign payments made by U.S. corporations

Table 3.2. Most Corrupt Nations[192]

1	Cameroon
2	Paraguay
3	Honduras
4	Nigeria
5	Tanzania
6	Indonesia
7	Colombia
8	Ecuador
9	Venezuela
10	Russia

to secure favorable foreign government action. Following a voluntary disclosure program, the SEC reported that more than 450 companies had admitted to collectively making more than $400 million in questionable payments.[193] Among the corrupt conduct uncovered in the investigation was the payment of $22 million by Lockheed Aircraft Co. to foreign officials.[194] This was of particular concern because Lockheed had recently received a $250 million loan provided by the Emergency Loan Guarantee Board to prevent its bankruptcy.[195] This caused some embarrassment on Capitol Hill, which had, in effect, subsidized Lockheed's corruption.

Public outcry led to congressional response and presidential action. "Congress viewed corporate bribery of foreign officials as unwise from a business standpoint, unethical, inimical to the principles of free and fair competition, and a threat to the conduct of the nation's foreign policy."[196] On December 19, 1977, President Jimmy Carter signed the Foreign Corrupt Practices Act into law, making the United States the only country to pursue criminal prosecution against its national corporations for engaging in bribery of foreign officials.[197]

Overview of the FCPA

The FCPA can be analyzed in two parts: the requirements relating to accountability, and the provisions addressing antibribery concerns. The teeth of the FCPA are contained in antibribery provisions which relate to appropriate behavior of "issuers" and "domestic concerns" when conducting business within a foreign country. The accounting requirements were included as part of the Act at the insistence of the SEC, and provide both a guide for corporate compliance and a means for governmental review. These requirements benefit both parties involved, and failure to comply with the accounting requirements does not ensure criminal liability or presume corrupt behavior.

The accounting provision requires an issuer of securities: (1) to maintain books, records, and accounts which, in reasonable detail, accurately reflect transactions and dispositions of assets and (2) to devise and maintain a system of internal accounting controls.[198] These internal controls are to provide reasonable assurances that transactions entered into are in accordance with management's authorization.[199] The "reasonableness" of the assurances and details required to be provided within the accounting provision are those which would "satisfy prudent officials in the conduct of their own affairs."[200] These accounting provisions are primarily a means by which corporations can establish their compliance with the FCPA. Criminal liability is imposed only for knowing circumvention or failure to follow internal accounting controls.[201] Willful violations, by providing false and misleading statements with respect to any material fact, may lead to fines of $1,000,000 and imprisonment of not more than 10 years against an individual, and a fine of up to $2,500,000 against the business entity in violation.[202]

The heart of the FCPA lies in the prohibited practices sections, which effectively define the scope and reach of the Act.[203] The FCPA prohibits payments to: (1) foreign officials, (2) foreign political parties, or (3) any intermediaries, while knowing that the money will be offered directly or indirectly to government officials.[204] A person's state of mind is "knowing" with respect to conduct, a circumstance, or a result, if such person is aware of such conduct or circumstances, or that such result is substantially certain, or that such person has a firm belief of the existence of a circumstance or the substantial certainty of a result.[205] Conscious disregard of a known fact, or willful ignorance of questionable conduct will not insulate a company or its agents from liability under the Act.

The 1988 Amendment to the FCPA provided limited exceptions and affirmative defenses to foreign payments. It distinguished payments made for the facilitation of routine governmental actions from those illegal payments made to influence governmental policy decisions. Routine governmental actions are those ministerial activities which are ordinarily and commonly performed by a foreign official.[206] These facilitating payments, also called "grease" payments, are legal payments made to foreign officials to expedite the processing of permits, licenses, or other governmental documents and do not violate the FCPA.[207] While the distinction depends on whether the payment made to the foreign official is to influence a discretionary act or to prompt and expedite a ministerial one, in practice where you draw the line is wholly unresolved.[208] The Amendment also allows exceptions for reasonable and bona fide expenditures relating to the promotion of business activities.[209]

Enforcement of the Act is shared by the SEC and the Department of Justice (DOJ). The SEC investigates violations by issuers and may request an independent investigation by the Department of Justice. The DOJ investigates violations by domestic concerns. Both agencies may seek injunctive relief in cases not involving criminal prosecution.[210] In any cases which involve criminal prosecution, penalties may range from not more than $2 million, to $10,000 and not more than 5 years imprisonment for willful violation by officers of an entity in violation.[211]

Penalties under the FCPA

Penalties for failing to comply with the FCPA depend on the type of violation (record keeping and accounting or illegal payments) and the willfulness of the misconduct. Record keeping and accounting violations may lead to civil penalties of $100 per day during the period the company fails to comply.[212] A willful violation of the provision, or willfully and knowingly making false or misleading statements in filed reports, may lead to criminal penalties of not more than $1,000,000, or not more than 10 years imprisonment, or both. When the violation involves the business entity generally, (the statute reads, "except when such person is a person other than a natural person"), a fine not to exceed $2,500,000 may be imposed.[213] The Securities and Exchange Commission handles enforcement of the accounting provisions, and compliance is required by all entities registered therewith.

Enforcement under the FCPA for illegal payment violations is split between the Department of Justice, which enforces violations by domestic concerns under §78dd-2, and the SEC, which supervises enforcement of issuers under §78dd-1. The levels of penalties for violations under each section are identical.[214] Entities are subject to fines of not more than $2,000,000 and civil penalties of not more than $10,000.[215] The officers, directors, employees, agents, and shareholders acting on behalf of the entities, who willfully violate the sections, are subject to fines up to $100,000, or 5 years imprisonment, or both. Violations without the willful requirement are subject to civil penalties up to $10,000.[216] Any penalty imposed, either civil or criminal, on those acting on behalf of the entity, may not be paid by the company, directly or indirectly.[217]

Enforcement of antibribery legislation

The expansion of U.S. corporations into emerging markets brings with it the rise in FCPA enforcement efforts. Since the law's enactment in 1977, only 16 bribery prosecutions under the FCPA have been brought. However, in 1997, at least 75 cases were under investigation by the DOJ.[218]

The increase in enforcement efforts has resulted from the changing face of the international political and business environments. The increase in U.S. foreign investment; the emergence of markets in the former Soviet Union, Eastern Europe, India, and China; the increase in infrastructure projects; the privatization of government enterprises; and the U.S. push to get foreign countries to prosecute corruption all have led to increased enforcement at home.[219] This increased enforcement heightens legal risks to corporations trading overseas and compels those businesses to focus on their internal procedures to ensure compliance.

The reach of the FCPA includes all foreign subsidiaries held by U.S. companies, as well as any project consortiums, joint ventures, or partnerships entered into with foreign corporations. Compliance falls on the shoulders of the corporation, and due diligence in the investigation of potential partners is crucial to avoid liability. This starts with investigating all business transactions entered into with foreign partners. It is necessary to determine that all participants in a transaction have no connection with foreign government officials. This includes all agents, consultants, primary shareholders, and management of the foreign corporation. Because an employee's knowledge can be imputed to the company, due diligence in the investigation of partners is the best way to ensure compliance and avoid imputed knowledge to the company with regard to any illegal payments that may be uncovered. Adequate preliminary investigation is critical to prevent, detect, and deter FCPA violations.[220]

The increased efforts by the DOJ to investigate and prosecute FCPA violations has led to greater risks for corporations that do not scrutinize the background of their foreign partners.[221]

The concerted push by the developed nations, and the principal role of the U.S. in speaking out against corruption, has led to increased enforcement at home. The negative effects of corruption in stunting economic growth and dissuading foreign investment, as well as destabilizing transitional governments, is now widely acknowledged. Corruption pervades many emerging economies and has the effect of depressing the population and their quality of life. However late the efforts for multilateral enforcement, recognition of the problem and its negative impact provide hope for change.

Practical effects of antibribery legislation

The significance of the antibribery convention and the practical effect of the FCPA on U.S. business opportunities worldwide is a matter of dispute. Sources at the U.S. Department of Commerce estimate that American companies lost more than 100 international contracts, valued at $45 billion, in 1994 and 1995 because of foreign companies' willingness to pay bribes.[222]

These findings were disputed by an Institute for International Economics study which concluded, "there is little evidence that the FCPA has a major negative impact on overall U.S. exports." The study noted that U.S. share of all world exports in 1994 was 12%, and that U.S. firms account for 21% of world exports of power-generating equipment and 40% of the global market for aircraft and parts, both of which are notoriously corrupt sectors. While admitting that being the world's lone superpower and part of the most desirable market has helped U.S. exports, the study also noted that the FCPA has forced U.S. companies to become more efficient in the marketing of their products, rather than relying solely on the quality of their bribes.[223]

The OECD Convention on Combating Bribery of Foreign Public Officials

The Convention on Combating Bribery of Foreign Public Officials in International Business Transactions, signed by members of the Organization for Economic Cooperation and Development (OECD) along with five other nonmember states, is a significant step toward leveling the field of international contracting. The Convention provides that each signatory country "shall take such measures as may be necessary to establish" criminal proscriptions against the bribing, and the aiding and abetting of such bribery, of foreign officials by any of its citizens. Punishment shall be "effective, proportionate and dissuasive criminal penalties...comparable to those applicable to the bribery of the 'signatory country's' own public officials."[224]

While this represents a major effort toward leveling the field, the convention's vague, open wording, required to persuade multinational participation, provides some glaring weaknesses in its enforcement. First, it fails to require specific accounting standards necessary to expose off-the-books payments. Second, it does not include the prohibition of payments to political parties which can sometimes serve as an indirect means of bribery to officials, especially in cases where one party is closely tied to all governmental decisions. Next, it fails to demand changes to the tax provisions of countries, such as France and Germany, that allow foreign corporations to deduct bribes as necessary business expenditures in connection with competition in "new markets." Finally, the strength of the convention lies in the willingness of the signatory states to enforce the adopted standards. However significant this collective effort to address the problem of corruption may prove, it follows by 20 years U.S. efforts to alleviate the same problems, and appears to lack the force necessary to expose and prosecute violators.[225]

An aggressive timetable for implementation was proposed. All signatories to the treaty were to present it before their parliaments by April of 1998. It was hoped, although highly unlikely, that the convention would be approved by a majority of participants by the end of 1998, giving effect to a global prohibition on bribery and corruption.[226]

Other international organizations are also pledging support in the campaign to stamp out corruption. The International Monetary Fund (IMF) and the World Bank, major lenders to emerging markets, are committed to adopting stricter standards and conditions when lending money. They have pledged to pay greater attention to the rule of law, the efficiency and accountability of the public sector, and the institutions which will manage economic growth and commerce in countries which are provided with assistance.[227] The World Bank has instituted rules that allow it to investigate corruption in projects associated with its US$19 billion per year development program and has vowed to blacklist corrupt companies and governments.[228] If access to these funds is contingent upon open and transparent economic practices, the countries seeking aid, which are generally emerging economies where corruption is most widespread, will be forced to implement reform. These heightened standards could be effective in promoting change within historically corrupt governments.

Progress for OECD legislation

The OECD legislation, agreed to in late 1997, will take effect only upon approval of the signatory countries. For this "soft law" to take force, it must be ratified by the legislatures of five of the ten OECD countries with the largest export shares, and which represent by themselves at least 60% of the combined total exports of those 10 countries.[229] (These 10 countries are, in order of largest export share, the United States, Germany, Japan, France, United Kingdom, Italy, Canada, Korea, Netherlands, and Belgium-Luxembourg).[230] If, after December 31, 1998, the Convention has not taken effect, any signatory may deposit in writing its readiness to accept the provisions. The Convention shall then enter into force upon the declarations of two such signatories.[231]

Thus far, little progress has occurred in the way of meeting the December deadline. The only country to achieve ratification is Bulgaria, which was one of the five non-OECD signatories.[232] The U.S. Senate approved the convention on July 31, 1998. Seven other countries, including Germany, Japan, and Belgium, have placed before their legislatures draft bills that would ratify the terms of the convention.[233] Britain hopes to put a memorandum ratifying the convention before Parliament in October.[234]

The nefarious impact of bribery and corruption

There is no dispute about the negative effects that corruption and bribery have on international trade and the stability of developing economies. Bribery represents inefficient spending of funds that acts as an invisible tariff, inflating project costs and discounting the quality of work and the materials used. It stifles efficient free market trade by rewarding corruption and providing no incentive to produce quality work and materials. Corruption has the effect of subverting political processes and distorting policy making. It can encourage bureaucrats with discretionary authority to invest in public works projects that are not necessary or are ill advised. Corruption provides a disincentive for privatization of government-held entities that could be run more effectively and efficiently within the private sector. It enriches a small elite bureaucracy on the take, but discourages foreign investors who will not invest in economies where it is perceived that corrupt leadership and ineffective institutions unfairly distribute the spoils of economic prosperity. This stalls the development of countries that have the necessary resources to become an emerging economy and, therefore, impoverishes a nation.[235]

Fortunately, bribery may be dying a natural death. The collapse of communism has forced those governments to become more open and democratic. The emerging press in these countries serves as a watchdog for the people and requires that politicians be more accountable for their decisions. Also, the economic crisis in East Asia could have a positive effect, with the IMF and the World Bank committed to more stringent monitoring of those who benefit from economic aid.[236]

Corruption in the transport sector

Because of the magnitude of the contracts and the high level of foreign government involvement, the transportation sector continues to be plagued by corrupt practices. What follows is a survey of recent reports detailing allegations of corruption uncovered in public works projects.

- In Sri Lanka, an investigation has been requested by the United National Party to look into possible corruption in the sale of a 40% stake in Air Lanka to the United Arab Emirates. UNP officials claim they have documented evidence that shows both airline officials and cabinet ministers obtained illegal gain.[237]

- In Indonesia, investigations by Commission V of the House have been commenced to look into the US$800 million markup in the construction of Pertamina's Export-Oriented (Exor) Refinery in Balongan, Indramayu, West Java.[238]

- In Thailand, Pradit Pattaraprasit, Deputy Minister of Transport and Communications, has launched an investigation into the state railway. He charges that illegal land deals are costing the government 500 million baht (US$10 million) a year. The investigation will focus on 22 plots of land leased to private entrepreneurs at no cost. These lands are in turn being leased to small businesses, without the consent of, or compensation to, the railway. One government official said that during the previous administration, at least 50% of the government's budget was being siphoned off by corruption.[239]

- In Hong Kong, 12 people were arrested on suspicion of bribery and corruption in connection with the construction of the multibillion dollar Hong Kong International Airport at Chek Lap Kok.[240] Four people (two engineers, one construction foreman, and one industrial diver) have been charged with allegedly covering up the use of substandard material during the construction of facilities connected to the new airport's railroad.[241]

- In Zimbabwe, the contract for a new airport in Harare was delayed for years under allegations of bribery and corruption, and was awarded to Air Harbour Technology despite its offering a bid US$30 million higher than the lowest bidder. In 1991, the government signed a contract with Aeroports de Paris, but revoked the agreement because the new airport design lacked any features "uniquely Zimbabwean." The local representative of Air Harbour Technology, Leo Mugabe, is the nephew of Zimbabwean President Robert Mugabe.[242]

Summary and Conclusions

The development of new airports has always required strong political leadership. As we have seen, the strength and longevity of political leadership in Singapore and Malaysia made it possible for a focused approach to airport development and economic growth. In contrast, weak and short-lived leadership in Thailand left the new Bangkok airport project floundering for decades.

Earlier projects tell the same story. In 1941, Mayor LaGuardia announced plans at a site rejected in 1930, the Idlewild golf course, on the edge of Jamaica Bay, New York. The city purchased, filled, and stabilized from December 1941 until June 1945 the 4600 acres of what became Idlewild International Airport, later renamed John F. Kennedy International Airport.[243] In 1958, President Dwight Eisenhower put an end to the endless studies and contentious debate surrounding the location of a second airport to serve Washington, D.C., by selecting the "political dark horse" of Chantilly, Virginia, as the site of Dulles International Airport.[244]

The amount of revenue a new airport project requires is enormous, often tempting politicians to hold out their hands, and stuff money in their pockets. In the United States, political rewards are often tendered in "set-asides" to "minority contractors" by minority politicians, whose racial minorities comprise political majorities in the cities which elect them. We have seen how Hispanic Denver Mayor Peña and African-American Denver Mayor Webb steered contracts toward minority contractors. Atlanta Mayor Maynard Jackson, also an African-American, halted construction at Atlanta International Airport until black-owned companies were awarded more than $70 million in construction contracts.[245]

The political influence, and potential corruption, of cities and their mayors might be diluted if semiautonomous independent regional airport authorities are created to build and operate airports. An airport serves not just the citizens of a city that builds it. An airport serves as an O&D catchment area for a larger metropolitan area, and of course, as a connecting point for through traffic. Allowing a city to control a region's airport necessarily disenfranchises those citizens who reside beyond the city's limits. A region's population might be able to exact political revenge on a mayor who rewarded his cronies with lucrative patronage contracts were they not disenfranchised by living outside the city's limits. Moreover, an airport authority might shield the airport manager from the whims of the mayor and city council. In the United States, most airport directors serve at the discretion of the mayor, who can remove him or her at will. This necessarily politicizes the process of airport management, and militates against objective decision making.

The contracting process also needs to be cleaned up. The extra money it takes to get a contract can flow to the political reelection committee, or to the politician's wallet, or those of relatives. Whatever the reason, issuing the contract to other than the lowest bidder drives up the cost of construction; issuing a contract to a less-qualified contractor means the work may be shoddy. The problem is widespread. Legislation to subdue corruption has been far from universal or effective. Perhaps the financial crisis which began in Asia in 1997 will inspire reform.

Examples can be drawn from around the world. For example, allegations of corruption have fallen on the issuance of landfill contracts at the New Bangkok International Airport,[246] and issuance of contracts at the new airport at Trinidad and Tobago.[247] In Zimbabwe, a contract for upgrading the international airport was let to a Cyprus company whose representative in Zimbabwe was President Robert Mugabe's nephew.[248] In the United States, the FBI investigated allegations of corruption in the issuance of concession contracts at Atlanta Hartsfield International Airport, resulting in bribery convictions against two City Council members.[249] Federal authorities also investigated allegations of understating lease revenue at Baltimore/Washington International Airport.[250]

Bribery and corruption are manifestly antithetical to the public interest, for they result in a poorer product at an inflated cost. Yet large infrastructure projects such as airports too often elicit one of man's most unsavory traits—greed.

We have seen airline opposition help derail new airport projects (as at Chicago and Minneapolis), or fall on deaf ears (as in Denver). We have also seen new airports advanced to inspire economic confidence in a politically volatile period (as at Hong Kong), to jump-start an economy in recession (at Denver), or to further a booming economy (as at Kuala Lumpur). Increasingly, environmental issues influence whether and where a new airport will be built, and how it will be configured.

Politics are an integral part of our everyday life. They play an indispensable role in the interactions within any group. They can aid or impede the progress of any activity depending on the will of the politician and those the politician serves. Nowhere is this more apparent then in the public works sector, where economic outcomes are vast, parties are divided, and both support and opposition are firmly set. Often politicians play the hand that tips the balance, sealing the fate of the project and setting a course for the future of the community from which there is no retreat.

While politicians like to think of themselves as altruistic visionaries sacrificing their personal goals for a life of public service, in fact, they are often puppets on a string manipulated by special interests. The days of the town meeting and representation by the common man are over, even in so-called democratic governments. Election to political office requires "compromise," too often being the interests of an influential and motivated constituency. As these case studies reveal, even where the goals within the community are aligned, rarely does the political process provide an effective and efficient means to the common end. Infighting, corruption, sedition, and cronyism plague public works projects, and somewhere in the mix the broader interests of the community can be lost.

Governments are beginning to recognize the corrosive effects that corruption in the political process can have in displacing the public interest. The United States, while by no means immune to corruption, was the first to make a concerted effort toward accountability of its corporations. The collective conscious of the global community is beginning to take note, and recent legislation by the OECD has resolved to place worldwide prohibitions on corruption. While these efforts provide some degree of hope, it is important to remember that rarely do even the purest ideals escape the political process unsoiled.

4

Airport Finance

"There is an inevitability in managing an airport: It typically must be operated 24 hours a day, 365 days a year; it requires continual maintenance, upgrading and expansion to reflect a growing transportation market; and most importantly, these obligations must be funded."[1]
—LOUIS A. TURPEN, CEO, GREATER TORONTO AIRPORTS AUTHORITY

"At US$8–14 billion a copy, how many [new "green field"] airports can the industry afford to build over the next 40 years? Even if the airport financing problem can be solved, it is doubtful that environmental and public opposition in the industrialized regions can be surmounted."[2] —MANFRED SCHÖLCH, DEPUTY CHAIRMAN, FLUGHAFEN FRANKFURT

In Chap. 4, we examine:

- The sources of capital available to less developed nations
- The inadequacy of traditional funding sources (such as AIP, GARBs, and PFCs) as a means of meeting the future air infrastructure needs in the United States
- The effect of federal legislation on airport revenue diversions within the United States
- U.S. Department of Transportation policies regarding excessive airport fees
- The means by which the world's new "green field" airports are financed
- The nature of GARBs with respect to airline "use and lease" agreements and bond underwriting agreements
- The costs, benefits, and motivations for privatization

- Efforts within the United States to facilitate civil aviation development at current and former military airports
- Sources of airport revenue
- ICAO recommendations concerning airport and user charges
- The necessity for accounting and control procedures

Introduction

Airport economics requires evaluation of financial issues at two levels. First, an airport seeking to expand its facilities, or a governmental entity seeking to build a new airport, must raise sufficient capital to finance such infrastructure development from public or private sources, or a combination of both. *Capital costs* consist of the component costs (e.g., labor, materials, and equipment) of construction of the airport and its component parts. Second, once built, an airport must earn sufficient revenue to pay its operating expenses and retire its debt. Such *operating costs* include expense items such as interest and depreciation or amortization on debt, taxes, and maintenance and administrative costs, such as salaries, power, and repairs. In addition to government grants and subsidies, the airport turns to its tenants—the airlines, concessionaires, parking—and the passengers they serve to finance its costs. This chapter focuses on both of these issues.

Issues surrounding financing of intermodal facilities at airports, and surface transport links to and from them, are discussed in Chap. 9. Suffice it to say here that those who object to public subsidization of transit tend to ignore public subsidization of other governmentally owned transport systems, such as public roads and highways, the waterway system, seaports, airports, and the air traffic control system. Here, we examine issues of airport finance.

Sources of Capital

Sources of capital for airport development include governmental or international organization loans and grants, commercial loans from financial institutions, equity or debt (typically, bonds) from commercial capital markets, including private investors, banks, investment houses, and fund pools, and the extension of credit from contractors and suppliers. Commercial loans typically incur the highest interest rates, though such rates may be reduced by governmental loan guarantees. Airports must also evaluate the amount of foreign capital needed, for debt often will be needed to be repaid in that foreign currency, and therefore subject to both competitive internal needs for foreign currency and currency revaluations, favorable and unfavorable.[3]

Foreign governments may be willing to provide capital to airport projects in less developed nations, out of a sense of altruism, or with the purpose of promoting trade and commercial relations between the two nations, or exporting technology and equipment from firms domiciled in the lender nation. Some nations have developed economic and social development programs in various parts of the world, providing loans on preferential terms, or supplies, equipment, and technology. Examples include the following:

- Belgium: General Administration for Cooperation and Development
- Canada: Canadian International Development Agency
- Czechoslovakia: Ministry of Foreign Affairs
- Denmark: Danish International Development Agency
- France: Central Fund for Economic Cooperation
- Germany: Ministry of Economic Cooperation
- Italy: Department of Cooperation
- Japan: Overseas Economic Co-operation Fund
- Netherlands: Foreign Ministry
- Norway: Norwegian Agency for International Development
- Russian Federation: Ministry of External Economic Relations
- Spain: International Cooperation
- Sweden: Swedish International Development Administration
- United Kingdom: Overseas Development Administration
- United States: U.S. Agency for International Development[4]

Specialized export-promoting agencies (e.g., the Export Development Corporation of Canada, the Export-Import Banks of Japan and the United States, COFACE of France, HERMES of Germany, and the Export Credits Guarantee Department of the United Kingdom) may also be able to make direct loans or guarantee private loans, or insure the risk assumed by domestic firms providing goods and services for airport development.[5]

Several international bank and fund organizations have been established to aid developing nations by assisting in financing and execution of projects, particularly infrastructure projects, which foster economic development. These include the following:

- International Bank for Reconstruction and Development and its affiliates, the International Development Association and the International Finance Corporation
- African Development Bank
- Asian Development Bank
- Caribbean Development Bank
- Inter-American Development Bank
- European Union Development Fund
- Japan Overseas Economic Cooperation Fund
- Organization of Petroleum Exporting Countries Fund for International Development
- Arab Bank for Economic Development in Africa
- Islamic Development Bank
- Saudi Fund for Development
- Abu Dhabi Fund for Arab Economic Development
- Kuwait Fund for Arab Economic Development
- Arab Fund for Economic and Social Development[6]

In each instance, a loan or grant will be made to a governmental agency, or to a private entity having the support and guarantee of the government. Hence, the government must designate the project as a high priority for development in order to receive such assistance.[7]

The United Nations Development Programme (UNDP) provides developing nations with expertise in planning and executing airport projects, including feasibility and cost-benefit analyses, master planning, and construction. Funding for minor equipment may also be obtained from UNDP, though the principal role of the agency is to provide expertise rather than capital.[8]

Public Sector Financing: The U.S. Example

The United States accounts for between 40 and 50% of the world's air transport movements, the single largest airline and airport market in the world. It is for this reason that we examine how the United States has struggled with the knotty problem of airport finance.

Capital infrastructure needs

Infrastructure investment in the United States declined significantly beginning in the early 1970s, coinciding with enormous expenditures in financing the Vietnam War and the sharp increase in fuel costs following the Arab Oil Embargo of 1973. In the ensuing years, public infrastructure investment was halved, so that by the mid-1990s, the United States was investing less than any G-7 nation, and at only one-third the rate of Japan.[9]

The U.S. Federal Aviation Administration (FAA) predicts that passenger enplanements will grow 60% by the year 2003.[10] Airport development and capital reconstruction needs in the United States are between $6 billion and $10 billion annually, of which 60% is eligible for federal support.[11] But airport capital requirements are growing faster than air traffic. Between 1983 and 1992, U.S. enplanements increased 4.5% annually; airport capital costs increased 10.4%, and airport operating costs increased 12.6% annually.[12]

The estimated $10 billion costs far exceed the U.S. aviation trust fund and passenger facility charge authority of $11 billion over 5 years (a time period when U.S. airports will need $50 billion). Adding airport bonds and other local revenue to the total meets only half the airport development needs in the United States. Without sufficient capital, the number of seriously congested airports (airports experiencing more than 20,000 hours of annual delay) will grow from the present 22 to 32 in less than a decade. These delays cost airlines more than half a billion dollars a year, and passengers billions more in opportunity costs.[13]

Funding sources

Historically, airport funding in the United States has come from two sources: (1) federal ticket taxes [or Airport Improvement Program (AIP) funds] from the Airport Trust Fund collected on every airline ticket purchased in the United States and (2) tax-free general airport revenue bonds (GARBs) issued by municipalities. Often, 80% of the capital for the airport project comes from AIP grants, while the remaining 20% is raised by municipalities in GARBs.[14] Between 1982 and 1994, some $16 billion in AIP funds was allocated by the FAA for improvements at eligible airports.[15] Between 1946 and 1996, the U.S. government granted more than $24 billion in grants to airports.[16] In 1990, Congress created a federally authorized but locally collected program of airport passenger facility charges to supplement public capital needs.

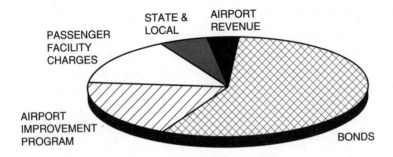

4.1 *This chart reveals the sources of capital for U.S. airports and their relative importance.*

In 1996, the 3304 U.S. airports that composed the national airport system spent about $7 billion on capital development. More than 90% of that came from three sources—airport and special facility bonds ($4.1 billion), the Airport Improvement Program ($1.4 billion), and Passenger Facility Charges ($1.1 billion).[17] Figure 4.1 provides a breakdown of the funding sources of U.S. airport capital expenditures.

Airport and Airway Trust Fund

The Airport and Airway Improvement Act of 1982[18] established the framework for federal financing of U.S. airport development and improvement projects. The Airport and Airway Trust Fund is the depository for airport infrastructure revenue under the Airport Improvement Program. Such revenue is derived from airline ticket taxes, fuel taxes, airway bill taxes, and other taxes and fees. Prior to 1997, the aviation taxes deposited in the Airport and Airway Trust Fund included a 10% ticket tax, a 6.25% cargo waybill tax, a $6 international departure tax, a 15 cent per gallon general aviation gasoline tax, and a 17.5 cent per gallon general aviation fuel tax.[19] Legislation passed in 1997 shifted much of AIP funding away from the traditional 10% tax imposed on each ticket sold. The new taxing structure is as follows:

Commercial

- Ticket tax of 9% in FY 1998, 8% in FY 1999, and 7.5% in FY 2000 through FY 2002

- Segment charges per passenger of $1.00 in FY 1998, $2.00 in FY 1999, $2.25 in FY 2000, $2.75 in FY 2001, and $3.00 in FY 2002

- International departure and arrival taxes of $12

- Frequent flyer award tax
- $0.043 commercial user fuel tax (formerly the deficit reduction tax)
- 6.25% cargo waybill tax

Noncommercial

- $0.193 aviation gasoline tax
- $0.218 aviation jet fuel tax[20]

In the late 1980s, the Airport and Airway Trust Fund (created by the Trust Fund Code of 1981) had accumulated some $15 billion. Some criticized the failure of Congress to spend it as a means of masking the size of the federal deficit (with unspent money in the trust fund, less would have to be borrowed in the private capital markets). Moreover, the AIP was tapped to pay for FAA operations.[21] Between 1992 and 1996, AIP funding declined by 23%, from $1.9 billion to $1.45 billion.[22] By 1998, Congress had funded AIP at $1.7 billion, about $800 million below the authorized amount. The AIP not only funds airport infrastructure improvement and expansion, it funds 70% of the Federal Aviation Administration's annual budget (the other 30% coming from the general fund).[23] To make up an anticipated capital shortfall, a special commission headed by former Congressman Norm Mineta recommended funding at $2 billion annually over 5 years.[24] (See Fig. 4.2.)

The AIP provides funding for airport planning and development projects that enhance capacity, safety, and security, and mitigate noise. Some 3300 airports have been designated by the FAA as eligible for AIP funding. Most funds are allocated according to a statutory entitlement formula based on passenger enplanements, with set-aside categories for designated types of airports and airport projects.[25] The FAA has discretionary authority to distribute the remaining funds (approximately $300 million of the $1.46 billion available in 1997).[26] The AIP allocation formula is set forth in Fig. 4.3. The largest U.S. airports get only about 10% of their funding through AIP, while the smaller airports as a group acquire about half of their funding from AIP.[27]

Passenger facility charges

It is anticipated that the AIP will be able to finance only 20 to 30% of the total capital needs of airports.[28] As a consequence, U.S. airports will become increasingly reliant on other sources of capital to

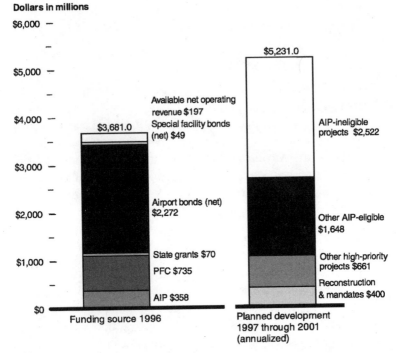

Dollars in millions

4.2 *The U.S. General Accounting Office anticipates a funding shortfall for large-hub airports, which will be ineligible for AIP funding, of more than $2.5 billion between 1997 and 2001.*

U.S. General Accounting Office.

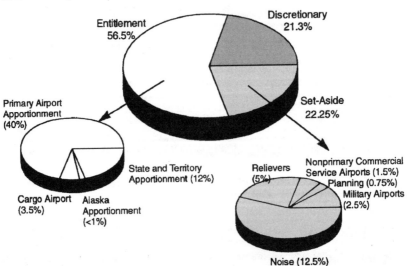

4.3 *This chart reveals the allocation method for the FAA's Airport Improvement Program.* U.S. General Accounting Office.

cover what likely will be a reduced federal ability to finance infrastructure improvement and expansion.

Recognizing the need to generate local discretionary sources of capital, Congress in 1990 created the Passenger Facility Charge (PFC) program, which allowed airports to impose a $1, $2, or $3 charge directly on each boarding passenger (up to a maximum of $12 per passenger) for FAA-approved projects, where the airport is willing to surrender half of its AIP funding. By 1994, the FAA had approved applications by about 100 airports to generate $6.4 billion in PFC revenue.[29] Between 1992 and 1996, the FAA had authorized 244 airports to collect $12.5 billion in PFCs.[30] By the late 1990s, PFCs were generating more than $1.1 billion annually for approved projects at more than 200 airports.[31]

The PFC program differs from the AIP funding program in a fundamental way. While AIP funding at a local airport must be directed and approved by a state transportation agency, the PFC funds flow directly to the airport proprietor, eliminating the state's oversight on how the money shall be spent. This has led to criticism on grounds that the local focus thereby transcends regional needs.[32] Congressman Henry Hyde (Republican, Illinois) has lamented that he supported legislation authorizing PFCs in the hope that a third airport would be built at Chicago, only to see Chicago O'Hare Airport hoard PFC revenue for itself. The PFC surcharge had been created in 1991 to support construction of new airports, but this purpose has been thwarted.[33]

Federal mandates

Federally mandated programs and regulations affecting airports are numerous and costly. They include:

- Part 139 airport certification
- Wildlife and endangered species
- Hazardous waste
- Farm and land conservation
- Wetlands, floodplains
- Scenic rivers
- Coastal barriers
- Historical preservation
- Soil contamination cleanup
- Disabilities access

- Security
- Noise
- Air quality
- Water quality (e.g., storm water runoff)

Sometimes federal regulatory requirements conflict. For example, FAA safety requirements insist on aircraft deicing during freezing rain or snowy weather. But EPA prohibits soil contamination with deicing fluid. Similarly, airports simulate crashes, firefighting, and rescue with actual burning fire, which must be extinguished; but air pollution regulators object to the smoke.

The airport industry associations estimate that federally mandated programs cost airports some $1 billion a year. The Federal Aviation Administration Authorization Act of 1994 makes the cost associated with the Clean Air Act, the Federal Water Pollution Control Act, and the Americans with Disabilities Act eligible for AIP funding.[34]

Revenue diversion

While on the one hand, Congress has been concerned with assisting airports with their need to secure adequate funding, on the other, it has been alarmed at the possibility that airports might use airport revenue for nonairport purposes. Airports account for between 4 and 6% of airline industry costs, and a diversion of revenue "downtown" could only worsen their financial condition.

The Anti-Head Tax Act of 1973 authorized airports to collect only reasonable and nondiscriminatory rental charges, landing fees, and other service charges.[35] The Airport and Airway Improvement Act of 1982 provides for a fee and rental structure which would make the airport as self-sustaining as possible, and states that such charges must be reasonable and be used only for airport purposes.[36] Specifically, it provides that, as a condition precedent for receiving federal funds, airports must agree that their facilities "will be available for public use on fair and reasonable terms and without unjust discrimination."[37]

Air-side revenue streams include landing fees, fuel taxes, and maintenance and cargo facility leases. Land-side revenue streams include terminal rents and gate leases, concessions, parking fees, and PFCs. All revenue generated by a public airport, as well as local fuel taxes, must be used for legitimate airport purposes—the capital or operating costs that directly and substantially relate to air transport. In *Northwest Airlines v. County of Kent,* the U.S. Supreme Court held that an airport charge is reasonable "if it (1) is based on some fair approximation of the use of the facilities, (2) is not excessive in relation to the benefits conferred, and (3) does not discriminate against interstate commerce."[38] Under this analysis, airports were "given wide latitude in selecting a particular rate methodology and fee structure."[39] Shortly after this decision, the city of Los Angeles announced it intended to use airport revenue to cover city operating expenses, including police protection for the city at large.

The Federal Aviation Administration Authorization (FAAA) Act of 1994 provides that airport charges, fees, or taxes be used for airport or aeronautical purposes.[40] When an airport accepts federal grants for airport improvements it must assure that the airport will be available for public use on fair and reasonable terms without unjust discrimination.[41]

The FAAA Act also authorizes the Secretary of Transportation to determine the reasonableness of airport fees, though the Secretary is not to set the level of the fee.[42] The legislation explicitly affirms that different rate methodologies may be employed. Strict time deadlines are imposed. The DOT Secretary must determine whether a significant dispute exists within 30 days of the filing of a complaint by an airline or a request by an airport.[43] If he concludes affirmatively, a DOT Administrative Law Judge (ALJ) must issue a decision within 60 days. The Secretary must issue a final decision within 120 days after filing. If the Secretary has not issued a decision within 120 days, the ALJ's decision becomes the final decision of the Department.[44]

The DOT issued a Policy Regarding Airport Rates and Charges in February 1995.[45] It provides that airports are not to charge more than aeronautic costs on a break-even basis. The rate base is to be valued according to historic cost, rather than fair market value.[46] DOT has ordered refunds of airport fees deemed to be excessive.[47] A DOT Administrative Law Judge ordered Los Angeles International Airport to refund $7.2 million to tenant airlines which the airport had collected for fire and police services, debt service, and accounting errors.[48] In 1997, the FAA suspended $63 million in grants to Los Angeles–owned airports until it returned some $31 million it diverted from airports to other projects the preceding year.[49]

Allegations of the use of airport revenue for nonairport purposes are abundant. The DOT Inspector General found Westchester County,

New York, spent nearly $24 million on nonairport projects in the early 1990s, while in Hawaii, $64 million was spent on a dog track next to the airport. The only category of airports found not guilty of revenue diversion were independent port authorities, not subjected to rule by mayors and city councils.[50]

One way around the diversion prohibition is for an airport to forgo federal assistance. San Francisco International Airport, which receives no direct tax dollars, purchases $60 million in services from the city, and transfers 15% of its concession revenue (about $20 million) to the city each year.[51]

State and local financing

Most states also provide financial assistance to airports, usually in the form of matching funds for AIP grants. States fund their contribution through such vehicles as aviation fuel and aircraft sales taxes, highway taxes, bonds, and general fund appropriations. In 1996, the states provided $285 million to U.S. airports.[52] Occasionally a local municipality or outside developer may provide financial assistance to an airport. Usually, such assistance is in the form of operating subsidies for smaller airports.[53]

Public Sector Financing at the World's New Airports

New "green field" airports are being built around in the world. This section provides a comparative analysis of the means whereby airports are financed at several of these venues.

Munich's Franz Josef Strauss Airport

Munich's airport was built by the Munich Airport Authority. The Free State of Bavaria provided 51% of the funding, the government of Germany 26%, and the city of Munich 23%.[54] All three governmental institutions must approve improvements, and all three audit the airport's expenditures.[55]

By 1991, the airport was anticipated to cost DM1 billion (US$590 million) for land and DM7.5 billion (US$4.41 billion) for construction. Had it been built 20 years earlier, in 1971, the full cost for land and construction would have been DM4 billion. The additional cost represents the delay caused by environmental opposition and litigation.[56]

Revenue of DM600 million (US$380 million) was anticipated for the first full year of operation. However, high mortgage and interest payments will cause gross losses for several years.[57] The new airport reported a net deficit after depreciation of DM100 million ($64.5 million) in 1993 and DM95 million ($61.3 million) in 1994.[58]

Osaka's Kansai International Airport

The Kansai International Airport Company was formed in 1994 to supervise construction and manage the airport.[59] Kansai International Airport was Japan's first airport to have been built and operated by a private firm and funded commercially.[60] Kansai was originally projected to cost 1 trillion yen, or about US$7.7 billion.[61] The national government was to invest US$546 million, and local governments and private investors invested US$137 million, with US$6 billion anticipated to be borrowed.[62]

But Kansai cost 40% more than was originally projected.[63] Along the way, the United States insisted that more American firms be involved in the construction process, and alleged some measure of bid rigging in favor of Japanese construction firms. The dispute caused something of a diplomatic rift. But steering construction contracts to higher-cost local firms no doubt also contributed to KIA's high costs. Ultimately, the government of Japan provided two-thirds of the funding, local governments provided one-sixth, and about 1000 local businesses contributed one-sixth.[64] KIA is saddled with US$15 billion in debt (and rising), and needs about US$2 million a day to service debt.[65] It lost more than US$200 million in its first 6 months of operation.[66]

Cost overruns at Kansai, stimulated principally by the fact that parts of the island on which the airport was built began to sink during construction, required the airport to impose charges of about US$24 a ton, or nearly US$9673 as a landing fee for a 747,[67] nearly the same as Narita, the world's most expensive landing fee airport.[68] Jetway and baggage handling fees are also among the highest in the world, as are rental rates on concessionaires (it was anticipated a cup of coffee might cost $10).[69] Air-side charges (e.g., landing fees) are 10% higher than the world's most expensive airport, Tokyo Narita, while land-side charges (e.g., concessionaire rentals) are up to 300% higher.[70] Office space and baggage facility rental fees are 331% and 750% higher, respectively, at KIA than at Narita; cargo office, first-class lounge, and check-in counter space is 332%, 347%, and 330%

higher, respectively.[71] The passenger airport departure tax is US$26 (the world's highest); the automobile toll across the two-layer truss airport bridge is US$17, while a cargo vehicle pays US$120; a taxi ride from Osaka station costs US$190;[72] and parking a car for 3 days at the airport costs US$145.[73] Thus, KIA is an extremely expensive airport at which to do business.

Debt service will cost 200 million yen (US$2.04 million) a day.[74] The airport was projected to sustain a US$400 million loss during its first year of operation[75] and take 30 years to pay off its debt.[76] That is quite different from projections made in the 1980s that the airport would reach break-even by 1998, and pay dividends by the year 2003.[77] For the fiscal year ending March 1998, KIA sustained an operating profit of nearly 27 million yen, but a net loss of 24 million yen after various expenses, including nearly 51 million yen in interest expenses.[78]

Funding for the second-phase expansion will be divided between local and national governments and the private sector. For site construction, the local and national governments will provide 30% of the cost in capital contribution and 25% in interest-free loans, while the private sector will finance the remaining 45% with loans. The facilities will be funded with an economic contribution provided by the national government and private sector of 30% in investment and 70% in government-guaranteed bonds.[79]

Denver International Airport

The history of airport cost overruns in the United States did not begin with Denver International Airport, though Denver took them to new heights, appropriate for a city which sits a mile above sea level. Early statements by Denver Mayor Federico Peña pegged the cost at $1.5 billion, of which half a billion dollars would come from the federal government, half a billion would come from the sale of the 7 mi^2 of the existing airport (Stapleton International),[80] and half a billion would come from bonds. DIA's ultimate cost was $5.3 billion. The last "green field" airport in the United States before DIA was Dallas/Fort Worth International, which was originally estimated to cost $200 million, but by 1970 had grown to $500 million.[81]

In Washington, D.C., Transportation (DOT) Secretary Sam Skinner had come under criticism for allowing the airport trust fund to accumulate some $15 billion. Skinner was looking for a "demonstration project," and was persuaded to view the proposed site on his way to

Alaska to deal with the oil spill of the *Exxon Valdez*. Skinner quickly became a DIA proponent, and DOT agreed to help finance the project in the amount of a $500 million Airport and Airway Trust Fund grant. By late summer 1989, the first federal funding installment ($60 million) was received, the FAA approved the final Environmental Impact Statement, and groundbreaking on the project began on September 28, 1989. The major part of the funding, however, would still have to come from the sale of bonds (which, upon completion, amounted to more than $4 billion sold since the first $704 million issue on May 8, 1990).[82]

The massive cost overruns at DIA were caused by several factors. United Airlines demanded several midstream design changes of magnitude as a price for signing up for Concourse B. These included an automated baggage system which the city decided should go in all three air-side concourses, requiring major redesign and reconstruction of the main terminal, already under construction. The airport planners at DIA also chose to stretch the technological envelope, without installing back-up systems, most notably on the baggage system and the underground train system linking the remote concourses to the main terminal. Back-up and contingency planning, even intermodal transportation planning, was notoriously poor at DIA. The inability of the baggage system to perform as designed gave United Airlines the excuse it needed to delay DIA's opening until the second hub carrier at Denver, Continental Airlines, announced it would eliminate its Denver hub. All this created enormous delays and increased costs.[83]

Macau International Airport

The $912 million airport was financed through an agreement with the Banco Nacional Ultramarine, the Banco Comercial de Macau, and financial institutions in New Zealand and Australia.[84] Landing fees will be an average of the five busiest airports in the region.[85] It is anticipated that the landing fee for an Airbus A320 will be US$730; the passenger departure tax will be US$16.[86]

The Macau Airport Company (CAM), which holds the 25-year concession on the airport, is owned 51% by the Macau government, 33.3% by Hong Kong's STDM (owned by Stanley Ho, who also owns all nine Macau casinos, the hydrofoil shuttle from Hong Kong, and the sea terminal, and accounts for 25% of Macau's gross domestic product), and 22% by other investors.[87]

One concern in both Macau and Hong Kong was reversion of the colonies to China in 1999 and 1997, respectively. The 25-year concession granted CAM was an effort to assure some stability. Mainland Chinese approval of airport financing of MIA was without incident, unlike the reservations Beijing raised with respect to Hong Kong's new airport.[88] Moreover, with Hong Kong reverting 2 years earlier, Macau is able to assess how China's ascendancy to political dominance affects the economic environment. And MIA will be operational before reversion, while Chek Lap Kok opened only after Hong Kong reverted to China.[89]

Kuala Lumpur International Airport

The Kuala Lumpur International Airport at Sepang is managed by the Public Works Department of the Malaysian federal government.[90] The airport was financed primarily by government-guaranteed unsecured bonds. KLIA also had access to a 61.5 billion yen Japanese government loan.[91]

Hong Kong International Airport at Chek Lap Kok

The new Hong Kong International Airport at Chek Lap Kok cost some US$8.7 billion.[92] However, with the additional suspension bridges, tunnels, highways, high-speed rail, and ocean terminal facilities being built, the total cost was US$19.9 billion.

The new airport was originally planned with the Provisional Airport Authority (PAA) working in conjunction with the departments of Civil Aviation, Highways, and Territory Development, as well as the Mass Transit Railway Corporation, the Hong Kong Legislative Council, the Airport Committee of the Joint Liaison Group (with China), the Airport Consultative Committee, and airlines.[93] Although 100% owned by the Hong Kong government, the PAA was established as an independent corporation operating under commercial principles.[94]

To pave the way for financing the airport terminal, the Hong Kong legislature converted the provisional body that oversees airport construction into an independent corporate body, named the Airport Authority. The Mass Transit Railway Corp., a semigovernmental body that runs the subway system and is building a railway to the airport, was already incorporated. The bill followed an agreement between the United Kingdom and China concluded in the summer of 1995 which resolved the dispute on finance support for the airport and its railway.[95] China had objected to what it perceived to be excessive reliance on borrowed capital, which would cause a heavy burden of indebtedness after it assumed sovereignty over the British colony in 1997.[96] China also resisted on grounds that it had neither been consulted on the desirability of a new airport, nor been included in the decision-making process.[97] It objected to what it perceived to be an excessively costly "gold-plated" facility devised by the British, for whom China has had long-standing antipathy.[98] Moreover, Beijing was unhappy about Governor Chris Patten's proposals for enhanced democratic representation in the Hong Kong local government.[99] Chek Lap Kok was originally scheduled to open in July 1997, but the lack of a financial agreement with China slowed the schedule. The airport opened in July 1998.[100]

As noted above, only 40% of the price tag of about US$20 billion is for the airport itself. The remaining funds are being spent for the 4500-ft rail and roadway suspension bridge, a 21-mi airport express rail line, and a third harbor tunnel for Hong Kong Island. Approximately 30% of costs will be raised from tenants, franchises, and such accounts as air traffic control. Aeronautical charges, about triple those of Kai Tak, provide 37% of the airport's total commercial and aeronautical revenue.[101] Landing charges are HK$2600 (US$330) per aircraft up to 20 tons, and HK$74 per additional ton. Parking charges for aircraft are imposed in 15-minute units, with terminal building frontal parking stand fees at HK$183, terminal building remote stands at HK$147, cargo apron stands at HK$117, and maintenance apron stands at HK$94. Lower fees are imposed between midnight and 7:00 A.M.[102]

The Hong Kong government put up US$4.5 billion in cash for the airport and US$2.9 billion for the airport railway. The airport and railway each borrowed US$1.4 billion from banks, to be repaid in 2001, four years after the airport was scheduled to open.[103] The Chinese government insisted that three-fourths of the airport debt be retired by its opening date.[104]

The airport includes major private investment opportunities, with 90 ha of commercial land available and 76,000 m² of floor space available for lease in the passenger terminal building. The Hong Kong Sky Mall has 30,000 m² of retail space, with five retail zones and 140 shops both air side and land side.[105]

Oslo Airport Gardermoen

Oslo's new airport at Gardermoen was financed with equity capital from the Norwegian Civil Aviation Administration, conventional loans amortized over a 30-year period, and proceeds from the sale of Oslo Airport Fornebu. Commercial activities and airport charges will cover airport operating costs, note repayment, and interest on equity capital. The airport has 13,000 m² of commercial space. Commercial activities are anticipated to cover 35% of total operating income. It is anticipated the airport will produce a 7% return within a few years of its opening.[106]

Seoul's Inchon International Airport

The South Korean government is spending 25% of its national budget for infrastructure improvements to close the gap between demand and supply.[107] In 1996 alone, the South Korean government approved finance plans for five major infrastructure projects totaling US$25 billion. The projects are designed to ease South Korea's overburdened transport system, which has been blamed for a steep rise in production and distribution costs. Logistical costs in South Korea were estimated to be 16% of GNP.[108]

Korean Airports Authority will raise US$2.36 billion (47%) from operating profits, loans, bonds, and sale of land at the prepared site. The Korean government will provide US$2.38 billion in subsidies (48%). Private sector investment will provide US$240 million (5%).[109] KAA funding will be used for airport facilities. The government's funding will be used principally for airport access and site preparation. Private sector investment will be used to construct the cargo terminals and other secondary support facilities.[110]

The Korean Ministry of Construction and Transportation designated seven major projects at the new Inchon International Airport as Social Overhead Capital projects to be built with US$2 billion in private sector financing on a build-operate-transfer (BOT) basis. Four of these projects were inaugurated in 1996: (1) the Airport Exclusive Access Expressway, which includes a US$1.4 billion, 4.4-km double-deck suspension bridge; (2) a US$62 million cargo terminal; (3) a US$37 million cogeneration power plant; and (4) a US$108 million aviation fuel farm.[111] To make them more attractive, the government has offered tax benefits and a guaranteed rate of return on investment.[112] Lee Hank Koo, president of Daweoo Research Institute in Seoul, praised private sector funding on grounds that "Because businessmen are profit-oriented, their choices will help the government select lucrative projects and curtail unnecessary investment that may be prompted by political motivations."[113]

The Asian financial meltdown and soaring construction costs caused the South Korean government to amend its financial plan in 1998. The cost of building Inchon International Airport had risen from an original estimate of US$3 billion, to US$4.4 billion in 1995, to US$5.8 billion in 1998. Though the government of Korea had originally committed to providing 40% of the capital for the airport, it announced an intent to sell a third of its stake to foreign airlines, and to consider allowing foreign management to run the airport.[114]

Private Sector Financing

Early airport construction was financed by general obligation bonds backed by the "full faith and credit" of a governmental unit and secured by taxes collected by it. The industry was in its infancy, and airports were not capable of generating sufficient revenue to finance infrastructure costs. Since World War II, general airport revenue bonds (GARBs) have replaced general obligation bonds as the preferred means of financing new airport construction, expansion, or improvement. GARBs are paid off by revenue generated by the facility they financed.[115] More than 95% of all airport debt issued at U.S. airports since 1982 has been in the form of GARBs, though special facility bonds secured by revenue from the indebted facility (e.g., hangar or maintenance facility) are sometimes issued.[116] Both GARBs and general obligation bonds historically have been tax-exempt (as industrial revenue bonds), allowing states, municipalities, and airport authorities to lower the long-term costs of capital financing.[117]

GARBs typically run for a 25- to 30-year term (as opposed to general obligation bonds which run for 10 to 15 years) and usually pay higher interest than general obligation bonds. Airlines typically stand behind the revenue bonds with "use and lease agreements," pledging to make up the difference in revenue shortfalls by paying higher landing fees.[118] The quid pro quo for the residual funding agreement historically has been a long-lease term for gates, a "majority-in-interest clause" giving airlines a say (often an effective veto) over airport expansion, and a return of excess revenue collected, often in the form of lower landing fees. This has resulted in some criticism, for the dominant airlines are given a means of thwarting airport efforts to build infrastructure to accommodate new entrant airlines, which might be able to discipline the incumbent's high fares. (See Chap. 11.) The danger is not only competitive control, if not monopolization, of public resources, it is reliance on airlines which, under deregulation, have been a financially unstable bunch.[119] As of 1990, majority-in-interest clauses were in effect at 36 of the 66 largest U.S. airports.[120] Today, about half of the largest airports in the U.S. rely on airlines to back airport revenue bonds.[121] In a compensatory lease arrangement, the airport takes the risk, and the surplus revenue stays in the airport.

The cost of private capital typically is higher than public capital, though it can be ameliorated by governmental guarantees and insurance. The competitiveness of airport bonds in the market can be gauged by the bond ratings by the major investment houses, the interest rate, and the default ratio. Tax exemptions on the purchase price or interest can also stimulate investor interest in airport bonds.[122]

Between 1982 and 1996, U.S. airports issued $53.6 billion worth of bonds, of which $17.3 billion was to refinance existing debt.[123] Airport revenue bonds are the most important single source of capital available to small and medium-size airports. Of the average $6 billion of total annual airport capital spending in the United States, revenue bonds accounted for $3.5 billion annually in "new money" and $1.6 billion in "refunding" or debt restructuring. Airport debt at the 22 largest airports in the United States almost doubled, growing from $445 million in 1988 to $880 million in 1994.[124]

Bond underwriters have a fiduciary responsibility to exercise "due diligence" on the issuance of bonds, taking reasonable care that all material facts are disclosed, and having a reasonable belief that the bonds will meet their assigned interest rate. Theoretically, highly speculative projects should not find financing because of the enormous legal risk underwriters assume. In reality, however, the bond market does not function in that manner because the securities industry has been largely unregulated and driven by commissions and transactions—in a word, profit. Municipal bonds issuance is a tremendously profitable multibillion dollar business. Companies that issue them earn profits in two ways—commissions on sales and capital appreciation on bonds purchased at a discount and later sold at the market. Thus, deals get done, not because they should be, but because they can be.[125] Moreover, while municipalities and airport authorities tend to issue construction contracts on the basis of competitive bidding, the issuance of bond underwriting agreements tends not to be handled on a competitive bidding basis. This gives bond underwriters incentive to express their appreciation for the business in the form of political contributions to the elected officials of cities who steer business their way. In a largely unregulated environment, elected officials can effectively extort campaign contributions under a "pay to play" philosophy.[126] As noted in Chap. 3, the U.S. Securities and Exchange Commission has adopted rules attempting to circumscribe campaign contributions by financial institutions which float municipal bonds.

Scrutiny of debt by the credit-rating agencies, such as Moody's and Standard & Poors, and investors encourages airports to be efficient and market-oriented in their operating and investment decisions.[127] Ratings assess the financial soundness of the project—the ability of projected cash flow to meet the financial obligation that has been incurred. The higher the rating the lower the cost of capital.[128] The exemplary financial performance of airport bonds has earned them the status of premium-grade investments in the tax-exempt municipal bond market, thereby reducing the cost of capital.[129] Ultimately, the institutional investors drive the market, for they have enormous amounts of cash to invest.[130]

Another private sector funding mechanism is the build-operate-transfer (BOT) approach, whereby the contractor commits to financing, construction, operations, and maintenance for a specified number of years (known as the "free use period"), after which it transfers the facility over to the government.[131] That leads us to a discussion of privatization.

Privatization

Though publicly owned, many airports long have had some measure of private operation. As noted above, private capital finances various levels of airport development throughout the world, which itself encourages market-driven management. In many nations, private companies perform essential airport services, including private airlines, caterers, concessionaires, and contractors. For example, less than 3% of the people who work at the three major airports operated by the Port Authority of New York and New Jersey (i.e., Kennedy, LaGuardia, and Newark) are public employees.[132]

Most of the world's airports are owned and operated by their national governments. In the United States, airports are owned and operated by municipal, local, or state governments or regional airport authorities. But worldwide, this is changing. Louis Turpen, CEO of the Greater Toronto Airports Authority, summed up the financial imperatives driving the historical trend toward airport privatization:

> *Historically, governments funded and drove airport decision making. In the years leading to deregulation, airlines, via the mechanism of the long term lease, assumed the primary role as guarantor of an airport's financial stability and in great measure dictated airport development. But airlines are faced with an increasing dilemma. Though still wanting to dictate the direction of the airports they serve, competitive forces are eroding airline profits and their ability to continue as guarantors of airport revenues. They simply can't afford it. As a result, airlines continue to look for substitute guarantors, preferably ones that are silent partners such as the passenger, via the "head tax" mechanism, to preserve their position of influence without incurring the cost. So do airports. Instead of recognizing that they themselves are now the guarantors, airports continue to seek financial assistance from government and the airlines. Such assistance is not forthcoming. This situation has created a vacuum, and a new breed of guarantors have surfaced—active investors who are quite willing to provide money which oftentimes the airport should not spend—in exchange for an ownership position.*
>
> *Privatization is a result for the need to service a voracious appetite for capital dollars....*[133]

At the dawn of the twenty-first century, the trend toward privatization is robust. Airports have been sold to private investors or turned over to private operators in nations as diverse as Argentina, Australia, Austria, Bolivia, Canada, Chile, Colombia, Denmark, Ecuador, Germany, Greece, Mexico, South Africa, and the United Kingdom. But actually, privatization is a generic term which means different things to different people—to some, it is the outright sale of assets, while to others, it is turnkey or joint venture arrangements, the lease of assets, private investment, private management, and/or corporatization.[134] In recent years, at least 47 nations have moved toward further privatization, from selling minority shares in individual airports, to inviting private investors to build runways or terminals, to selling major airports outright.[135]

The motivations for privatization are varied. Privatization offers governments a short-term alternative to raise capital for new airport infrastructure, or to cash out the public investment. Nations with a high social welfare burden and a declining ratio of taxpayers to recipients are enjoying a one-time benefit by "cashing out" their investment in major infrastructure industries—telecommunications, broadcasting, energy, and transportation—thereby postponing the day of reckoning when either taxes must be raised, social welfare programs cut, or both. Elsewhere, free-market ideology dominates, with the privatization, liberalization, and deregulation of airlines taking governments into uncharted territory.[136] Privatization also relieves nations of the burden of heavy capital investment, giving airports direct access to the market for debt and capital.[137]

While private developers usually bear a higher cost of capital vis-à-vis the government, and lack the government's eminent domain powers, private firms, driven by a profit motive, often produce a product (here, airport services) with fewer employees, enhanced innovation and marketing acumen, and greater economy and efficiency.[138] The privatized British Airports Authority has proved that real estate and concessions can be developed into a significantly enhanced revenue stream. Nonetheless, airports are a monopoly bottleneck, and unless regulated, have the ability to extort monopoly rents from their customers (primarily the airlines).[139]

Sir Walter Raleigh observed that he who controls the seas controls the trade; he who controls the trade controls the wealth; and he who controls the wealth controls the world. These days, airways have replaced the oceans, and airports have replaced seaports in importance. Airlines are too numerous to be profitable in mature markets. But airports are the bottlenecks through which passengers and high-valued cargo must flow. Some would argue that they are "natural monopolies" whose infrastructure cannot be replicated to provide a competitive alternative, because of cost, land, and environmental restrictions.[140] Thus, it would be imprudent to privatize them without regulatory supervision of rates and charges imposed upon carriers.

The United Kingdom became the first major entrant into the land of airport privatization, with its sale of British Airports Authority (BAA), which controls seven major airports, including London's Heathrow, Gatwick, and Stansted, in 1987, in a US$2.5 billion public share offering. The government continued to provide oversight of airline access, airport charges, safety, security, and environmental protection, and to hold veto power over airport investment or divestiture. BAA has been consistently profitable. Despite fee caps and US$782 million in infrastructure improvements, it earned a profit of US$455 million in 1995.[141]

In 1995, the Mexican government passed laws allowing private operations of its 58 airports, including 50-year renewable leases.[142] One group won the right to operate nine airports in southeast Mexico for US$116 million. It plans to invest US$160 million upgrading runways and terminals over 5 years.[143]

Among the new green field airports being built around the world, Berlin Brandenburg International Airport is the first to depend almost entirely on private funding, with a 74.9% private stake in Berlin Brandenburg Flughafen. The successful bidder will be granted a concession to plan, build, own, and operate the new airport. Construction is expected to begin in 2002 and be completed in 2007. When the new airport is open, Berlin's existing airports at Tegel, Tempelhof, and Schoenefeld will be closed.[144]

With the proposed privatization of Amsterdam's Schiphol Airport, officials recognized a need to establish an independent regulator to approve landing fees. Allowing the private owners of the airport to set fees at will would likely encourage them to maximize the wealth interest of shareholders, to the injury of the airlines and their customers. Regulating fees would encourage airport owners to develop nonaeronautical revenue sources.[145]

Governments which have privatized airports have adopted one of four regulatory approaches—rate of return regulation (e.g., Spain, France, Greece, and the Netherlands), rate of return price caps (e.g., the United Kingdom), aeronautical price caps (e.g., Australia, Austria, Denmark, and Mexico), and limited governmental oversight (e.g., Canada, New Zealand, and the United States).[146] The general principles in the United States which govern airport fees, rates, and charges require that they be "fair and reasonable," not "unjustly discriminatory," and make the airport "as self-sustaining as possible."[147] As revealed in Table 4.1, Simat, Hellieson & Eichner, Inc., assessed the strengths and weaknesses of each of the four alternatives.

From the perspective of the airport owner or operator, revenue can be subject to high levels of market and regulatory risk. Government regulation always poses the risk that owners will not be allowed to earn a reasonable return on investment, even where increases in fees have been contractually agreed. One potential remedy to the vulnerability of airport revenue streams to changing economic and

Table 4.1. Strengths and Weaknesses of Airport Regulatory Approaches

	Rate of return	Rate of return price cap	Aeronautical price cap	Government oversight
Predictable aeronautical prices	Moderate	Moderate	Strong	Weak
Predictable airport profits	Strong	Moderate	Weak	Weak
Improving airport operating efficiency	Weak	Moderate	Strong	Weak
Ability to attract investment capital	Strong	Moderate	Moderate	Strong

regulatory conditions is to have the airport company enter into a management contract with the local government, allowing the government to collect the revenues and pay the company a set management fee (adjusted for inflation and currency revaluations).[148]

In 1989, Albany County, New York, sought to lease or sell its airport to a private consortium as a means of eliminating the county's operating subsidies to the airport. Though the county eventually dropped the proposal, in 1992 President Bush issued Executive Order 12803, which provided that federal agencies should assist state and local governments in their efforts to privatize infrastructure assets,[149] notwithstanding the natural monopoly bottleneck that airports pose. In 1995, Orange County, California, sought to sell John Wayne Airport to generate capital for the county's bankruptcy.[150] Yet U.S. law makes privatization problematic, since revenue of federally financed airports lawfully cannot be used for nonairport purposes.

In 1996, the U.S. Congress directed the FAA to establish a limited pilot program under which some of these constraints would be eased. The first two airports to participate in the pilot privatization program were Brown Field near San Diego and Stewart International Airport in New York state.[151]

Some cities have contracted out the management of their airports to private firms. In 1992, Pittsburgh International Airport contracted with BAA to manage its retail facility, which has more than 100 retail outlets. Per-passenger retail spending increased 250%. The Indianapolis Airport Authority signed a 10-year management contract in 1995 under which the private firm guarantees at least $32 million in cost savings and increased revenue, 70% of which will be passed on to the airport authority, which will reduce charges to the airlines, and 30% of which will go to the private contractor.[152]

Privatization is touted as a means of reducing political influence in infrastructure development, creating new sources of local revenue, and by injecting a profit motive, inspiring efficiency. Airlines fear the prospect of increased rates and charges.[153]

Military Conversion

Another means of expanding capacity is military conversion. The Aviation Safety and Capacity Expansion Act of 1990 established a Military Airport Program (MAP) to facilitate civil aviation development at current and former military airports located in congested metropolitan areas. In order to be eligible, the airport must be (1) a current or former military airport, (2) able to be converted to either a public-use commercial or a reliever airport, and (3) capable of enhancing airport and ATC capacity in a major metropolitan area and reducing existing or projected flight delays at commercial airports having more than 20,000 hours of annual delays.[154] Funding was set at not less than 1.6% of AIP funds.[155]

Congress also created a reliever airport set-aside fund to reduce congestion at commercial airports and provide additional capacity for general aviation. The largest conversion to date has been at Austin, Texas, where Bergstrom Air Force Base, located only 8 mi from the Texas state capitol building at Austin, was converted into a civilian airport, replacing the city's existing airport. (See Fig. 4.4.) Funding was authorized at up to 5% of AIP funds.[156] Another partial conversion involved Mather Airport near Sacramento, which consumed 2875 acres of the 5800-acre U.S. Air Force base that closed in 1993.[157]

4.4 *Opened in 1999 with 25 gates, the new Austin-Bergstrom International Airport was a former U.S. Air Force base. The airport includes a 500,000-ft² terminal and a second 9000-ft-long runway parallel to the existing 12,250-ft runway.* City of Austin.

Operating Costs and Revenue
ICAO regulation of airport charges

At the outset, the template of lawful rates and charges must be understood. The Chicago Convention on International Civil Aviation of 1944 provides that among the principal purposes of the International Civil Aviation Organization (ICAO) is to "avoid discrimination between contracting States."[158] More specifically, Article 15 of the Chicago Convention requires that "every airport in a contracting State which is open to public use by its national aircraft shall likewise...be open under uniform conditions to the aircraft of all the other contracting States" and that airport and air navigation charges imposed on foreign aircraft shall be no higher than those imposed upon domestic aircraft.[159] Though a State may recover its costs by assessing fees for air navigation, it may not charge a fee solely for the privilege of flying into, out of, or over its territory. All charges should be published and communicated to ICAO.[160] Airport and air navigation charges and fees may be reviewed by the ICAO Council upon complaint of a contracting State.[161]

ICAO's Council has issued a series of recommendations dealing with various aspects of airport and user charges. The Council expressed a general principle in favor of assessing fees in a manner in which "users shall ultimately bear their full and fair share of the cost of providing the airport." Cost should include the full economic cost, including depreciation and interest, but allowing for all revenue, aeronautical and nonaeronautical. When the fees are set, airlines are not to be charged for facilities and services they do not use or otherwise not properly allocable to them. Landing charges should be based on aircraft maximum permissible takeoff weight. ICAO has also approved a cost-based formula based on separate en-route/in-flight and terminal/approach charges, adjusted for aircraft weight and distance flown. Others have suggested additional factors should be considered, such as the time of day, level of airport congestion, and airspace utilized.[162] Two types of charges—security charges and noise-related charges—should be designed to recover no more than the relevant costs of providing security and noise-abatement equipment and services. In contrast, other charges may produce sufficient revenue to exceed direct and indirect costs by a reasonable margin.[163] Of course, airport and air navigation fees and charges may not discriminate between domestic and foreign carriers.[164]

Finance methodologies

In covering operating costs, airports tend to use one of three approaches. The most popular is the *residual cost,* or "cash register," approach, which seeks to balance total costs with total revenue. Once the airport's costs have been determined, nonairline revenue is subtracted from total expenditures to determine what additional revenue is needed to break even. Airline specific fees are then set to make up the remaining deficit.

A second approach is the *cost of service,* or "multiple cost center" method. The airport is divided into cost centers, and fees and charges for each cost center are set at a level to cover the costs allocated to it. A third method is the *public subsidy* approach, under which the difference between cost and revenue is subsidized by the airport.[165]

Once an airport is operating, it must generate sufficient revenue to retire debt and cover operating expenses. Airports generate revenue from landing fees and terminal leases, concessions (e.g., parking fees), departure taxes and passenger facility charges, and other sources (e.g., advertising and fuel sales). Airport operating revenue funds the airport's operating expenses, debt service, and sometimes nonoperating expenses, such as capital development (under a "pay-as-you-go" financing scheme).[166]

Airport revenue falls into two broad categories—revenue derived from air traffic operations, and revenue derived from ancillary (non-aeronautical) operations.[167] *Air traffic operations* are a major revenue stream. These include aircraft landing and parking charges, passenger and cargo charges, and leases of airline hangars and gates. User charges account for between 50 and 65% of total revenue for an airport.

As of 1998, the airports with the highest landing fees for a Boeing 747 were:

1 New Tokyo International Airport (Narita)
2 Kansai International Airport (Osaka)
3 Hong Kong International Airport
4 Frankfurt Main Airport
5 New York John F. Kennedy International Airport
6 Singapore Changi International Airport

7 Paris Charles de Gaulle Airport

8 Seoul Kimpo Airport

9 Washington Dulles International Airport

10 London Heathrow Airport

Ancillary, or *nonaeronautical, activities* include concession fees (e.g., rentals and profit-sharing arrangements with concessionaires such as restaurants and shops), revenue derived from rental of land, premises, and equipment (e.g., hotels, airline cargo space, kitchens, and office space rent), income derived from the airport's shops and services (e.g., baggage handling and parking), and various fees charged to the public.[168] At the largest airports in the United States, 20% of revenue is derived from airline landing fees, 40% from parking and concessions, 20% from terminal leases, and 20% from other sources.[169] A worldwide survey of airports conducted by Airports Council International revealed that 54% of airport revenues come from aeronautical sources (such as landing fees, aircraft parking, lighting, and airbridge charges) and 46% is derived from nonaeronautical sources (such as concessions, parking, rental car facilities, and advertising).[170]

Fees imposed upon airport concessions may be based upon (1) bids by tender, (2) assessment of market value, (3) the annual costs of the building and land, or (4) a combination of the above. The primary basis used by most airports for selecting concessionaires is by public tender, though some airports do not necessarily accept the highest bid, allowing such factors as standards of service and competitive prices to play a role in the selection process. In determining market value, airport operators often compare the value of premises of similar character in the vicinity, taking into account the nature of the activity, the size of the market, and the volume of transactions. In determining the costs of the building and land, full costs are usually taken into account, including maintenance, operating and administrative expenses, and capital costs (depreciation and interest).[171]

Concession fees may be variable or fixed. Variable fees are usually stated as a percentage of sales, or less commonly (because of difficulties of monitoring and auditing profit), a percentage of net profit. Some airports impose an increasing percentage as the volume of business increases. Most airports that use variable fees also stipulate a minimum payment. Fixed concession fees are usually applied to those activities likely to yield only modest profits (e.g., barbers, bookstores, flower stands, newspaper vendors, slot machines, and taxis). Some airports divide space into different zones, charging higher fees for more desirable locations.[172] Airports should also take care to ensure that retail prices charged by concessionaires are fair and competitive.[173] Some do so by placing a ceiling of no more than, say, 10% higher than prices charged in the central business district of the city the airport serves.

The airport as shopping center

Though retail sales outlets and restaurants have been located in airports since the 1920s, and expanded robustly with duty-free shops in the 1960s and 1970s, they have reached new levels with privatization, beginning with the British Airports Authority's (BAA) privatization of U.K. airports in 1987. At Heathrow and Gatwick Airports, retail sales increased significantly. At Heathrow, sales totaled $500 million annually, or about $25 per enplaning passenger.[174] Pittsburgh's new terminal, opened in 1992 and operated by BAA, was designed to maximize retail opportunities. Some airports today look more like suburban shopping centers than like airports. Examples include Toronto's Trillium International Terminal and Osaka's Kansai International Airport.[175] BAA is planning an aviation theme park at Gatwick.[176]

The new Hong Kong International Airport forces passengers to pass by or through its sprawling Sky Mall, from which it plans to earn a third of its revenue. At Dubai, the airport is adding an additional 9000 m^2 of duty-free shops, and a duty-free warehouse to serve charter flights. Frankfurt expects to double its commercial revenue after it opens a casino in 1999.[177] Robert Brugemann observes:

Airports are becoming more like downtowns every year, boasting new or expanded security forces, fire stations, central heating and cooling plants, hotels, medical centers, welfare organizations, restaurants, retail outlets, miniature department stores, chapels, banks, museums and other exhibition spaces, bowling alleys, sex cinemas, antique stores, and, at Milwaukee, an antique book store. In addition to the employees and passengers, there are activists leafleting, religious groups making conversions, pickpockets plying their trade, and a contingent of homeless people using the airport for shelter. The airport, like the shopping center, is becoming a central public space of our cities today, and this perhaps explains the surprising ways in which the architectural design of large shopping centers and airport terminals is starting to converge.[178]

Several reasons have been advanced to explain why airports are handing over more space to retailers:

- *A good economy.* People spend more on impulse purchases when money isn't so tight.

- *Customer convenience.* People prefer to get errands done whenever possible, including during lulls while flying, instead of doing them on their personal time.

- *More air travelers.* The growing number of airport passengers has created a more appealing retail market.[179]

Airport Accounting and Control Practices

Each airport should establish appropriate financial accounting and control practices (in accordance with recognized accounting rules, standards, or conventions), not only to ensure that its economic resources are properly and lawfully deployed, but also to give management essential data to operate the airport, and existing or potential lenders a basis on which to make their investment.

Financial accounting refers to the system in which income and expenses are recorded to present a comprehensive financial picture.[180] Typically, the airport will periodically (monthly, quarterly, and annually) produce a profit and loss statement and a balance sheet. The profit and loss statement summarizes the revenue and expenses over the period, with the difference being the profit or loss. The balance sheet summarizes the assets and liabilities, with the difference being an increase or decrease in the airport's net worth over the period.[181] The airport should also produce a periodic budget, with a subsequent explanation of positive or negative variances from budget.[182]

Financial control refers to the system of monitoring financial performance to ensure that expenses comport with plan, and income flows correspond to budget. Financial control is a three-step process: (1) comparing actual income and expenses with plan; (2) determining whether income or expense variances from plan are a problem of the budget, management of the airport, or external factors; and (3) what corrective action should be, and can be, taken.[183] Careful accounting and control can also thwart fraud or embezzlement, assuring that the public's resources are well spent. Internal and external auditing should be performed to assure that the financial data are accurate, and to identify waste and embezzlement. Law enforcement should be vigorously pursued against corruption.

Assessing Alternative Costs

Although site selection is the subject of the next chapter, it is imperative that in assessing alternative sites, or alternative expansion projects within an existing airport, or even alternative public works projects of which new or expanded airports are but one, the costs and benefits of the various alternatives be evaluated. If all sites or projects are otherwise of equal merit, the balance tips to the one with lowest cost. Since resources are scarce, airport projects should be considered against the relative merits of competing proposals for public financial resources.[184] Only in this way can the public interest be best advanced.

Summary and Conclusions

The capital requirements needed to finance airport projects are enormous. As Chap. 1 revealed, between US$200 billion and US$300 billion in airport projects are under way or contemplated around the world. Typically, airport projects are financed with a mix of debt and equity, the equity usually coming from the government treasury. Airline financial instability created by deregulation and liberalization enhances airport financial risk. Though there is a movement to airport privatization in certain parts of the world, local and national governments typically occupy center stage. Privatization also potentially poses problems of monopolistic abuse if unregulated.

Airports must generate sufficient revenues to cover operating expenses and service debt. The modern trend is to reduce reliance on aeronautical charges imposed upon airlines, and seek improved concession and rental revenue streams. It is here that privatization offers its most promising contribution. Increasingly, the financial opportunities are turning airports into shopping centers with runways.

5

Airport Planning

"Only with the development of the new Dallas/Ft. Worth Airport has an air center at last been planned and constructed that promises to achieve the full potential offered by air transportation."[1]—WES WISE, MAYOR OF DALLAS

"If you don't know where you're going, any road will get you there."[2]—BRUCE BAUMGARTNER, AVIATION DIRECTOR, DENVER INTERNATIONAL AIRPORT

In Chap. 4, we examine:

- The need for comprehensive systems planning at the national, regional, state, and local levels
- The changing face of global trade and its effect on efforts to develop both enduring and malleable air management systems which can quickly and efficiently accommodate the demands of the region
- The planning process itself, including the planning organization which is to advise, coordinate, and formulate policy, while considering need and securing acceptance
- The necessary factors which must be considered in determining the needs of the community, the market, and the air industry
- Useful factors in considering requisite airport capacity
- Economic factors which should be considered in projecting future regional airport demand
- Recent air traffic statistics
- Alternative methodologies used in determining capacity and forecasting demand

- Essential components of an airport master plan
- A brief look at airport planning at several of the world's new airports

Introduction

The capacity of an airport is constrained by the weakest link in the chain of (1) airway capacity, (2) runway capacity, (3) apron capacity, (4) terminal capacity, and (5) surface access capacity.[3] Bottlenecks anywhere along the path of the aircraft, the surface transportation vehicle, the pedestrian, the freight or mail create obstructions to efficiency and impose economic and noneconomic costs. An airport's maximum capacity is defined by the maximum capacity of its runways, gates, terminal facilities, baggage handling capacity, trains, curb space, roads, or parking, for example. Congestion at any point along the path can back up movements at any earlier point along the path. What good is a prompt landing if there is no gate at which to park the aircraft? What good is a prompt departure from a gate if the queue at the runway is 30 aircraft? What good is a timely arrival if the bags take an hour to work their way to the carousel? What good is an expeditious airport if the roads leading to and from it are mired in bumper-to-bumper automobile gridlock?

Airports therefore must plan to meet capacity requirements with infrastructure growth, and they must do so from a comprehensive perspective, taking into account all the elements of movement, any one of which can destroy the efficiency of the whole. Unfortunately, where demand for new airport infrastructure is high, the physical and environmental impediments, as well as political opposition for expansion or new airport development tend also to be high.[4] Congested airports tend to be located in crowded metropolitan areas. Thus, urban areas typically need additional airport capacity to satiate passenger and cargo demand, but urban airports are hemmed in by development, making it particularly difficult to expand beyond their existing boundaries, at least in terms of land-consumptive runways, taxiways, terminals, hangars, and cargo warehouses.

Airport planning must also be flexible, recognizing that the evolution in demand and technology will mandate changes in the airport's design. New airports typically have long planning horizons. A mere decade was consumed between Denver's decision to build a new airport and its completion. Osaka's Kansai International Airport opened 26 years after initial site selection for a new airport had begun. Munich's Franz Josef Strauss Airport opened 38 years after it was originally conceived. More than 4 decades after the need for a second airport at Bangkok was recognized, construction still is not completed. Thus, planning must evolve as demand and technology evolve. And planning must incorporate modular designs and sufficient space to accommodate demand and technology-driven expansion once the airport is open, or new infrastructure has been built.

Louis Turpen notes that an airport should first attempt to identify what it is—an origin-and-destination facility, or a connecting hub, or an international gateway, or a regional end point—and develop the facilities to support the needs so identified. Airport planning must be performed within a strategic framework, requiring strategic rather than tactical reactions. Most important, the organization must reflect the cornerstones of the airport business:

- Safety and security
- Customer service
- Environmental sensitivity
- Financial responsibility[5]

This chapter provides an overview of the planning process. Other chapters focus on various aspects of airport planning more closely. Airport finance, a critical component of the planning process, is examined in Chap. 4. Site selection and environmental planning are discussed in Chap. 6. Air-side and land-side design issues are reviewed in Chaps. 7 and 8, respectively. And intermodal access is addressed in Chap. 9.

The Air Traffic Control System

Systems planning examines the need for and relationships between various kinds and sizes of airports serving the overall aviation transport system. This may require coordination by national, state, and local governmental institutions, each assessing its respective geographic dimensions of the total equation. In the United States, the Federal Aviation Administration issues a periodic National Plan of Integrated Airport Systems.[6]

Airway and airport inefficiencies contribute to delay, congestion, fuel consumption, environmental pollution, and a thinner margin of safety. Air traffic control delays are projected to cost the airline industry more than a billion dollars annually. Four priorities have been identified by airline leaders—hardware replacement, software improvement, acceleration of the national route program, and development of satellite (Global Positioning System) navigation.[7]

In the United States, much criticism has been levied at the FAA's slow pace of air traffic control (ATC) modernization, particularly the sluggish pace of replacing automation equipment. In 1983, the FAA estimated the cost at $2.5 billion; by 1994, the cost was $7.6 billion, and the project was 8 years behind schedule. Sixty-four projects totaling $3.8 billion had been completed; 158 projects remained.[8]

The FAA has put the blame on burdensome procurement rules and proposed reforming them by giving integrated product teams broad authority to award contracts.[9] The U.S. General Accounting Office disagreed, criticizing the FAA as having underestimated the technical complexity of the systems being developed, and having provided inadequate oversight of contractors. Part of the problem appears to lie in the fact that the FAA had seven different administrators in 10 years.[10]

Air traffic congestion also appears to be a growing problem in east Asia. Traffic growth has been attributed to four factors:

1 Off-shore investment by Japanese, Korean, and Taiwanese companies, whereby the manufacturers of North Asia seek cheaper production locations in Southeast Asia; this has created an "axis shift" of trunk routes from Southeast Asia toward Europe, toward a north-south alignment, connecting the major cities of Pacific Asia.

2 The emergence of China as a major market and production location, particularly around Hong Kong and the Pearl River Basin, and Shanghai.

3 The rising levels of personal income in the region.

4 The rapidly expanding global role of cities in the region, as global trade patterns make intercontinental air transport essential.[11]

As a consequence, a relatively narrow, 300-km traffic corridor has been created, with Japan and Korea on the north, and Sumatra and Java on the south, reinforced with the great circle routings to Europe and North America. The growth of traffic on this narrowly circumscribed geographical corridor has major implications for the management of air space in the region.[12] As noted elsewhere:

It is possible that air space management problems could be serious, as the approach paths to the facilities in the Pearl River Delta will overlap. The difficulty with improving air space management is that it calls for international cooperation and agreement. The European experience has shown that is difficult to achieve....In short, although capacity additions and improvements are undoubtedly needed in several locations, the region will need to move quickly to more sophisticated regional approaches to air space management to keep up with the rate of growth of air traffic in the region.[13]

Air traffic management is usually within the province of the federal government, with standardization and coordination provided internationally by the U.N. International Civil Aviation Organization (ICAO). Nonetheless, the safety and efficiency of air traffic flows is also of concern to airports. Airport planners, therefore, need to take account of airspace flows, proximity to other airports, height of off-site buildings within the approach corridors, and historical meteorological data.[14]

National, State, Regional, and Local Plans

Airport planning is performed at several different levels, reflecting the need to integrate a safe and efficient air transportation system despite the differing jurisdictional lines. In the United States, planning is performed at four levels: (1) the 10-year National Plan of Integrated Airport Systems, continuously updated and published biennially by the Federal Aviation Administration; (2) Statewide Integrated Airport Systems Planning, prepared by state DOTs or aviation agencies; (3) Regional/Metropolitan Integrated Airport Systems Planning prepared by regional or metropolitan planning agencies; and (4) Airport Master Plans, prepared by individual airports.[15]

The Planning Process
Public input and acceptance

In most communities, airport planning transcends technical engineering and design issues. It is a complex and politically sensitive public process. Many different airport users and diverse interests must be accommodated. Legal (principally environmental) restrictions influence decision making. As we saw in Chap. 3, political considerations must be accommodated. The business community and the press can also be highly influential in molding governmental and public opinion.[16] Several constituencies must be involved early and throughout—the politicians, the various governmental agencies, the tenants, the business community, and the general public.[17] Their involvement avoids unnecessary surprises, and helps build consensus.[18] Therefore, the airport planning process should be characterized by consultation and cooperation between various constituencies.

The planning organization should seek the advice and input of interest groups prior to and during the preparation of the airport master plan.[19] A *master plan* is the comprehensive and detailed concept for the ultimate development of an airport, both in terms of aviation and nonaviation uses, and the use of land adjacent to it.[20] The process should be undertaken in a way that ensures that the plan thereby produced will receive acceptance by the appropriate governmental officials and the general public.[21]

The planning organization

In the preplanning stage, an organization is established to undertake the study, develop a work program, and provide a means for financing the project.[22] The organization should establish policy that is acceptable to the airport community, bring together for advisory and coordinating purposes the relevant aviation and nonaviation interests, and provide a process that is both technically sound and responsive to aviation policy and the coordination of the various constituencies. Thus, the planning organization should perform three principal functions: (1) policy formulation, (2) advice and coordination, and (3) technical planning.[23] Failure to do this properly may result in fragmented public support for the master plan's recommendations, unrealistic recommendations unacceptable to the aviation community, and a completed study of little utility and difficult to implement.[24] For complex projects, formal policy, technical, and review committees meet regularly. Ideally, they open their meetings to the public. Frequently, once the project has been properly scoped, consultants are hired to provide data, plan development, alternatives assessment, and other assistance.[25]

In the United States, the Airport and Airway Improvement Act of 1982 and the Airport and Airway Safety and Capacity Expansion Act of 1987[26] require that 0.5% of airport development funds be dedicated to systems planning, including assessing the need for new infrastructure and the type of facility best able to meet that need. Once a systems plan is developed and the community planning process is begun, specific proposals for new airports or expansion projects are considered under what is termed "project planning" or "master plan development." Five basic phases are involved—needs assessment, facilities assessment, facilities design, environmental assessment, and financial planning.[27] Each should be done on a short-term (0 to 5 years), intermediate term (6 to 10 years), and long-term (20 years) planning horizon.[28]

Needs assessment and demand forecasting

Needs assessment usually requires forecasting of anticipated aviation activity. *Forecasting* requires an expert judgment, or estimate, of future air traffic and demand. Such forecasts are based on the assumption that assessment of historical data and trends (e.g., aircraft operations, enplaning passengers) may have a predictive relationship vis-à-vis events in the future. An array of aviation, socioeconomic, and demographic information will form the basis of the forecast.[29] Forecasters must analyze such information as historical trends in aircraft movements, passenger and cargo volume, population and economic growth characteristics of the region, national and international traffic, geographic factors, and airline industry dynamics, including competition with respect to pricing and frequency, and government regulation.[30] More airline competition typically translates into lower fares, and because of the price elasticity of demand for air travel, more demand. Conversely, less airline competition typically translates into higher fares, and less demand for air travel.[31] Also examined are demand/delay relationships, and the capability of existing airports to satiate present and projected future demand with existing capacity.[32] Projections of the mix and type of aircraft and volume of movements are essential to identify the aircraft which will drive the geometric and structural design of the runways, taxiways, tarmacs, and terminals, and the navigational aid requirements of the airport.[33]

Though forecasting is an extremely difficult task, airport authorities, central governments, commercial airlines, and aircraft manufacturers rely on their forecasts for planning purposes.[34] The purpose of forecasting is not to predict the future with precision, but to provide data that can be useful in reducing uncertainty. If overly optimistic forecasts prompt investments in airport infrastructure too early, then premature capital costs and unnecessary operating expenses can be incurred. On the other hand, if overly pessimistic forecasts dissuade infrastructure expansion, efficiency costs can be high. Thus, the purpose of forecasting is to provide a framework for gauging the timing of airport investments in a way which minimizes forecasting error costs in either the excessively optimistic or pessimistic direction.[35]

Though historical annual and seasonal data are useful, peak demand defines capacity needs.[36] Thus, the annual capacity capability of an airport measured in passengers or volumes of cargo and mail is a less helpful number than the airport's capacity on a peak day at a peak hour. Therefore, forecasts are most useful when converted into peak period data (based on what ICAO calls the "typical peak hour," or the thirtieth or fortieth busiest hour over the course of a typical year) for aircraft movements, and passenger, cargo, and mail throughput. The U.S. Federal Aviation Administration recommends the "design hour," or the peak hour of the average day of the busiest month.[37] Aircraft movements are a useful point of departure in assessing runway, taxiway, apron, and air traffic requirements. When coupled with airline and airport employee and accompanying visitor ("meeters and greeters") data, passenger, cargo, and mail throughput define terminal and intermodal transport requirements.[38] Specifically, the following data are useful predictors of requisite airport capacity:

1 Annual throughput of international and domestic passengers, cargo, and mail, categorized by scheduled and non-scheduled airlines and general and military aviation, and by arrivals, departures, transit, and transfer/trans-shipment

2 Typical peak hour aircraft movements and throughput of passengers, cargo, and mail

3 Average day of peak month throughput of passengers and aircraft movements

4 Number of airlines serving the airport, their local and network size, and their route structure

5 Types of aircraft serving the airport

6 Number of aircraft to be based at the airport, and their base and line maintenance requirements

7 Intermodal surface transportation connections between the airport and the surrounding metropolitan area

8 Number of visitors and airline and airport employees by category, including segregation of passengers into origin-and-destination and connecting categories

9 Historic trends in passenger, freight, mail, and aircraft traffic volume

10 Demographic, population, and economic growth characteristics of the region, including the types and levels of business activities, and hotel and motel registrations

11 Geographic factors affecting transport requirements, including distance from other population centers

12 Intramodal and intermodal competition[39]

Numerous forecasting techniques have emerged, including forecasting by judgment; trend extrapolation; market share models; econometric models such as multiple regression or logit models for trip generation; trip distribution and modal choice analysis; trend projection; and linear, exponential, and logistic curve extrapolation.[40] Nonetheless, forecasting remains an extremely subjective process that can result in widely differing predictions depending on the assumptions made and techniques used.[41] The U.S. Federal Aviation Administration (FAA) uses a step-down process whereby national enplanement forecasts (based on multiple regressions involving anticipated levels of gross domestic product, airline yields, and revenue passenger miles) are employed to predict airport enplanements.[42] Nonetheless, the FAA recognizes the shortcomings of its methodology:

> *[I]n general, [econometric] models and equations are simple portrayals of a complex system. They cannot account for a number of political, social, psychological, and economic variables and for all the interrelated actions and reactions that eventually lead to a particular set of results. It is particularly important, therefore, that the initial model results are reviewed, revised, and adjusted to reflect the analysts' best judgment of the impacts of the events occurring or expected to occur during the forecast period.*[43]

Forecasting is more of an art form than a science, and as an art form, more impressionism or surrealism than realism. During the first two decades following World War II, aviation forecasters tended to underpredict actual passenger volumes, for the enormous growth in air traffic during the 1950s and 1960s emerged as a result of unanticipated technological advances, particularly the emergence of jet aircraft, which enhanced speed and capacity and lowered costs. But since 1970, forecasters have tended to overpredict demand.[44] (See

Fig. 5.1.) Moreover, liberalization of economic regulation (e.g., international "open skies") and deregulation of airlines have made the task of predicting air transport trends enormously more difficult.[45] For example, massive infrastructure improvements have been made in cities like Nashville, Tennessee, Raleigh/Durham, North Carolina, San Jose, California, and Colorado Springs, Colorado, for hubs which were created by particular airlines, and subsequently abandoned by them. In 1990, the Federal Aviation Administration predicted that by 1995, Denver's new airport would be served by three hub carriers and handle 66 million passengers. But the forecast was off by more than 100%. Though in 1986 Denver's Stapleton International Airport handled 34.7 million passengers, in 1995 the new Denver International Airport handled only 32 million passengers.[46] (See Fig. 5.2.) Not until 1998 did Denver's traffic reach a new level of about 37 million passengers.[47] In other words, forecasting air transport demand is a little like predicting the weather—the weather man is right more often than not, but the variables are too many for him to be able to predict the future with precision, and the further out on the horizon the weather man attempts to look, the less likely he is to be right.

One must also recognize the exceptionally fluid and fickle nature of air travel demand. Air travel is a derived demand product, meaning that people consume air travel as a means to an end—people travel to an airport to fly to a business meeting, a vacation, or a visit to friends and relatives. Demand is highly cyclical, depending on the time of day, day of week, and season, and broader macroeconomic market fluctuations.[48] International, regional, and local air traffic is influenced by economic, demographic, technological, commercial, and political forces; freight traffic is also influenced by tariff and quota changes, as well as currency fluctuations.[49] Hence, macroeconomic trends and the nature and composition of the local traffic mix (business and pleasure) must also be integrated into the forecast. Among those broader economic factors affecting demand forecasting are:

1 *Economic growth and changes in industrial activity.* In addition to national and regional economic activity, forecasting should be tailored to local economic characteristics and trends.

2 *Demographic patterns.* The size and composition of the area's population, including its population, age, educational, and occupational distribution, are important.

3 *Disposable personal income.* The higher the disposable personal income, the greater likelihood that the area will enjoy higher levels of consumer spending on air travel.

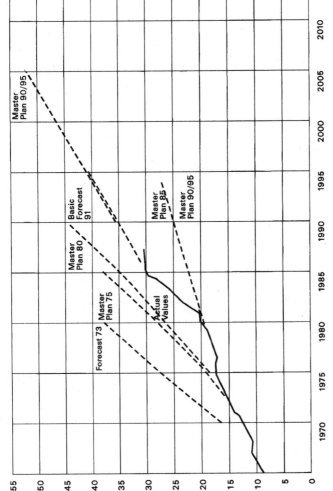

Air Traffic Development and Forecasts
Millions of Passengers

5.1 *Predictions of passengers flowing through Frankfurt Main Airport were overly optimistic in the 1970s and early 1980s, and overly pessimistic in the mid-1980s, vis-à-vis the demand that ultimately materialized. By the 1990s, the trend lines were drawn between the two extremes. Forecasting is an extremely difficult task, and only time will tell whether the projections are now right.* Flughafen Frankfurt AG.

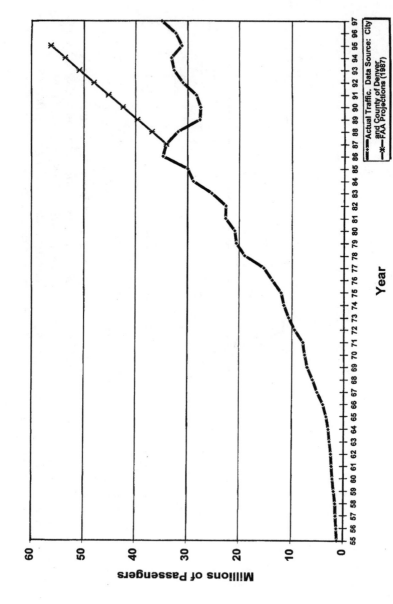

5.2 *In 1987, the U.S. Federal Aviation Administration predicted Denver's new airport would serve 56 million passengers annually by 1995. The actual figure was only 31 million passengers, 45% less than the FAA projected.*

4 *Geographic attributes.* The geographic distribution and distances between population centers may affect the type of transportation services required.

5 *Other external factors.* These include such things as changes in fuel prices, the regulatory environment, taxes, fees, and currency restrictions.

6 *Local aviation actions.* Demand for aviation can be affected by such locally determined factors as ground access, support services, user charges, and plans for future development.[50]

One must also be concerned about issues of capacity and delay. *Capacity* refers to the processing capability of a facility over a period of time. When capacity becomes saturated by demand, delays occur. Alternative concepts of capacity include *practical capacity,* which corresponds to tolerable or reasonable levels of delay, and *ultimate capacity,* which is the maximum number of aircraft, passengers, cargo, or mail the facility can process. Runway capacity is typically defined in terms of flight arrivals and departures per hour. Factors that determine ultimate capacity include the layout and design of the runway system, air traffic control procedures, and environmental and regulatory conditions of the airport.[51] Terminal capacity is typically defined in terms of the number of passengers it can reasonably accommodate per hour. Cargo and mail capacity is defined in terms of the number of parcels the facility can reasonably process per hour.

Air traffic volume

Passenger demand has grown significantly since the dawn of commercial aviation. (See Fig. 5.3.) Compared to 1996, the number of passengers carried on airlines worldwide increased by 5% in 1997 to 2,705 billion. The volume of freight increased by 8% to 55.71 million tons. Africa registered the largest increase in passenger traffic (10% growth), followed by Latin America and the Caribbean (8.2%).[52] Tables 5.1, 5.2, and 5.3 identify the world's busiest airports on the basis of passenger volume, cargo volume, and aircraft movements, respectively, as well as their growth rates.

Facilities assessment

Facilities assessment involves comparing the forecasts of future demand with existing capacity. It should attempt to determine the capacity of the aircraft, passenger, cargo, and ground vehicular

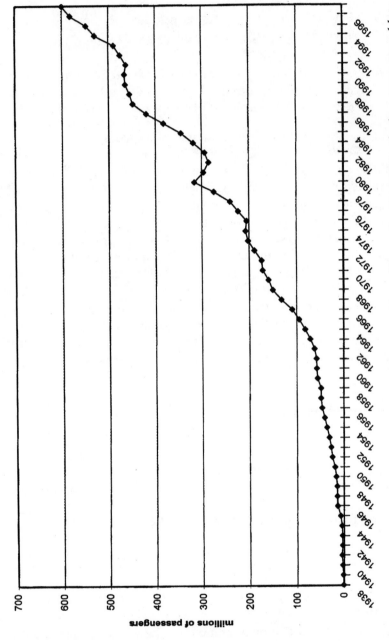

5.3 *Though U.S. passenger demand slowed in the early 1980s and early 1990s, the long-term trend has been upward.*

millions of passengers

Table 5.1. Total Air Passengers in 1997[53]

		Number of passengers	Percent change compared to 1996 (negative in parentheses)
1	Chicago (ORD)	70,294,601	1.6
2	Atlanta (ATL)	68,205,769	7.7
3	Dallas/Fort Worth (DFW)	60,488,713	4.2
4	Los Angeles (LAX)	60,142,588	3.7
5	London (LHR)	57,974,931	3.8
6	Tokyo (HND)	49,302,268	5.9
7	San Francisco (SFO)	40,499,947	3.2
8	Frankfurt (FRA)	40,262,691	3.9
9	Seoul (SEL)	36,757,280	5.9
10	Paris (CDG)	35,293,661	10.9
11	Denver (DEN)	34,972,936	8.3
12	Miami (MIA)	34,533,268	3.1
13	Amsterdam (AMS)	31,569,977	13.6
14	Detroit (DTW)	31,520,656	4.6
15	New York (JFK)	31,228,956	0.3
16	Newark (EWR)	30,866,374	6.0
17	Phoenix (PHX)	30,536,061	0.5
18	Las Vegas (LAS)	30,305,553	0.0
19	Minneapolis/ St. Paul (MSP)	29,070,480	4.9
20	Hong Kong (HKG)	29,020,369	(3.9)

infrastructure.[56] The facilities assessment process should produce an inventory of the existing physical plant, its condition and useful life, and land use on and near the airport.[57]

When demand exceeds capacity, delay results, causing airlines and their passengers to lose productivity and efficiency. The concept of *practical capacity* corresponds to "reasonable" and "tolerable" levels of delay, while *ultimate* or *saturation capacity* refers to the maximum number of aircraft an airport can handle given constant demand.[58] As noted above, capacity typically is calculated in units of operations (flight arrivals and departures) per hour. Factors that most

Table 5.2. Total Cargo in 1997[54]

		Total cargo	Percent change compared to 1996 (negative in parentheses)
1	Memphis (MEM)	2,233,490	15.5
2	Los Angeles (LAX)	1,872,528	8.9
3	Hong Kong (HKG)	1,813,122	14.0
4	Miami (MIA)	1,765,827	3.3
5	Tokyo (NRT)	1,738,795	6.9
6	New York (JFK)	1,661,400	1.6
7	Seoul (SEL)	1,567,639	15.1
8	Frankfurt (FRA)	1,514,278	1.1
9	Chicago (ORD)	1,407,589	12.1
10	Singapore (SIN)	1,358,044	12.1
11	Louisville (SDF)	1,345,318	(1.8)
12	Anchorage (ANC)	1,289,625	9.4
13	London (LHR)	1,268,621	10.5
14	Amsterdam (AMS)	1,207,282	7.3
15	Newark (EWR)	1,048,954	9.5
16	Paris (CDG)	1,028,602	5.1
17	Taipei (TPE)	913,520	14.7
18	Atlanta (ATL)	864,474	8.0
19	Dayton (DAY)	812,440	6.1
20	Dallas/Fort Worth (DFW)	810,621	4.7

strongly influence capacity are number, layout, and design of runways; air traffic control procedures; and environmental controls.[59] An assessment of the existing infrastructure may involve an inventory of such things as:

- Runways, taxiways, and aprons and related marking and signing
- Passenger and cargo buildings and other terminal buildings and areas, by function
- General aviation buildings and areas, by function; fire fighting and rescue buildings; federal facilities

Table 5.3. Total Aircraft Movements in 1997[55]

		Total movements	Percent change compared to 1996 (negative in parentheses)
1	Chicago (ORD)	883,758	(2.8)
2	Dallas/Fort Worth (DFW)	851,101	0.4
3	Atlanta (ATL)	794,621	4.4
4	Los Angeles (LAX)	781,492	2.3
5	Detroit (DTW)	541,216	1.1
6	Miami (MIA)	533,084	(0.3)
7	Phoenix (PHX)	521,689	(1.0)
8	Oakland (OAK)	519,504	6.4
9	St. Louis (STL)	516,889	0.6
10	Minneapolis/ St. Paul (MSP)	491,273	1.8
11	Boston (BOS)	482,488	5.8
12	Denver (DEN)	477,384	7.4
13	Las Vegas (LAS)	473,127	(0.5)
14	Philadelphia (PHL)	465,692	11.2
15	Santa Ana (SNA)	464,750	(0.9)
16	Newark (EWR)	461,786	2.4
17	Pittsburgh (PIT)	457,410	1.2
18	Long Beach (LGB)	450,512	(6.3)
19	Charlotte (CLT)	444,679	(0.4)
20	London (LHR)	441,449	0.1

- Surface access to the airport, including vehicular circulation and surface access
- Aviation fuel and aircraft servicing systems
- Utilities, including water, gas, electric, telephone, drainage, and sewage
- Proximity of airports to one another, and their influence on flight patterns[60]

Land use on the airport and real estate adjacent to it must also be reviewed, particularly to determine airway obstructions and compatibility of use to noise levels above the 65 day-night average sound level (Ldn) contour.[61] If projected demand exceeds capacity, decisions must be made as to whether, and how, the existing airport can be expanded, or whether, and where, a new airport should be built.[62]

Where more than a single airport serves the community, or where a new airport is being developed, their aggregate capacity should also be evaluated. For example, the Frankfurt Airport Company entered negotiations for purchase of the former U.S. Air Force base at Hahn, located about 100 km west of Frankfurt Main Airport, as a potential site for freight capacity to relieve demand at Main.[63]

Environmental assessment

Also important are *environmental considerations,* which mandate developing an airport plan that is compatible with surrounding land use and developmental objectives. Most developed nations, and some developing nations, have promulgated environmental legislation which affects the airport planning process. For example, in the United States the National Environmental Policy Act of 1969 raised up environmental factors as an essential consideration of airport planning, requiring an Environmental Assessment and Environmental Impact Statement for most major airport projects using federal funding, and mandating that environmental impacts be considered early and throughout the planning process.[64] Among the elements to be considered are

- Air and water quality
- Solid waste generation and disposal
- Floodplains, wetlands
- Endangered/threatened flora and fauna
- Biotic communities
- Parklands/recreational areas
- Historic/architectural/archaeological/cultural resources, and prime and unique farmland[65]

We discuss environmental issues in greater detail in Chap. 6.

Facilities design

Facilities design requires the preparation of an airport layout plan, which is a graphic depiction to scale of the existing and ultimate airfield configuration, schematic terminal design, land use plan, and the intermodal transport connections.[66] Objectives of the plan include optimization of efficient aircraft operations and passenger flows, accommodating surface transport connections, and avoiding environmental degradation.[67]

Designing an airport is like designing a city, for it must have all the essential functions of a city. Typically, an airport must be designed to include the following facilities and functions:

- Runways
- Taxiways
- Aprons
- Aircraft hangars and maintenance facilities
- Aeronautical navigation facilities
- Aviation lighting facilities
- Aircraft fuel facilities
- Passenger terminals
- Customs facilities
- Immigration facilities
- Quarantine facilities
- Catering facilities
- Airline offices
- Meteorological facilities
- Communications facilities
- Electric power supply facilities
- Gas supply facilities
- Heat and cooling facilities
- Sewage treatment facilities
- Waste disposal facilities
- Water supply facilities
- Baggage handling facilities
- Air cargo facilities
- Postal facilities
- Rescue and fire fighting facilities
- Police facilities
- Automobile parking facilities
- Automobile rental facilities

- Taxi, bus, and van plazas
- Rail terminals
- People-mover systems
- Hotels

We will discuss design issues in greater detail in Chaps. 7 and 8.

Financial planning

Financial planning involves assessing the capital needs of the project, identifying public and private sources of capital, and projecting the revenue streams necessary to cover such costs. Sound forecasting of anticipated traffic development and infrastructure capacity are essential to any airport development project and its financing.[68] Once forecasts have been made and facility requirements determined, capital costs and subsequently recurrent costs can be estimated. Traffic forecasts can also be useful in predicting income from both primary sources (e.g., landing fees, gate rentals, hangar rents) and secondary sources (e.g., concessions).[69] Revenue streams typically are landing and parking fees, gate and hangar rentals, ground handling charges, aviation fuel and oil concessions, fixed-based operator rentals, and various terminal concessions (including shops, restaurants, and hotels).[70] Economic feasibility should be determined for each component of the airport.

Economic impact surveys consist of traffic forecast information, and the growth in economic activity—direct, indirect, and induced—anticipated therefrom. Such surveys are important in securing financing from foreign governmental sources, particularly development banks and funds, which weigh the impact of infrastructure development on national economic development. Direct economic impacts result from airport-related activities, such as services provided to airport users (e.g., freight forwarders, taxis, and hotels). Indirect economic impacts arise from the purchase of goods and services and investments made by those enterprises which produce direct impacts. Induced economic impacts result from the purchase of goods and services and investments made by individuals employed by enterprises linked directly to airport activity. The five key indicators of direct, indirect, and induced activity are employment, personal incomes, business revenue, tax revenue, and capital investment.[71] We will examine economic impacts of airport activities in greater detail in Chap. 10.

In most nations, the federal government provides most of the financing of new airports, and thereby has enormous influence over the project's development. In the United States, for example, though airports are typically built by local governments, federal financing ensures that the Federal Aviation Administration plays a key role in negotiating an agreement on an airport plan which satisfies national air system requirements. Private capital, in the form of bonds, also plays an important role in airport infrastructure finance. We discuss financial issues in greater detail in Chap. 4.

Alternative planning methodologies

The International Civil Aviation Organization recommends a somewhat different planning process. It suggests that the initial planning process begin with development of general policy objectives (e.g., designation of time frame, planning horizon, and geographic limits of the planning area) and a study design. After these are completed, technical planning begins with an inventory of existing infrastructure. Then a forecast of demand is made in order to ascertain future capacity requirements, with alternatives to expansion also examined.[72] Major airlines also engage in a sophisticated strategic planning process that assesses where they are and where they want to be, including which airports they want to serve.[73]

At the heart of any planning process is the assembly of sufficient, comprehensive, and objective data upon which rational decision making can be based. According to ICAO:

> One of the problems of airport planning is that basic facts and principles have not been presented comprehensively. This is especially true in respect to passenger facilities. Formal analysis is essential for any reasonably satisfactory future development. Therefore, basic facts need to be stated so that they can be challenged and tested throughout the world and, if found incorrect, replaced by others which can be similarly tested until a faultless body of data is compiled. The deductions made and the principles established should be similarly arrived at by analysis so that the present situation of conflicting "options" is replaced by data....[74] The master plan should evolve through consideration of all the factors which affect air transport and which will influence or impinge on the development and use of the airport throughout its working life.[75]

The data collected should not only address the airport's physical facilities, they should also measure utilization, volume, and composition of traffic, the price of transportation, the financial condition of the airlines using the airport, and government transport and environmental policy, law, and regulation.[76]

Finally, the objectives of the planning process should include providing for the orderly and timely development of an airport adequate to meet the present and future air transportation, safety, efficiency, and environmental needs of a region, integrating aviation into a comprehensive seamless intermodal transportation system, and promoting the establishment of an effective governmental organization capable of implementing the master plan in a systematic fashion.[77]

The Airport Master Plan

An airport *master plan* consists of a comprehensive conception of the long-term development of an existing airport, or creation of a new airport and land adjacent thereto.[78] It should reflect a current assessment of what exists and what is required, and the research and logic which served as the foundation for plan development. The basic documents consist of a plan report and set of drawings.[79] As noted above, the process for its creation involves collecting data, forecasting demand, predicting facility requirements, and determining plans and schedules.[80]

Using demand-capacity analysis, air-side capacity should be calculated and compared with aircraft demand forecasts, and landside capacity should be calculated and compared with passenger demand forecasts to determine the need, identity, and timing of infrastructure investment.[81] The master plan should be a guide for development of both the aviation and nonaviation physical facilities of the airport, the development of adjacent land areas, the determination of environmental effects of construction and operation, and the establishment of access requirements.[82] The plan itself will be of considerable interest to a wide spectrum of groups, including for example, private citizens, interest groups, airport users, airlines, concessionaires, governmental agencies, and the press.[83]

The essential components of a master plan are:

1 An inventory of the physical facilities of the airport and airspace infrastructure and nearby airport-related land uses.

2 A demand forecast for short, intermediate, and long terms to determine the necessary capacity for airport facilities.

3 An assessment of the capacity of the airport to satiate projected demand in terms of air-side capacity (e.g., number and dimensions of runways, taxiways, and aprons) and land-side capacity (e.g., terminal building space, parking, and surface access), and the delay imposed by inadequate capacity.

4 Site selection when the capacity of the existing airport is inadequate, or where a decision has been made to build a new airport.

5 Consideration of existing and potential environmental impacts as well as appropriate mitigating measures.

6 Simulation (sometimes with computer models) of airport operations in order to assess the merits of development alternatives.

7 Evaluation of the cost-effectiveness and financial feasibility of various alternative concepts and solutions.

8 Preparation of drawings of an airport layout plan (consisting of the airport boundary, runway configuration, and areas reserved for land-side facilities), a land use plan (showing areas reserved for terminals, maintenance, cargo facilities, general aviation, and other areas within the airport boundary, as well as recommended off-airport land uses, considering safety and noise), a terminal area plan (revealing the various terminal area components and their relationships), and access plans (showing major highway and rail routes from the airport to the central business district).

9 Plan implementation, which includes schedules, costs, and sources of revenue for airport development.[84]

Master plans typically begin with a statement of strategic goals or objectives, providing general direction. Ideally, planning attempts to identify issues from the perspective of both the airport in a system, and the airport as a system. The Master Plan for Salt Lake City International Airport sets forth several objectives which would be appropriate for any major airport:

1 Develop an integrated airport system that balances airfield capacity with terminal, parking, access, cargo, and other airport facilities' capacities (including development of an on-airport land use plan that effectively uses all airport property, and a plan for access/curb layout that minimizes terminal area congestion).

2 Plan an airport system which balances the Authority's responsibility to develop facilities to meet aviation demand with local and state transportation and environmental needs [including

encouraging the use of high-occupancy vehicle (HOV) modes and rail service, if appropriate].

3 Plan for a world-class terminal complex that is easily adaptable to changing airline service patterns (including a concept that is adaptable to expansions or reductions in airline hub and point-to-point service).

4 Maintain the high level of compatible land use that exists around the airport today (including minimization of adverse noise impacts).

5 Develop an airport that supports local and regional economic goals while providing the flexibility to accommodate new opportunities and shifts in development patterns (including keeping costs within acceptable limits and establishing an efficient airport layout integrated with existing transportation infrastructure).[85]

Among the issues which will likely surface during this process are growth in air operations and passengers, the potential need for a new airport, the role of existing and contemplated airports, the possibility of capacity expansion, ground access needs, relocation of roads, power lines and buildings, and airspace obstructions and landfill problems.[86] As discussed in Chap. 2, if a new commercial airport is to be built, decisions must be made regarding the role of the existing airport: (1) Will it supplement the existing airport, emphasizing a specific type of traffic (as Montreal's Mirabel supplements Dorval Airport, or Washington's Dulles supplements Reagan National Airport)? (2) Will it fully replace the existing airport (as Denver International replaced Stapleton Airport)? (3) Will it replace the existing airport for all but general aviation operations? Moreover, if current demand does not warrant building a new airport, the new site can be selected and preserved, or land banked, for future use.[87]

Finally, decisions on the timing of airport infrastructure expansion should undergo a cost/benefit analysis and alternatives assessment. A comparison of annual delay with and without the proposed infrastructure improvement produces a theoretical delay reduction in units of time. When multiplied by aircraft operating costs and passenger opportunity costs, this total can be compared with the cost of annual debt amortization, and the maintenance and operational costs of the new infrastructure investment to arrive at a cost/benefit assessment.[88] A review of alternatives should include an assessment of the consequences of doing nothing, the provision of reliever airports for general aviation, and the investigation of potential sites for a new airport.[89]

The Business Plan

An airport is a business. Actually, an airport is an amalgamation of 20 or more separate and distinct businesses. Prudent airport managers focus on developing a comprehensive business plan which attempts to improve product lines, satiate consumer needs, and thereby maximize revenue. Airport managers should also be prepared to invest economic and human resources in those lines of business with the highest potential gains. In other words, airport managers need to understand that an airport is a business (actually a combination of businesses), and run the enterprise as such. Because airports are natural monopolies, and have historically tended to be government-owned enterprises, their managers have tended not to think in these terms.

Privatization of airports, of course, naturally induces this process, as the airport owners come to grips with the fact that their facilities must turn a profit. Maximization of shareholder value is the predominant motive among privately owned enterprises. Even government entities that control publicly owned airports and wish to avoid privatization may seek to improve the efficiency and economy of their operations, while maximizing revenue with new and innovative marketing and business investments.

At most airports, the potential for increased revenues is vast. An essential ingredient of its realization is airport management that understands the potential benefits of revenue enhancement, knows how it can be accomplished, and is willing to invest the human and economic resources to achieve it. It may be useful to supplement internal resources with fresh market ideas by bringing in entrepreneurial management personnel from the private sector whose acumen and skills have been honed by a competitive environment. Of course, government pay scales are such that this may be difficult to achieve, for salaries of skilled managers tend to be bid up by competition. Public enterprise usually skirts around the problem by hiring a consultant team to address specific issues, though this can be less than satisfactory, since consultants do not have to live with the results of what they recommend.

The first step in the process of developing a business plan should be to identify the customers served by the airport, and determine how they might be served better. Some airports think of airlines as their customers. Airlines are actually tenants, while airports are landlords. The true customer is the passenger or the shipper of goods, rather than the transport provider.

Airport management should identify the businesses that operate, or should operate, on the airport property. Airlines lease gates, hangars, baggage space, offices, employee lounges, business and frequent flier lounges, ticket counters, and so on. Aircraft maintenance, deicing, and catering facilities on the airport property may serve several different airlines or be airline-dedicated. Fuel facilities typically serve several different airlines. Facilities typically exist for air cargo and postal collection, storage, and distribution. Hangars and service facilities may exist for general aviation airplanes and jets. On the land side, the terminal may house a multitude of concessions, including newsstands, pharmacies, food courts, restaurants, pubs, souvenir stands, and tax- and duty-free shops. Many are in the entertainment business, with movie theaters, museums, gambling, kids' play sets, and panoramic viewing lounges in their facilities. Some airports have whole hotels in or adjacent to their main terminals. Parking garages may be a significant source of revenue. Fees may be imposed on buses, vans, limousines, and taxis which serve the airport. Space typically is leased in the terminal and on the airport property to automobile rental firms. Space may also be dedicated to security operations and government services such as customs and immigration.

Once the existing business lines are identified, the second task is to sort through them and determine which are providing the highest profit margins, and which have the potential to provide higher margins if better managed or expanded. Benchmarking with other lessors of space in the city or other airports allows a determination of whether the revenue derived per square foot or as a percentage of sales is at an appropriate level. Any particular lessee or concessionaire may be over- or undercharged by the airport. Appropriate adjustments should be made as the leases expire. Moreover, as concession leases expire, alternative lessors that may provide more revenue, or enhance customer service, should be explored.

The volume of passengers, "meeters and greeters," vehicles, cargo, and mail should provide some rough sense of the revenue potentially realizable with more astute marketing. In its market analysis, the airport should evaluate what services other airports provide as a benchmark against which to measure whether a different blend of concessionaires and lessees should be recruited to enhance its smorgasbord of services to the traveling public. Actually, since airports are becoming more and more like shopping centers, airports should also look to them to determine what types of goods and services might be offered passengers passing through the terminal and its concourses. In a sense, an airport is much like the lessor at a shopping center, though providing more services than a typical real estate developer. Nonetheless, viewing the opportunity as a potential real estate development is an appropriate way to view the airport property. Passengers who use the airport, and the airport's tenants, should be surveyed to determine what additional services they would like to have.

Passengers, visitors, and employees should be surveyed to determine their income levels, needs, preferences, and factors influencing their purchases. What types of facilities do they prefer, what do they think of the quality and price of goods and services offered? Past, current, and projected traffic volumes, including a breakdown into origin-and-destination, connecting, domestic, and international passengers, the number of meeters and greeters, as well as the average time spent by each of these categories in the airport, are all useful marketing data that can be used to find the optimum mix of concessionaires and airport provided services.[90]

From all that, the airport's managers should be able to determine the markets it is not serving or is underserving. Providing more service usually requires an expenditure of capital. An evaluation of space should be made to determine whether existing space is being dedicated to highest-value use, and whether additional revenue could be realized in excess of cost by the creation of more space for additional business lines. For example, where space in a terminal or concourse is limited, it may be possible to consolidate the food court and move it up a level to make room for other concessions. People who are hungry usually have no difficulty finding vendors, feel little inconvenience in moving up an escalator, and appreciate an amalgamation of food choices in one space. In contrast, someone passing through an airport is less likely to move up an escalator to another floor to purchase a postcard or a souvenir key chain or T-shirt. Thus, food services need not be placed in heavy-traffic corridors, while other vendors may need to be if sales are to be sufficient to cover leases.

In determining where to invest its resources, airport management should identify the most promising markets not being served or being inadequately served and the cost to capture those markets, and calculate potential returns on investment. This will allow it to prioritize investments. Ultimately, these can be incorporated into the airport master plan and capital improvement program.[91]

Airport Certification

Airports the world over are usually either owned or certified by their federal governments. U.S. airports serving aircraft having more than 30 seats must be certified by the FAA under Part 139. The regulations require that the airport develop a certification manual which sets forth comprehensive operating procedures. The FAA provides oversight on airport safety, security, hazardous materials, aircraft fire fighting, and rescue.[92]

The U.N. International Civil Aviation Organization (ICAO) also issues international standards and recommended practices which influence airport design and operation. ICAO's Annex 14 ("Aerodromes") sets forth international standards for airport certification, operation, and safety.[93]

Planning at Asia's New Airports

Airport construction can be an extremely complicated endeavor. For example, the Dubai airport project required at least 21 different design and execution programs, from concourse and control tower development to terminal renovations.[94] Let us examine planning at several of the world's new Asian airports:

Osaka's Kansai International Airport

Site surveys and technological planning for a new Kansai International Airport began in the mid-1960s. The Kansai International Airport Company was created in 1984 to build and manage the new airport. Construction began in January 1987. Landfill was completed in late 1991.[95] Kansai International Airport initially fell short of traffic projections, but exceeded cost projections.

Seoul's Inchon International Airport

A feasibility study for a new Inchon International Airport was conducted in 1989 and 1990. The site at Yongjongdo was selected in 1990. A master plan was finalized in 1992.[96] Groundbreaking took place in November 1992. The Korean Airport Construction Authority was created to construct the new Inchon International Airport on

September 1, 1994, under the terms of the New Airport Construction Promotion Law, enacted the preceding August.[97] The first phase of the airport's construction began in 1992.[98] The initial phase is anticipated to be completed in 1999, and the airport is scheduled to be operational in the year 2000, at a cost of $4.98 billion.[99]

Indonesia's new airports

In Indonesia, the Ministry of Communication, which oversees the nation's airports, prepared an Integrated Air Transport Study assessing Indonesia's airport needs. It identified four major airports (i.e., Solo, Lombok, Medan, and Padang) which it deemed essential to facilitating the government's policy of expanding the social and economic resources of the country.

The first step in the process was the performance of comprehensive feasibility studies for each airport. The principal issues addressed in the feasibility studies were the following:

1 Establish a sound engineering and financial database for the facilities required to meet future aviation demands for the air service area as well as provide the necessary information for potential investors to assess the financial risk of developing and operating the airport facilities.

2 Provide an airport plan that will improve the on-time flight performance.

3 Establish a plan that will provide better daily utilization of the scheduled airline fleet.

4 Prepare a terminal plan that will provide increased security for air travelers.

5 Provide an airport layout plan that is in compliance with ICAO design requirements.

6 Provide a schedule of development projects that utilizes the limited available airport development funds in the most cost-effective manner.

7 Provide a plan having development projects that will be attractive to private investors.[100]

Note that the Indonesian feasibility study plan emphasizes satisfaction of the efficiency needs of the carriers and the attractiveness of the project to private investors, two constituencies which do not always receive a high priority in other airport development projects. Cost containment, safety, and security are other major goals of this plan.

Alternatives to Airport Expansion

Of course construction of additional land-side or air-side capacity is not the only solution to capacity inadequacy. Other alternatives to building new airport infrastructure to accommodate demand include enhancing use of existing facilities via better rationing (e.g., higher peak period landing fees, to move demand to less congested parts of the day), and improvements in navigational and aircraft technologies (e.g., larger and STOL aircraft, as well as the introduction of the future air navigation system).[101]

More efficient use of airport resources can be achieved if the large peaks and valleys of aircraft takeoffs and landings are spread more evenly throughout the day, particularly at congested hub airports. One could argue that it is not an inadequacy of poured concrete that creates congestion, it is the decision of airlines to schedule takeoffs and landings in banks, particularly at hub airports. As an analogy, patrons of fine restaurants often have long waits for tables during supper, while the restaurant has empty tables most of the rest of the day. Many restaurants offer a discounted lunch menu to fill that capacity during the noon hour. However, efforts to impose *peak period pricing* (higher landing charges during periods of highest demand, and lower fees during periods of lower demand) of airport resources to flatten the demand curve somewhat have met with fierce political resistance from the general aviation community. That constricts the menu of remedies, most often requiring billions of dollars for incremental improvements in airport capacity.

Nonetheless, the U.S. Department of Transportation has explicitly approved the concept of peak period pricing, saying, "A properly structured peak period pricing system that allocates limited resources using price during periods of congestion will not be considered to be unjustly discriminatory. An airport proprietor may, consistent with the policies expressed in this policy statement, establish fees that enhance the efficient utilization of the airport."[102] Massport has long considered imposing peak period landing fees at Boston Logan International Airport, where airlines schedule flight arrivals and departures at levels higher than Logan's 120 movements per hour capacity.[103]

Moreover, airports that have *common-use facilities,* which require airlines to share common ticket counters and gates, need to build less infrastructure than airports with *dedicated ticket counters and gates.* Thus, a particular airport ticket counter or gate might serve Lufthansa at 9:00 A.M., Singapore Airlines at 10:30 A.M., Varig at 12:00 P.M., and Delta Air Lines at 1:30 P.M. Though many airlines prefer dedicated facilities, the result is that such facilities go vacant for long periods of the day. At major U.S. hub airports, one could roll a bowling ball down a concourse after a hub rotation and hit nary a soul. Such vacancy is wasteful of limited public resources.

Yet another means of reducing airport and airway demand is to shift passengers to surface modes, particularly for relatively short hauls. Though surface transportation is the subject of a later chapter (Chap. 9), suffice it to say that buses, railroads, and ocean and river ferries ought to be examined as alternatives to air transport in congested corridors.[104] Such alternative modes should be linked to the airport in a way to allow seamless intermodal transfers.

Summary and Conclusions

Airport planning is a little like science fiction. It requires creativity and vision tempered by objectivity and prudence. It involves amassing data from every conceivable source and then, on the basis of these data, offering a best guesstimate of future events. The size and cost of these projects define the importance of the planning phase. A new airport must be capable of meeting the immediate and future needs of the air industry, the passengers, and the community it serves.

The planning process is characterized by the open, transparent flow of information between all parties. Because of the size of the project, consultation and cooperation between the community and the planners are imperative if the plan is to receive general acceptance. Another component of the planning process, demand forecasting, reduces uncertainty by providing estimates to be used as a timetable for future construction and a framework around which the airport master plan is developed. Planning also aids operations at the regional and national level, by providing input in the development and subsequent revises of an integrated air traffic system.

The mantra for airport planning should be: practical, functional, flexible. A new airport must be practical. It must meet the needs of those it serves. It must be functional and efficient, capable of handling increases in air traffic during peak periods, while weathering the fiscal storms that invariably come with slower ones. And finally, it must be flexible. The airport master plan should consider everyone's interests. It should provide for the community's present needs, while preparing for its future exigencies.

Several overriding considerations should govern airport planning. A cost/benefit analysis must be made to determine whether public resources should be devoted to airport development rather than alternative public projects, and if an airport is to be built, that economic resources are wisely spent. Airports must be designed to ensure sufficient flexibility and expandability to meet evolving needs, and airport plans should be tailored to emphasize local problems and prospects. As an example, opening in 1994 with five runways and three remote concourses, Denver International Airport was designed in modular form to accommodate twelve runways, five concourses, and a two-fifths expansion of its main terminal as traffic growth warrants infrastructure expansion.[105] According to ICAO, "The most efficient plan for the airport as a whole is that which provides the required capacity for aircraft, passenger, cargo and vehicle movements, with maximum passenger, operator and staff convenience and at lowest capital and operating costs."[106]

Finally, it must be remembered that a master plan is nothing more than a guide for airport development. It should not provide mandates for specific improvements; it should only set forth the alternative improvements which may be undertaken. While setting forth a direction of development, it should not detail precisely how the development should be manifested.[107] Flexibility is necessary to guard against the negative consequences of change in uncertain environments, for the only constant is change. Given the fact that human affairs create unexpected events, and therefore unanticipated difficulties, the relative inflexibility of major infrastructure projects such as airports inevitably increases their risk. Preparing for the unexpected in airport development might be addressed by developing a series of "go/no go" checkpoints, whereby at specific points in the decision process the situation will be reevaluated and decisions made on the basis of new information and existing conditions.[108]

6

Site Selection and Environmental Planning

"Noise is the sound of commerce. If you want commerce you will have to put up with some noise."[1]—Donald D. Engen, Administrator, U.S. Federal Aviation Administration

In Chap. 6, we examine:

- The legal, political, environmental, and geographical considerations of site selection, airfield design, and use and zoning of land surrounding an airport
- Environmental issues such as noise, atmospheric emissions, water pollution, global warming, ozone depletion, and sustainable development
- Issues of land availability within reasonable proximity of the central business district of the city an airport will serve

Introduction

Early airports were built away from the cities they served, on inexpensive land, and where a minimum number of obstructions allowed maximum flexibility and safety in flight operations. Small aircraft flying infrequently posed little objection on grounds of noise. But the growth of air transport in terms of size and range of aircraft, thrust of engines, and frequency of takeoffs and landings, coupled with the expansion of cities to engulf airports, has caused the airport's needs for land and the aircraft's bombardment of noise to collide with the interests of surrounding landowners. As Edward Gervais, chief of airport planning at Boeing put it, "Most current airports have grown up from the DC-3 days, and now they're surrounded by residences and businesses."[2]

Airports are therefore challenged by the need to acquire sufficient airspace for access, sufficient land for ground operations, all within a potentially hostile political environment.[3] The problem is exacerbated by the fact that aviation is the fastest-growing mode of transport. That has raised the profile of environmental issues such as noise, land use, air and water pollution, climate change, and energy efficiency.[4]

Building a new airport, and selecting a venue for it, stems from a decision that the existing airport cannot be expanded adequately to accommodate anticipated aviation demand.[5] In determining whether a new airport should be built, and assessing which of the potential sites should be chosen, the U.S. Federal Aviation Administration has summarized the salient criteria as these:

> *The principal considerations for comparison of new sites to the existing airport will be airspace and airspace capacity, airfield and ground access costs (including value of time), aircraft operational costs, environmental impacts, financial feasibility, and long-term viability. Consideration also must be given to alternative roles for the existing airport and alternative transfer times to a hypothetical new airport.[6]*

We begin our review of the issue with the fundamental question of finding land sufficient in size and suitable in location for airport development.

Land

Airports consume vast quantities of land. Growing airports have a seemingly insatiable thirst for more land. Not only do airports require land for runways (typically about 2 mi long each), terminals (some of the largest public buildings ever), concourses, hangars, cargo facilities, kitchens, parking, and highways, they restrict land use in their flight paths. For safety's sake, approaches must be clear of office towers, water towers, and smokestacks. For sanity's sake, the approaches must be clear of residential housing. Few cities have available, or reasonably priced, land within reasonable proximity of their central business districts for new airports. Airports that are built in an urban area frequently find themselves hemmed in by surrounding development.[7] (See Fig. 6.1.)

As an example of an airport's insatiable thirst for land, witness Amsterdam's Schiphol Airport, hub to KLM Royal Dutch Airlines. In 1920,

6.1 *Boston's Logan International Airport serves 25 million passengers annually on a site consisting of only 2400 acres. To the north, south, and east, Logan is hemmed in by water. To the west, it is bordered by the neighborhoods of Boston. Growth in passenger and cargo volume must be accommodated within its existing boundaries.* Massport.

the airport consisted of a single field in the Haarlem-Mermeer polder. By 1945, it was a couple of runways in a small corner of the polder. By 1967, the central buildings and four runways had consumed about a quarter of the polder. It began another expansion project in 1994 to grow to five runways and an infrastructure of roadways and railways by 2005.[8]

By the mid-1990s, as the Dutch government considered privatization of its 75.8% stake in Schiphol, it began to face up to the question of what to do about the government-imposed capacity limitations of 44 million passengers per year (Schiphol handled 27.8 million in 1996), 3.3 million annual tons of cargo (it handled 1.1 million tons in 1996), and day and night noise restrictions. Three alternatives were identified:

1 *An overflow model.*   Combine Schiphol's current runway system with a nearby overflow airport that would handle charters, cargo, and low-cost carriers.

2 *A tandem model.*   Schiphol would handle only origination-and-destination passengers, while connecting traffic would be handled by an airport built on a man-made island in the North Sea near Lelystad.

3 *A remote runway model.*   Schiphol would handle origin-and-destination passengers and cargo for all of the Netherlands, while connecting traffic would operate out of a nearby airport on a man-made island connected by a high-speed people mover.[9]

In selecting a potential site for a new airport, the number of runways and size of terminal and other buildings should be estimated to project an overall contour of the airport, which will be useful for initial site screening purposes.[10] Prudent airport planners acquire more land than is necessary to satiate current demand, so as to allow room for future expansion, and restrict use of surrounding real estate via zoning and covenant restrictions. Ideally, the airport acquires control (if not outright ownership) of all land use falling within the airport's 65 Ldn contours, leaving such land unoccupied, or devoting it to aviation-related activities, such as long-term parking, a rental car campus, or air cargo facilities, and such additional land as may be necessary for future airport expansion.[11] So as to avoid the inflationary impact of land speculation and arbitrage, once a site has been selected, land should be purchased expeditiously.[12] Table 6.1 reveals the size of several of the world's new airports.

Table 6.1. Relative Size of Major Airports

Airport	Hectares	Acres
Macau International	190	469
Oslo Gardermoen	270	667
Osaka Itami	317	783
New York LaGuardia	275	680
Osaka Kansai	510	1,262
Paris Orly	629	1,552
Tokyo Narita	680	1,680
Tokyo Haneda	894	2,208
Shanghai Pudong	949	2,344
Tokyo Narita (completion of Phase 2)	1,065	2,631
Tokyo Haneda (completion of Phase 2)	1,100	2,717
Seoul Inchon	1,174	2,900
New York Kennedy	1,195	2,952
London Heathrow	1,197	2,957
Brussels Airport	1,245	3,075
Hong Kong International	1,248	3,084
Munich Franz Josef Strauss	1,387	3,427
Guangzhou New International	1,453	3,589
Singapore Changi	1,663	4,108
Amsterdam Schiphol	1,750	4,323
Jakarta	1,800	4,446
Kuala Lumpur International	1,850	4,570
Salt Lake City International	2,884	7,123
Paris Charles de Gaulle	3,104	7,667
Shanghai Pudong (full build-out)	3,198	7,899
Second Bangkok International	3,200	7,904
Seoul Inchon (full build-out)	4,743	11,715
Seoul Inchon International	5,615	13,869
Dallas/Fort Worth	7,203	17,791
Kuala Lumpur International (full build-out)	10,121	24,998
Denver International	13,760	33,987

The amount of land an airport will require is influenced by (1) the performance characteristics and size of aircraft anticipated to use it, (2) the expected volume of traffic, (3) meteorological conditions (including average temperatures and prevailing wind speeds and direction), and (4) the elevation of the site (with higher-elevation airports like those at Denver and Johannesburg requiring longer runways than sea-level airports).[13]

Location

The three most important considerations for real estate development are location, location, and location. The same is clearly true for airports as well.[14]

Airport site selection requires an in-depth analysis of alternative locations, considering such features as physical characteristics of the site, the nature of surrounding land use development and flight path obstructions, atmospheric conditions, land availability and its cost, ground access, the compatibility of surrounding airspace, and the site's proximity to aeronautical demand.[15] Each potential site should be systematically evaluated, deleting those with clear deficiencies in areas of construction cost, topography, airspace, ground access, and environmental impacts.[16]

In assessing the physical characteristics of a site, topography is important, for runways need to be reasonably horizontal and unimpeded by hills or mountains in their flight paths. Earth and rock may need to be blasted and removed if a hilly area is selected for the airport site. Drainage is also important, and a well-drained site is preferable to one that gathers water.[17] Soil types also must be examined, for some expansive soils have a tendency to buckle, causing runways and buildings to crack.

Unobstructed airspace is also important to the safe and efficient operation of an airport. Therefore, new airports should not be located in a venue likely to interfere with the flight approaches of aircraft using other airports. High terrain, trees, and structures (e.g., skyscrapers, radio and television towers, smokestacks) should also be avoided.[18] An example of an airport with difficult approaches is Toncontin International Airport in Tegucigalpa, Honduras, where an aircraft must land between camelback mountains on one side and hillside dwellings so close passengers can see people eating inside their homes; the landing strip has no radar, runway lights, or instrument landing system, and has a deep ravine at the end of the runway.[19] Another airport where the aircraft approach passes uncomfortably close to buildings is San Diego's Lindbergh Field, where passengers can look out their windows and peer into office buildings and watch secretaries typing away at computers.

Prevailing winds may cause industrial smoke to limit visibility, and such areas also should be avoided. Areas which fall within the migratory patterns or nesting sites of birds, particularly large birds such as swans and geese, should be avoided. Meteorological data may reveal that certain areas are susceptible to high winds, turbulence, fog, or high rainfall, which obstruct visibility or create turbulence, and these should be avoided too.[20] Fog has a tendency to settle in lowlands where there is little wind.[21]

Airport siting decisions have two primary, sometimes conflicting, dimensions—avoiding blasting land inhabitants in the flight paths with politically intolerable levels of noise, and finding suitable undeveloped land within reasonable distance of the central business district (CBD) of the city it will serve so that it can conveniently be used by its inhabitants. With regard to proximity to the CBD, and access to the new airport by the passengers who will use it, to facilitate public transport, airports such as London's Gatwick were placed near existing rail corridors, while Frankfurt Rhine-Main was placed near the intersection of two autobahn corridors. Both were built in 1936.[22]

Paradoxically, airports need to be located near population centers and surface transportation corridors so that people (including passengers, shippers of airfreight, and airline and airport employees) can use them conveniently. Yet the runways should be aligned so the flight paths do not cross over heavily populated areas. This requires compromise between these two conflicting principles. Building an airport too far from an urban area defeats the objective of reducing door-to-door transit times. Therefore, it is important to obtain sufficient land at the runway ends, or regulate the land use under the flight paths via zoning, so as to mitigate adverse noise impacts on the human population. It is also important to obtain sufficient land to satiate future capacity needs.[23] "Land banking" can reduce long-term costs while minimizing future adverse environmental impacts.[24]

Noise is a relatively more serious political problem in developed vis-à-vis undeveloped countries. Airports and their flight paths should be located away from residential areas and schools, wherever possible. If possible, a buffer zone around the airport should be created to minimize conflict. Zoning of surrounding real estate to avoid such future uses should be imposed. Delineation of noise contours should identify which areas are most likely to be blasted by noise, and these should be zoned for light industrial, commercial, recreational, or agricultural activity (so long as they do not attract birds) rather than residential housing.[25]

Multigenerational efforts to deal with crowding and congestion at Chicago are instructive as to the serious problems posed by location, and the tension between centrality and periphery. Established in 1922 (and rebuilt in 1927) outside the developed area of the city, Chicago's Municipal Airport (later renamed Midway Airport after the Pacific naval battle of World War II) soon found itself hemmed in by residential and commercial development and unable to expand.[26] A 1941 study reviewed several alternative locations of a new airport that would be close to the Chicago CBD, including on a man-made island or polder in Lake Michigan, in the warehouse district on Chicago's south side, near the west side slums, or on stilts above a rail yard. A 1946 study recommended an area 2 mi south of the Chicago Loop, which would have required clearing of 242,000 blighted or near-blighted dwelling units. Mark Bouman summarized the problems with these alternative sites, saying, "the virtues of centrality were also its undoing: being close to industrial districts meant coping with smokestacks and smog; being close to the commercial district meant coping with high land costs and tall buildings; being in Lake Michigan near the Loop meant dealing with fog, spray, and trick winds; and being close to any site of economic value meant paying exorbitant costs for land acquisition."[27] Ultimately, Chicago built O'Hare International Airport north of the city in 1963.

Contemporary efforts to build a third airport at Chicago, or a second airport at Minneapolis, have been thwarted by airlines not wanting to face new competition. Other cities have also failed in their efforts to build new airports, including New York (which wanted to build a fourth airport in the Great Swamp, New Jersey) and London.[28]

In order to find adequate land at reasonable prices and diminish political not in my back yard (NIMBY) opposition, newer airports have been built at greater and greater distances from the CBD (see Table 6.2). But growing suburban sprawl creates the same problems for peripheral locations that central locations suffer from—difficulty in land acquisition, high costs, and political (NIMBY) opposition.[29] Paradoxically, population increases create more demand for air transportation service.

Table 6.2. Airport Distances from Central Business Districts

Airport	Miles	Kilometers
Hong Kong Kai Tak (1929)	3	4.8
Geneva Cointrin	3	4.8
Brussels Melsbroek	4	6.3
Osaka Kansai (1994)	4	6.3
Frankfurt Rhine/Main (1936)	5	8.1
Salt Lake City International	5	8.1
Oslo Fornebu	6	9
Detroit City	6	9
Athens Hellinikon	6	10
Denver Stapleton (1929)	7	11.3
Vancouver International (1930s)	7	11.3
Munich Riem (1939)	7	11.3
Manila Ninoy Aquino	7	12
New York LaGuardia (1939)	8	12.9
Austin Bergstrom	8	12.9
Zurich Kloten	8	12.9
Atlanta Hartsfield (1925)	9	14.5
Paris Le Bourget	9	14.5
Paris Orly (1961)	9	14.5
Chicago Midway	9	14.5
Paris Charles de Gaulle (1974)	12	19.4
London Croydon	12	19.4
New York John F. Kennedy (1958–1962)	15	24.2
London Heathrow (1946)	15	24.2
Chicago O'Hare (1955)	15	24.2
Hong Kong International Airport (1998)	15	24.2
Munich Franz Josef Strauss (1992)	16	25.8
Toronto International Malton (1937)	17	27.4
Kansas City International (1968–1972)	17	27.4
Dallas/Fort Worth International (1965–1973)	17	27.4
Detroit Metropolitan	19	30.8
Los Angeles International (1930)	20	32.5

Table 6.2. Airport Distances from Central Business Districts (*Cont.*)

Airport	Miles	Kilometers
Houston Intercontinental (1967)	22	36
Bangkok Nong Ngu Hao (2003)	23	37
Hong Kong International (1998)	23	37
Denver International (1994)	24	38.7
Washington Dulles (1958–1962)	27	43.5
London Gatwick	27	43.5
Stockholm Arlanda (1962)	28	45
Oslo Gardermoen (1998)	29	47
Detroit Willow Run	31	50.2
Buenos Aires International	32	51.6
Seoul Inchon (2000)	32	51.6
Buenos Aires Ezeiza	32	51.6
Montreal Mirabel (1975)	40	64.5
Tokyo Narita (1978)	41	66.1
Kuala Lumpur International (1998)	41	66.1
Los Angeles Palmdale (proposed, but abandoned)	45	72.5
Chicago Kankakee (proposed, but abandoned)	50	81

Some airports have resolved the land and noise issues by filling coastal land in the ocean, albeit at enormous expense. Examples include Osaka's Kansai International Airport, Macau International Airport, Hong Kong International Airport at Chek Lap Kok, and Seoul's Inchon International Airport. A flight path over the ocean bombards no homes with noise, and requires no displacement of existing commercial, industrial, or residential buildings.

Other airports have attempted to buffer noise and improve the visual aesthetics by planting large volumes of trees around the airport perimeter. Washington's Dulles International Airport planted 1.5 million trees, for example. Milan's Malpensa 2000 project envisages planting a million trees and erecting other noise barriers, as well as undertaking other environmentally friendly projects, such as dedicated takeoff and landing runways to avoid overflying the most densely populated areas, installing an underground pipeline directly from a refinery to reduce truck traffic on the highways and the aprons, and assuring that the water table under the airport remains pollution free.[30] (See Fig. 6.2.) At Munich, more than a million shrubs and about 5000 large trees were planted.[31] Trees not only muffle noise, but offer an aethestically pleasing visual horizon to an airport. In selecting the types of trees to be planted, such factors as which trees are most likely to reduce noise year-round, how much the trees cost to buy and maintain, and which types of trees and shrubs are unlikely to attract birds that may pose a flight hazard should be taken into account.

Sustainable Development

Transportation has been described as the world's most serious environmental villain. Concerns over profligate consumption of nonrenewable resources (i.e., fossil fuels), global warming, ozone depletion, and acid rain, as well as air pollution, urban sprawl, congestion, and safety, have warranted careful examination of the role of transport in making the planet less habitable. As one source put it, "As a result of decades of careless development practices, our generation is now confronted with the reality of irreversible environmental damage."[32] The concept of sustainable development first emerged in the 1980s out of the "green," or environmental, movement, with a holistic, comprehensive, long-term view of humans' impact on life on this planet.

An early definition of sustainable development was "development that meets the needs of the present without compromising the ability of future generations to meet their own needs."[33] Thus, sustainable development focuses on conservation of the earth's nonrenewable resources, and avoiding pollution which threatens the life or health of present or future generations of human and other life forms. Human activity should be pursued in the least damaging manner, so that the environment is left in a state as good as, or better than, we found it—so that, in essence, our grandchildren and their grandchildren will have a decent life. Sustainable development was embraced as a global mission by the U.N. Conference on Environment and Development, known as the "Earth Summit," that met in Rio de Janeiro in 1992. At Rio, 180 nations signed an agreement to roll back greenhouse gas emissions to 1990 levels by the year 2000. At Kyoto, the

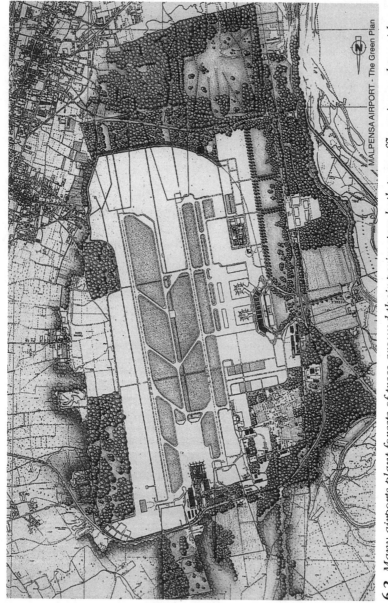

6.2 *Many airports plant forests of trees around their perimeters both to muffle noise and provide a more aesthetically attractive appearance. In one of the most ambitious such projects, Milan's Malpensa 2000 project, envisages the planting of 1 million trees.* Aeroporto Milano Linate.

MALPENSA AIRPORT - The Green Plan

industrialized nations agreed to reduce greenhouse gas emissions below 1990 levels by 2008–2012.[34]

Sustainable transportation has been defined as "Transportation that does not endanger public health or ecosystems and meets mobility needs consistent with (*a*) use of renewable resources at below their rates of regeneration and (*b*) use of non-renewable resources at below the rates of development of renewable substitutes."[35] Thus, sustainability does not insist that all transport should come to a halt. It attempts to promote a transportation system that is least offensive in terms of its consumption of the earth's resources and pollution of the earth's environment. According to one source, "Transportation needs must be met without generating emissions that threaten public health, global climate, biological diversity, or the integrity of essential ecological processes."[36]

Environmental Impacts of Transportation
Environmental feasibility

Strict environmental laws have been promulgated in many developed nations of the world. Moreover, in democratic nations, political acceptability of airport expansion is essential if the project is to move forward. Both factors converge to make environmental feasibility (including mitigation of adverse environmental impacts) of an airport project as important as economic or engineering feasibility.[37]

Much scientific inquiry and public policy debate have recognized the enormous problems that transportation poses in both the consumption of nonrenewable resources and the discharge of unsavory emissions. As the largest single source of these problems, the automobile and other forms of surface transport have been the focus of sustainability analysis. In contrast, relatively little has been said or written about the role air transport plays in global emissions.[38]

Yet of all the modes of transport, aviation is uniquely global. The typical turbofan jet engine burning kerosene produces unburned hydrocarbons, soot, carbon monoxide, and nitric oxide (NO, one type of nitrogen oxide).[39] Hydrocarbons produce smog; carbon monoxide takes oxygen out of the blood system; and nitrogen dioxide (NO_2) produces excessive nutrients in bays and estuaries (40% of the NO_2 entering Chesapeake Bay, for example, comes from air). Although air transport contributes a relatively small share

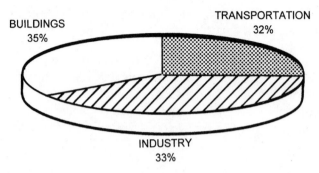

6.3 *Transportation accounts for nearly a third of all U.S. carbon emissions, a significant contributor to global warming.* NASA.

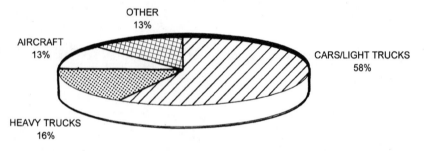

6.4 *Among transportation vehicles, aircraft account for 13% of total emissions, and are the only direct source of pollutants in the earth's upper atmosphere.* NASA.

of total pollutants (about 2.6 to 2.7% of global carbon dioxide emissions, for example),[40] it is the *only* industry which discharges harmful emissions (such as nitrogen oxides and carbon dioxide) directly into the upper atmosphere, thus contributing more profoundly to global warming and ozone depletion. Moreover, of all modes of transport, commercial aviation is growing fastest—outpacing any other form of transportation. This makes aviation of growing concern to sustainability in the twenty-first century. (See Figs. 6.3 and 6.4.)

As the major consumer of nonrenewable energy resources (i.e., fossil fuels), transportation is among the world's most prominent polluters [accounting for a third of the world's carbon dioxide (CO_2) emissions], with the automobile being the single worst culprit. Yet many predict that emission controls will result in a decrease in global automobile emissions over time, as has already occurred in the industrialized nations.

Noise

Noise has been the most prominent of environmental concerns about the aviation sector since inauguration of the jet age in the late 1950s. Jet engines produce noise as fuel ignites and exhaust gases and turbine blades strike the surrounding air. Aerodynamic noise is produced as air is displaced by the profile of the aircraft.[41] Solutions have included land use planning (zoning) around airports, sound-proofing homes in flight paths, altering flight paths to minimize noise impacts, imposing flight curfews at night, and mandating quieter stage 3 engines on aircraft. Prudent airport planning requires the measurement of aircraft noise, land use planning, and aircraft noise abatement procedures, such as quieter engines and home insulation or residential area condemnation and relocation.[42]

Assessing noise impacts requires quantification of noise, including frequency, pitch, time of day, and number of intrusions. Several metrics of aircraft noise have been developed.[43] The U.S. Federal Aviation Administration (FAA) has adopted a noise threshold of 65 decibels (dB) DNL as the trigger for unacceptable noise levels. That standard has been criticized by environmentalists on grounds that it is based on an averaging of noise, rather than a loud single event such as a passing aircraft, and that noise levels significantly lower than the threshold of 65 dB are annoying to many people.[44]

As an alternative, California and several European governments have adopted the community noise equivalent level (CNEL), which imposes a 5-dB penalty during the hours of 7:00 P.M. to 10:00 P.M., in addition to the DNL's 10-dB nighttime penalty. Environmentalists have argued that the threshold should be 55 dB CNEL rather than 65, and that single event noise rather than averaging should be taken into account by using the single exposure level (SEL) in conjunction with the CNEL.[45]

Typically, airports and governmental agencies in developed nations embrace a multitude of methods for reducing noise pollution. For example, in 1990, Amsterdam's Schiphol Airport became one of the first airports in the world to formulate an Environmental Policy Plan. The comprehensive plan specifies 24 action items, from installation of a noise monitoring system to an environmental protection system directed at promoting the use of public transport. In 1967, the Kosten unit (Ke), named for Professor C. Kosten, was adopted to measure aircraft noise. The 35 Ke zone surrounding the airport has been reduced from 42,000 homes in 1979, to about 15,000 in 1990, with the use of quieter aircraft and better planning of runways and flight paths. Many houses near Schiphol have been insulated against noise. By 2015, the 35 Ke zone will contain 10,000 homes, an absolute maximum. Night flying must meet the 26 LAeq standard, which means that the annual average bedroom noise levels during nighttime (11 P.M. to 6 A.M.) must not exceed 26 dB, while daytime flying must not exceed the 40 Ke level. (LAeq stands for equivalent continuous sound level, which is useful for describing a time period that consists of occasional short bursts of noise sandwiched between relatively quiet periods.) A fifth runway is being constructed at Schiphol to steer flight paths away from population centers.[46]

Aeroports de Paris also employs a multitude of innovative mechanisms to reduce noise impacts. Aircraft noise is monitored carefully by noise monitoring equipment at strategic points around Paris. At Paris Orly Airport, strict curfews on aircraft takeoffs and landings are imposed between 11:30 P.M. and 6:00 A.M. Aircraft landing fees are

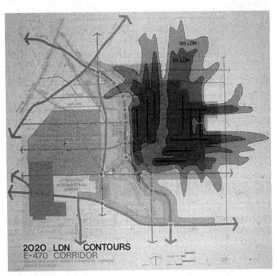

2020 LDN CONTOURS
E-470 CORRIDOR

6.5 *Typically, airport planning involves computer mapping of noise contours to predict where the most severe noise impacts will be felt.* Denver International Airport.

graduated according to noise emissions, causing noisier aircraft to pay higher taxes. Noise contour maps are drawn up to identify regions where no new construction is permitted. Flight paths are directed around residential areas.[47] (See Fig. 6.5.) Both to buffer noise and improve the aesthetic appearance of the airport property, a major tree-planting project is under way south of Charles de Gaulle Airport.[48] Recognizing the need to keep the community informed about what the airport is doing to try to reduce noise bombardment, and to encourage dialogue with the community, Aeroports de Paris established an Environmental Resources Center to act as a contact point and meeting place with the community, and to display information on technology and pollution.[49]

The greenhouse effect

The greenhouse effect is a natural phenomenon that warms the earth, enabling it to support life. Without it, the temperature of our planet would be a frozen $-18°C$, rather than the current average of $+15°C$. Light from the sun penetrates the atmosphere, warming the surface and oceans of the planet. Much of this heat is reradiated back out again in the form of infrared radiation. Because infrared rays have a longer wavelength than visible light, they can be absorbed by certain atmospheric gases, labeled *greenhouse gases.* This absorption warms the atmosphere, which in turn, radiates some of that heat back again to Earth.[50]

Among the most prominent of nonrenewable resources being consumed at a vigorous rate are fossil fuels, particularly oil. During the twentieth century, world energy consumption has increased more than 12 times, while per capita energy consumption increased 3.7 times.[51] Fuel consumption by transport increases at the rate of 2.6% a year.[52] The combustion of fossil fuels releases significant pollutants into the earth's atmosphere; these include greenhouse gases, such as carbon dioxide, methane, nitrogen oxides (NO_x), volatile organic compounds, and unburnt hydrocarbons. Greenhouse gases are those which are particularly effective in absorbing longer-wavelength radiation beyond the visible light spectrum, and trapping it in the atmosphere.[53] Transportation accounts for 45% of all volatile organic compound emissions.[54]

In earlier centuries, the pollution created by smaller human populations could be handled by the natural cleansing processes of rivers, seas, and the atmosphere—the carrying capacity of the planet was vast. But as the human population of the planet reached into several billions, the ability of natural forces to cleanse human emissions became strained, while, at the same time, human projects like deforestation reduced the earth's carrying capacity. Thus, human activity has significantly increased the volume of greenhouse gases in the earth's atmosphere. Some estimate that human activity has doubled methane concentrations in the atmosphere, and increased carbon dioxide by about 30%.[55] Atmospheric concentrations of carbon dioxide have increased from 280–285 parts per million in the year 1800 (when coal was the primary fuel), to 350 parts per million today (with petroleum as the primary fuel).[56] Unfortunately, carbon dioxide has an effective lifetime of centuries in the atmosphere.[57] In assessing the global impact of fossil fuel use, the consensus of 2000 top meteorologists and other experts was that the balance of evidence suggests these fossil fuel emissions have had "a discernible human influence on global climate."[58] According to Professor Ulrich Schumann, aircraft carbon dioxide emissions "add linearly to emissions from other sources" and contribute to global warming.[59]

As the principal discharge of combustion, carbon dioxide is believed to be the single most significant factor contributing to global warming. Carbon dioxide traps the sun's heat, increasing the planet's surface temperature; also contributing are nitrogen oxide, carbon monoxide, water molecules, chlorofluorocarbons, and methane.[60] Some contend that the condensation trails of ice crystals flowing from jet engines at high altitude may also exacerbate the greenhouse effect, although others deny this.[61] Water trails are believed 10,000 times less damaging than other greenhouse gases, helping form contrails and cirrus clouds, whose shadows cool the earth, but by blocking the planet's infrared emissions, also warm it by trapping heat in the atmosphere.[62]

Because of fossil fuel emissions, transportation is by far the world's most vicious polluter. Among sources of CO_2 emissions, transport has grown in both absolute and relative terms. In the United States, despite severe environmental regulation of automobile emissions, CO_2 emissions emanating from transportation sources rose 79% between 1965 and 1992, from 229 mmt of carbon to 408 mmt, a rate of growth outpacing any other source.[63] By some accounts, the combined effect of aviation emissions of CO_2 and NO_x could represent 10% of human-created global warming by the end of the twenty-first century, because the demand for air transportation may outstrip technological remedies.[64] Only about half the carbon dioxide emissions are absorbed by forests, oceans, and such; the rest stays in the atmosphere as a greenhouse gas.[65]

Transport also accounts for 43% of NO_x emissions in the United States, and 60% in Europe.[66] As we shall see, nitrogen oxide depletes ozone at higher altitudes, while paradoxically, below a level of 12 km, NO_x increases the amount of ozone, acting as a potent greenhouse gas. In a 1998 study published in *Science* magazine, two dozen scientists concluded that aircraft emissions of nitric oxide interact with sunlight in the upper troposphere to produce ozone, resulting in the formation of more greenhouse gas than previously thought.[67] In Europe, tropospheric ozone has increased 500% since 1970; concentrations are increasing 1 to 2% per year in the northern hemisphere.[68]

Since the mid-nineteenth century, global average temperature has increased about 1°F (0.5°C), and sea level has risen between 4 and 8 in (between 10 and 20 cm, approximately). By 1998, the earth's surface temperature had reached its highest level since people first began to measure it in the mid-nineteenth century. The 10 warmest years on record occurred after 1983.[69] The warming patterns are unlike those which might be expected from natural variability, and no alternative solar or volcanic causes have been identified, suggesting that the culprit may well be man.[70] Though some authorities contend that the warmer climate can be explained by normal variation, the dominant view among climate scientists is that at least some of the contemporary warming is caused by the trapping of solar heat because of emissions of industrial gases such as carbon dioxide.[71] It has been estimated that if we continue at our current rate of carbon dioxide pollution alone, average global temperatures will rise by 1.5°C to 4.5°C over the next 40 to 50 years. If average global temperature rises by 3°C, the Antarctic and Greenland ice caps would partially melt, and ocean water would expand, raising the sea level between 30 and 150 cm. A complete melting of the global ice caps is a process that would likely take several hundreds of years.[72] Global warming will also cause climatic changes and alter crop production.

Depletion of the ozone layer

Between 12 and 30 mi above the surface of the earth, the ozone layer shields the planet from harmful cancer-causing ultraviolet radiation.

Since 1967, the ozone layer over the equator has decreased by 3%, and over Europe and North America by 10%. Satellites have discovered an ozone "hole" appearing during the springtime polar vortex of the Antarctic stratosphere, while more moderate ozone depletion has been found at the midlatitudes, where most of the earth's population resides.[73] By 1996, the hole covered 7 million mi^2, nearly as large as the combined area of the United States and Canada.[74] Each 1% decrease in the ozone layer can lead to a 3 to 6% increase in skin cancer.[75] For human beings and other animals, increases in ultraviolet radiation can also cause immune system suppression, increased sunburns, cataracts and epidermal lesions, and reduced vitamin D synthesis. In plants, it can inhibit the process of photosynthesis, reducing agricultural productivity.[76]

Airlines are the second largest consumer of petroleum as a fuel, behind highway transport (i.e., automobiles and commercial trucks), and ahead of rail and water modes.[77] Commercial aircraft produce approximately 3150 mg of CO_2 and 1240 mg of H_2O per kilogram of aviation fuel burned. Air transport is responsible for but a small fraction of the earth's pollution, accounting for 2 to 3% of global carbon dioxide and nitrogen oxide emissions. But as noted above, aircraft are the *only* source of nitrogen oxide emissions in the upper atmosphere, which may represent a threat to the ozone layer.

Chlorofluorocarbons are the primary threat to the ozone layer; but the release of oxides of nitrogen into the upper atmosphere creates a series of chain reactions in which ozone molecules in the stratosphere are converted via photochemical dissociation into oxygen molecules.[78] Scientists are also concerned that nitrogen oxide emissions above cloud level may represent a threat to the ozone layer, which shields the earth from harmful ultraviolet radiation.[79] Approximately 20% of aircraft emissions occur at stratospheric cruise altitudes.[80] The current state of scientific knowledge offers no definitive proof of cause and effect, for the chemical interactions and climatic conditions are quite complex, and occur at an altitude in which they are notoriously difficult to measure. But ministers from 26 OECD member states issued a consensus statement that "the balance of evidence suggests a discernible human influence on human climate....The environmental impacts of rapidly increasing air transport are also of concern."[81]

NO_x emissions below cloud level are also washed to the earth as acid rain. Acid rain can have a deleterious effect on forests and wetlands.[82] Nitrogen oxides and other volatile organics combine to form tropospheric ozone, more commonly known as *urban smog,* which can irritate lungs, reduce resistance to infection, and aggravate heart disease, asthma, and bronchitis.[83]

Urban air pollution

The profound growth of population and its urban concentration has created localized concerns over carbon monoxide, ozone, and hydrocarbon emissions, as well as particulates. Particulates are fine particles that go deeply into the lung when inhaled. Temperature inversions can trap ground-based pollution, creating health concerns. Aircraft and airports are significant polluters. Burned and unburned jet fuel aerosols contain several carcinogenic organic compounds, including benzene and formaldehyde. For example, pollution from takeoffs and landings at Los Angeles International Airport alone is the equivalent of hydrocarbon emissions of 300,000 automobiles, and the NO_x emissions of one million automobiles.[84] Chicago O'Hare International Airport's 900,000 flight operations rank it as one of the top 3 to 5 toxic pollution sources in the state of Illinois.[85]

In 1993, aircraft at U.S. airports produced 350 million pounds of volatile organic compounds (VOCs) and nitrogen oxides (NO_x during their landing and takeoff (LTO) cycles, more than twice their 1970 levels. An airport's arriving and departing aircraft can create as much, if not more, ground level VOCs and NO_x than their industrial neighbors.[86]

Surface contaminants

Various chemicals abundant at airports, such as oil, kerosene, and aircraft deicing fluid (particularly ethylene glycol), and other hazardous and toxic substances such as solvents and metals have a potential to cause environmental damage to soil, groundwater, or surface water—and ultimately the human environment—if they stray off the airport property.[87] Prevention is far more cost-effective than cleanup. For example, Massport spent $61 million to clean up contaminated soil and groundwater at Boston Logan International Airport. It was estimated that there had been 2500 spills over 3 decades. Consuming a million gallons of fuel a day inevitably results in some spills. Massport found 31 discrete areas of jet fuel in the ground at Logan.[88] Certain chemicals, such as aircraft deicing fluid, can be collected and recycled.

Recycling

Several airports have become proactive with respect to recycling, and have established programs to reduce the amount of waste products leaving the field. As noted, deicing fluid is recaptured and recycled at many airports. Osaka's Kansai recycles waste water. Munich Strauss and Boston Logan have waste containers segregated for glass and cans, paper, and other trash so that the recyclable garbage can be collected.

Projected Growth of Air Transport and Its Emissions

Air transport appears to be growing faster than the other modes of transport. Passenger air transport grew 260% between 1970 and 1990, while air cargo grew 220%.[89] Many project that global air transportation will double over the next 10 to 15 years, with developing nations growing at a faster rate than developed nations.[90] Airbus predicts that the world's airlines will purchase 15,000 new passenger jets as global air travel triples over the next 20 years.[91] McDonnell Douglas predicted the need for 18,000 aircraft over the next 2 decades, with only 52% of today's active passenger fleet being retired.[92]

Moreover, the per person-kilometer carbon dioxide contribution of aviation is between 4 and 8 times that of travel by automobile, more than 10 times that of travel by bus, and 22 times that of electric-powered train.[93] For airfreight transport, CO_2 emissions are 20 times per tonne-kilometer greater than for a medium-sized truck, and 240 times greater than slow rail.[94] Thus, the tradeoff for speed appears to be pollution.

Carbon dioxide, methane, nitrogen oxide, and ozone have all increased dramatically in concentration over the past half-century as population, industry, and transportation have exploded.[95] Some sources ascribe to air transport a 24.3% share of transportation-related emissions which could potentially affect the climate, and project an increase in energy use by this mode of 180% by the year 2005. The International Civil Aviation Organization predicts a 65% increase in fuel consumption between 1990 and 2010. The combustion of any fossil fuel produces nitrogen. Others predict that technological improvements will allow nitrogen oxide emissions to hold constant even while fuel consumption (now 180 million tonnes) doubles.[96] But still other sources calculate that carbon dioxide emissions from global subsonic aviation may rise from 554 million tonnes in 1990 to 957 million tonnes by the year 2015. By the year 2100, aviation could account for 14% of the world's anthropogenic carbon dioxide emissions.[97]

Environmental Law and Regulation
The U.S. statutory regime

The United States has promulgated an extensive body of legislation dealing with aircraft noise and emissions that has had a profound influence on airport planning, design, and operation. Environmental factors must be considered carefully in the expansion of an existing airport or the development of a new one. Studies should be made of the impact of airport construction and operation on air and water quality, noise levels, industrial waste, and wildlife, and efforts should be made to mitigate the adverse environmental consequences wherever possible.[98]

The most common environmental problem airports pose is noise. Noise and other environmental impacts influence siting decisions. Adverse noise impacts can be minimized with land acquisition, runway realignment, or changing a runway extension from one end to the other.[99]

The Clean Air Act of 1963 was Congress' first effort to address the problem of air pollution. Congress first dealt with aircraft noise in the Aircraft Noise Abatement Act of 1968,[100] which authorized the FAA to set noise control and abatement standards for aircraft.

Comprehensive federal environmental regulation began with the National Environmental Policy Act of 1969,[101] which, with companion legislation, established the Environmental Protection Agency (EPA), and required that an environmental assessment (EA) and environmental impact statement (EIS) be prepared, the latter for any "major federal action significantly affecting the quality of the human environment." The EA determines whether potential impacts are significant, explores alternatives and mitigation measures, and provides essential information as to whether an EIS must be prepared. The EA focuses attention on potential mitigation measures during the planning process, at a time when they can be incorporated without significant disruption.[102] If the FAA concludes that there are no significant adverse environmental impacts, or that with appropriate prevention or mitigation efforts they will be minimized, it issues a "finding of no significant impact" (FONSI). If, however, it concludes the impacts are significant (which is sometimes the case in a major airport project), the FAA prepares an EIS.[103] The EIS must include an assessment of the environmental impacts, evaluate reasonable alternatives, and suggest appropriate mitigation measures.[104] It must review such issues as the impact of the project on noise, air quality, water quality, endangered species, wetlands, and flood plains. However, the thrust of the statute is process; there is no mandatory obligation to implement mitigation measures, even if they are feasible.[105]

These environmental requirements were explicitly affirmed for airports in the Airport and Airway Development Act of 1970. Such legislation has required that environmental factors be considered in both site selection and design. Airport master plans ordinarily must consider the following:

- Changes in ambient noise levels
- Displacement of significant numbers of people
- Aesthetic or visual intrusion
- Severance of communities
- Effects on areas of unique interest or scenic beauty
- Deterioration of important recreational areas
- Impact on the behavioral pattern of a species or other interferences with wildlife
- Significant increases in air or water pollution
- Major adverse effects on the water table[106]

As an example of the regulatory labyrinth through which airports must pass to proceed toward development, consider this single sentence from Salt Lake City Airport Authority regarding a major terminal and airfield expansion: "The current expansion has been in the planning process for nearly fifteen years and has included two Master Planning efforts, an FAR Part 150 document (an airport noise compatibility planning study), a Capacity Task Force Document, a Draft Environmental Assessment, an Expanded Environmental Assessment, and an Environmental Impact Statement as well as numerous smaller studies and documents."[107]

As noted above, Congress amended the Federal Aviation Act in 1968 to require the FAA to prescribe standards for noise measurement and abatement.[108] The FAA promulgated regulations thereunder for aircraft certification.[109] The Noise Control Act of 1972 gave the EPA the mandate to take an active role in the formulation and evaluation of noise standards, including aircraft noise, coordinating noise regulation with the FAA.[110] The statute explicitly allows citizen suits against any person alleged to be in violation of any noise control requirement. The EPA also regulates aircraft emissions, though the FAA was given veto power over any aircraft emission standards which might jeopardize safety. The FAA also was given authority to review flight and operational procedures to determine how they might be molded to mitigate adverse noise impacts.[111]

The Quiet Communities Act of 1978[112] provided for federal funding and technical assistance for a noise control program to be administered by state and local governments. The Aviation Safety and Noise Abatement Act of 1979[113] focused on reducing the impact of noise by establishing a system for airport noise compatibility land use planning.[114] Under it, the FAA has promulgated extensive Airport Noise Compatibility Planning Regulations.[115] That statute, and the Clean Air Act of 1963, confer jurisdiction on the EPA and FAA to monitor and regulate aircraft engine noise and exhaust emissions. Airlines are required to comply with all applicable noise control regulations and exhaust emission standards.

In 1969, the FAA promulgated regulations requiring implementation of noise abatement technology on aircraft.[116] Under these regulations, all Stage 1 aircraft were phased out from the U.S. fleet by 1988. The Airport Noise and Capacity Act of 1990 shifts authority away from airports to the FAA, requiring that airlines phase out Stage 2 aircraft by December 31, 1999.[117] In 1991, the FAA promulgated regulations requiring airlines to reduce (by modification or retirement) the number of Stage 2 aircraft operated by 25% by December 31, 1994, by 50% by December 31, 1996, by 75% by December 31, 1998, and by 100% by December 31, 1999.[118] Under this schedule, the entire U.S. fleet will be 100% Stage 3 by the turn of the century. However, a carrier may apply for a waiver from these requirements if 85% of its fleet is compliant by July 1, 1999, and it has a plan for becoming fully complaint by December 31, 2003.[119] The European Union has also adopted a program for phasing out Stage 2 aircraft over 7 years, beginning on April 1, 1995.

But the problem of mandating less noise from jet engines is that it may result in worse emissions, for the technology that reduces the decibel rate of engines requires higher temperature burn, which produces more pollution. Conversely, some technological improvements can reduce both noise and emissions. For example, air traffic control modernization, particularly including satellite navigation, will result in less circuitous flight paths, less congestion, and therefore less fuel burn and noise.[120]

The 1977 Clean Air Act Amendments established the National Ambient Air Quality Standards. The combined impact of this legislation, as well as the 1990 Clean Air Act Amendments and the 1991 Intermodal Surface Transportation Efficiency Act, is that nonattainment can mean ineligibility to receive federal matching funds for new transportation projects, such as airports and highways. One source noted, "To the extent that the growth of an airport leads to growth in flights, and the emissions from those flights, the administrative provisions of the Clean Air Act may act as a de facto limit on the size and operations of an airport in a given district that has not yet attained its air quality goals."[121]

Section 404 of the Clean Water Act gives the U.S. Army Corps of Engineers jurisdiction over wetlands management. Since 1989, the U.S. government has embraced a "no net loss" policy toward wetlands, requiring wetland loss be mitigated by upgrading wetlands elsewhere. This policy helped derail Chicago's proposed new airport at Lake Calumet, and will likely drive other U.S. airport projects upland.[122]

One relatively obscure piece of legislation that may impact older airport development is the Historic Preservation Act of 1966,[123] which requires that before federal funds are spent, account must be taken of the effect the project will have on any "district, site, building, structure, or object that is included in or eligible for inclusion in the National Register."[124] Some airport facilities, such as the Marine Terminal at New York LaGuardia Airport, are on the National Register.

Local noise regulation and federal preemption

The United States government vested plenary power in itself over navigable airspace in the Air Commerce Act of 1926. Under the Federal Aviation Act of 1958, navigable airspace includes areas more than 1000 ft above land as well as the airspace in the vicinity of airports needed to ensure safety in aircraft takeoff and landing.

Article 1 §8 of the U.S. Constitution vests in the Congress the power to regulate interstate commerce; inconsistent state or local laws are struck down as preempted by federal law. Local governments have been preempted from exercising their police powers to promulgate noise abatement requirements which affect aircraft flight patterns, or to impose curfews on unwilling airport proprietors.[125] However, they may exercise their police, land use, and zoning powers to regulate the location, height, and size of structures (for example, to prohibit the erection of a skyscraper at the end of a runway), so long as the regulation is for a health or safety purpose unrelated to the regulation of noise or the use of navigable airspace.[126]

While noise is of national concern, aircraft noise around airports is a highly localized political and legal problem. Local governments and their airport sometimes find themselves in the cross hairs of litigation objecting to aircraft noise. State nuisance and inverse condemnation laws are sometimes used by individuals seeking airport noise abatement. A property owner may allege that property has been taken without just compensation in violation of the Fifth Amendment to the U.S. Constitution.[127] Monetary damages may be awarded.[128]

The flying of an aircraft directly over private property can constitute a "takings" for which just compensation is required under the Fifth Amendment of the U.S. Constitution if the noise and vibration caused by the overflights significantly limit the utility of the property to its owner and cause its value to diminish.[129] Some state courts have held that flights from airports may violate common law doctrines of trespass or nuisance. Financial liability lies with the airport proprietor.[130]

In *United States v. Causby*,[131] the U.S. Supreme Court concluded that continued low-altitude military flights destroying plaintiff's poultry business constituted a "takings" requiring compensation under the Fifth Amendment of the U.S. Constitution, but that comprehensive federal regulation made the airspace a public highway above a certain altitude, for which no complaint could succeed on trespass grounds. Noting the conflicting rights of landowners to the airspace in the immediate reaches of their land, and the need of overflying aircraft for access, Professors Prosser and Keeton have urged, "A privilege to use air space for overflight of any height could be recognized so long as the exercise of that privilege did not unreasonably interfere with the use and enjoyment of the land surface."[132]

So as to minimize legal liability and political discomfort, numerous local airports have taken action to reduce aircraft noise or mitigate its effects, including access or use regulations or restrictions.[133] Airport proprietors may exercise their proprietary powers to control noise by promulgating noise abatement and curfew regulations, provided that such regulations are fair, reasonable, and not discriminatory, and do not unduly affect the free flow of interstate commerce.[134] For example, some airports impose flight curfews (prohibiting takeoffs and landings during certain late evening hours), prohibit the landing of Stage 2 aircraft, or establish perimeter rules prohibiting nonstop flights beyond a specified radius.[135] However, "overbroad, unreasonable and arbitrary" regulations will be struck down by the courts as imposing an unreasonable burden on interstate commerce.[136] Noise restrictions must be fair, reasonable, and nondiscriminatory, and intended to serve a legitimate public purpose.[137]

Though airport proprietors may regulate use of the airports they control, efforts of local municipalities to regulate the flight of aircraft have been struck down as preempted by federal law. However, reasonable zoning ordinances that merely regulate or restrict airport location or ground operations or assure compatible land uses within the vicinity of the airport have been deemed not federally preempted and within the policy power of the government as appropriately related to health, safety, or general welfare goals. Paradoxically, without zoning, land around the airport perimeter may become high-density development because the land is not suitable from a market perspective for low-density use. Airport zoning may restrict land use so as to, for example, restrict the height of structures in the aircraft approach paths to assure safety.[138]

Other means of avoiding inverse condemnation litigation include land use planning and zoning around airport perimeters. Airport planners must project the "noise footprint" that will fall on surrounding land by virtue of aircraft operations, with an assumption that an impact above 65 Ldn is incompatible with the reasonably quiet use of residential real estate. Zoning such land for industrial or agricultural use, for example, can be an effective means of reducing legal and political problems. An even more effective, albeit expensive, means of accomplishing the same goal is an outright purchase of all land that falls within the 65 Ldn noise footprint, resorting to condemnation powers under eminent domain, if necessary, or to purchase "avigation easements" over surrounding land.[139]

Siting Decisions at the World's Major Airports

The utility of an airport is largely influenced by its proximity to the business and residential areas of the metropolitan area it will serve. Although surface transport modes typically serve a city's downtown, airports are placed on the periphery of a metropolitan area, because of both the enormous amount of land an airfield requires and the environmental impact of noise from aircraft.

Frankfurt's Rhine-Main Airport

As noted above, Frankfurt's Rhine-Main Airport was built in 1936, near the intersection of two autobahn corridors, making it conveniently accessible to surface transport. The decision to build a new runway (Runway West 18) was made in 1965, but it was not constructed until 1981. The principal objection was that 5 acres of woodland had to be cut down.[140]

The number of runways is driven by existing and projected future takeoffs and landings. Airline use of an airport drives demand. For example, Lufthansa's Jurgen Weber has demanded a new runway at Frankfurt (Lufthansa's primary international gateway), which in his view has already reached its limit. Though conceding that the airport regularly hits hourly capacity limits on aircraft movements during peak demand periods, the Frankfurt Airport Company insists demand will not exceed full capacity until 2005.[141]

Stockholm's Arlanda Airport

Built in 1962, Stockholm's Arlanda Airport was located some 45 km north of Stockholm's CBD, in a location sufficiently remote and rural that it was anticipated that the airport could be expanded to accommodate future needs. Unfortunately, Luftfartsverket, the Swedish civil administration (which oversees 14 civilian and 5 military airfields in the nation), though participating in regional planning, was less than successful in its attempt to prevent home construction in the flight paths. As a consequence, new town sites have sprung up during the more than 3 decades since the airport was built, resulting in a number of operating restrictions being imposed upon the airport. Arlanda took over domestic operations from the existing airport, Bromma, in 1962, and international operations in 1983. By 1997, Arlanda was handling nearly 14 million passengers annually.[142]

Arlanda originally was built with two runways—a 2500-m east-west runway and a 3300-m north-south runway. By 1989, planning began for the construction of a third runway. Siting of the runway would take into account maximum peak capacity efficiency, minimum taxiing distances, and a smaller noise footprint, and ideally would be parallel to one of the existing runways. In Sweden, such a project is subject to a two-step process. First, under the Swedish Natural Resources Law, the government determines whether the project is preferable to feasible alternatives. Second, the Swedish Environment Protection Licensing Board evaluates the application, predicts the noise footprints, and establishes the tolerable pollution limits. To mitigate noise on the 14,000 people living in the FBN 55-dB noise level, the two-runway configuration has been limited to 275,000 movements per year, while the three-runway configuration is limited at 372,000 movements per year. (FBN is the Swedish-language abbreviation for aircraft noise level.) Actually, the third runway will allow a reduction in noise exposure to the nearby population. Restrictions have also been placed on CO_2 and NO_x emissions. Underground reservoirs have been protected from fuel and deicing fluid runoff.[143]

Munich's Franz Josef Strauss Airport

As early as 1954, the Munich Airport Authority was contemplating a replacement for Riem Airport.[144] Political leaders and airport officials began to study the need for a new airport seriously after a midair collision over Munich in 1960.[145] Thus, safety, noise, and capacity considerations drove the need to build a new airport and retire the old one. Yet noise, pollution, ground traffic congestion, and other environmental concerns pose the most formidable obstacles to new airport development.

In 1963, studies began to determine the best location for a new airport. Twenty sites were evaluated. In 1966, the list was reduced to two—Hofolding Forest and the Erding-North/Freising area. In 1969, the Bavarian government selected the latter for the new airport. Some hoped the new airport would be open for the 1972 Munich Olympic Games, but the approval cycle was laborious, and permission to construct was not conferred until 1979. Nearly 250 public meetings were held, all breaking at 5:00 P.M. so the farmers could go home and milk the cows, and some with such vociferous "not in my backyard" opposition that the police had to be called in.[146]

Construction began in November 1980, but was stopped after 5 months because of litigation, which forced the airport planners to reduce the size of the airport (from 19 to 5.4 mi²) and its terminal.[147] In Germany, judges are free to rule on all details, including terminal size.[148] Not until 1986 did the Federal Administrative Court in Berlin render a final legal decision to go forward with the new airport.[149]

Opposition came primarily from environmentalists and local politicians who resisted perceived noise which would disrupt rural calm. But airport planners gave environmental concerns a major priority from the start. Approximately 5000 trees were planted. Grassy vistas, flowers, landscape, and field and wetland nature conservation areas surround the airport, while 70% of the airport itself has been "greened" with plants and vegetation.[150] Palm trees were erected in the glassy lobby of the airport hotel, designed by Helmut Jahn, though they grow naturally nowhere in Germany.[151] Two-thirds of the airport's 1500 ha are green and have a year-round cover of vegetation.[152]

Once construction resumed, it took 7 years to complete building and testing. Strauss opened on May 17, 1992, some 38 years after it was originally conceived.[153] Munich's new airport was originally planned for 5066 acres and three runways; litigation reduced it to 3427 acres and two major runways.[154] At 5.4 mi^2, Strauss is one-tenth the size of Denver International Airport. Nonetheless, the new airport is 5 times the size of the airport it replaces. Half the land was set aside for future development.[155]

Munich Riem blasted about 200,000 people with noise. The figure at Munich Strauss drops to a few hundred. The size of Strauss ensures that the area subjected to a constant noise level of 75 dB(A) (zone 1 under German law) lies almost exclusively within the airport's boundary. Flight restrictions prohibit pilots from using reverse thrust to enhance braking unless necessary for safety reasons. No more than 38 aircraft movements are permitted between 10:00 P.M. and 6:00 A.M. The landing fee schedule favors the use of the quietest aircraft currently available, those which conform to ICAO's Chapter 3 requirements. Nearly 10% of Riem's cost was in noise abatement, waste management, landscaping, and measures to keep the water, soil, and air clean.[156]

Osaka's Kansai International Airport

One of the fundamental problems confounding public officials in Japan is the practical inability to condemn land under eminent domain power, and purchase it for fair market value. Traditional farmland is valued by its owners well beyond its economic value, and many farmers resist all efforts to purchase their property. Moreover, as is true in most industrialized countries, Japan is quite noise-sensitive. Efforts to expand the existing airport at Osaka were resisted by local residents for the same reasons people everywhere resist additional airport infrastructure—noise and fear of accidents.[157] Thus, the land, safety, and noise problems could be resolved only by building an airport out at sea.

Japan's Ministry of Transport began to study the possibility of building a new international airport in the Kansai region in 1968. The Cabinet approved the initial Kansai airport plan in 1984. Noise pollution was a driving force for the decision of the Ministry of Transport to authorize construction in 1986.[158] Construction began in 1987, and reclamation of the enclosed area began in 1988. Construction of the passenger terminal began in 1991 as the bridge trusses were completed. Osaka's Kansai International Airport (KIA) opened on September 4, 1994.[159]

To understand why Osaka's new airport was built on an artificial island, one must understand the geography and development, as well as the politics, of the country. Japan is a volcanic archipelago, whose relatively small areas of flat land have long since been developed into agricultural, industrial, and residential uses. Thus, there is little flat undeveloped land within close proximity to an urban area in need of an airport.

Second, one must recall the history of construction of Tokyo's Narita Airport. Shortly after construction was announced in 1966, violent clashes between 15,000 riot police and 3000 farmers supported by leftist militants resulted in the deaths of four police officers, attacked by left-wing protesters armed with bamboo poles and gasoline bombs. The result was that only one runway could be built, and it withstands a takeoff or landing every 2 minutes throughout the day, while about 100 protesters maintain a 24-hour vigil outside the security fence. Two families, owning 6 ha, refused to sell the land desperately needed for a second runway.[160]

The experience taught airport officials conciliation. Although under Japan's Airport Maintenance Law, Class A international airports, like Narita or Kansai, need not consult with local municipalities, they have begun doing precisely that. In order to build a 2000-m second runway by 2000, Narita proposed to compensate residents not covered by the Aircraft Noise Hazards Control Law, to recover green space lost during construction, and to help farmers who received land in exchange for expropriated parcels.[161]

Political problems have also arisen at Kobe. Kobe's municipal assembly unanimously approved construction of a new airport on landfill in Osaka Bay in 1990. Though the project originally had wide support, the city was devastated by an earthquake in 1995, causing severe budgetary problems. Airport opponents pointed to Osaka, alleging that city is near bankruptcy after the enormous sums it spent in building KIA. By 1998, a third of Kobe's eligible voters had signed a petition demanding a referendum on the airport.[162]

Hong Kong International Airport

The new Hong Kong International Airport (HKIA) at Chek Lap Kok was located on a site partly off shore. Takeoffs and landings occur in over-water locations so as not to blast residents with noise, as did Kai Tak Airport, which HKIA replaced. In building the new airport, rare white dolphins were discovered nearby. A marine reserve was established to protect this rare species.[163]

Oslo Airport Gardermoen

Discussions about the need for a new airport at Oslo began in the mid-1960s, but abated as traffic softened in the 1970s. Demand grew in the 1980s, rekindling interest in a new airport. Gardermoen, 45 km north of Oslo, was seen as a possible location, for it already had military and charter aircraft capability.

But in 1988, against the advice of the Ministry of Transport and Communications, the Norwegian Parliament selected Hurum, about 50 km southwest of Oslo, as the site, and preliminary surveys began. Debate raged in the media and the government, and in 1990, meteorological and flight operations reports revealed that Hurum would be an unsuitable site for an airport. The Norwegian Parliament reversed itself, and Gardermoen was surveyed for its suitability. In 1992, the Parliament formally endorsed Gardermoen as the site for the new airport.

Between 600 and 700 people lived on the proposed airport site. This required acquisition of 255 houses, 25 farms, and 16 industrial properties. The airport authority adopted the following environmental policy:

> *The new main airport is to be environmentally sound, designed and operated so that the negative effects on its surroundings are reduced. Environmental considerations are to be given the same status during planning, construction and operation as functional, technical and financial considerations.*[164]

The airport also established a "good neighbor" program, whereby nearby homes were purchased or soundproofed, communications lines opened, neighborhood planning encouraged, and local governments given financial assistance to aid them in dealing with health and social problems created by the airport.

Seoul's Inchon International Airport

Located 32 mi west of Seoul at Inchon, the new Seoul International Airport was built on landfill adjacent to Yongjong Island. Because of its noise-free isolation from inhabited areas (takeoffs and landings at the new airport are over the ocean), Inchon International Airport is Korea's first 24-hour international hub.[165] Inchon Mayor Choi Sun-ki observed, "In the past, we have focused on industrial and economic development, but now we are facing serious environmental problems, so we have placed environmental reform as a priority for our city's policies."[166]

Summary and Conclusions

It is important to reserve land for future airport development. The government of Thailand set aside ample land 25 years ago within proximity of Bangkok for future airport development. Munich attempted to set aside 17 mi², which unfortunately was whittled down to 5.4 mi² by the German courts. Denver can also be praised for having the foresight to set aside an enormous land mass (53 mi²) for future expansion of DIA.

Airports should be located so as to minimize adverse noise impacts. Of course, this lesson is closely related to the preceding one. Several Asian airports (e.g., Kansai, Macau, Hong Kong, and Inchon) have been built on the ocean, not only because of the dearth of suitable level land, but because of the noise impact on dense population clusters. Landfill on sea bed is among the world's most expensive and complicated engineering feats. But once constructed, 24-hour-a-day takeoffs and landings may be possible.

In selecting among alternative sites for a new airport, planners should engage in a cost-benefit analysis of each potential site across three dimensions—operational, social, and financial. With respect to the operational characteristics of each site, issues such as land availability, airspace availability, the effect of any restrictions (e.g., topographical, meteorological) on operational efficiency, and both short- and long-term potential capacity rise to prominence. With respect to the social dimensions, planners should consider proximity to passenger and cargo demand centers, adequacy of surface transportation access to the CBD and suburban residential areas, potential noise problems, and current land use and the need for zoning control measures in areas surrounding the airport. Finally, each will have differing economic characteristics, requiring a cost-benefit analysis.[167]

A systemic approach should be adopted which analyzes modes on the basis of sustainability. A transportation system which promotes sustainable development would contain three attributes: (1) it must be environmentally sound, (2) it should be efficient and flexible, and (3) it must be safe and secure. Each of these criteria contains three elements—technology, planning and policy, and ethics.[168] Airport location and design should be as environmentally benign as possible. As theologian Deitrick Bonhoeffer observed, "the ultimate test of a moral society is the kind of world it leaves to its children."[169]

7

The Air Side

Airfield Design and Construction

"In all my years of flying and testing aircraft, there have been a few times I was glad to see solid ground....I always appreciate having a place that pilot and aircraft can call home."[1]—CHARLES "CHUCK" YEAGER, BRIGADIER GENERAL, U.S. AIR FORCE

"Imagine constructing Charles de Gaulle Airport, Phase 2 of Singapore's Mass Rail Transit, the Golden Gate Bridge, Shanghai's Wan Pu Bridge, the Kennedy Expressway in Chicago, two Boston Harbour tunnels, housing for the 20,000 residents of Bend, Oregon, and reclaiming enough land for 16 Disneylands all at the same time. Imagine further that you are doing it all in one of the world's most densely populated cities, and in a political climate of incredible complexity."[2]—BECHTEL CORPORATION, COMMENTING ON ITS CONSTRUCTION OF THE NEW HONG KONG INTERNATIONAL AIRPORT AT CHEK LAP KOK

In Chap. 7, we examine:

- The historical evolution of airfield planning and design, and the technological catalysts therefor
- Fundamental principles of airfield design, including runway, taxiway, apron, and terminal layout
- Airfield and air navigation technologies
- Construction of airfields, including landfill

271

Introduction

This chapter addresses design and construction of the air side of the airport. The *air side* includes all areas where aircraft may take off, land, taxi, or park, including runways, taxiways, and aprons, and all areas providing aircraft service, such as fuel farms, deicing pads, maintenance hangars, and so forth. The air side is the "movement area of an airport, adjacent terrain and buildings or portions thereof, access to which is controlled."[3] We begin this chapter with a historical review of the metamorphosis of terminal and airfield configurations over time.

The next chapter (Chap. 8) examines design of the land side of the airport, focusing on the terminal and its various appendages. The *land side* begins at that area of an airport where the passenger loading device (typically a "jetway") at a gate connects with the passenger terminal or concourse, and proceeds through the passenger building, cargo facilities, and the ground access system.[4]

Terminal and Airfield Configuration: A Multigenerational Analysis

Many of the first-generation airports were military facilities, which after the Great War (World War I) were wholly or partially converted to civilian operations, with airlines operating out of former military hangars and barracks. For example, the city of Amsterdam purchased a military airfield to serve passengers arriving for the 1928 Olympics, in what later became Schiphol Airport.[5] In Asia, many of the first-generation airports were also postwar military facilities, but of World War II vintage, though this happened in Europe as well.[6] Between the world wars, Orly Airport near Paris had been the site of dirigible hangars, more than 300 m long. Orly was taken over by the U.S. Army during World War II, then returned to the French government in 1946. By the end of that year, a wooden air terminal had been erected. Today, Orly is France's second busiest airport.[7] Similarly, London's Stansted Airport was originally George Washington Field, built by U.S. Army 817th Engineer Aviation Battalion in 1942.[8] At each of these former military airfields, additional purpose-built passenger and freight facilities gradually were added to house the different airport functions—administration, passports and immigration, customs, weather, communications, and airlines.[9]

Crossing vast oceans in nascent aeronautical technology, the original transcontinental commercial aircraft were flying boats, capable of landing on water should the aircraft experience engine failure. Thus, another type of first generation landing field was water. Flying boats typically landed on bays. At New York's LaGuardia Airport a marine terminal was built to serve arrivals and departures of Pan Am "Clipper" aircraft, named after the clipper ships of an earlier era. Pan American World Airways was expanding its operations across oceans, and flying boats were then the safest means of transport. At Wake Island, a barren rock between Midway and Guam, the airline dynamited the lagoon to rid it of coral so its flying boats could land safely.[10]

The second generation of airports was characterized by the arrangement of buildings around an airfield according to a recognizable plan. The runways were grassy fields drained of moisture, with only the area immediately adjacent to the terminal and hangar (the apron) paved. A few airports in this era were illuminated, to allow flying at night.[11] The first municipal airport and terminal was built by the German government at Königsberg (now Kaliningrad) in East Prussia, to allow it to be connected via air to the rest of Germany after East Prussia became geographically isolated as a result of ceding Danzig to Poland at the end of World War I. Berlin's Templehof Airport was built the following year.[12]

By the 1930s, it became apparent that airports were being boxed in by buildings located along the periphery, leaving inadequate room for expansion. (See Fig. 7.1.) The French engineer A. B. Duval developed a concept of a wedge-shaped building zone projecting from the edge to the center of the airport, so that 80% of the airport periphery could remain undeveloped. That concept was adopted in the 1930s at Lyon, France, Birmingham, England, and Helsinki, Finland.[13]

The fourth generation is represented by London's Heathrow and Paris' Orly airports, where the passenger buildings were located on an island in the central part of the airport, with runways grouped in constellations around the terminal. (See Fig. 7.2.) The island is dominated by a passenger terminal building allowing aircraft direct access, and a central core of buildings, roads, and automobile parking garages, with a tunnel or underpass giving access to the island.[14]

Adopted in the 1950s, when gate concourses were added to central terminal buildings,[15] the fifth-generation airport is typified by London's

7.1 *Early airports were landing strips surrounded by buildings, which, over time, began to hem them in.* Copenhagen Airports.

7.2 *Paris Orly Airport was among the first to locate its terminal in the middle of the airfield so as to enhance aircraft access.* Photo by G. Halary. Aeroports de Paris.

Gatwick and Amsterdam's Schiphol airports, with a rectangular land-side building with "pier fingers" jutting from it.[16] (See Fig. 7.3.) Movable sidewalks transport passengers long distances from the ticketing area in the central terminal, out along the concourses to their gates,

7.3 *Amsterdam Schiphol Airport exemplifies the pier-finger terminal design, with enclosed concourses jutting into the air side, where aircraft are parked on both sides. The advantage of this design is that it allows an entire airport terminal to be housed under a single roof. The disadvantage is that it can create air-side congestion and delay as moving aircraft try to maneuver around parked aircraft.*
Amsterdam Airport Schiphol.

where they enter the aircraft (parked in rows on the aprons at the edge of the air side) via a jetway that protrudes from a finger. The fundamental concept was to minimize distances between land side and air side.[17]

The sixth-generation airport emerged in the late 1950s and early 1960s, and is typified by New York Kennedy and Los Angeles International airports, where independent satellite unit terminal buildings, typically dominated by a single airline, with a single multiple-use international terminal at one end, are located around a central highway corridor in the middle of the airport surrounded by runways, and connected together by bus or rail lines.[18] (See Fig. 7.4.) This airport design is efficient for intracarrier connections, but creates inconvenience for passengers needing to transfer between carriers.

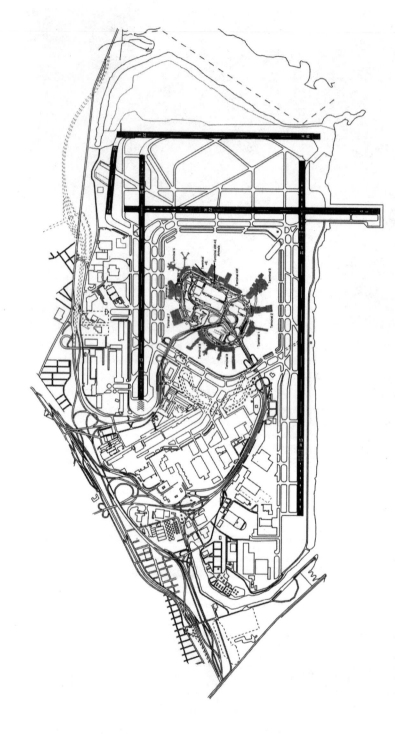

7.4 *New York Kennedy International Airport reflects the type of airport in which several independent terminals operate between a central transportation/parking campus and runways.* Port Authority of New York and New Jersey.

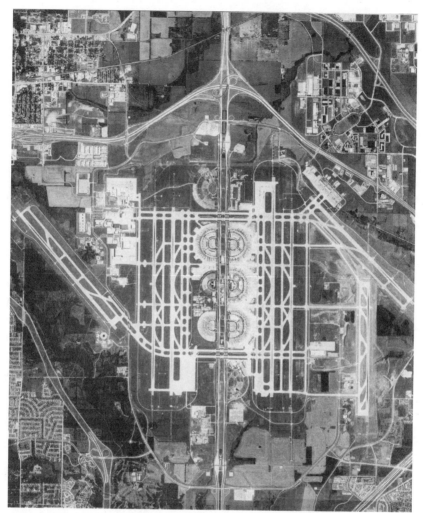

7.5 *In the 1970s, the dominant focus of airport designers was accommodating the local origin-and-destination traveler, by placing parking garages close to gates, facilitating expeditious check-in and departures from the airport.* Dallas/Fort Worth International Airport.

The seventh-generation airport, the "O&D Airport," is typified by Dallas/Fort Worth International, Kansas City International, and Rio de Janeiro airports, with terminal buildings placed within convenient proximity of parking. (See Fig. 7.5.) In each case, curved terminals wrap around parking areas, while aircraft are parked around the outside of the curved buildings. Thus, minimum walking distance is required between parking, check-in, and departure gates. While of

7.6 *The designers at Denver International Airport perfected the notion, earlier embraced at Atlanta Hartsfield, of remote parallel air-side concourses connected to the main terminal and land-side facilities by underground train.* Denver International Airport.

great convenience to origin-and-destination (O&D) passengers, the design is deficient for the connecting passenger, who must often walk vast distances between arrival and departure gates. Efforts have been made to resolve this difficulty by providing people-mover systems (pioneered at Tampa in 1970), typically belowground trams moving between clusters of gates.

The eighth-generation airport, the "hub connecting" airport, is typified by Atlanta Hartsfield and Denver International airports, whereby remote satellite concourses are surrounded by the airfield and connected to the main terminal building (surrounded on both sides by car parks) via a belowground people-mover system.[19] (See Fig. 7.6.) Orlando International Airport reflects another derivation of this approach, with circular pods connected to the main terminal by aboveground monorails. Many airports use trams to link terminals and concourses, including Atlanta, Birmingham, Dallas/Fort Worth, Denver, Gatwick, Hong Kong, Houston, Las Vegas, Miami, Newark, Pittsburgh, and Seattle-Tacoma.[20] Tampa's and Orlando's trams are aboveground. (See Fig. 7.7.) Trams at Dallas/Fort Worth and Kuala Lumpur are partly aboveground and partly subterranean. Washington Dulles tried another innovation, though mercifully not followed elsewhere—access to the remote midfield concourse is via oversized bus-like elevated lounges built by Chrysler.[21] Chicago O'Hare added an underground movable sidewalk showered in pastel lights and tinkle-bell music between its land-side terminal and United Airlines' remote

7.7 *Remote air-side terminals, like these at Tampa International Airport, are sometimes linked to the main terminal by aboveground trams, enhancing the aesthetic appeal of the passenger experience.*
Photo by David Lawrence. Hillsborough County Aviation Authority.

concourse. Many airports transport passengers from the terminal to remote parked aircraft in buses which roll across the tarmac, or by buses from one terminal pier to another, as at Detroit's Wayne Airport.

Configuring the airport requires assessing the number and orientation of the runways relative to the terminal. The number of runways is dictated by the volume of aircraft movements, while the orientation of runways is driven largely by prevailing winds, the size and shape of the perimeter of the airport property, and land use restrictions in the airport's vicinity. The terminal building should be located as close to the runways as possible to provide for efficient aircraft movements between them.[22]

Airfield Design

Known as *landing fields,* the original airports were simply grass fields, allowing landing and takeoff from any direction within 360°. The original aircraft were so light in weight that their pilots had to point them directly into the wind in order to effectuate a safe takeoff or landing; hence, 360° radius flexibility was imperative. Cinders

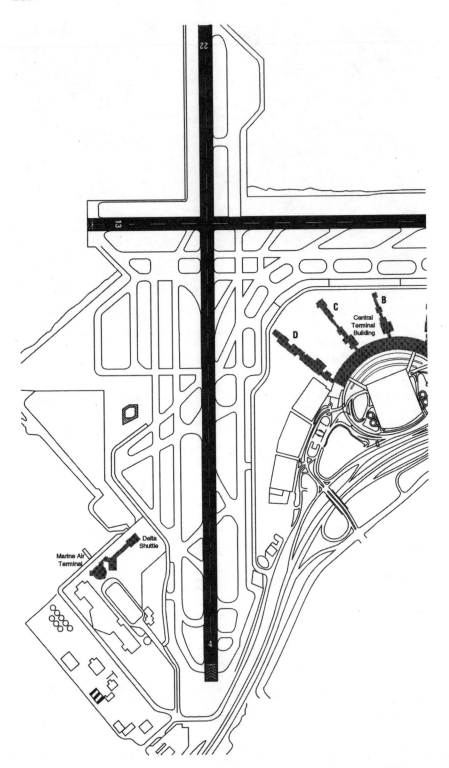

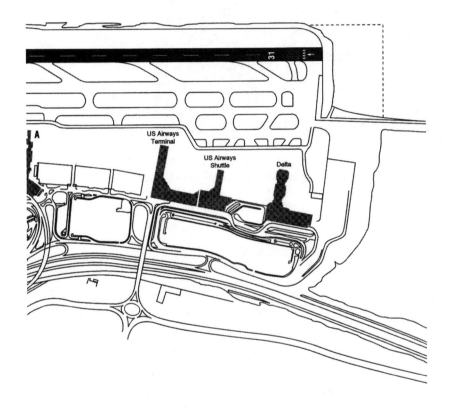

7.8 *LaGuardia Airport was one of the first designed in which runway layout was deemed as important as design of the terminal building. Note that it also contains a marine air terminal for the early Pan American Clipper aircraft.* Port Authority of New York and New Jersey.

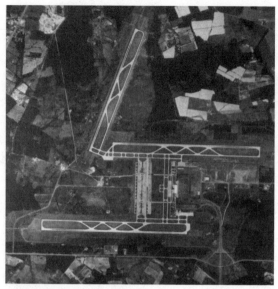

7.9 *Washington Dulles International Airport was one of the first airports designed for the jumbo jets, with the Boeing 747, Douglas DC-10, and Lockheed L-1011 coming on line in the late 1960s and early 1970s. Note the widely separated, staggered runways, a unique concept when this aerial photograph was taken, in the mid-1970s.* Metropolitan Washington Airports Authority.

were added at Chicago Municipal Airport (later renamed Midway) in 1926 to reduce the problem of water and mud on the field. Elsewhere, gravel, crushed rock, and ashes were added to the airfield.

These were followed by hard-surfaced runways, initially in the shape of a cross. As aircraft became heavier, they became less affected by crosswinds, and more likely to sink into the mud on unsurfaced air fields. In Europe, the Halle-Leipzig Airport added a concrete runway for takeoffs in 1926, the world's first hard-surfaced runway. Newark built the first hard-surfaced runway in the United States in 1928. The first concrete runway was built the following year at Dearborn, Michigan. In 1936, Stockholm's Bromma Airport became Europe's first airport to utilize a full system of hard-surfaced runways. Influenced by the advent of the DC-3 in 1936, New York's LaGuardia Airport, built in 1939, and Washington National Airport, built in 1941, were two of the first U.S. airports in which the layout of the runways and taxiways was considered as important as the design of the terminal building. (See Fig. 7.8.) LaGuardia was originally designed with four runways, allowing takeoffs and landings from eight directions. Washington's Dulles International Airport, opened in 1962, was the first airport designed for modern jet transports. (See Fig. 7.9.) Paris' Charles de Gaulle Airport, opened in 1974, would be designed with the jumbo jets in mind, with widely separated staggered runways, allowing parallel approaches at different altitudes. The Brisbane Airport, opened in 1988, Denver International Airport, opened in 1995, and Hong Kong International Airport, opened in 1998, all were designed for the next generation of super jumbo aircraft not even yet built.[23]

As aircraft technology matured, airfields had to be designed to accommodate the larger, heavier aircraft, with higher-thrust engines capable of lifting increased weight off the ground. This generally required longer runways built with a surface capable of withstanding the loads, though more advanced aircraft have abated the trend toward longer runways somewhat.[24] As air travel demand increased and airlines responded with more flights, runway saturation required more runways, spaced farther apart, allowing simultaneous takeoffs and landings from multiple directions.

As observed in Chap. 5, forecasts should attempt to predict the number of aircraft movements, type of aircraft, nature of the traffic, and other essential criteria essential in determining the number, layout, and dimensions of runways, taxiways, and aprons. Because of the vast amounts of land which runways, taxiways, and aprons consume, as well as the land use restrictions on nearby real estate necessary to assure safe and environmentally inoffensive takeoffs and landings, the runways and taxiways are the essential starting point for designating the airport layout.[25] In other words, in designing an airport, the runway and airfield layout should be done first. However, these airfield components, and their possible alternative layouts, cannot be considered in isolation. They should be considered in conjunction with the other essential infrastructure of an airport (e.g., passenger terminals, cargo buildings, aircraft maintenance hangars, parking) to select the optimum integrated schemes essential for efficient operations, and identify those areas where compromise may be required.[26] Moreover, as always, sufficient land should be set aside to accommodate future growth, including adequate space for adding or lengthening runways as demand increases or technology changes. (See Fig. 7.10.) Clearly, airports are land-consumptive endeavors.

Runways and taxiways should be located so as to provide adequate separation between flying aircraft and reduce delay in landing, taxiing, and takeoff. They should be staggered, so that parallel aircraft

DENVER INTERNATIONAL AIRPORT
AIRFIELD / TERMINAL LAYOUT

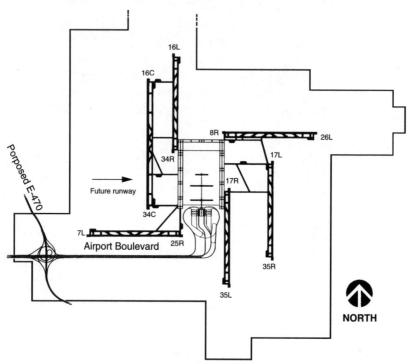

7.10 *At 53 mi²* (*about the size of the District of Columbia*), *Denver International Airport has sufficient land for 10 runways and multiple concourses.* Denver International Airport.

approaches can be made at different altitudes. Taxiways should be placed so as to provide the shortest possible distance from the terminal to the ends of the runways, and be sufficiently abundant, adequately sized, and at proper angles so as to allow landing aircraft to exit the runway as quickly as possible. Adequately sized aprons should be located adjacent to runways to allow several aircraft to park in a queue while awaiting takeoff, with sufficient space to bypass a parked plane. The terminal itself should be located to minimize distances to the takeoff ends of the runways, and to shorten taxiing distance for landing aircraft as much as possible.[27]

Alternative airport layout plans are also developed to enhance efficiency of airline operations, keeping taxi distance and runway crossings to a minimum, given meteorological conditions, capacity requirements, noise and land use constraints, and air traffic restrictions. The airfield should be designed with an eye to terminal configuration and intermodal transport rights-of-way and their location.[28]

In designing an airfield, several alternative runway configurations assist planners in assessing noise and other environmental impacts on surrounding land.[29] Computer models can graphically identify noise contours to identify areas which will be saturated with noise, thereby enabling planners to avoid flights over residential areas. For example, airport planners in Hong Kong evaluated 120 runway configurations before coming up with a runway design capable of handling 47 flights per hour in Phase I, compared with Kai Tak's low-thirties maximum capacity per hour.[30]

Runways, Taxiways, and Aprons

Runways have several essential elements—structural pavement sufficient to support projected aircraft loads, shoulders capable of resisting erosion due to jet blasting and capable of handling maintenance equipment, a runway strip which surrounds the runway and shoulder, a blast pad adjacent to the runway ends, a runway end safety area, a stopway, and a clearway consisting of an undeveloped zone beyond the blast pad to protect against aircraft over- or undershots.[31]

7.11 *Passenger loading bridges facilitate the speed and comfort with which passengers can board and depart an aircraft parked on an apron adjacent to the terminal.* Photo by Stefan Rebscher. Flughafen Frankfurt.

Taxiways are designed to provide aircraft with surface egress and ingress to runways. Exit taxiways should be designed to minimize runway occupancy time by aircraft which have landed. Rapid-end taxiways are those which have an angle of 25 to 45°, allowing high-speed exit from the runways.[32] Taxiways are designed to facilitate efficient aircraft movements between runways and terminals or hangars.[33] *Aprons* are paved air-side areas in which aircraft are parked for loading and unloading passengers, mail, or cargo, fueling, parking, or maintenance.[34] (See Fig. 7.11.)

Runway configuration must be planned to integrate with existing airspace limitations imposed by existing air uses (such as air traffic patterns created by nearby airports) and obstructions to navigation such as topography and buildings or other urban structures (such as radio or water towers), and prevailing meteorological conditions (e.g., wind, fog, rain) and electromagnetic interference.[35] Hong Kong's Kai Tak Airport had numerous physical and man-made obstacles standing in the straight-line flight path, over an extremely dense population base. This required serious circuitry and banking in the approach flight path, including a 47° turn to line up with the runway. (See Fig. 7.12.) Mercifully, Kai Tak was closed in 1998 when Hong Kong International Airport opened. The new airport has straight approaches over the ocean, and adequate safety areas at the ends of the runways to provide a margin of safety should an aircraft over- or undershoot the runway.

Runways should also be spaced so that they are sufficiently apart to accommodate parallel instrument landings during periods of inclement weather. The FAA prefers spacing of 4300 ft between runways for simultaneous instrument landings. At airports such as Cleveland Hopkins and San Francisco International, the parallel runways are simply too close together to allow simultaneous parallel landings during periods of inclement weather. At San Francisco, there is only 750 ft between the runways.[36]

Runways, taxiways, aprons, and terminals must be designed to accommodate the aircraft which will use them. In addition to volume of movements, consideration also must be given to aircraft weight and mass (which determines the thickness of pavement), wingspan and fuselage length (which influence the width of runways, taxiways, and aprons, and the configuration of the passenger building), aircraft turning radii (the distances from the center of the rotation to the wing tips, nose, and tail of the aircraft), passenger and cargo capacity (which influences the size and design of the passenger terminal), and takeoff length (which determines the length of runways). Runway length is also affected by temperature (the higher the temperature, the longer the runway required), altitude (the higher the airport, the longer the runway required), surface wind (headwinds shorten necessary runway length; tailwinds lengthen it), slope (an uphill grade requires longer runways than a downhill grade), and surface condition (water, snow, and slush make longer runways necessary).[37] Drainage and slope must be adequate to remove water, snow, and slush from the surface. To reduce the potential for hydroplaning and improve braking, the runway surface is typically grooved in a transverse direction.[38]

Safety, noise, topography, and land availability are among the principal concerns in runway layout. To the extent possible, runways should be oriented so that aircraft do not fly over heavily populated areas, both for safety and environmental reasons. They should also be oriented toward the prevailing wind when it blows consistently from a particular direction. Aircraft have difficulty taking off and landing when the *crosswinds* (winds at right angles to the aircraft) are excessive.[39] The maximum allowable crosswind depends on the size of the aircraft, its wing configuration, and the condition of the runway surface.[40] The standard minimum usability of a runway with respect to crosswinds, as established in ICAO's Annex 14, must be 95% with crosswind of 20 knots on runways of 1500 m or longer, 13 knots on runways between 1200 and 1500 m, and 10 knots on runways less than 1200 m long. The determination of the appropriate direction of a runway is performed by using a *wind rose,* a series of concentric circles cut by radial zones drawn to the scale of wind magnitude on polar-coordinate graph paper.[41] Additional runways, laid at different angles, may be necessary to accommodate aircraft during periods of strong crosswinds; however, since they are only to be used under high headwinds, their length may be shorter than the primary runway.[42] Aircraft use of runways and airspace adjacent thereto must include sufficient separation so that smaller aircraft are not caught up in the wake turbulence or wake vortex of larger aircraft.[43]

Airports should also be located away from concentrations of birds, such as nesting estuaries or garbage dumps. For example, Israeli aviation officials had to place a curfew at Tel Aviv's Ben Gurion International Airport between 1:00 and 3:00 P.M. because the number of birds feeding at the nearby dump is heaviest during the early afternoon.[44]

There are essentially four types of runway configurations—single runway, parallel runways, intersecting runways, and open-V runways.

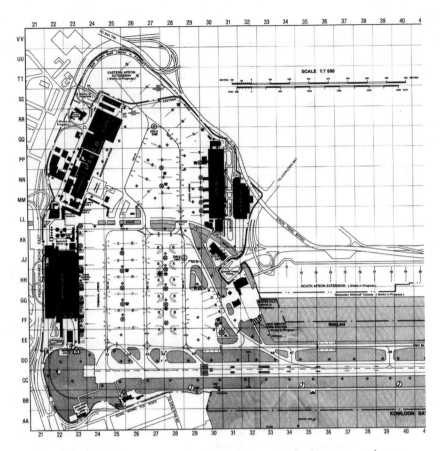

7.12 *Physical and man-made obstacles made flight approaches to Hong Kong's Kai Tak Airport among the most treacherous in the world.* Airport Authority Hong Kong.

The simplest is a single runway, which can handle between 50 and 100 aircraft movements per hour under visual flight rule (VFR) conditions, and between 50 and 70 movements per hour under instrument flight rule (IFR) conditions, depending upon the types of aircraft and navigational technology available. The capacity of parallel runways depends on their spacing. Some runways are only 700 ft apart. Where the distance between them exceeds 4300 ft, they can be operated independently under IFR conditions. Parallel runways can handle between 60 and 200 operations per hour under VFR conditions, and between 60 to 125 operations under IFR conditions, depending on runway spacing and navigational equipment. Intersecting runways are usually built when strong crosswinds come from more than a single direction, or when the airfield land perimeter will not permit parallel

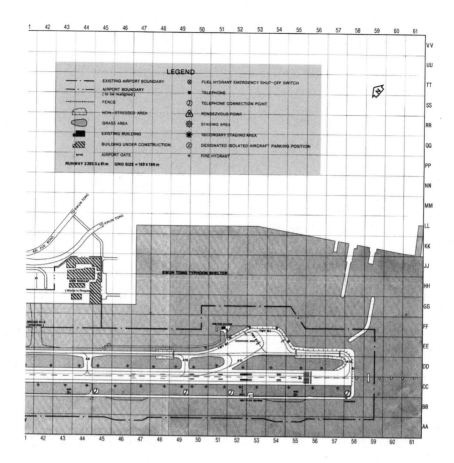

runways. When wind is light, both runways can be used simultaneously. The farther the intersection is located from the takeoff end of the runway, the lower its capacity. Open-V runways are those which, from the air, resemble the letter V, and these too are often built because of shifting strong crosswinds. When winds are light, both runways can be used. Maximum capacity is enjoyed when operations are directed away from the base of the V (this is called a *diverging pattern*); capacity is minimized when operations are toward the base of the V (this is called a *converging pattern*).[45]

Generally speaking, a parallel runway layout is preferred to intersecting runways, though terrain, noise constraints, or flight obstacles may make parallel runways infeasible. London's Heathrow Airport, originally a military airfield on the outskirts of London, and opened to commercial aviation in 1946, is laid out with six runways in a Star of David configuration, so as to provide simultaneous takeoffs and landings from as many as six different directions, depending upon the wind.[46] In contrast, Ronald Reagan Washington National Airport, built on landfill in the Potomac River, has three crossing runways. San Francisco International has four. Intersecting runways create both capacity limitations and potential safety problems.

Larger airports include parallel one-way taxiways to alleviate airfield congestion and delay. Taxiways should be laid out in a way to minimize distance between the terminal and the ends of the runway. Taxiways should also be placed at several points along the runway, and angled, so as to facilitate high-speed exit by landing aircraft—these are known as *exit taxiways or turnoffs*. This frees the runway up for another landing.[47]

Aprons serve terminals, and therefore must be planned in conjunction with them. Among the objectives to be taken account of in siting aprons are: (1) providing minimum distance between runways and aircraft stands; (2) allowing freedom of aircraft movement to avoid delay; (3) reserving adequate area for demand-based expansion and advances in aircraft technology; (4) achieving maximum efficiency, safety, and user convenience; and (5) minimizing adverse environmental effects.[48] The apron size is influenced by the number of aircraft stands it needs to accommodate, the present and future aircraft mix, and their dimensions and parking configuration, as well as ground service and service road requirements.[49] Ideally, an apron will be sufficiently large to allow an aircraft to back out and turn without blocking a taxiway.

Several methods of deplaning passengers exist, from parking the aircraft on the apron and walking the passengers across the tarmac to the terminal building (where they are processed through immigrations or customs, and retrieve their baggage) at smaller airports, to having the aircraft pull up to a jetway which connects to the aircraft for convenient passenger "deplaning" (see Figs. 7.11 and 7.13):

1 *Simple concept.* Aircraft are parked on the apron angled either nose-in or nose-out, for self-taxi in and out. Passengers walk across the tarmac to the terminal building. Though perhaps appropriate for small, low-density airports, having passengers walking around the tarmac creates security and safety concerns, and is less convenient for passengers and their baggage.

2 *Linear concept.* Aircraft are parked side-by-side nose-in along one side of the terminal. When fully boarded, they are pushed out, consuming less apron space aside either wing of the aircraft, but more at its tail.

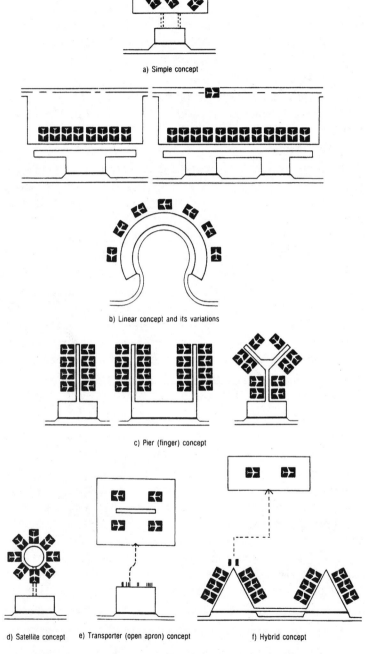

a) Simple concept

b) Linear concept and its variations

c) Pier (finger) concept

d) Satellite concept   e) Transporter (open apron) concept   f) Hybrid concept

7.13 *Potential terminal concepts are* (a) *the simple concept,* (b) *the linear concept,* (c) *the pier (finger) concept,* (d) *the satellite concept,* (e) *the transporter concept, and* (f) *the hybrid concept.* International Civil Aviation Organization.

3 *Pier (finger) concept.* Aircraft are parked side-by-side nose-in along both sides of a terminal pier which juts into the apron area. This can be an efficient means of boarding connecting passengers, for walking distances between gates is shortened. However, the pier finger concept creates inefficiency for aircraft as it restricts movements into and out of dead ends.

4 *Satellite concept.* Aircraft are parked all the way around a satellite terminal remote from the main terminal, and connected to it by surface or underground transport. The remote terminal concept allows for more efficient airfield movements by taxiing aircraft.

5 *Transporter concept.* Aircraft are parked at a remote apron (sometimes called a remote stand), and passengers board buses to take them to the terminal. The advantage is that aircraft may be parked closer to the runway, reducing taxiing and improving aircraft flexibility. However, passengers, baggage, and cargo must be moved longer distances.

6 *Hybrid concept.* A hybrid consists of any combination of the above. For example, some international airports use the transporter concept to augment capacity during peak periods when all gates are occupied.[50] An example of the hybrid concept is the new Kuala Lumpur International Airport, which uses the linear and pier (finger) concepts at its main terminal, and the satellite concept at its remote terminal, with the two linked together by an above- and belowground rail line.

Nose-in parking coupled with passenger loading bridges (jetways) appears to be the preferred parking configuration for most modern high-volume airports, for it consumes less apron area and less aircraft ground time as a result of efficient movement of passengers and efficient positioning of ground service equipment. It is also superior in terms of passenger safety, convenience, comfort, and security, than its alternatives. Its major drawback is that it often requires a tractor (or fuel-consumptive reverse engine thrust) for departure.[51] Additional aprons must be designed for the airport to accommodate such areas as the cargo terminal, maintenance terminal, parking, holding and deicing bays, general aviation, and helicopters.[52]

Runways, taxiways, and aprons are paved with layers of concrete and/or asphalt with different densities, strength, and smoothness. For example, at Kansai International Airport, the runways and taxiways are composed of an asphalt concrete pavement, while aprons are composed of a prestressed concrete pavement, and other areas are composed of plain concrete pavement. Each includes several layers to ensure durability and longevity.

Let us examine a specific runway/taxiway/apron project which increased takeoff and landing capacity by 77%. By improving its runways and taxiways, Brussels Airport was able to increase the capacity of its three runways from 45 movements per hour to 80 per hour, enhancing speed, operational efficiency, and safety. It did this, not by increasing the number of runways, but by adding 1.2 million m^2 of high-speed exits and adding taxiways and platforms, equivalent to about 60 km of highway.

At Brussels Airport, taxiways were hardened with asphalt, while aircraft stands and platforms were laid with concrete. The choice was driven by the physical characteristics of each material, and the visual signal given to pilots—that black zones are movement areas and white zones are parking areas. However, some airports have used concrete, rather than asphalt, in their runways (see Fig. 7.14). At Brussels, the taxiways were composed of a subfoundation of 20 cm of stabilized sand, 20 cm of lean concrete, and 26 cm of four layers of asphalt. (See Fig. 7.15.)

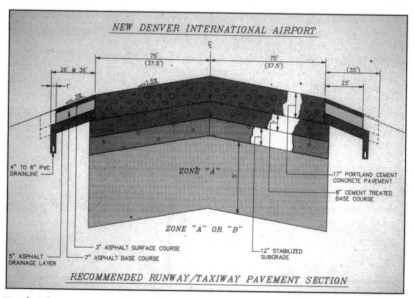

7.14 *The runways at Denver International Airport include multiple layers of various grades of cement framed by asphalt.* Denver International Airport.

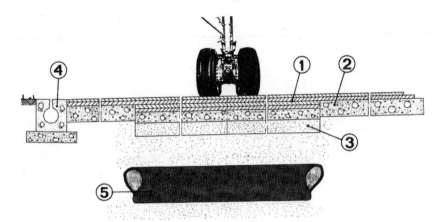

7.15 *Taxiways at Brussels are composed of (1) asphalt surface, (2) foundation in lean concrete, (3) subfoundation in stabilized sand, (4) circular groove, and (5) drain.* Brussels Airport Authority.

Taxiways are 30 m wide, with a shoulder of 10 m to protect the engines of the largest aircraft. Asphalt offers relatively good resistance against wheel ruts, allows smoother curves than concrete, and can be repaired quickly. In order to reduce the number of longitudinal joints, three asphalt machines operating side by side laid the top layer. High-speed exits were added to expedite aircraft movement off runways. The exits were treated with a kerosene-resistant antislip layer.

The aircraft stands at Brussels Airport were built with concrete, with a 20-cm subfoundation of stabilized sand, a 20-cm foundation of lean concrete, and a 35-cm layer of cement concrete. (See Fig. 7.16.) The top layer was blended in a way to resist deicing products and frost. An underground pipeline was installed, connecting the airport stands with fuel storage tanks in Antwerp, thereby eliminating the need for fuel trucks on the tarmac or to and from the airport.

Runways and taxiways were fitted with lateral and axis beacons and lights to guide the pilot from touchdown to the gate. (See Fig. 7.17.) In accordance with ICAO standards, the taxiways were equipped with blue lateral beacons and red stop bars (at angles perpendicular to the direction of air travel) at intersections, and green beacons on the axis lines. By lighting up only the beacons relevant to an aircraft's navigation of the airfield, the pilot can be led step-by-step from touchdown to the gate under a "follow the greens" system. To provide nighttime illumination, floodlights were installed atop masts surrounding the concourses.

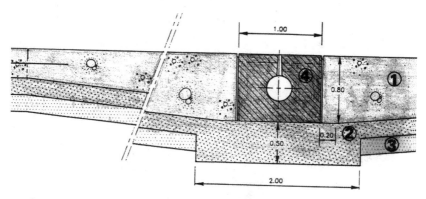

7.16 *Aircraft stands at Brussels are composed of (1) a concrete surface, (2) a foundation of lean concrete, (3) a subfoundation of stabilized sand, and (4) a circular groove.* Brussels Airport Authority.

At Brussels, the maximum slope of taxiways was 1.5% and of platforms 0.5%, which creates special drainage demands. Circular water drainage grooves are laid throughout the runways, taxiways, and aprons, connected to drains at junction chambers every 50 m. The airport is also equipped with separate drainage systems for wastewater and rainwater, the latter connected to an oil separator to reduce contamination by oil or kerosene. Tunnels are placed under the taxiways to expedite ground vehicle movements between the passenger terminal and cargo areas. To facilitate maintenance, a labyrinth of large underground tunnels makes cables and pipes accessible.[53] (See Fig. 7.18.)

Similarly, Frankfurt Main Airport increased aircraft movements from 62 per hour in 1992 to 76 per hour in 1998. It was able to reduce the vehicle congestion on the ramp by moving services underground. With enhanced technology, high-speed exits, and approach-departure enhancement rules, Frankfurt achieved simultaneous operations on closely spaced runways, reduced separation between aircraft, and reduced runway occupancy time, all of which has enhanced capacity without expanding the airport perimeter. According to airport Deputy Chairman Manfred Schölch, "We will build up and down, because we can't expect to expand beyond current airport fence lines."[54] Aircraft safety and efficiency dictates that in designing an airport, the runways be laid out first. Runways and taxiways define the airport. Only after the airfield is laid out should the terminal be designed, for the terminal design must integrate with the airfield.

Finally, with respect to runway maintenance and repair, the most economical way to lengthen the life of the pavement is to rehabilitate

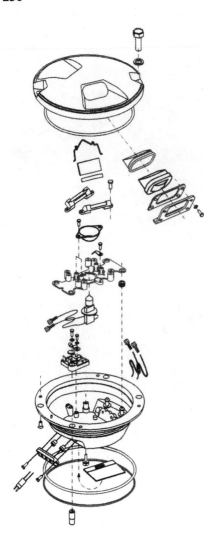

7.17
A diagram of a runway axis beacon, which leads aircraft to their destination. Brussels Airport Authority.

the runway while it is still in good condition. Waiting until serious problems arise can increase costs because more expensive methods of repair must be used.[55]

Navigational Aids

In order to maintain adequate separation between aircraft so as to en-sure safety, governments the world over coordinate their supervision of aircraft under uniform standards established by the U.N. Interna-tional Civil Aviation Organization. Air traffic control is typically di-

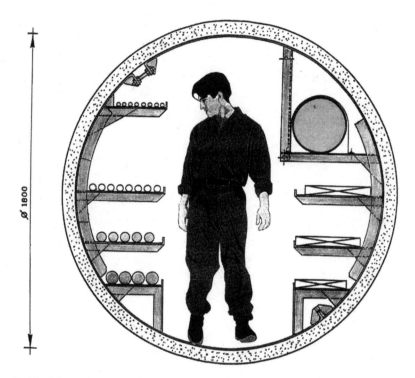

Ø 1800

7.18 *Though initial costs are high, underground utility tunnels sufficiently large to allow a worker to make repairs or lay new conduit or pipes eliminate the need to dig up airfields, and therefore are lower-cost in the long-term. Here, cables are on galvanized steel shelves, while a water pipe is to the right.*
Brussels Airport Authority.

vided into three areas: (1) area control, (2) approach control, and (3) aerodome control. *Area control* is designed to ensure adequate separation of aircraft once an aircraft has left the airspace controlled by an airport and is en route to another airport. *Approach control* (or terminal radar control) gives approaching and departing aircraft radio instructions within the approach control area (the airspace extending like a staircase from the airport). Airport surveillance radar monitors location and altitude of aircraft under terminal radar control. *Aerodome control* consists of clearance given to an aircraft landing at or taking off from an airport in a control zone. This is performed from the air traffic control tower located at the airport.[56] (See Fig. 7.20.)

Airports must be designed to integrate with the air traffic control system, to provide navigational aids for approaching aircraft, and to control taxiing aircraft and vehicles moving about the airport. Enhanced navigation technology can improve the safety and efficiency of takeoffs and landings. For example, upgrading ground equipment enabled Paris to join Dallas and Denver as cities where three simultaneous parallel aircraft approaches are possible.[57] At St. Louis Lambert International Airport, installation of a $9 million precision runway monitor allows high-speed, high-resolution monitoring of incoming aircraft to enhance safety and capacity at the airport's parallel runways, only 1300 ft apart.[58]

An airport's instrument landing system (ILS) allows air travel to proceed safely regardless of poor visibility. The ILS consists essentially of a localizer beacon and glide path transmitter at the outer and middle marker beacon. The localizer beacon informs pilots of the appropriate direction for landings through a glide path. The glide path transmitter informs the pilot of the appropriate descent.[59] Figure 7.19 reveals the potential location of several of these systems vis-à-vis the approach glide path and runway.

Approach lighting at the ends of the runways will require advance planning and purchase of land for installation and clearance of obstacles in the approach area. Preplanning also includes installing sufficient duct capacity in the runways and taxiways during initial construction. Adequate duct capacity will enable runway lights to be upgraded to Category III technology without tearing up the runways.[60] Table 7.1 reveals the ICAO categorization of instrument landing systems.

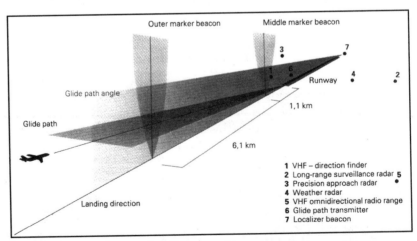

7.19 *This drawing depicts the typical location of air navigation technologies and the flight path leading to the runway.* Flughafen München.

Table 7.1. International Civil Aviation Organization Instrument Landing System Categories

Category	Decision height	Runway visual range	Remarks
I	60 m (200 ft)	800 m (2600 ft)	
II	30 m (100 ft)	400 m (1200 ft)	
IIIA	0 m	200 m (700 ft)	Visual observation required for operation on the runway or taxiway
IIIB	0 m	50 m (150 ft)	Visual observation required for operation on the runway
IIIC	0 m	0 m	No operations depend on visual observation

The navigation aids a major airport is likely to have, and their acronyms, are the following:

1 Instrument landing system (ILS)/microwave landing system (MLS)

2 VHF omnidirectional radio ranges (VOR) or nondirectional radio beacon (NDB), which work as an intercept point at which the aircraft should intercept the ILS course; when used in conjunction with an ILS, the NDB is called a *compass locator* (CL)

3 Distance measuring equipment facilities (DME) (generally collocated with VOR or ILS or MLS); terminal DME provides approaching aircraft with information as to the distance to the touchdown point; guide slope (GS) gives approaching aircraft information regarding their angle of descent

4 Collocated tactical air navigation systems and VOR (VORTAC)

5 LLZ localizer, which provides approaching aircraft with guidance information to the centerline of the runway

6 Middle marker (MM) identifies a point 900 m from the end of the runway; an inner market (IM) identifies a point 400 m from the end of the runway

7 Far field monitor (FFM) checks the ILS as to accuracy

8 Radars—approach, secondary, and surveillance type[61]

Airport Control Tower and Control Center

The airport control tower should have a clear and unobstructed view of the entire movement area, including runways, taxiways, aprons, and parking spaces, and of air traffic in its vicinity. (See Fig. 7.20.) Shorter towers may be useful to manage aircraft ground traffic around aprons. The area control, or flight information center, should be in reasonably close proximity to the airport control tower and sufficiently large to accommodate its personnel and equipment.[62]

At 33 stories, Denver International Airport has the highest air traffic control tower in the world. Some airports can take advantage of a nearby hill to place a tower, thereby lowering cost. The airport in Phuket, Thailand, is an example of this.

Safety Infrastructure

Though at most airports, deicing of aircraft is performed at the gates, aircraft deicing pads should be located as close to the departure runways as possible, so that aircraft can take off immediately after deicing fluid is applied. Deicing fluids typically include glycols, thickener, and corrosion inhibitors, which present low-level environmental concerns if they contaminate the soil or ground or surface water. Therefore, the deicing pads should collect and recycle the deicing fluid, so that it does not become an environmental hazard. Fire, crash, and rescue facilities should be located on the airfield, as close to the center of the runways as possible, to minimize response time during emergencies.[63]

Aircraft Fuel Facilities

Commercial aircraft consume vast quantities of fuel. The design and location of fuel facilities should adhere to the highest principles of safety, environmental prudence, and aircraft service efficiency. Fuel should be stored as close to the aircraft fueling area as possible, which is usually at their parking position at stands near the terminal building, located near the fuel intakes on the wings of the aircraft. Pipelines running from a central storage area linking pits at the aircraft stand avoid both excessive storage near the passenger terminal and fuel trucks on the tarmac.[64]

Cargo and Mail Facilities

Cargo is transported in the belly of passenger aircraft, in combination aircraft (frequently in containers), and in all-cargo carrier

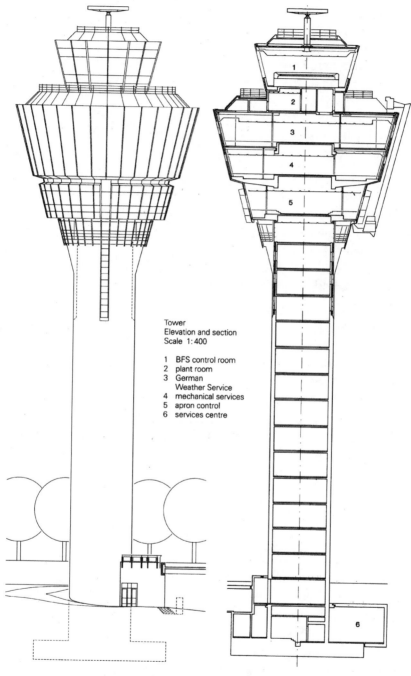

Tower
Elevation and section
Scale 1:400

1 BFS control room
2 plant room
3 German
  Weather Service
4 mechanical services
5 apron control
6 services centre

7.20 *An airport air traffic control tower must be sufficiently tall so that air traffic controllers can see all parts of the landing field. It should have separate windowless rooms for controllers seated at radar screens.* Flughafen München.

freighters. It is recommended that all-cargo aircraft be separated from combination aircraft. Typically, combination aircraft are parked on the aprons adjacent to the passenger terminal building, while all-cargo aircraft are parked near cargo terminals. The flow of cargo and accompanying documents to and from, and between, aircraft should be smooth and cover the shortest possible distance.[65] To facilitate intraline cargo connections between cargo and combination carriers, cargo facilities should be located in reasonable proximity to the passenger terminal.[66]

Many airports have devoted resources to facilitating cargo. Airfreight carriers have established cargo hubs at airports in cities such as Dayton, Indianapolis, Louisville, Manila, and Memphis. Frankfurt Main Airport built a 90-ha cargo area ("Cargo City South") on the south side of its airport, allowing it to handle 2.75 metric tonnes of cargo by 2010. It is also served by a rail line.[67] Table 7.2 reveals the world's airports with the largest cargo volumes.

Most new airports include dedicated air cargo facilities. For example, the new Hong Kong International Airport includes an automated three-story airfreight center, with 150 independent freight stations for processing and storing cargo. Total available space is 120,000 m². Ceilings are sufficiently high to allow two 9-ft containers to be stacked on one another.[69] Manila also became a major airfreight cen-

Table 7.2. Airport Air Cargo Volume[68]

Airport	Million tons
1 Memphis	69.1
2 Los Angeles	1.72
3 Miami	1.71
4 New York Kennedy	1.63
5 Tokyo Narita	1.63
6 Hong Kong	1.59
7 Frankfurt	1.50
8 Louisville	1.37
9 Seoul	1.36
10 Anchorage	1.27
11 Chicago O'Hare	1.25
12 Singapore	1.21

ter in 1995 when FedEx leased a hangar from the Philippine Air Force at Villamor Airbase adjacent to Manila International Airport.

Postal facilities are also essential components of airport infrastructure. Typically, the government operates a mail and package sorting facility on or near the airport, while combination passenger/freight air carriers operate sorting facilities which funnel bundled mail and packages onto the appropriate departing or connecting flight. A serious problem at many older airports is the inadequacy of space on the airport campus for postal facilities. Locating the mail facility off of the airport property creates surface transportation problems, for mail then has to be trucked on and off the airport for sorting. In the United States, one of the most advanced automated airline sorting facilities is at Pittsburgh International Airport, while some of the more primitive such systems exist at Detroit, Houston, and Phoenix airports.

The Bidding and Contracting Process

Major contracts for most airport projects are let on a competitive bidding process. Usually, eligible foreign bidders team up with local contractors in a consortium to bid the project. Governments and airport authorities utilize different procedures to evaluate the bids.

For example, in bidding the new Inchon International Airport, the Korean Airport Construction Authority (KOACA) took into account three factors: (1) prior construction record was given a weight of 20% in the prequalification round; (2) technological superiority of the proposed work was given 50%; and (3) price was given 30%. Bidders scoring 70 points or more in the prequalification round were considered in the final selection round, where the sole criterion was price. KOACA employs a fixed-price approach to contracting, whereby the successful bidder agrees the cost will not exceed its specific bid price.[70]

In Japan, the Fair Trade Commission warned 31 Kansai companies, several of which were involved in the construction of Osaka's Kansai International Airport, to stop the traditional practice of *dango,* or bid rigging. In 1994, the Japanese Construction Ministry issued an "action plan" requiring that public works projects exceeding a certain budgetary threshold be subject to competitive bidding under the guidelines established by the World Trade Organization.[71]

Blind grading should be incorporated into the contracting process. It is unlikely that an American firm would have received the contract to design Seoul's terminal, that an Italian firm would design Osaka's, or that a Japanese would design Kuala Lumpur's, without blind grading (i.e., judging proposals on merit alone, irrespective of nationality of the contestant). The superior designs at the lowest cost may be available abroad because of foreign experience, expertise, and economies of scale. Of course, a nation attempting to stimulate job creation can insist that a certain percentage of the labor force be local.

Airfield Construction and Technology

In order to build airfields, mountains have been flattened (e.g., Charleston, West Virginia, and Kuala Lumpur, Malaysia), lakes partially filled (e.g., Chicago Meigs), wetlands drained (e.g., Amsterdam

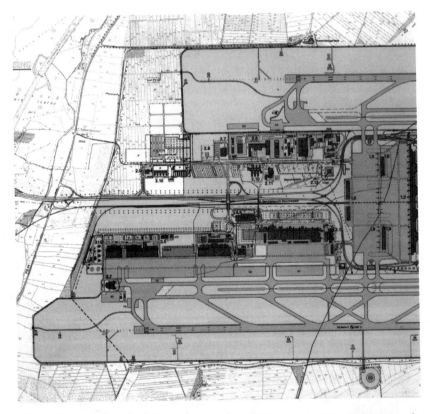

7.21 *The airfield at Munich's Franz Josef Strauss Airport consists of two staggered parallel runways surrounding a terminal complex.*
Flughafen München.

Schiphol, New York Kennedy, and the Second Bangkok International Airport), bays filled (e.g., Singapore Changi, Osaka Kansai, Seoul Inchon, and both Hong Kong's Kai Tak and Chek Lap Kok), and enormous quantities of earth moved (e.g., Denver International). Sizable airport drainage ponds also have been created to remove water from the runways (e.g., Chicago O'Hare).[72] Let us examine the engineering problems confronted in building several of the world's new airports.

Munich's Franz Josef Strauss Airport

Munich's new Franz Josef Strauss airport has two staggered east-west runways 4000 m in length (13,120 ft). (See Fig. 7.21.) Alfons Wittl of the Munich Airport Authority noted, "Two new runways are a minimum for any new airport. Munich 2 will be open at all times in winter. While

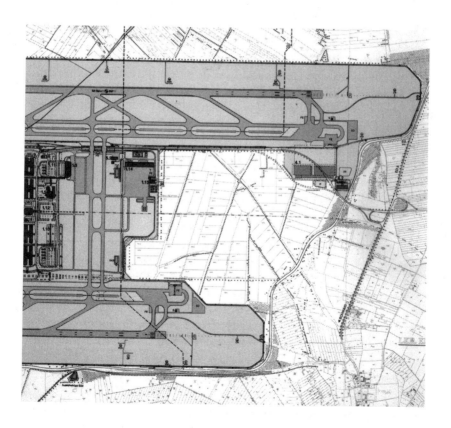

one runway is being cleared, the other can remain open." A third, short, general aviation and corporate jet runway was planned for the north end of the airport, as well as a separate general aviation terminal.[73] The spacing and length of the runways are designed to disperse the noise and eliminate the need for aircraft to reverse engine thrust on landing. Their location is designed to avoid having aircraft fly over nearby towns.[74] The maximum capacity of the two parallel Category IIIC runways is 275,000 aircraft movements a year (about 170,000 were anticipated for its first year of operation).[75] Each runway can accommodate a takeoff or landing every 2 minutes under instrument flight rule conditions.[76] Like the airport it is replacing, Strauss will have noise curfews, restricting takeoffs and landings to between 6:00 A.M. and 10:00 P.M., but allowing 28 additional operations per day for Stage 3 aircraft between 5:00 and 6:00 A.M. and 10:00 A.M. and 12:00 P.M. A year after it opened, Strauss handled 63 aircraft an hour during peak periods, up from 32 at Riem.[77]

At Strauss, the two 4000-ft staggered, parallel runways are equipped with Category IIIA and B technology for landings during inclement weather. High-speed taxiways allow aircraft to exit quickly after landing. The airport incorporates an automated computer-controlled system developed by Siemens for aircraft parking which eliminates the need for "follow me" vehicles or human aircraft directors, and thereby reduces delays caused by aircraft waiting for gates.[78] Apron controllers in the 197-ft tower illuminate green lights in the center of the taxiway to guide pilots to their assigned gates. Amber lights mark the point at which pilots will turn into their gates, and a sign confirms the assignment by relaying the flight number. Loops laid in the apron detect the nose wheel location.[79] Fifty-four parking stands are equipped with electronic docking systems.[80]

Safety is enhanced during winter with a computerized early-warning system with 39 data-gathering points identifying crucial ice and snow surface conditions. The deicing stations are positioned to deice a running aircraft in 4 to 5 minutes just before it enters the runway.[81] Half of the deicing solution used on aircraft is recaptured and recycled.[82] In fact, in order to assuage environmental interests, nearly 40% of all airport refuse is recycled.

Osaka's Kansai International Airport

Japan is essentially a chain of volcanic islands whose flat land has long been consumed by business, manufacturing, or agricultural uses.

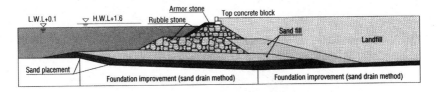

L.W.L+0.1 H.W.L+1.6 Armor stone Top concrete block Rubble stone Sand fill Landfill Sand placement Foundation improvement (sand drain method) Foundation improvement (sand drain method)

7.22 *At Kansai International Airport, a man-made island was created with landfill surrounded by a stone barrier in Osaka Bay. Water was removed from the clay bed by the sand drain method.* Kansai International Airport.

When Japan's Ministry of Transport began surveying land for a new airport at Osaka in 1968, six of the eight potential sites involved sea reclamation, while a seventh involved leveling an island. The final site was selected in 1984; construction began 3 years later.[83]

Built on an artificial island in Osaka Bay, the new Kansai International Airport (KIA) was the world's most expensive airport completed up to that time (it has since lost that distinction to Hong Kong International Airport). The water in which the island was constructed has a depth of 18 m, with a soft clay layer underlying the sea bed. Creating the island involved a landfill of 511 ha, requiring the movement of 178 million m³ of soil and rock, the equivalent of 70 Great Pyramids at Giza, Egypt.[84] To strengthen the sea bed foundation, sand piles were driven into the alluvial clay layer so as to remove water from it.[85] One million pilings were sunk to firm up the undersea clay floor upon which the airport was built.[86] (See Fig. 7.22.)

Kansai took 7½ years to build. KIA was originally scheduled to open in the spring of 1993, but during construction, it was plagued by terrorist attacks and a costly 18-month delay caused by technical problems.[87] The airport's construction was also impeded by the fact that parts of the man-made island began sinking because of the soft seabed. Sand was driven into the seabed to strengthen it prior to the time the landfill was added in to slow the process.[88] The unanticipated work delayed the opening by 18 months and raised the cost of the airport by 40% over initial projections. As a consequence, Kansai levies some of the highest airport charges in the world.[89]

At the airport's opening, the reclaimed land upon which the airport sits was sinking at the rate of 1 in every 25 days.[90] Initially 49 ft above sea level, by 1998 it was only 16 ft above sea level, and sinking at the rate of more than 1 ft per year. If subsidence continued at

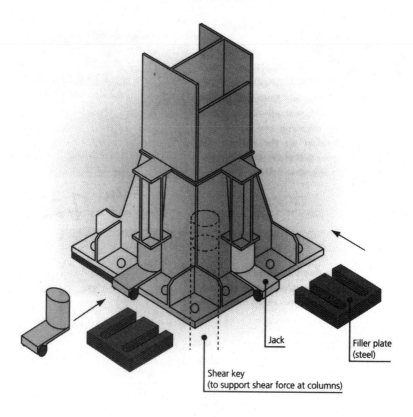

Jack

Filler plate
(steel)

Shear key
(to support shear force at columns)

The height of the column pedestals can be
adjusted using the compact jack with rollers.

7.23 *Each pillar at the passenger terminal at Kansai International Airport has a hydraulic jack which can be adjusted to compensate for soil subsidence.* Kansai International Airport.

that rate, the airport would be submerged in less than 15 years.[91] But KIA officials assured your author the rate of subsidence will slow and even stop before, like Atlantis, it sinks into the sea.

To compensate the weight of the terminal building, nearly twice its weight in earth was removed from the island. A 2.5-m-thick layer of approximately 360,000 tons of iron ore was added as a foundation to keep the ground properly balanced.[92] As a further precaution to keep the terminal from cracking because of uneven settlement, subsidence monitors were located throughout the building.[93] To compensate for uneven settlement, each of the terminal's 900 pillars can be individually adjusted by powerful hydraulic jacks (each with a maximum thrust of 300 tons), which can be individually lowered or

7.24 *To build the island on which Kansai International Airport sits required 178 million m² of earth and sand.* Photo by Yasuhiro Takagawa. Renzo Piano Building Workshop.

raised so that steel filler plates can be inserted to lock the position.[94] (See Fig. 7.23.) Walls in the buildings are hung from the ceiling, while doors are fixed to the floor, so that they can be adjusted as the ground subsides unevenly. Cable and wire were laid with slack to accommodate settling. Air conditioning ducts also include flexible joints for this purpose.[95] The air traffic control tower is a well-braced facility, with counterweights to compensate for earthquakes, which frequently occur in the Japanese archipelago.

The man-made island on which the airport sits is shaped like an aircraft carrier.[96] (See Fig. 7.24.) The island houses the airport terminal, a runway, a ferry port, a train station, a police station, two fire stations, a hotel, shops, offices, and restaurants. It is linked to the mainland by a 3.75-km double-deck truss bridge, with a highway on the upper deck and rail lines on the lower deck.

On the airfield, Kansai International Airport installed some 500 sensors, and another 200 along roads and rail links, able to detect hazardous objects. This allows management to field a security staff of only 600 officers, compared to Narita's 3600.[97]

Plans are to expand the island to add two additional runways.[98] Phase 2 involves adding a 545-ha island on which a 4000-m-long runway and a satellite passenger terminal will be built. This will place the passenger terminals between the parallel runways. This is a vast improvement over the airport's original configuration, which would have placed the parallel runways side by side. Spacing them allows enhanced runway utilization, and placing the terminal in the center of the airport enhances aircraft ground movement efficiency. Though construction of Phase 2 is scheduled to begin in 1999, the reclamation project must await completion of an environmental impact statement.[99] Phase 3 would add a 3500-km crosswind runway, set at an angle from the other two.

Macau International Airport

The master plan for the new airport at Macau was developed by Flughafen Frankfurt Main AG, Airconsult, in conjunction with Aeroports de Paris.[100] An artificial island was built adjacent to the island of Taipa (one of two islands adjacent to the peninsula of Macau),[101] by hydraulic sandfilling,[102] to expand the tiny Portuguese colony's usable real estate by more than 20%.[103] Four hundred hectares reclaimed between the islands of Taipa and Coloane became the venue for the airport and airport-related businesses.[104] Land reclamation includes the runway and much of the area for future ramp

7.25 *Though the new Macau International Airport required sea reclamation, costs were held down by using workers from mainland China.* Civil Aviation Authority of Macau.

expansion.[105] (See Fig. 7.25.) The runway has been lengthened and the ramp area increased since the initial plan, to accommodate anticipated cargo business.[106] By virtue of the fact that the airport is in the bay, noise impacts on residents are anticipated to be modest, allowing 24-hour departures.[107]

Macau International Airport (MIA) was built by the Sociedade do Aeroporto Internacional de Macau, otherwise known as the Macau Airport Company (CAM), a German/Portuguese consortium of Siemens and Soares de Costa,[108] which holds a 25-year franchise to develop and operate the airport.[109] Construction began in January 1992, but a number of design changes delayed offshore financing.[110] The truly amazing thing about MIA is its modest cost, in part attributable to using low-cost labor from mainland China.

Kuala Lumpur International Airport

Thirty-thousand workers from 50 different countries cleared 10,000 ha of bush and rubber trees near Sepang, 50 km south of Kuala Lumpur, to prepare a site for the new Kuala Lumpur International Airport.[111] (See Fig. 7.26.) Unsuitable soils led to the decision to open

7.26 *Construction of the New Kuala Lumpur International Airport involved 30,000 workers from 50 different countries. Here is the view of the airfield from the main terminal, with the remote terminal on the horizon. They are linked by a tram, in the foreground, which is aboveground at both the main and remote terminal, but goes underground beneath the taxiway.* Photo by P. S. Dempsey.

the airport with only one runway. Forty percent of the second run-way lies on peat swamp and marine clay. The design and construction of the airport's main terminal, contract pier, and baggage handling system were performed by a joint venture of four Japanese companies, Perspec, at a cost of $670 million.[112]

Hong Kong International Airport at Chek Lap Kok

The new Hong Kong International Airport at Chek Lap Kok is the twentieth century's largest civil aviation project.[113] When completed, the 4.82 mi^2 island site was 4 times the size of Kai Tak Airport, twice the size of Osaka's new Kansai Airport, and nearly the size of the Kowloon Peninsula or London's Heathrow Airport.[114] The island is large enough for two 3800-m-long (12,464-ft) runways with a 1525-m separation.[115] The runways are capable of handling the next generation of 84 × 85-m mega aircraft, which will carry up to 900 passengers. The project also includes the world's largest railway suspension bridge.[116] Of the airport's 1248 ha, 938 were reclaimed from the sea and 310 from leveling Chek Lap Kok and Lam Chau Islands.[117] Site preparation cost US$1.26 billion and involved the largest fleet of dredgers the world has ever seen.[118] An avalanche of mud, rock, and sand was moved at the

7.27 *Hong Kong International Airport at Chek Lap Kok required leveling two islands and the largest fleet of dredgers ever assembled to fill the sea in order to build an airfield sufficient for two parallel runways and terminal and cargo facilities.* Airport Authority Hong Kong.

rate of 10 tons a second for 30 months. For a while, the airport site was the world's fourth largest open pit mine.[119]

The two islands account for a quarter of the airport's surface area, with the rest reclaimed from the sea between them. Unfortunately, soft marine mud, in some places 27 m thick, lay above the bedrock. The consortium that won the contract—Japan's Nishimatsu, Britain's Costain, America's Morrison Kundsen, Holland's Ballast Nedam, Belgium's Jan De Nul, and China's Harbour Engineering—opted to engineer the problem out, rather than engineer around it. This meant assembling the world's largest fleet of dredgers to suck the mud and clay out and dump it in deeper water. A layer of sand was laid on top. Then granite was blasted from the islands and laid on top of that, most forming the foundations for the runways and taxiways. At completion, the island was to be 6 to 7 m above sea level, and 347 million m³ of material was moved—about 400,000 m³ of rock, soil, mud, and marine sand per day, or 10 tons per second.[120] Twelve kilometers of sea walls also were built to protect the island against typhoons.[121] (See Fig. 7.27.)

Seoul's Inchon International Airport

Sea reclamation for Seoul's new Inchon International Airport was done by Hyundai Construction Co. It involved filling in the sea between

7.28 *Construction of the airfield at Inchon International Airport required filling in the sea between Yongjong and Yongyu Islands, near Seoul.* Korea Airports Authority.

Yongjong and Yongyu Islands, which was at a depth of about 16 ft.[122] (See Fig. 7.28.) To ferry construction workers to Yongjong Island from nearby Yuldo Islet, small passenger boats were rented.[123] When Inchon opens in 2004, its two runways will be able to handle one flight every 30 seconds, 24 hours a day, or 27 million passengers a year. A third runway is scheduled for 2005–2006, and a fourth in 2020. Inchon also plans to build an entire community surrounding the airport, with apartments, schools, and office buildings.[124]

Runways at the new Seoul International Airport at Inchon will open with a Category IIIA instrument landing system (ILS), with a Category IIIB system to be in place in the second phase of the airport's development.[125] Inchon International Airport will include an Integrated Communications Center which will integrate voice, data, and video information in a central computerized distribution system, and oversee takeoffs and landings, taxiing and gate pushbacks, baggage distribution, and security.[126]

Second Bangkok International Airport

The Second Bangkok International Airport is being built 30 km southeast of Bangkok in a floodplain. Historically, the area had been swamp land drained for use in fish farming and growing hydroponic plants. Dutch consultants were commissioned to design dikes and levies to remove the water, something the Dutch have been doing

7.29 *Construction of the airfield at the Second Bangkok International Airport required draining the "Swamp of the Cobras" and building dikes to keep the water out.* Photo by P. S. Dempsey.

for hundreds of years, to make the site suitable for an airfield. (See Fig. 7.29.) At full build-out, sometime after 2020, the new airport will be capable of handling 100 million passengers annually. (See Fig. 7.30 on pp. 316 and 317.)

Summary and Conclusions

The design of an airport should begin with the design of the airfield. After the airfield is laid out with a view to the safety and efficiency of aircraft operations, the ground facilities should be sited, preferably between the runways, though with sufficient land to allow for future expansion.

Safety and efficiency should be the guiding objectives of airfield design. Airfields should have runways appropriately spaced and staggered, pointed into the wind, and of sufficient length to accommodate aircraft anticipated to land there. Taxiways should allow expeditious movements to the terminal which, at larger airports, should be located in the center of the field, between parallel runways. Air and ground navigation technologies should allow safe, efficient, and expeditious movements from the airways to the runways, and from the runways to the terminal. Sufficient land should be reserved for future infrastructure which will be needed to accommodate demand increases over time.

Construction remains a critical part of the process of airport development. According to airport architect Curt Fentress, "The most difficult and burning issue is schedule and budget."[127] Getting it built on time and on budget remain two of the most difficult and important tasks to be accomplished.

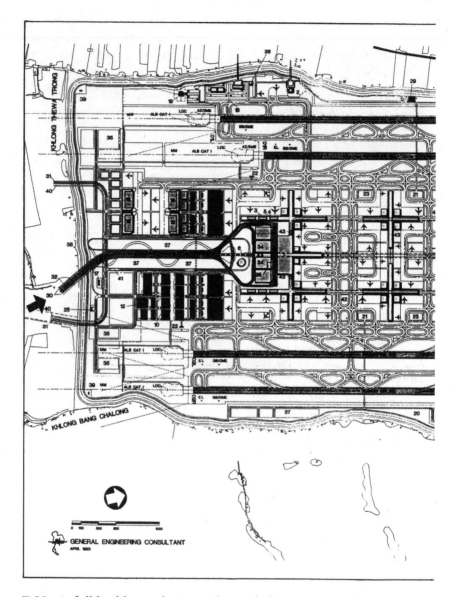

7.30 *At full build-out, the Second Bangkok International Airport will have four parallel runways surrounding several terminals, capable ultimately of handling 100 million passengers per year.*
Second Bangkok International Airport.

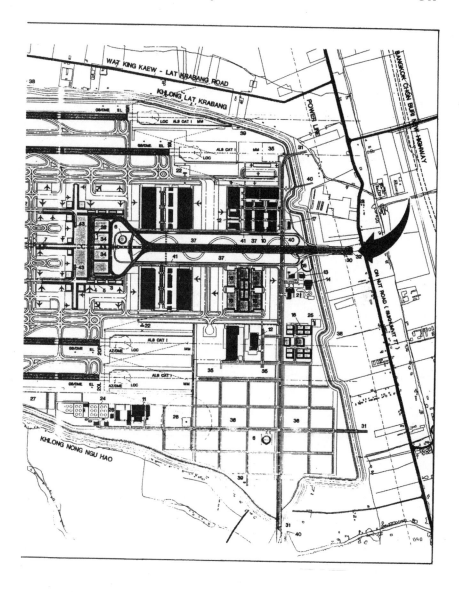

8

The Land Side

Terminal Design

"If ever a place of work justified the pride of those who built it, with their brainpower and their hands, it was this, this meeting place of earth and sky."[1]—GENERAL CHARLES DE GAULLE, FRENCH HEAD OF STATE, AT THE OPENING OF THE SOUTH TERMINAL AT PARIS ORLY, FEBRUARY 24, 1961

"Airports present gateways of opportunity to migrant populations, escape to holiday-minded couples, familiarity to frequent-flying executives. For a world in motion they are home, which reminds us that despite being inherently alienating (often their features resemble each others' more closely than those of the cities beyond their sliding doors), they are equally sources of connection. Their scattered confines link together a global village."[2]—NASID HAJARI, JOURNALIST

In Chap. 8, we examine:

- Fundamental principles of terminal design
- The different types of passenger terminals and their utility
- The functional elements which must be accommodated in terminal design, including ticketing, baggage handling, customs, immigration, security, and boarding
- The need for people-mover systems to move passengers over long distances
- The need for clarity of signage and graphics

- The design of terminals to accommodate shopping, hotels, and commercial activity
- The need for design that accommodates the needs of the disabled
- The desirability for an airport to be not only functionally efficient, but also aesthetically pleasing to those who use it
- The desirability of adopting proven technologies to run complex essential systems
- The need to design essential systems with adequate back-up in case they break down
- How airports attempt to rectify design flaws

Introduction

As an integral part of the transportation infrastructure, airports are all about getting from here to there, a brief pause in a longer journey. Hopefully, that can be done as effortlessly and safely as possible. Fundamentally, an airport should be designed to move passengers, cargo, mail, aircraft, and surface transportation vehicles efficiently, expeditiously, and pleasantly—at least cost and with less hassle. Intramodal and intermodal connections should be as seamless as possible. The essential functions which must be performed at an airport include passenger ticketing, baggage check and pickup, customs, immigration, security, boarding and deboarding, and in the case of cargo and mail, tendering or picking up with appropriate air cargo waybills and documentation. The airport should be well designed and well signed to enhance rather than hinder efficient traffic flows.

In a sense, airports are little different in function from the railway terminals they replaced. At a rail station, a departing passenger needs to purchase a ticket and board a train from the appropriate platform. At an airport, a departing passenger needs to be ticketed,

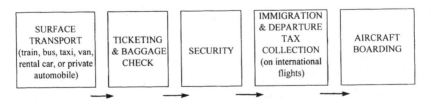

8.1 *Departure functions.*

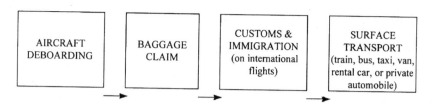

8.2 *Arrival functions.*

receive a boarding pass, check luggage, and board an aircraft at the appropriate gate. On international flights, the passenger will also pass through immigration and customs, at departure and/or on arrival. At arrival, a passenger needs to collect bags and exit the terminal. At properly designed intermodal facilities, a passenger has the opportunity to both board and exit a train and an aircraft. Simply put, the departure functions are shown in Fig. 8.1 and the arrival functions are shown in Fig. 8.2.

Along the way, various shopping opportunities may be present, so that a departing passenger may purchase a gift or a magazine, get a shoe shine or haircut, or be entertained, for example. An airport can be designed to tend to the passengers' other needs as well—eating, drinking, services, or pharmaceutical products, or just going to the toilet. Some airports are essentially shopping centers surrounded by runways. Others have integrated hotels into the property and a few into the terminal building. The airport terminal might also be designed as a pleasing or dramatic architectural statement. Some airports are designed to provide a glimpse of the local or regional culture. British architectural critic Deyan Sudjic contends that airports are the contemporary substitute for the public square of old, a place where crowds congregate and strangers cross paths.[3] Thus, airports serve a myriad of purposes. The best do so harmoniously.

Early architectural designs of airports owe their origins to the mode of passenger transportation which aviation largely replaced—the railroads. Early airport terminals were fashioned after the railroad Union Stations of the day, aircraft hangars after the rail yard shed, and aircraft interiors after the Pullman railroad cars.[4] The word *terminal* itself was borrowed from the railroad industry. The word *hangar,* borrowed from the French, means a shed for hay, open on one side and built into the side of a farmhouse.[5]

Principles of Terminal Design

The essential purpose of an airport terminal building is to interchange passengers, their luggage, and freight and mail between surface transportation (i.e., automobile, taxi, bus, truck, or rail) and air transportation modes in a comfortable, convenient, expeditious, and economical manner.[6] As described in the immediately preceding chapter, the location of a terminal building and air cargo and postal facilities must be coordinated with the design and location of the runways, taxiways, and aprons.[7] From a time and fuel-consumption perspective, aircraft operations are more efficient if the terminal is located as close to the runways as possible. However, care must be taken to ensure that neither expandability, flexibility, nor safety are sacrificed at the altar of short-term aircraft efficiency.[8] Passenger buildings should also be integrated with parking areas and roads so as to minimize walking distances between curbside and the ticketing/check-in/baggage areas.[9] Total walking distances and travel and processing time for passengers from curbside to aircraft boarding should also be reasonable, with people-moving systems (e.g., moving walkways or automated trams) available to traverse longer distances.[10] Time is also an important consideration in designing ticketing and baggage-claim facilities, for surveys of passengers reveal that time spent there is the most important characteristic in airport preference.[11]

Designing a terminal building, or buildings, requires consideration of a number of variables, including the (present and future) capacity of the facility, growth potential, types of aircraft, intermodal transport access (including automobile curbside needs), the layout geometrics of the airfield, commuter facilities, general aviation facilities, customs, and other buildings on the site, including maintenance hangars, cargo facilities, kitchens, parking garages, and car rental facilities.[12] Efficiency of aircraft, automobile, transit, rail, and passenger flows is also an important goal. Safety and security are overriding considerations.

The overriding policy of terminal design should be simplicity. The terminal should be designed so that the passenger encounters simple, obvious, unobstructed flow routes. This is the theme that was embraced, for example, at the $2.4 billion expansion at San Francisco International Airport (SFO). The architectural mission was all about simplicity. "Make it as hassle-free as possible," according to Jason Yuen, former director of planning at SFO. Architect Craig Hartman designed the great hall at the international terminal to be one in which passenger flow is to be as "intuitively clear" and direct as possible, rather than relying on signs.[13]

To aid in achieving simplicity, functions should be separated. If facilities such as car parks, hotels, and control towers are integrated into the terminal design, the flow plan tends to be distorted and expansion flexibility is hampered. Units or modules within the terminal building should be arranged in the simplest manner possible so that passengers can easily comprehend what is located where, and to make it possible to expand the areas as future capacity needs arise.[14]

Where there is abundant land, terminals tend to be designed to be long, horizontal facilities, though departures and arrivals may be handled on different levels. In high-density areas, where land is scarce and expensive, terminals tend to be vertical, or stacked, with various functions handled on multiple levels.

As noted in Chap. 5, in determining what should be included in an airport, it is useful to begin with an assessment of its size and capacity, its integration with the airfield and ground transportation, the type of airport it is predominantly likely to be [e.g., origin-and-destination (O&D) and/or connecting], what is likely to be the mix of its traffic base (e.g., business travelers, leisure travelers, or cargo; international or domestic; departure, arrival, transit, transfer, and/or general aviation), what types of carriers will serve it (e.g., domestic or international, O&D or network, scheduled and/or charter), and what type of aircraft will arrive and depart (e.g., wide-body, narrow-body, commuter, air taxi, and/or general aviation), for each has different needs.[15] Some terminals designed for O&D efficiency, such as Kansas City International Airport, have been unable to attract sustained hub traffic, though several airlines (i.e., Braniff, Eastern, and TWA) have tried, albeit unsuccessfully. Others designed with separate airline terminals, such as Los Angeles International or New York Kennedy, make interline transfers awkward. (See Fig. 8.3.)

In determining the needs the terminal should serve, planners should consult with the major tenants—the airlines, concessionaires, general aviation users, as well as the relevant governmental agencies, including air traffic control, customs, immigration, and agricultural inspection, for all will play a role in airport operations.

8.3 *Los Angeles International Airport was designed with a series of independent terminals surrounding a central core with parking garages, a control tower, and other futuristic buildings.* Photo by P. S. Dempsey.

An airport terminal must be designed for numerous, sometimes conflicting, purposes and objectives. As Koos Bosma observed:

> *Linear continuity, separation of passenger flow, prevention of intersecting paths, optimum signage, and separate stationary areas (lounges, restaurants, waiting areas, and shops) make considerable demands on the architectonic setup of airport terminals. The logistics of an airport, of course, are geared to getting passengers through the terminal as fast as possible; nonetheless, the commercial factor of their "dwell time" must also be shaped.*[16]

> *The question that…planners and architects have been addressing…is whether essentially static architecture can shape such unstable surroundings. After all, the tension between the continuing changes in the transport process and the architectural monumentality of the transfer centers is considerable. Nevertheless, there are a number of fixed programmatical demands for an airport: safety and convenience, walking distances, parking and access, to name a few. These demands still allow for substantial design leeway, although this too is restricted by current views on*

other criteria—like flexibility and scope for expansion, centralization and decentralization—which are inherent in airport design and determine what type of airport a country or municipality decides to build.[17]

Principles of flow route design for airport terminals include attempting to achieve, to the extent possible, the following:

1 Routes should be short, direct, and self-evident; they should not conflict with or cross the flow routes of other traffic.

2 Changes in the levels of pedestrian routes should be avoided wherever possible.

3 Passengers should be able to proceed through the terminal without staff instruction; signage should be frequent, clear, and understandable throughout the passenger flow path.

4 Departing passengers should be able to check their baggage at the earliest possible point.

5 Each flow should be unidirectional; milling spaces (areas where passengers engage in random movements) should be adjacent to, and not a part of, the flow routes.

6 Free flow between surface and air transport should be interrupted as little as possible; government authorities and airlines should interrupt flow only as necessary, with a view to maximizing passenger convenience and security and efficiency and economy of operations.

7 Passengers should not be forced to pass through the same type of control more than once; agencies with responsibility for customs, immigration, and agricultural control should attempt to integrate their passenger review.

8 The last control function through which the passenger should pass should be security; a sterile buffer zone should exist between the security control point and the aircraft.

9 Visual continuity should exist between one functional area and the next, so as to draw the passenger naturally through the flow system; visual blockages are confusing, and disrupt efficient passenger flow; for example, the airline ticketing and baggage-check counter should be visible immediately upon entering the terminal (intuitive wayfinding).

10 Features which cause hesitancy, such as ambiguous or nonexistent signage, flow routes which appear wrong, and multidirectional junctions should be avoided.

11 The speed of the passenger flow should be timed to match that of the other systems, such as baggage flow and aircraft turnaround time; the capacity of the flow should match the overall capacity of the airport.

12 To maximize efficiency in the use of airport facilities, joint airline use should be encouraged, particularly in such areas as ticketing and gate assignment.[18]

Again, passenger flow patterns should be smooth, clear, and obvious. Automobile curbside and rail facilities should be as close to baggage check and ticketing counters as possible. At Osaka's Kansai International Airport and Hong Kong International Airport, for example, both the automobile curbs and rail stations are immediately in front of the terminal building. Once inside the terminal building, the passenger should immediately encounter well-signed ticket counters. Ronald Reagan Washington National Airport fails in this regard, for the immediate view of departing passengers is through glass onto the airfield. Though the view is visually stimulating, passengers must search left and right for ticket counters, and search again for awkwardly placed arrival and departure gate screens.

Once having been ticketed and having checked baggage, departing passengers should then enjoy direct and unobstructed flows through security (and, at international airports, immigration facilities) to departure gates with ample space in which to sit while awaiting departure. Where distances are significant, passenger movements should be expedited with movable sidewalks and/or trains. Along the way, in areas which will not obstruct flow, shopping and restaurants, as well as airline lounges, should be available for passengers who need them.

The arriving passenger should encounter this same level of unobstructed flow in reverse. Passengers departing from aircraft should find signs clearly directing them to transfer aircraft, baggage claim, customs and immigration, and surface transportation alternatives (taxis, rail lines, parking garages, and automobile rentals). Baggage claim areas should be well marked as to which arriving aircraft's bags are at which carousel, and the carousels and the space around them should be of sufficient size to accommodate waiting passengers. Domestic and international passengers must be segregated until the latter pass through customs and immigration. After that, they can enter an area where "meeters and greeters" are awaiting their arrival (usually the other side of a clear glass wall, where they can be seen), and proceed out of the terminal to surface transport. Again, ideally such surface transportation departure areas will be located immediately outside the terminal. At larger airports, arrival and departure flows are segregated on different levels of the airport.

Finally, research reveals the following criteria have been identified as the most important in assuring the comfort and convenience of the passenger in the transfer between air side and land side:

- Availability of seats (a sufficient number of seats in waiting areas to handle passenger volume)
- Walking distances between gates (for connecting passengers) and between gates and ticketing, baggage claim, and surface transportation
- Accessibility to concessions and services
- Orientation (availability of information)
- Waiting time
- Occupancy (density)[19]

These, then, are the qualitative indices on which terminal designers should focus.

Types of Terminals

In Chap. 7, we identified the different types of terminal designs that have emerged and their integration with the airfield. Here, we repeat the categories, but focus on their land-side characteristics.

1 *Simple concept.* The passenger terminal consists of a single-level common waiting and ticketing area with several exits onto a single apron area.

2 *Linear concept.* This is an extension of the simple concept, with the building repeated in a linear extension to provide more apron frontage, gates, and space. When the terminal is extended so far as to make walking distances long, people-mover systems can be added.

3 *Pier (finger) concept.* Typically, a main central building is used for ticketing and baggage, while boarding is taken care of at the gates along both sides of an extended corridor (the pier, or finger). Examples include Minneapolis/St. Paul International Airport and Cleveland Hopkins International Airport.

4 *Satellite concept.*   Midfield satellite terminals with boarding gates at which aircraft are parked are connected to the main terminal with a people-mover system. Examples include Sea-Tac International Airport and Atlanta Hartsfield International Airport.

5 *Transporter concept.*   Aircraft are parked at a remote apron, to and from which passengers are transported by surface vehicle (typically a bus). Examples include Washington Dulles International Airport.

6 *Unit terminal concept.*   Adopted at larger airports (e.g., New York Kennedy and Los Angeles International), individual self-sustained module units are built around a loop of connecting roads. Each unit provides complete passenger handling and aircraft parking positions.[20]

One other way to categorize airports is to divide them into those with centralized terminals versus those with decentralized terminals. Centralized airports consolidate ticketing and baggage claim in one terminal, while decentralized airports disaggregate these functions among multiple processing facilities. Orlando and Seattle/Tacoma airports were the first in the United States to embrace a central terminal complex design, with remote satellite boarding concourses. New York Kennedy and Los Angeles International Airports are examples of decentralized airports, with multiple terminals, each with their own ticketing and baggage claim facilities.[21]

Location of Terminals

Airport passenger terminals should be located within reasonably close proximity of existing and future surface transportation links, on the land side, and runways and taxiways, on the air side.[22] This will reduce passenger walking distances to ticketing and aircraft taxiing distances on the airfield. To enhance efficiency of aircraft movements, terminals should be surrounded by runways.

Ticketing and Check-in

Ticketing and baggage-check positions should be immediately obvious to passengers as they enter the terminal. Their layout is influenced by the principles discussed above—simplicity, unobstructed and clear passenger flow, and minimum distance between terminal entry and aircraft boarding.[23] Prominent signage is extremely important to lead customers directly to their airlines.

The number of counter positions required for ticketing should be determined by the airport in consultation with the airlines that will use them. In the United States, airlines usually embrace an exclusive use approach both to ticket counters and gates. Elsewhere, airports insist on a multiple-use approach, whereby any particular ticket counter or gate will be occupied by Carrier A at 9:00, and Carrier B at 10:30, and so on throughout the day as each carrier's peak needs arise. For example, Munich's new Franz Josef Strauss Airport embraces the multiple-use approach. The multiple-use approach results in a far more efficient use of airport facilities. Generic common use terminal equipment (CUTE) allows individual airlines to access their host computers and thereby share counters.[24]

Several different concepts have been embraced. Check-in may be via the centralized check-in concept (whereby passengers and baggage are processed in a central area, usually the departure area of the terminal), split check-in concept (e.g., ticketing and boarding pass issuance is handled at one location and baggage check at another), and gate check-in concept (whereby ticketing, boarding passes, and baggage check are all handled at the gate). Check-in counters may be linear, flow-through, or islands.[25]

Baggage Handling

Passengers should be free to check their baggage at the earliest point upon arrival at the terminal. Curbside checking is convenient, but security is enhanced with checking at the ticket counter, where a boarding pass is issued, the ticket collated with passenger identification, and the passenger is profiled and asked several security-related questions. Baggage processing should be timed to correspond with passenger flows in speed and capacity. Checking of baggage for security purposes should be performed prior to boarding. To avoid a tragedy like the explosion in the cargo hold of Pan Am flight 103 over Lockerbie, Scotland, each checked bag should be collated with a boarding passenger. As explained above, signage (with up-to-date information correlating flight arrivals with baggage positions on television monitors) is also important to lead the passenger to the proper baggage carousel or other luggage area. Milling can be reduced by placing bags on rotating carousels or belts. The baggage

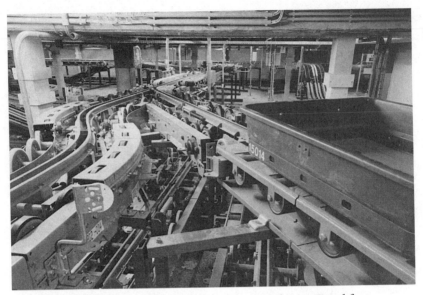

8.4 *The baggage conveyor system at Terminal 2 at Frankfurt International Airport is among the world's most sophisticated and reliable.* Flughafen Frankfurt.

area should be conveniently located near the departure doors of the terminal building, where passengers can easily access public transport, car rentals, and taxis. At larger airports, departure should normally be handled on a different floor from arrivals.[26]

Frankfurt Airport operates the world's first and largest automated baggage system, spanning 56 km of guideways. (See Fig. 8.4.) It is connected to the two terminals with a high-speed transfer tunnel.[27] Baggage carts can be used from the parking levels to check-in.[28] Munich's airport planners learned from the experience at Frankfurt, which in the early 1970s was the first airport in the world to install the individual cart-and-track baggage system. Early on, Frankfurt had serious problems with the baggage system. Munich chose a baggage system with more conventional and reliable "tried-and-true" technology, and made it a priority that the baggage system would be working on opening day. The Dutch firm Vanderlande Industries installed the same system it had installed at Amsterdam's Schiphol Airport, a system of conveyor belts with saucer-shaped wood containers attached to the belts holding bags. Munich insisted the system be completed a full year before opening, so that it would have a full year of testing to work out the bugs. As a consequence,

Munich's baggage system worked perfectly from day one. Unlike Denver International Airport (which gave the maintenance contract to a firm other than the company that built it so as to further expand the minority contracts), Munich insisted that Vanderlande maintain the system for its first 5 years of operation.[29]

According to Munich airport operations director, Peter Trautmann, "We only need three things to work perfectly. The first was the baggage system, the second was signage and monitors, and the third was the telephone system."[30] They therefore insisted that all critical systems be tested for months prior to opening. As noted, the Dutch-built conveyor belt baggage system was tested a full year before the new airport opened. Telephone books were put in old luggage purchased from thrift stores to provide weight.[31]

On the other hand, Denver International Airport (DIA) had grave difficulty getting its 26-mi, $218 million automated baggage system to work properly, and after several opening dates were missed, opted for a primitive tug-and-cart system as the primary system for two of its concourses, and a back-up system for its third.[32] Three years after DIA opened, its automated baggage system served only Concourse B's outbound operations,[33] though United Airlines planned to spend another $22 million to upgrade it (by installing new software and making mechanical changes) to handle inbound and transfer baggage.[34] Kuala Lumpur abandoned an electronic baggage system after seeing problems with a similar one at Denver International Airport.[35] Yet the Second Bangkok International Airport is slated to install a destination-coded vehicle system like Denver's.[36]

Kansai also opted against installing the computerized baggage system that plagued DIA. Osaka's Kansai International Airport installed a three-dimensional tilt tray sorter system. Once flight data are inputted onto bar-coded tags, the bags go down a spiral conveyor set at a 14° angle to reach the lower level where the tilt-tray sorter distributes bags to aircraft. Each of the two lines has the ability to process 3500 pieces per hour. Tags that are unsuccessfully read by scanners are forwarded to a makeup conveyor where they are manually processed. The total length of Kansai's conveyor line is about 5000 m.[37] (See Fig. 8.5.) The baggage system installed at Kansai was tested repeatedly, and was working smoothly at opening.[38]

Bar coding was developed by the airline industry in the mid-1980s to correlate the passenger's bag with the proper flight. It uses a 10-digit

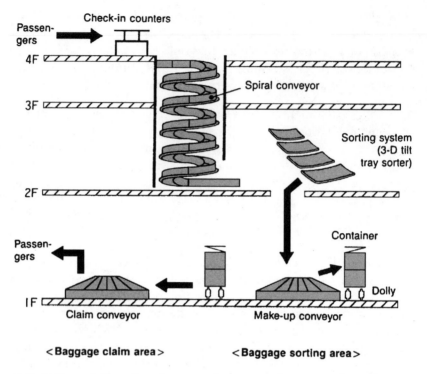

8.5 *At Kansai, bags flow downward on a spiral conveyor from the fourth floor check-in area to a 3-D tilt-tray sorting system.* Kansai International Airport.

key number, known as a baggage source message (BSM) on a tag known as a "license plate." Scanners attempt to read the tag as it passes by so as to route it to the appropriate departing flight or arriving baggage claim carousel. For direct flights, accuracy is around 95%; but for connections, accuracy is only between 60 and 75%. Bags which are not successfully read by scanners are routed to an area where they can be read manually by airline employees.[39]

At Hong Kong International Airport, the baggage system can handle 19,200 pieces of luggage per hour. Conveyor belts bring bags to large rotating carousels, stopped by a sensor whenever another bag is on the carousel so as to avoid stacking and jamming of luggage. At larger airports, baggage retrieval areas typically are movable belts (such as at Chicago Midway) or rotating carousels (such as at Atlanta Hartsfield). (See Figs. 8.6 and 8.7.)

Passenger movements are expedited when the airport has an abundant supply of free carts available in the baggage-claim area, as exists, for example, at the new Hong Kong International Airport. Some

8.6 *Belts, like these at Chicago Midway Airport, move bags past waiting passengers.* Photo by P. S. Dempsey.

8.7 *Using gravity to hold the bags steady, tilted carousels, such as these at Atlanta Hartsfield International Airport, rotate luggage past large crowds of deplaning passengers.* Photo by P. S. Dempsey.

airports, like Denver International, charge outrageous fees ($1.50 at DIA) to rent a cart. Passenger convenience is enhanced when the carts are designed like mountain goats to move up and down escalators, as they are at Frankfurt and at Kansai International Airport.

People-Mover Systems

Major airports tend to be vast in size, with significant distances between various functions. Where possible, airports are designed to minimize distances between passenger arrival and aircraft boarding areas. But this is not always possible. The volume of activity at an airport may require scores of ticket counters, concourses, gates, and baggage carousels, each consuming large areas of space. The need to have efficient airfield operations by placing gates near runways may require remote satellite terminals. Gates may be insufficient in number to handle the number of arriving and departing aircraft, requiring remote aircraft parking stands on aprons. The need always is to move passengers quickly and comfortably within the airport.

Within a terminal or concourse, pedestrians can walk. Walking can be expedited with the installation of movable sidewalks, preferably enclosed, shielded from inclement weather. Movable sidewalks in subterranean locations can be made more aesthetically attractive with the judicious use of lights, as at Munich and Chicago O'Hare. One severe handicap of a curved terminal building, as at Dallas/Fort Worth, is that it is incompatible with movable sidewalks, which require a straight-line dimension. Thus, the vast distances between gates at Dallas/Fort Worth International Airport cannot be reduced via movable sidewalks.

Trams can be installed to move passengers longer distances along concourses. For example, at Hong Kong International Airport, an underground driverless train system provides guide movement of passengers along the 750-m central concourse, supplemented with 2.5 km of movable sidewalks. At Kansai International Airport, an aboveground tram moves passengers along the mile-long concourses extending from the sides of the terminal. Seattle/Tacoma has a satellite transit system which operates on three loops—two which link the main terminal to a long concourse and a satellite terminal, and one which provides a linear shuttle within the main terminal.[40]

Between floors of a building, passengers can be moved by stairs, escalators, or elevators. Elevators must be of sufficient size to handle

8.8 *At airports with an inadequate number of gates, passengers are often shuttled between the terminal and their aircraft via buses, which drive on the tarmac. Typically, they are wide-bodied buses with few seats, like these at Hong Kong International Airport.* Photo by P. S. Dempsey.

several passengers with baggage carts, an oversight at Munich's Franz Josef Strauss Airport, which has aesthetically pleasing glass-walled, but small, elevators throughout the terminal.

Passengers must sometimes be moved to and from aircraft parked on the tarmac. At smaller airports, passengers simply walk between the aircraft and the terminal. At larger airports, this may pose both safety and security concerns. Typically, they are transported in wide buses, such as at Hong Kong or Zurich. (See Fig. 8.8.) Some airports have buses to shuttle passengers between concourses of a terminal, such as at Detroit. Others use buses to shuttle passengers between their various terminals, such as Boston. In such instances, the sidewalk areas, or islands on which pedestrians stand, should be sufficiently wide to handle the volume of passengers and luggage which accumulate to wait for buses. Many airports use buses or vans to shuttle passengers between the airport and the automobile rental campus or local hotels.

Many airports are designed with remote terminals, facilitating flight connections, for aircraft can park on all sides of the terminal. In 1962, Washington Dulles International Airport introduced vehicles that can be raised and lowered, an approach not widely followed, though they do appear here and there, such as at Toronto Pearson

8.9
These aboveground trams link the main terminal at Tampa International Airport with its remote satellites. Hillsborough County Aviation Authority.

International Airport. The lounges are 54 ft long, 16 ft wide, and 17.5 ft high, and can accommodate 102 passengers, 71 of them seated.[41] Atlanta International introduced an underground driverless tram with a parallel pedestrian tunnel with a movable sidewalk. Denver International copied Atlanta's concept, but left out the pedestrian tunnel in a fit of myopia which (as we shall see at the end of this chapter) makes the airport largely dysfunctional when the train breaks down. Plans for Inchon International Airport call for eventually building two terminals with sufficient capacity to handle 100 million passengers per year. They will be linked by an Intra Airport Transit people-mover system capable of carrying 14,000 passengers per hour.[42]

Trams can be aboveground (as at Tampa, Orlando, and Kansai—see Fig. 8.9), belowground (as at Seattle, Atlanta, Denver, and Hong Kong—see Fig. 8.10), and both above- and belowground (as at Kuala Lumpur and Dallas/Fort Worth—see Fig. 8.11). Placing them aboveground is more aesthetically attractive, for looking at palm trees and lakes, as in Orlando or Tampa, is far more pleasing than looking at the dark concrete walls of an underground tunnel. Yet at Denver International Airport, as a price of signing up for gates on Concourse B, United Airlines insisted that the glass on the 365-ft-long pedestrian bridge linking Concourse A to the main terminal be painted black so as to obscure the magnificent view of the Rocky Mountains. Mercifully, the airport engineer chose to violate that nefarious contractual provision, and United was too embarrassed to insist upon it.[43]

8.10 *Subterranean automated people-mover vehicles, such as these at Seattle/Tacoma International Airport, allow for efficient and unobstructed surface aircraft movements.* <small>Photo by Don Wilson. Port of Seattle.</small>

8.11 *Meshing aircraft efficiency with passenger aesthetics, the people-mover system at Kuala Lumpur International Airport is aboveground at the main and remote terminals, but dips belowground under the aircraft taxiway.* <small>Malaysia Airports Berhad.</small>

Signage and Graphics

Flight information should be clearly and ubiquitously visually presented to passengers throughout the terminal. Some airports, like Frankfurt, Los Angeles, Hong Kong, and Washington Dulles airports, display flights on enormous boards (similar to those used at some European railway stations) at prominent locations in the main terminal. (See Fig. 8.12.) Others use television monitors to display flight arrivals and departures. (See Fig. 8.13.) Airports should be careful not to allow airlines to monopolize screen displays. For example, at Minneapolis airport, it is relatively difficult to find an arrival or departure monitor which lists flights other than those of Northwest Airlines. Several airports also clutter displays with multiple code-sharing listings, as do Los Angeles International and Denver International.

The principal purpose of airport signage is to move the traveling public through the airport's labyrinth of roadways and corridors ef-

8.12
Large railway-type boards, such as these above the check-in counters at the Bradley International Terminal at Los Angeles International Airport, provide flight information from great distances throughout the terminal. Photo by P. S. Dempsey.

8.13 *Flight information display systems (FIDS), such as this one at the new Kuala Lumpur International Airport, provide up-to-date flight arrival and departure information, giving passengers the appropriate flight number, gate, and baggage carousel.* Malaysia Airports Berhad.

ficiently. Clear, concise, comprehensible, and conspicuous signage is essential to lead passengers quickly and effortlessly to their destinations. Signs serve the purposes of guiding the passengers in the direction they wish to proceed (to and through the airport), and of informing the passengers what alternative services are available (such as concessions, telephones, and rest rooms) and, from a regulatory perspective, what is expected of them (such as customs and immigration requirements).[44]

The process of developing appropriate signage should involve interviews of passengers to discern what their needs are, and how clear or confusing the signs are, for signs are hung for the service of the passenger. Note too that signage is an evolutionary process, which must be amended, supplemented, and adjusted as the airport itself evolves.

Signage should provide the passenger with general information at first, and more specific information closer to the point of decision. Signage begins at the vehicular entry points to the airport. The highways leading into the airport should identify which roads lead to arrival and departure terminals, where particular airline ticket counters

8.14 *Large legible signs leading into the airport complex, such as these at New York Kennedy International Airport, should clearly identify the venue of airline arrivals and departures.* Photo by Evan Auster. Port Authority of New York and New Jersey.

and baggage carousels are located, and where parking and rental-car returns are located. (See Fig. 8.14.) Signage too small and too late, as at Munich, only serves to confuse automobile drivers and congest airport roads, and make them less safe. Where an airport is served by multiple airlines, adding code-sharing affiliates which do not serve the airport to the list only serves to clutter signs and confuse passengers. Moreover, with the growth of code-sharing alliances, listing them all becomes unmanageable. The better approach is to list only carriers which actually serve the airport on signs leading to it, and provide additional information within the terminal as to code-sharing affiliations. After all, most aviation regulatory authorities require that the airline or travel agent issuing the ticket inform the passenger of the identity of the carrier that will actually be performing the flight. Where airlines insist on engaging in such manifestly deceptive practices as code-sharing, it should be their responsibility to undo the confusion they have created. Where there is more than one terminal, color-coding of signs helps passengers to find their way to the appropriate airline.

The introduction of airline deregulation and regulatory liberalization has created enormous volatility in terms of the identity and number of air carriers serving airports. This has placed significant demands on airports to keep their airline identity signs current. Though more expensive than static signage, dynamic signage can easily be changed to reflect changes in the identity of the airlines serving the airport, often with a few keystrokes to a keyboard. Airports should undergo a life-cycle cost analysis to determine whether the initial cost of dynamic signage is more or less than the anticipated changes and replacements which will need to be made over time to static signs. Airports can supplement visual displays with radio transmissions informing those arriving at the airport of where to check-in, as is done at Dallas/Fort Worth.

Within a parking garage, dynamic signage can direct drivers to those areas with the most available spaces. Signage is also important to help drivers to locate their automobiles once parked. Such parking garage signage can be alphanumeric (e.g., Level 3, Row 27, East), though color-coded thematic graphic symbols are easier for patrons to remember than letters and numbers. For example, the parking lots at Disney World are labeled according to unforgettable cartoon characters (e.g., Chip, Dale, Goofy, Minnie). Whenever in doubt, look to Disney as the model of how to move large volumes of passengers comfortably and expeditiously, with the least amount of confusion or congestion.

Signage for transit riders can begin with the transit vehicle, alerting the passenger to which airlines are closest to exit points. This can be supplemented with oral information over speakers in the vehicle. Once the passengers step off the vehicle, signs should lead them into the terminal and to the airline they seek with ease. The train bringing passengers to and from Hong Kong International Airport has television monitors embedded in the seats to introduce visitors to the airport and the intermodal connections at the rail terminals.

Within the terminal, signs should lead passengers vertically to the level they need, and horizontally to the place on that level where they need to be. Arriving passengers typically need to know where ticketing, baggage check, and boarding gates are located, as well as any internal transit vehicles or movable sidewalks linking terminals. Departing passengers typically need to know where baggage claim and transportation are located.

At Hong Kong International Airport, baggage carousel locations for arriving flights are displayed both on large boards in the main terminal, and at overhead television monitors at the immigration counters.

Arriving and departing flights can be listed on television monitors, or as in railway stations and several of the world's busiest airports, on large boards. The advantage of the large board displays is that they typically list the 20 or 30 flights closest to departure, in order, so that a late-arriving passenger can quickly glance at the board to learn the time of departure and proceed directly to the gate. Code-sharing, however, clutters the boards with duplicative listings of the same flight, thereby shoving upcoming departures off the boards. Another advantage of the large boards is that passengers can wait in the comfort of a lounge area until their aircraft is near departure. Similarly "meeters and greeters" can remain seated in the lounge until the board informs them the flight is close to arrival. This reduces congestion of passenger corridors. The Bradley International Terminal at Los Angeles International Airport has several well-placed large boards to inform passengers of upcoming departures, though unfortunately, it is cluttered with duplicative code-sharing listings.

Some airports seem to serve the interests of their principal tenants to the detriment of passengers seeking to use the services of their competitors. Flight displays at Minneapolis/St. Paul International Airport are notoriously poor for passengers seeking information on flights other than those flown by Northwest Airlines. Similarly, Atlanta Hartsfield International Airport lists the flights of its dominant tenant, Delta Air Lines, more prominently than those of its competitors.

The new Hong Kong International Airport took the signage issues seriously:

> *Signage has been an important part of the terminal's design strategy. The building is open and spacious, with abundant light and clear orientation. The signs will comprise contrasting white lettering on a vivid blue background, which will be very visible within the interior's mostly neutral color schemes. The signs will be written in both English and original complex Chinese characters. They will include easily identified pictograms and will be backlit so a person with average vision can read them from 30 to 35 meters.*

> *The flight information display system (FIDS) boards will also be highly effective. Three meters high and eight meters long, the liquid crystal display (LCD) readouts will be visible from the entrance ramps in the Departures Hall. They will contain about three hours of flight information.[45]*

Yet signage in the rail terminals linked to the new Hong Kong International Airport could be improved. Other airports have had more difficulty with the issue. When Frankfurt Airport opened a new terminal in 1994, lack of sign clarity confused passengers, thereby making it difficult for them to get around. At Munich, parking lot signs were so small passengers couldn't find their cars.[46]

Clarity is enhanced by adopting a uniform system of sign styles and sizes, consistent terminology, recognizable symbols, and consistent coloration for standard functions.[47] As a rule of thumb, a sign with lettering 1 in high can be seen 50 ft away; 2-in signs can be seen 100 ft away, and so forth. International airports should have signs in the most prominent foreign languages. Though signs should be easy to find, they should not be too plentiful, for too many signs can create clutter, a "barnacle syndrome," which confuses the passenger. Information booths and television kiosks can also serve to help the confused passenger.

Security

Terrorist hijacking of aircraft and bombing of airports in the 1970s required an airport design focusing on security. These events caused airport designers to embrace the bottleneck principle. Arrival and departure halls are sometimes placed on different floors, with a strict segregation between secure areas (after passport inspection and bodily searches) and open areas.[48] Typically, passengers and carry-on baggage are screened by technological devices or dogs for weapons and explosives prior to being admitted to the boarding area of the airport.

Several international agreements have been concluded [many under International Civil Aviation Organization (ICAO) auspices] to deal with aerial security, hijacking, and terrorism. These include the Tokyo Convention of 1963 (which required that a hijacked aircraft be restored to the commander), the Hague Convention of 1970 (which declared hijacking to be an international crime and required that the state to which an aircraft is hijacked exert jurisdiction over the hijackers and prosecute them, imposing "severe penalties"), and the Montreal Convention of 1971 (which added airport security to the international regime). The European Convention of 1977 for the Suppression of Terrorism provided that hijacking would not be deemed a political offense for purposes of extradition. The Bonn Declaration of 1978, an agreement of G-7 leaders, provided that flights would be ceased immediately to or from any nation which refused to prosecute or extradite a hijacker, or return the hijacked aircraft.[49]

The United States has also promulgated comprehensive legislation dealing with aerial terrorism and hijacking. The Antihijacking Act of 1974 imposed a penalty of 20 years imprisonment or death if a passenger is killed during a hijacking, and authorized the President to suspend the landing rights of any nation that harbors hijackers. The Aircraft Sabotage Act of 1984 imposes penalties of up to $100,000 or 20 years imprisonment, or both, for hijacking, damage, destruction or disabling an aircraft or air navigation facility. The Security and Development Act of 1985 authorized expenditures for enhancing security at foreign airports. The Air Traveler Protection Act of 1985 required the DOT Secretary to assess security at foreign airports, and notify the public if a foreign airport fails to correct a security breach.

The Aviation Security Improvement Act of 1990[50] established a Director of Intelligence and Security in the Office of the DOT Secretary, and an Assistant FAA Administrator for Civil Aviation Security, and gave the FAA responsibility to oversee security at major airports.[51] FAA security managers supervise security arrangements. The FAA carries out periodic threat and vulnerability assessments, and publishes guidelines on such topics as airport design and construction, screening of passengers and property, public notification of threats, security personnel investigation and training, cargo and mail screening, research and development activities, security standards at foreign airports, and international security negotiations.[52] FAR 107 deals with airports, 108 with airlines, 109 with freight carriers, and 129 with foreign carriers.

Congress has conferred comprehensive jurisdiction on DOT to regulate the screening of passengers and property, deal with threats to domestic and international civil aviation, establish security standards at foreign airports, issue travel advisories, require passenger manifests, provide airport construction guidelines, and foster an accelerated program of security research and development, including explosive detection.[53] In order to protect public safety, passengers can be charged with and convicted of federal crimes committed aboard aircraft,[54] including hijacking an aircraft, intimidating a flight crew member,[55] abusive sexual contact,[56] carrying weapons, loaded firearms, and explosives or incendiary devices aboard an aircraft,[57] and conveying false information and threats to commit a felony.[58]

The crash of TWA flight 800 near Long Island, New York, in 1996, inspired President Clinton to appoint Vice President Gore to head a commission to study safety and security in aviation. The Commission issued 31 security recommendations, of which the FAA was responsible for implementing 21. (See Table 8.1.) An FAA reauthorization bill imposed a host of new security requirements. The FAA must certify companies providing security screening and improve training and testing of security scanners, give priority to developing automated surveillance targeting systems, assess weapons and explosive detection technologies, work with airlines to develop passenger profiling systems, conduct employment investigations for individuals who exercise security functions with cargo and baggage, and coordinate joint threat and vulnerability assessments with the FBI. The NTSB must establish a program to provide family advocacy services after air crashes. Proposals to require matching of bags with passengers in domestic aviation were objected to by airlines because of the cost and delay which would be imposed at hub airports, though a test was begun to measure their efficacy and cost.

In the Omnibus Consolidated Appropriations Act of 1997, Congress authorized $144 million for the purchase of advanced security screening equipment for checked and carry-on baggage. The FAA planned to deploy 54 certified explosive detection systems to screen passengers' checked bags and 489 trace detection devices to screen carry-on bags by the end of 1997. But by 1998, the FAA had deployed only 13 explosive-detection systems and 125 trace detection devices.[59] At this writing, the only FAA-certified explosive-detection machine is the CTX 5000, developed by InVision Technologies, which uses computed tomography. But even as sophisticated a system as the CTX 5000 is only as good as the people operating it. This has required an emphasis on Screener Proficiency Evaluation and Performance (SPEARS), predicated on analysis of the three essential components of job analyses—selection, training, and performance monitoring.[60] The FAA spent $132 million on checked-luggage bomb detection devices capable of handling 225 bags an hour. But a 1998 DOT inspection, which audited 13 of the $1.3 million machines, found that nine handled fewer than 200 bags a day, some so few that their operators could not maintain their proficiency.[61]

New airport technology is also making immigration and security a more efficient operation. For example, at New York Kennedy and Miami International airports, magnetic stripe cards have been installed to

Table 8.1. Recommendations of the White House Commission Report on Aviation Safety and Security

No.	Recommendation	Lead agency
1	The federal government should consider aviation security as a national security issue and provide substantial funding for capital improvements.	FAA
2	FAA should establish federal mandated standards for security enhancement.	FAA
3	The Postal Service should advise customers that all packages weighing over 16 ounces will be subject to examination for explosives and other threat objects in order to move by air.	Postal Service
4	Current law should be amended to clarify Customs Service's authority to search outbound international mail.	Customs Service, Postal Service
5	FAA should implement a comprehensive plan to address the threat of explosives and other threat objects in cargo and work with industry to develop new initiatives in this area.	FAA
6	FAA should establish a security system that will provide a high level of protection for all aviation information systems.	FAA
7	FAA should work with airlines and airport consortia to ensure all passengers are positively identified and subjected to security procedures before boarding.	FAA
8	Submit a proposed resolution that ICAO begin a program to verify and improve compliance with international security standards.	FAA

9	Assess the possible use of chemical and biological weapons as tools of terrorism.	FAA
10	FAA should work with industry to develop a national program to increase the professionalism of the aviation security work force, including screening personnel.	FAA
11	Access to airport controlled areas must be secured, and physical security of aircraft must be assured.	FAA
12	Establish consortia at all commercial airports to implement enhancements to aviation safety and security.	FAA
13	Conduct airport vulnerability assessments and develop action plans.	FAA
14	Require criminal background checks and FBI fingerprint checks for all screeners and airport and airline employees with access to secure areas.	FAA
15	Deploy existing technology.	FAA
16	Establish a joint government-industry research and development program.	FAA
17	Establish an interagency task force to assess the potential use of surface-to-air missiles against commercial aircraft.	Department of Defense
18	Significantly expand the use of bomb-sniffing dogs.	FAA
19	Complement technology with automated passenger profiling.	FAA
20	Certify screening companies and improve screener performance.	FAA
21	Aggressively test existing security systems.	FAA

Table 8.1. Recommendations of the White House Commission Report on Aviation Safety and Security (*Continued*)

No.	Recommendation	Lead agency
22	Use the Customs Service to enhance security.	FAA
23	Give properly cleared airline and airport security personnel access to the classified information they need to know.	FAA
24	Begin implementation of the full bag-passenger match.	FAA
25	Provide more compassionate and effective assistance to families of victims.	NTSB
26	Improve passenger manifests.	DOT
27	Increase the number of FBI agents assigned to counterterrorism investigations.	FBI
28	Provide airport security training and antiterrorism assistance to countries whose airports are served by airlines flying to the United States	Department of State
29	Resolve outstanding issues related to explosive taggants and require their use.	ATF
30	Provide explosives detection training program to foreign, federal, state, and local law enforcement, as well as FAA and airline personnel.	ATF
31	Create a central clearinghouse within the government to provide information on explosives crime.	ATF

speed the separation of individuals to be interviewed by the Immigration and Naturalization Service (INS) from those who do not need to be interviewed. The encoded magnetic stripe includes basic identification information and a three-dimensional reading of a person's palm print. Upon entry to a gateway inspection area, the pass holder runs his or her card through a reader and provides a palm reading. Computers reconcile the card information and biometric reading to determine whether the individual should be separated for an interview. This reduces the number of INS inspections, and airline costs.[62]

One airport which has addressed the security threat of terrorism effectively has been Colombo's Bandaranaike International Airport in Sri Lanka. Since the early 1980s, the Liberation Tigers of Tamil Eelam, seeking independence for those regions of Sri Lanka inhabited by the Tamil minority, has been on a rampage of terrorism, attacking the military, blowing up banks, and in early 1998, bombing the country's dearest Buddhist religious shrine at Kandy. The airport has enhanced its security screening measures whereby air force soldiers screen all vehicles and individuals entering the airport several times; tourist coaches must be driven by airport-certified drivers; suspicious vehicles are isolated and sniffed by bomb-trained dogs; cargo, bags, and passengers are searched by sophisticated bomb-detection equipment; and bags are correlated with passengers. One design innovation was to eliminate the parking lot next to the terminal building and move curbside access to a location some distance away from the terminal, and to build a blast wall packed with sloping soil next to the curb to shield the terminal should someone try to bomb it.[63] Passengers and their luggage pass through metal detectors at curbside, before ticketing, and at the gates.

One other issue of security is baggage theft. Theft of laptop computers at security screening areas has occurred with increasing frequency, where one individual blocks the metal detector, shuffling through pockets for pens, glasses, and coins, while a partner in crime grabs the laptop computer coming off the scanning device belt and passes it off to a third who exits the airport. Another problem is baggage stolen off the baggage carousel. It is said that no bag makes it twice around the baggage carousel at the airport in Lagos, Nigeria. Both problems can be diminished with better security operations.

Customs and Immigration

The customs and immigration inspection areas should be arranged like a comb. Customs should be segregated into green zones (for passengers having dutiable goods to declare), and red zones (for all other passengers).[64] Passport clearance is enhanced when the inspection positions are homogeneous, allowing any passenger to use any position. Search and interview rooms should be located off to one side of the through-flow area.[65]

Shopping Areas

Though airports have long been venues for duty-free sales of alcohol, tobacco, perfume, handbags, neckties, and such, they entered a new level of commercialization in the 1990s, particularly after privatization of the British Airports Authority (BAA) demonstrated the revenue enhancement potential of shopping. Several new airports are essentially shopping malls surrounded by runways. In its seven British airports, BAA sells one-fifth of all the perfume in Britain and 800,000 neckties a year. At Pittsburgh International Airport, which BAA manages, it has more than tripled the revenue earned from an average passenger. Professor Bijan Vasigh notes that "Airports generate captive customers, and they have become acutely aware of that fact."[66] Noting that Heathrow Airport effectively functions as a great public square like Picadilly Circus or Trafalgar Square, one commentator observed, "The modern airport is not just a machine for flying from. It is a mini-metropolis of retail areas, business centers and hotels. It is also a national symbol."[67] Airports at Orlando and Kuala Lumpur have hotels embedded within them, while at Brussels, Dallas/Fort Worth, and Chicago they are a short walk away. Shopping abounds at Amsterdam's Schiphol Airport. At Frankfurt's airport there are peep show booths and casinos. Gambling machines are everywhere at airports in Las Vegas and Reno. Kuala Lumpur is planning a go-cart track and botanical garden near its airport. At Seoul's Inchon Airport, there will be a seaport, business and telecommunications center, and a "leisure-port," where travelers can enjoy yachting or fishing.[68]

Some airports, such as Hong Kong International, coerce passengers to walk circuitously through a shopping mall. Others, such as Ronald Reagan Washington National Airport, have shopping tucked in awkward places away from natural passenger flows.

Passenger Waiting Areas

To the extent possible, passengers seeking to move quickly from ticketing and baggage check to boarding gates should enjoy unobstructed flow over clearly defined and properly signed routes. But many passengers who arrive early, or are transferring between aircraft, will need space to stand or sit while they wait for their departure. These passengers should be directed to waiting areas, sited to the side, and clear of, the direct passenger flow to aircraft. Waiting areas should provide adequate amenities, such as flight information displays, toilets, telephones, and perhaps televisions, as well as concessions such as newsstands, cafes, and bars.[69]

Airlines also seek space for their club or first-class lounges for their preferred customers, particularly at airports at which they have dominant market share. For example, British Airways transformed its arrival lounges at London's Heathrow and Gatwick airports to add showers, changing rooms, breakfast, clothes pressing, and valet services for its first-class customers.[70] Japan Airlines has both a first- and business-class lounge at Tokyo Narita Airport, with the business-class lounge more opulent than many U.S. carrier domestic club lounges. Typically they offer cocktails, lounge chairs, televisions, telephones, magazines and newspapers, and conference meeting rooms, as well as various business amenities, such as fax machines, computer terminal modems, and telephone conference rooms.

Large common waiting areas can be located in the arrivals terminal area, or at individual gates. At the gates, space efficiency can be enhanced by combining the lounge area for a number of gates. Total space can be reduced by 20 to 30% by combining the lounge area of four to six gates.[71] Cleveland Hopkins International and Boston Logan International airports have designated areas filled with children's play equipment, a nice touch to relieve parents with high-energy but bored kids on connecting flights.

Aircraft Boarding Areas

Several different systems have been developed to accommodate passenger boarding of aircraft. Passengers can walk up boarding stairs or along a loading bridge, or be transported by bus or other vehicle to the aircraft. They can board from the tarmac or from an enclosed pier. Passenger loading bridges (jetways) are the most expeditious means of boarding, the most convenient for passengers (since they are shielded from the elements, weather, noise, and fumes), and most secure (since pedestrians are not wandering around the air side). The size and type of loading bridge should be able to accommodate aircraft different from those originally designated for it, for technology or market demand may require different-sized aircraft over time.[72] Some airports, such as Hong Kong International, have two loading bridges to allow expeditious loading and unloading of wide-body aircraft, such as the Boeing 747. One is typically used for first- and business-class boardings, while the other is for economy-class passengers. Others, such as Munich Strauss, have dual-level ramps which allow segregation of arriving passengers (many who will have to clear customs) and departing passengers.

Unfortunately, some airlines, such as Cathay Pacific, refuse to allow coach passengers to enter the "upper class," first- and business-class boarding jetway, no matter how constipated the coach bridge becomes. This is a particularly unfortunate practice when passengers are forced to stand in the heat and humidity of Hong Kong International Airport waiting to board an aircraft parked on the tarmac when the coach boarding stairway is moving at a snail's pace, and the "upper class" stairway is empty. Cathay's deboarding procedures are equally frustrating, as all coach passengers are forced to wait until the last "upper class" passenger deplanes, no matter how sluggish the departure. Though Cathay is well regarded by business travelers, its boarding and deboarding procedures leave coach passengers stranded.

Ronald Reagan Washington National Airport installed enclosed spiral ramps for gate-checked baggage, allowing bags to descend to the tarmac quickly and effortlessly, without forcing labor-intensive lifting by gate agents or "ramp rats." This is an efficient way to expedite processing of gate-checked bags.

Access by Disabled Passengers

At least one entrance to the terminal should be ramped so that a wheelchair can roll up it. The only suitable way to move disabled passengers between floors is via elevator.[73] Bathrooms and drinking fountains also should be designed to accommodate wheelchair-bound patrons.

The United States has promulgated legislation addressing accessibility of airports. Both the Rehabilitation Act of 1973 and the Rehabilitation, Comprehensive Services, and Developmental Disabilities Act of 1978 prohibit discrimination in any program or activity receiving federal financial assistance.[74] DOT regulations promulgated thereunder prohibit discrimination against the handicapped in most airports, except those served by smaller aircraft.[75]

Essentially, the regulations require equality of treatment for qualified handicapped people in terms of employment, access, or utilization of airports. Structural changes in facilities necessary to permit access by individuals with disabilities is required. Specifically, airport terminals "shall permit efficient entrance and movement of handicapped persons while at the same time giving consideration to their convenience, comfort, and safety."[76] The terminal design should also take into account their needs in location and types of elevators and escalators, minimizing any extra distance that wheelchair users must travel, compared to nonhandicapped passengers, to reach ticket counters, waiting areas, baggage handling areas, or boarding locations. Ticketing, baggage check-in and retrieval, boarding, telephones, teletypewriters, vehicular loading and unloading, parking, waiting areas, airport terminal information, and other public services such as drinking fountains and rest rooms—all must made accessible to the handicapped. Many of these requirements also have been imposed upon airlines which own, lease, or operate airport terminal facilities.[77]

Airlines were specifically excluded from the application of the Americans with Disabilities Act of 1990, because Congress had already enacted the Air Carrier Access Act of 1986,[78] which provides, *inter alia,* that "No air carrier may discriminate against any otherwise qualified handicapped individual, by reason of such handicap, in the provision of air transportation."[79]

Architecture and Aesthetics

Architectural critic Stefano Pavarini wrote, "Architecture's strength lies in the lasting, theoretically timeless, message it transmits; something built to last and not vanish into thin air. Architecture is, then, the bastion of existence, a message designed to last and prophecy of a story yet to be told."[80] Airport terminals have not always been distinctive in their architecture. Many are bland but functional industrial buildings. A few have attempted to replicate the sense of grandeur of the railway terminals they replaced. Yet only a few inspire the sense of awe of the railway terminals at Washington, D.C., New York, Philadelphia, or St. Louis. Stressing modern form in this most modern of transportation modes, several airport terminals constructed at the dawn of commercial aviation were art deco in their architecture—Washington National Airport and Chicago Midway, for example. (See Figs. 8.15 and 8.16.)

8.15 *Washington National Airport was at the forefront of the new art deco architectural movement at airports when built in 1940.*
Metropolitan Washington Airports Authority.

8.16 *Chicago Midway Airport, once one of the world's busiest, embraced an art deco design that will be leveled for a new terminal on the same site.* Photo by P. S. Dempsey.

But the truly distinctive terminals are a short list. They include Eero Saarinen's landing-bird TWA terminal at New York's Kennedy International Airport and his swept-arch terminal at Washington Dulles International Airport, as well as Helmut Jahn's glass-walled United Airlines remote terminal at Chicago O'Hare. Time will tell whether the new "green field" terminals are added to that list. This author is inclined to add Piano's winged terminal at Osaka, Kurokawa's forest-within-an-airport design at Kuala Lumpur, Foster's flying roof and glass-walled terminal at Hong Kong, and Fentress' snow-capped-peaks design at Denver and Korean garden design at Seoul to the list.

Note the nationality of these creative men is spread across the planet. Saarinen was born in Finland, Jahn in Germany. Both came of age while working in the United States. Piano is an Italian who designed a terminal in Japan; Kurokawa is a Japanese who designed terminals in Kuala Lumpur. Foster is an Englishman who designed a terminal in China. Fentress is an American who designed terminals in the United States and Korea. Such diverse nationality speaks well for the concept of blind grading of architectural proposals, and the willingness of governments to reach beyond their borders to embrace some of the most dynamic architecture on the planet.

An airport is the first and last impression many travelers will have of a city and its region. It therefore can be (and in the author's opinion, should be) a distinctive architectural and aesthetic statement of its venue. Albuquerque's airport is distinctly southwestern in design. Curt Fentress' design of the new Seoul International Airport is distinctly Korean. Kurokawa's design for the new Kuala Lumpur International Airport is distinctly Malaysian. Jakarta's airport is distinctly Indonesian in its rooflines. Bali's Airport is distinctly Balinese in the ornate Hindu sculptures which greet its visitors. Each is a statement of the architecture, art, history, and culture of the land to which it provides entry and exit. Denver International Airport conjures up images of Indian tepees sprinkled across the Great Plains to some, or the snow-capped peaks of the Rocky Mountains to others. Nantucket Memorial Airport evokes the style, sense, and culture of a New England village. Chattanooga's airport reminds one of its history as a nineteenth-century railroad hub. Ronald Reagan Washington National Airport's new terminal domes are reminiscent of the Thomas Jefferson Memorial (appropriate to a Virginia domicile) or the U.S. Capitol Building. Other airports convey a visual sense of flight, notably Osaka's Kansai International Airport, Hamburg Airport, Hong Kong International Airport, Salt Lake City International Airport, and Copenhagen Airport.

The fundamental challenge for architects is to evoke a sense of place (a visual icon, if you will, as important to a city as its principal municipal buildings) while satisfying the technical, technological, safety, and security needs of a major transport facility. Koos Bosma describes the effort to blend the vastness of terminal space design with functionality and aesthetic flair:

> *[T]he passenger terminals of the [newest] generation of airports—from the new facility at Stansted Airport north of London to the totally new Kansai International Airport in Japan—are designed as vast open spaces. But these spaces are to pass through, with ambiguous social connotations, in that they are collective spaces without a feeling of communality. The space itself calls forth associations with a transparent tube or station concourse: a mixture of street and interior. Architects strive to show the construction as a universal structure, but that structure also has to be unique for the location. The spaces are bathed in brilliant light, filtered through the transparent walls. The terminal roof, often easier for passengers to observe than the other elevations, becomes a fifth facade, and as such, is an essential part of the spatial composition.*[81]

Many airport terminals are architectural statements for the airlines which occupy them. Some of the most notable are Eero Saarinen

8.17 *Eero Saarinen's swept-wing design for TWA's terminal at New York Kennedy International Airport is among the most dynamic in architectural history.* Port Authority of New York and New Jersey.

and Associates' TWA terminal at New York Kennedy International Airport, completed in 1962, or the adjacent National Airlines terminal building designed by I. M. Pei and Associates, acquired by TWA subsequent to National Airlines' acquisition by Pan Am.[82] In his landmark building, Saarinen broke ranks with the traditional form of a long finger with airplanes parked in a line by building smaller satellite terminals surrounded by parked aircraft, connected to the main terminal building by passenger corridors.[83] Saarinen's main TWA terminal looks something like a large white landing bird with extended wings. (See Fig. 8.17.) The terminal consists of four vaulted domes with upward soaring curves rising 50 ft over the passenger areas.

Saarinen described his main terminal at Washington Dulles International Airport as "the best thing I have ever done." The design uses outward-leaning concrete pillars (68 ft high on the land side, and 50 ft high on the air side) to support a concrete roof with encased steel cables spanning 164 ft, with window walls of glass, steel, and aluminum. (See Fig. 8.18.) The American Institute of Architects gave it the First Honor Award in 1966.[84] Completed in 1962, the main terminal was only 600 ft wide, half of Saarinen's original design; it was expanded significantly in the 1990s.

8.18 *Saarinen described the main terminal he designed for Washington Dulles International Airport built in 1962, as "the best thing I have ever done."* Photo by Eric Taylor. Metropolitan Washington Airports Authority.

Paris Charles de Gaulle Airport opened with architectural fanfare in 1974, with a revolutionary pancake-shaped terminal design. To reduce walking times for passengers, the designers stacked different zones and services (arrivals, departures, restaurants, shops, offices, and parking garages) on top of one another in a central cylinder linked by underground walkways to seven "satellites" with access to the aircraft. Subsequent terminals were added on a modular basis, with the B terminal opening in 1981, A opening in 1982, D opening in 1989, and C opening in 1993. Each adhered to the basic concept, though introducing new technology, better natural light, and segregating departing and arrival passengers. Future 2E and 2F terminals will form two sides of an oval at right angles to the main terminal so that aircraft can be parked on both sides (under a pier finger arrangement). (See Fig. 8.19.) The two latest terminals embrace a "flying carpet" effect, which attempts to give passengers the feeling of walking through a large open space. Moreover, the outer walls do not actually touch the ground, giving passengers the impression the structure is floating in air. The terminals have been designed to minimize walking distances, to 45 m from entrance to check-in, and 100 m from check-in to embarkation.[85]

With the opening of Terminal 2 in 1994, Frankfurt Airport "blazed new trails in contemporary architecture," according to its literature (see Fig. 8.20):

> *[T]he most prominent features of this pace-setting concept are transparency, light and air. The overall impression is one of lavish space, with clear and simple orientation for passengers.*

> *A distinctive characteristic of the new passenger terminal at Frankfurt is the roof construction. It combines the technical-functional and the human aspect under a single steel-and-glass roof, thus providing a flashback reminiscence of the classicist architecture of Frankfurt's central railway station.*

> *A further architectural hallmark is the extensive glass surfaces of the hall with specially developed "intelligent" panes helping to create a pleasant indoor climate in all types of weather....*

> *On first entering the check-in hall, the passenger is emotionally attuned to the experience of flight. Extensive natural*

light and lavish green plants enhance the relaxing atmosphere, which imparts a sense of security and comfort to the passenger. All these features make the journey considerably more pleasant for the air traveler—and simplify work for staff at the airport.[86]

Frankfurt is also among the new generation of airports that have abundant shopping. The airport has more than 100 shops including a supermarket, duty-free shops, Harrods of London, and several German designer boutiques.[87]

British Airways wanted to offer its business travelers something new, different, and attractive at its Oasis lounge in Terminal 4 at Heathrow, where passengers can sleep, eat, or pray.[88] Completed in 1978, the Haj Terminal at Jeddah, Saudi Arabia, covered a large space with a tentlike fabric made of fiberglass coated with Teflon, suspended with steel cables.

Some airports offer passengers observation platforms from which they can enjoy a panoramic view of aircraft taking off, landing, or taxiing to and from gates. Observation platforms can be outdoors (as in Puerto Plata, Dominican Republic, for example), or indoors [as at Baltimore/Washington International Airport (BWI), for example]. BWI integrates an aviation technological exhibit into its observation area, displaying the marvels of flight, including a dissected Boeing 737 aircraft, complete with wings, tail, cockpit, engine, galley, and fuselage, all taken apart so that the passenger can see the inner workings of the aircraft. Several airports offer educational displays on flight, such as San Francisco International Airport, which provides a historical overview, going back to the days of ballooning. All this provides something to entertain and educate passengers waiting for an arriving or departing flight.

Many airport concourses are designed with generous walls of glass, allowing the passenger a view of the hustle and bustle of the airfield (such as the midfield terminal at Chicago O'Hare and the north terminal at Ronald Reagan Washington National Airport, for example) or the panorama of the surrounding landscape (such as Salt Lake City International Airport and Hong Kong International Airport, for example), or simply let in light (Fig. 8.21).

While a few airport terminals (like United's midfield terminal at Chicago O'Hare) are architectural statements of distinction in their

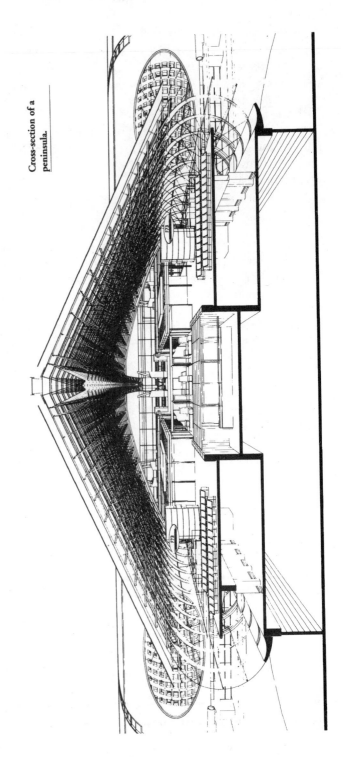

Cross-section of a peninsula.

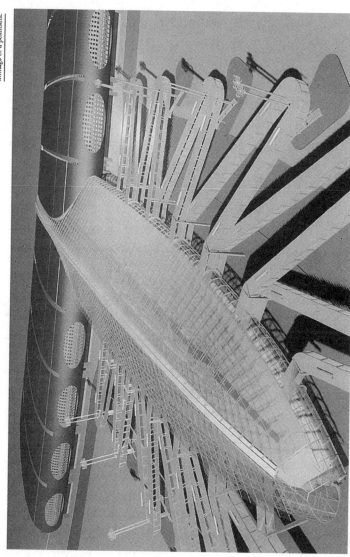

Computer-generated immage of a peninsula.

8.19 *Opened in 1974 among much architectural fanfare, Paris Charles de Gaulle Airport has undergone expansion that lives up to its tradition of bold aesthetics.* Aeroports de Paris.

8.20 *Many airport terminals, such as Terminal 2 at Frankfurt, are designed with an abundance of glass to allow light in and a view out.* Flughafen Frankfurt.

8.21 *Glass has become an integral part of many airports, such as the new north terminal at Washington Ronald Reagan National Airport.* Metropolitan Washington Airports Authority.

8.22 *A nod to the past, antique aircraft such as this biplane are hung in the lobbies of airport terminals and concourses throughout the world. This early British plane is at the new Hong Kong International Airport at Chek Lap Kok.* Photo by P. S. Dempsey.

own right, many are simply cheap industrial buildings. Art can dress up functional and relatively nondescript terminal buildings. For example, Baltimore/Washington International Airport departure terminal was once dominated by a huge stained-glass crab, letting everyone who passed know they were in Maryland, the state where they could feast on a marvelous pile of Chesapeake Bay crabs. Unfortunately, the glorious crab was removed, replaced by a multitude of photographs of Maryland. Several airports, including Hong Kong International Airport and Minneapolis/St. Paul's airport, have antique aircraft hanging from their ceilings, telling all who pass they are in the world of aviation. (See Fig. 8.22.) Dayton International Airport has a model of the Wright Brothers' original biplane in a display case, appropriate to the city whose native sons gave birth to flight. It was at Dayton these two bicycle repairmen designed and built the first powered aircraft, which flew at Kitty Hawk, North Carolina, in 1903. San Francisco International Airport has displays of its distinguished citizens sprinkled around the terminals—famous baseball players and such. Originally, the reconfigured Atlanta International Airport had a few abstract photographs and paintings crudely hanging above the escalators moving people up from the dark recesses of the underground tunnel linking its concourses, though the airport has since done an aesthetic makeover coincident with the 1996 Summer Olympic Games.

Instead of tacking up a few questionable art pieces as an afterthought, as Atlanta did, art can be (and again, in the author's opinion, should be) integrated into the design process, as was done in designing the new Denver International Airport (DIA). As an integral part of the DIA construction program, the airport art program was governed by a committee dominated by art "experts," in collaboration with Patricia Fuller, a former director of the National Endowment for the Arts "Art in Public Places" department. The idea was that the art committee would work with architectural teams so its work could be folded into the building's architecture.[89] Though with a squint of the eyes, Fentress' main terminal roof at Denver International Airport may evoke impressions of snow-capped peaks or Indian tepees rising above the plains, the art inside is not only qualitatively disappointing, in the author's view, it speaks relatively little about Colorado or the American West.[90] Though the political and intellectual elites have the power to decree what pleases them, more public input would have been beneficial. When Chicago was considering several alternatives for a new public library to be located on State Street, it allowed the general public to file past the architectural models and share their input with the individuals who made the selection decision. Had this been done at Denver, the art selected might have been of better quality. It might also have evoked a sense of the Colorado Rocky Mountain West, its rich history and beauty. Beauty is in the eye of the beholder, and the public has enormous common sense about what pleases it.

At Vancouver International Airport, the terminal designers commissioned Native American tribes as design consultants. The result is foliage, waterfalls, material, fabric, and art which convey a dramatic American Indian influence, which is the cultural heritage of British Columbia. The interior of the Bradley International Terminal at Los Angeles International Airport also conveys the local culture, with concession facades in the art deco architectural style of Beverly Hills and Hollywood. Koos Bosma described the psychological themes that should permeate airport architectural design:

> *Airport buildings not only supply comfort, but also satisfy emotional and symbolic needs. We unconsciously associate*

air travel with speed, adaptability, light, air, comfort, and service. It is the architect's job to reinforce that unconscious desire, with the use of plenty of glass, muted colors, comfortable furniture, and perfect treatment of the passenger. The space surrounding the flowing movement should radiate not only an atmosphere of comfort and luxury, but also one of reassurance and mild euphoria, against a backdrop of necessary facilities presented in a simple and efficient design. Transparency of organization and styling suggests minimal resistance. Architecture is also part of the airports' and airlines' marketing strategy. It is vital that airports be different, and that difference must be expressed in their architecture. The visitor must also be convinced that the socioeconomic disruptions caused by air traffic—dangers, disasters, pollution, noise, smell, traffic jams, jet lags—are outweighed by the advantages of the megastructure.[91]

Many larger airports consist of an eclectic collection of industrial buildings with little attempt at aesthetic integration or harmony. The arrogant failure of many architects to attempt to integrate their designs to blend harmoniously with the surrounding campus, a trend which began in earnest after World War II, is not unique to airports. Perhaps the most miserable example stands in the center of one of the world's most beautiful cities, Vienna, whose elegant baroque and gothic architecture of the Stepansplatz and Graben has been assaulted by Hans Hollein's Haas Haus, a futuristic clash of errant modernity in a place it clearly does not belong.[92] Airport authorities should insist that the architects who design their buildings not present a facade which so radically clashes with its surroundings. Architects whose hubris insists that they ruin the harmony of past architecture should be publicly flogged.

Terminal Design and Architecture at the World's New Airports

Munich's Franz Josef Strauss Airport

More than 100 architectural firms competed for the design of Munich's new Franz Josef Strauss airport. Its main terminal, designed in the mid-1970s, was heavily influenced by U.S. airport terminal design and its reliance on the automobile as the primary mode of landside access.[93] (See Fig. 8.23.) The style is a long linear blend of white

8.23 *Munich's Franz Josef Strauss Airport's 1970s design evolves from the sterile Bauhaus style which dominated postwar German architecture.* Flughafen München.

girders and glass.[94] One source described the aesthetics in these words:

> *A surprisingly sleek, yet incongruous apparition in concrete, steel and glass stretches into view as motorists near Munich Airport, sprawling spaciously over the moor near rural Erding. Cynics refer to it wryly as a "designer" airport....*
>
> *Nearly 80% of Munich passengers and visitors polled...responded favorably. Random remarks: "Ultramodern facility," "attractive, functional layout," "stunning architecture; clear, bright design features."*[95]

Four terminal modules have been built—one for domestic flights, one for Lufthansa's international flights, and two for charter and nondomestic flights.[96] Unfortunately, arrival and departure areas are on the same level, creating some confusion and consuming a massive expanse, requiring a long hike for pedestrians transferring between gates and those dropping off rental cars at the centralized rental return in the parking garage. The terminal building stretches 1 km (3170 ft) perpendicularly between the runways, nearly the same distance as DIA's Concourse B (3300 ft).[97] The terminal has 20 jetway gates, and an adjacent ramp can accommodate 28 aircraft, from which passengers will deplane and board buses to the terminal. All check-in, security, and waiting room areas are in close proximity (50 m, or 164 ft) from the curb to the nearest gate for departures.[98] The terminal is designed to allow passengers to make their connections within 35 minutes. It was designed to accommodate 15 million passengers a year.[99]

An open area south of the general aviation facility will allow room for future expansion by two additional terminal modules, which could provide capacity for an additional 6 million passengers a year.[100] It is anticipated that a second terminal will be built no later than the year 2005.[101]

Among the aesthetic touches in the terminal building is that large amounts of glass face the runways, which pose a cleaning problem because of the glass angle. Even the escalators have clear sides, revealing working parts. The elevators are made of glass, but they are too small to accommodate a glut of passengers with baggage. The airport also uses a new bench system (Tubis), made of an aluminum frame with perforated steel parts covered with upholstery.[102] Since the airport opening, changes have been made to correct oversights. More passenger seats were added. Duty-free shops were moved closer to passenger traffic. Food and drink vendors were added to the terminal's single arrival and departure level.[103]

In 1989, Professor Hans-Busso von Busse, the general planner for the passenger service area at Munich Strauss, presented an overall art concept developed in connection with the information and advertising concept. It embraced an "integrated air concept," including the interior and exterior. Works of art are sprinkled throughout the terminal.[104] The underground tunnel linking various parts of the terminal is also engulfed in colorful hues of bright light.[105] Willi Hermsen, Managing Director of Flughafen München, noted the effort of the airport to blend with its surroundings:

> The airport's visual design, which has earned several awards, is uniform and distinct. Taking the local landscape of Erdinger Moos, which is characterized by intensive agricultural land use, the architects, landscape planners and designers involved in the construction of the airport produced an urban landscape that ties in with the colors and forms of the countryside in a stylized manner. The land's characteristics,

its flatness, its straight lines, and its neatly grouped areas of
trees and greenery became elements in the design of a bright
and friendly "airport landscape."[106]

Osaka's Kansai International Airport

Before construction began, the developers of Kansai International Airport sought advice from six airport operators on design, construction, and operation. The six were BAA (formerly British Airports Authority), Aeroports de Paris, Amsterdam Airport, Frankfurt Airport, the Port Authority of New York and New Jersey, and Dallas/Fort Worth International Airport. After they studied the design concept, their input was incorporated into the detailed design plan.[107]

Paul Andreu of Aeroports de Paris established the airport's basic layout, including a 5600-ft-long concourse with a passenger terminal building at the center (Fig. 8.24). Renzo Piano and Building Workshop, one of 15 entries from the United States, Europe, and Japan, won the international design competition for the airport.[108] Piano described his first visit to the site, standing on a boat in Osaka Harbor, where the airport was to be built:

Architects are creatures of the land. Their materials rest on
ground....Aboard that boat we tried to think in terms of wa-
ter and air, rather than land; of air and wind, elongated,
lightweight forms, designed to withstand the earthquakes to
which the area is prone; of water, sea, tides; of liquid forms
in movement, energy, waves. Many of the ideas that shaped
the project were born that day on the sea.[109]

Italian architect Renzo Piano was one of the designers of the Georges Pompidou Center in Paris[110] and winner of the 1998 Pritzker Architecture Prize.[111] The ultramodern main terminal building at Kansai International Airport has a rolling, cloudlike roof,[112] shaped like the leading edge of a giant airfoil.[113] (See Fig. 8.25.) An architectural magazine described the aesthetics as follows:

Piano's design, which...was chosen mainly for aesthetic rea-
sons, employs a roof of aerodynamic curves reminiscent of
an aircraft fuselage. The curves are meant to assist ventila-
tion throughout the terminal. The other major features of the
design are strips of plantings—Piano calls them "valleys"—
running the length of the terminal. The theme of technology

and nature coexisting is reinforced with an extensive planting scheme for the man-made island.[114]

Architect Gordon Brown observed:

Like Saarinen [who designed the main terminal at Dulles, and the TWA terminal at Kennedy], Piano sought to translate the sensation of flight into an earthbound structure. The difference is that the Italian architect carries out on a truly massive scale. The boarding concourse, with a swooping, stainless-steel-paneled roof, is nearly a mile long.[115]

The terminal uses outside trams to serve a mile-long line of gates.[116] Since land for the airport was reclaimed, and therefore expensive, the airport was designed vertically, in four stacked layers. International passengers arrive on the first floor and depart on the fourth floor. The second floor is for domestic arrivals and departures, while the third floor contains the concession/shopping mall area (an idea contributed by Aeroports de Paris),[117] which includes 27 restaurants, 48 shops, and the automated guideway transit system.[118] KIA is Japan's only airport handling both international and domestic flights, creating an advantage for Japanese carriers funneling passengers onto their growing international networks.[119] The terminal has a capacity of 30.7 million passengers a year. Construction of Stage 2, to double its capacity, is scheduled to begin in 1999.[120]

Denver International Airport

The original Perez Associates design for the main terminal building was a glass-walled stepped pyramid over a massive central atrium whose interior evoked a sense of a great railroad terminal hall. Denver is a crossroads for rail lines, highways, and a major airline hub, so the salute to Denver's railroad past seemed appropriate. The Perez design also called for multiple levels—level 6 for departures, level 5 for baggage claim, public transport, taxis, and car rentals, and level 4 for private automobiles—to enhance the amount of curbside space.

But Denver's Mayor, Federico Peña, wanted a more distinct and dramatic architectural facade for the first new airport to be built in the United States in 2 decades (the last "green field" U.S. airport was Dallas/Fort Worth International), so he invited Denver architect Curt Fentress to submit a design. Peña had in mind something as memorable as the Sydney Opera House. Given 3 weeks to come up with

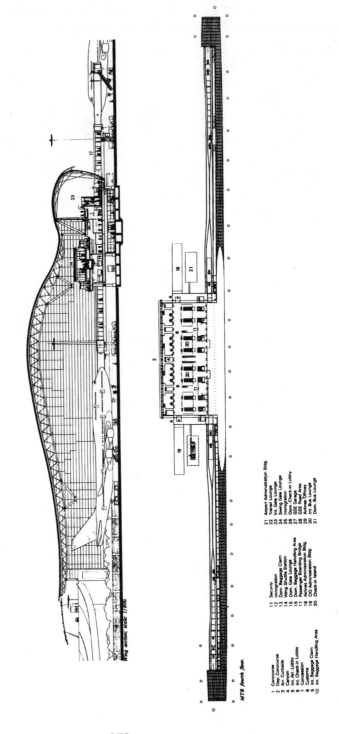

Wing section; scale: 1/800.

NTB fourth floor.

1   Concourse
2   Dep. Concourse
3   Arr. Curbside
4   Canyon
5   Int. Arr. Lobby
6   Int. Check-in Lobby
7   Concession
8   Customs
9   Int. Baggage Claim
10 Int. Baggage Handling Area

11 Security
12 Immigration
13 Dom. Baggage Claim
14 Wing Shuttle Station
15 Dom. Gate Lounge
16 Dom. Baggage Handling Area
17 Passenger Boarding Bridge
18 Airlines Administration Bldg.
19 CIO Administration Bldg.
20 Check-in Island

21 Airport Administration Bldg.
22 Transit Lounge
23 Int. Gate Lounge
24 Swing Gate Lounge
25 Immigration
26 Dom. Check-in Lobby
27 GSE Garage
28 GSE Road Area
29 Airlines Offices
30 Int. Bus Lounge
31 Dom. Bus Lounge

370

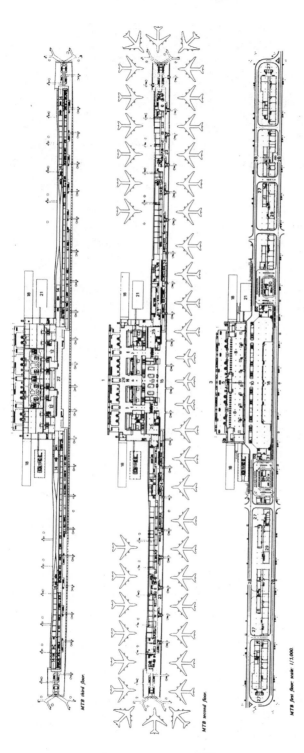

MTB third floor.

MTB second floor.

MTB first floor, scale: 1/5,000.

8.24 *Renzo Piano's terminal design at Kansai International Airport calls for long concourses extending from both sides of the terminal building, supplemented by a tram to whisk passengers to remote gates.* Renzo Piano Building Workshop.

371

8.25 *Renzo Piano's ultramodern roof line at Kansai resembles the leading edge of an aircraft wing.* Renzo Piano Building Workshop.

a design, Fentress folded a paper into a series of peaks and valleys, and proposed to drape the entire terminal in canvas—actually a translucent Teflon-coated synthetic material that could be stretched by cables from tall masts jutting up from the terminal. Such a roof adorns the Haj Terminal at the Jeddah International Airport in Saudi Arabia, stretching over several buildings like a great umbrella, providing shade from a blazing Arab sun. In the process, the 300,000 tons of structural steel in Perez' original design was reduced to 30,000 tons of steel in the Fentress design. DIA's terminal (with 3.5 million ft² spread over seven levels) cost less than $100 per square foot, compared to the United Airlines glass-walled remote terminal at Chicago O'Hare, which cost $370 per square foot.

Memorable it is, evoking images of American Plains Indian tepees, or, with a squint of the eyes, perhaps the snow-covered peaks of the Rocky Mountains and, from certain views, the cockpit bubble and windows of a Boeing 747 (Fig. 8.26). Denver sits on the Great Plains at the base of the Rocky Mountains, in a state with more than 50 peaks above 14,000 ft in elevation. Each of the 34 white fabric peaks is about 120 ft high. The interior is illuminated by sunlight passing through the fabric (admitting 11% of the sunlight) and, at several places, direct sunlight which slides across the floors of the enormous

8.26 *Curt Fentress' design for the roof of Denver International Airport has evoked images of tents, tepees, mountain tops, and clouds, among others.* Denver International Airport.

atrium as the day passes. Denver enjoys more than 300 days of sunshine a year, and the translucent roof enables the light to flow inward into the main terminal. But there is no consensus as to the image the roof evokes. Gordon Brown wrote:

> *Press accounts tended to peg DIA's white peaks as a reference to the surrounding mountains, but from both inside and outside, the fabric roof actually does little to remind the traveler of Colorado's rugged terrain. Mountains are the result of compression, while the terminal is clearly a product of tension. The feeling evoked by the roof is one of movement. In fact, if you look closely at its structure—its masts and rigging, and the awnings that look like bowsprits and a jib—as well as the wave-like patterns of the polished granite floor, what comes to mind is sailing.*[121]

Pavarini saw in Fentress' roof, "the nomad's tent, a symptom and symbol of departure and necessary condition for separation and parting, like the fate that awaits the whole of humanity."[122] But to others, DIA's sprawling tent evoked a different metaphor. One observer wrote, "Every time I approached DIA's main terminal I chuckle at its resemblance to an enormous white sow lying on its back, teats pointed skyward as it suckles United Airlines, contractors and bondholders."[123]

DIA's main terminal is named after aviation pioneer Elrey Jeppeson, a Colorado native. Unfortunately, a gaudy highway sign announces the nomenclature. The terminal's Great Hall is 6 times the size of New York's Grand Central Station. Fentress also designed the world's most impressive airport floor for the Jeppeson terminal, with colorful granite drawn from quarries around the world, cut by a now-bankrupt Italian firm to $3/8$-in sheets and assembled on a vacuum-impregnated, epoxy-infused, polypropylene-fiber mesh reinforcement.[124] It is far more lovely than the "art" that adorns DIA, commissioned by the airport's art commission.

Unfortunately, the concourses were designed to match the Perez stepped-pyramid design, with multistory atriums of glass and steel. The two designs were never integrated, leading to an eclectic architectural campus with no visual harmony between the main terminal and remote concourses. The exterior of the concourses is modern airport bland in appearance. But their interiors are visually interesting. Each of the three concourses has a 90-ft-tall, city-block-wide multistory atrium leading up from the trains connecting them with the main terminal, avoiding the subterranean claustrophobia of most airport train systems. Each of the concourse atriums is adorned with massive artworks that evoke a sense of travel. Though Denver devoted 1% of the hard construction costs to art, as noted above, much of it is of questionable quality.[125]

As we shall see later, not having a pedestrian tunnel parallel to the train tunnels created chaos when the train system became dysfunctional. Atlanta Hartsfield International Airport, upon which DIA was modeled, with remote concourses linked to the main terminal with an underground train and a parallel pedestrian tunnel with moving sidewalks, does not suffer from this cost-cutting myopia.

Macau International Airport

Macau International Airport includes a three-level, 54,000-ft^2 terminal facility.[126] The terminal includes only four loading bridges, with additional bus loading sites on the tarmac.[127] The terminal apron can handle 22 aircraft.[128] Arrivals enter on the ground floor, and departures enter on the first floor. The mezzanine above the ground floor is dedicated to airline lounges and concessions.[129]

Kuala Lumpur International Airport

Japanese architect Kisho Kurokawa came up with a "forest in a forest and forest in an airport" concept for Kuala Lumpur International Airport that incorporates tropical vegetation and a lush Malaysian landscape into the airport.[130] Ambrin Buang, a senior official at the airport, describes its design as "an atmosphere reflective of Malaysia and its people."[131]

A 15-year-old tract of jungle was planted in the center atrium of the remote terminal—an inverted glass cone with a forest core—bringing the views of the landscape from the airport perimeter directly into the heart of the airport. Forestry experts used wind tunnels to determine how best to cool the air in the arboretum. As a base for this patch of jungle, they laid 62,000 m^3 of soil and river sand on top of a waterproof membrane to protect the baggage system beneath it.[132] Mature trees were also planted around the main terminal.

The ceilings of the terminals add a fifth facade reflecting the Islamic influence of domes, a hyperbolic parabaloid structure in a modular design, coupled with the traditional Malaysian ceiling of timber (actually, the panels are made of metal for fire safety reasons, cut and painted to look like dark wood). (See Fig. 8.27.) The geometric floor pattern also reflects both the Islamic tradition of geometric patterns and the traditional straw mats of Malaysian homes. According to Hank Cheriex, one of the airport's principal architects, "The design reflects the symbiosis between the high-tech image of modern Malaysia, and traditional Malaysian identity."[133] One commentator described it in these terms:

> *Planners have taken a modular approach: A series of arching ceilings will be joined to form the terminal. Light will filter through skylights, and the terminal will be open and unmarred by columns or pillars....*

> *Kuala Lumpur's main building will boast a peaked and tented roof that connects to a linear series of gates. Its planners are relying on lush landscaping, lots of light and a naturalistic setting to create a forest in the airport.*[134]

Contracts for the design and construction, as well as operation and maintenance of the terminal at Kuala Lumpur International Airport, were given to a six-member Malaysian-Japanese consortium, led by United Engineers Malaysia Berhad.[135] The airport has 45 loading

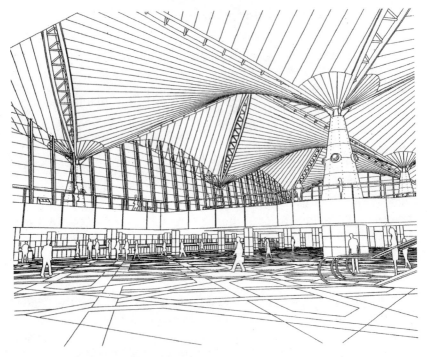

8.27 *The roof of the terminal at the new Kuala Lumpur International Airport is formed using hyperbolic parabolic shells, creating the abstract image of the Islamic dome, held up by a facade which looks like Malaysian timber.* Malaysian Airports Berhad.

bridges and 216 check-in counters.[136] The airport was the first in the Asia/Pacific region to install glass-walled passenger loading bridges, which give the passengers a view of the tarmac.[137] So, too, the people-mover system between the main and remote terminals is partially above ground so as to avoid the claustrophobia of the subterranean experience that exists at many airports.

Hong Kong International Airport at Chek Lap Kok

At Hong Kong International Airport at Chek Lap Kok (HKIA), between the two parallel runways lies an enormous passenger terminal, allegedly the largest enclosed space in the world but for a Dutch flower market. The initial terminal is a four-level, $^{1}/_{2}$-mi-long, 1-million-ft^{2}, Y-shaped building, with 48 air bridge gates, 288 check-in counters and a capacity of 30 million passengers a year.[138] About 8% of the terminal is devoted to 150 shops, restaurants, and services, exceeding the retail space at Singapore's Changi and Tokyo's Narita airports. The terminal has an energy-efficient roof, designed by Sir Norman Foster, who also designed the terminal at London's Stansted Airport. It is the roof's elevation, and progressive upward slope toward the terminal entrance, that gives HKIA a feeling of vast space, a sense of upward flight, and direction. (See Fig. 8.28.) One has the sense of having been swallowed by an enormous manta ray, looking up at the interior spine and rib cage as one proceeds deeper into the beast. The terminal's sheer scale is breathtaking. One source summarized its appearance:

> *Thanks to the roof, and the way the light comes through it, all the garish mess and complexity of an airport is sublimated to a successfully Fosterized interior, with a universal gray-green tinge, diffuse, subaqueous light and endlessly repeated shallow vaults which, all running in the same direction, help make sure you never get lost. The other big Foster move is to wall the building in glass, so that you can see the mountains, sea and/or aeroplanes from almost everywhere. The idea is simple enough, but it's something most airports don't give you.*
>
> *With its natural light, views out and simple shape, the Foster design is the opposite of the artificial, claustrophobic labyrinths that most airports are.*[139]

Another said, "Foster has roofed the main building with a billowing latticework of steel panels that bring to mind the gossamer batwing sails of a Chinese junk. The design, which floods the airy terminal with natural light, is monumental without being particularly alienating.[140]

Yet several design features deserve criticism:

1 The roof itself is a triumph of form over function. It could not support air conditioning units, so they were placed in the terminal's basement, obstructing airfield flows of tugs and carts and causing wasteful electrical energy consumption as colder air is forced unnaturally to rise.

2 An inordinate number of the 20,000 sheets of glass that wall the terminal are extraordinarily wavy, thereby distorting the view.

3 Nothing about the terminal's design (save perhaps the view out) gives one any sense of China. The terminal would fit as well in suburban London. Perhaps the design was intended to elicit the impression that Hong Kong is not *really* a part of the People's Republic of China.

8.28 *Norman Foster's floating-roof design for the new Hong Kong International Airport creates a spacious terminal walled with glass.* Airport Authority Hong Kong.

4 Entrances to the men's and women's restrooms converge in a space inadequate for either, let alone both, particularly for people lugging baggage.

5 Passenger flows are somewhat awkward: While arrivals proceed directly into the main terminal effortlessly from ramps connected to automobile and rail levels, and then to ticketing and check-in, passengers must then turn sharply left or right to the sides of the terminal to enter security and passport control. The straight-line route down the main pier has been obstructed by routing passengers circuitously through a shopping arcade. According to one source, "The much-lauded Norman Foster design of the terminal has to be questioned. It lacks 'flow' and the emphasis seems to be on marshaling passengers to the shopping mall."[141]

6 Connecting passengers arriving at the far end of the pier must hike all the way to the main terminal to change levels to the departure gates. If there are less circuitous connections, they are not signed.

The building can be expanded in 120-ft modular sections so as to expand capacity to 45 million passengers. A second X-shaped terminal is planned for the year 2010, to provide needed capacity until the year 2040, when the airport is anticipated to peak at 85 million passengers.[142]

Oslo Airport Gardermoen

From the outset, designers of the new Oslo Airport at Gardermoen recognized that the airport should be a statement of place—the nation through whose door a passenger enters. When in 1992 the Norwegian Parliament authorized construction of the new airport, it decreed, "The new main airport for the Oslo region shall be an outstanding example of Norwegian building traditions. Producing an integrated plan for artistic decoration shall be a required part of the architectural work for the main airport."[143] The airport was adapted to its surrounding landscape, protecting and enhancing existing vegetation. The main terminal features a Scandinavian design emphasizing Norwegian traditions and natural materials. The roof of the main terminal building is supported by the world's largest laminated wooden beams, while abundant glass allows daylight to penetrate the building. (See Fig. 8.29.) Airport shops and restaurants are also paying attention to architectural and aesthetic issues to accentuate the Norwegian flavor of the airport, using high-quality materials and offering Norwegian specialties on their menus.[144] The

8.29 *At Oslo Gardermoen, the airport design is Scandinavian in nature, employing local materials, with wooden beams, stone, and glass.* Illustration by Kurt Waltine. Oslo Lufthavn.

airport also devoted NOK18 million to artistic decorations around the terminal, showcasing contemporary Norwegian art.[145]

Airport planners should set forth a statement of principles which will govern design throughout the process. One of the most impressive such statements was created by the Oslo Airport Authority:

> *Oslo Lufthavn believes that it is of great importance to present uniform values for design and visual concepts for all aspects of the new airport, from the architecture to the design of printed material to landscaping and infrastructure. The airport will be an example of a modern hi-tech installation, with an emphasis on the best Norwegian construction practice and extensive use of natural materials such as stone, wood and steel. The new airport will be environmentally friendly and blend with its surroundings....*
>
> *The buildings at Oslo International Airport will reflect openness and the social equality of Norwegian society. They will have an open and friendly character designed to appeal to passengers, visitors and passers-by. The form and articulation of the individual buildings emphasize the following features of Norwegian construction traditions:*
>
> - *The building's primary functions will be clear and apparent.*
> - *Buildings with complex functionality should be divided into several parts, each of which with its own unique function.*
> - *Few, preferably local, materials should be used which will achieve a natural impression.*
> - *Logical construction principles will be visible in the design of the buildings.*
> - *New cultural impulses will be encouraged.*
> - *The degree of finishing on parts of the building should be graded, so that large areas are left rough and solid, while important construction and physical details are executed precisely and finished in detail.*
>
> *All buildings at the airport will reflect stringent requirements for safety, efficiency, and functionality. Each building will help create an image of Norway as a hi-tech society with a high level of education. They will also express our desire to conserve resources and protect the environment and show our proximity to nature.[146]*

Seoul's Inchon International Airport

The terminal at Inchon International Airport, the largest building in Korea, will be 530,000 m² (5.7 million ft²), with an expansion to 1.3 million m² by the year 2020.[147] It will sit in the middle of the four parallel runways ultimately planned. The terminal was designed to reflect Korean tradition, with the incorporation of pillars resembling Korean temples and the use of foliage resembling Korean gardens. (See Fig. 8.30.) The winning design was submitted by blind grading in 1992 by C. W. Fentress J. H. Bradburn and Associates of Denver, the same architectural team that designed the main terminal building at Denver International Airport.[148] Fentress spent several weeks wandering around South Korea to get a feel of the architecture of the nation before coming up with his design. At Seoul the "Great Halls" are "greened," with a hanging garden that is several stories high. As Curt Fentress observed:

> *The Korean Airport Authority wanted an airport that was reflective of the local culture, past as well as projected. Both [Denver and Seoul] sought gateways to their cities. Both wished*

8.30 *At Inchon International Airport, Curt Fentress came up with a design which, though modern, evokes Korean roof lines and an interior Korean garden.* Inchon International Airport.

to provide a definable and memorable image that rekindled the lost excitement of travel as well as furnishing a portal for arrival into a new place. In both cases, this desire produced the largest public works projects·in their respective countries within the last generation....

Combining the ideas of both temporary and permanent occupation revealed a design in Denver in which the roof top, the sky gate, became the visual gateway for arrival from both the ground and the air. The idea of using a tensile structure for Denver and the historical form of a sacred temple precinct in Seoul, replete with imagery derived from the colors, patterns and shapes of native animals, ecosystems, and even costume, to house a purely utilitarian plinth became economically plausible for some and architecturally essential to integrate these spaces. Thus, the airports assumed forms from which many things were intuitively communicated; at once contextual, regional, and instinctively derived by, of, and for human beings.[149]

Second Bangkok International Airport

In January 1995, the U.S. consortium Murphy Jahn/TAMS was chosen to design the terminal for the airport's first phase. But Thai politicians backing the bid of rival Aeroports de Paris pressured the airport authority to reverse its decision. This led to a diplomatic rift with the United States. The controversy over the contract, coupled with the refusal of some squatters on the airport land to move, placed the airport nearly a year behind schedule.[150] As noted in Chap. 3, governmental ambivalence caused more delay.

One issue of contention was the architecture. The "Terms of Reference" in the contract specified that "the facilities should reflect Thai character." The preliminary design submitted by the Murphy Jahn/TAMS team, including 1800 drawings, was not rejected within the rejection period provided under the contract. The Murphy Jahn design was for a modern glass-walled terminal. When the Thai government got around to objecting, it was too late. Some criticized the design as too modern, not adequately reflecting Thai culture, or the ego of "Helmut Jahn making a monument to himself." The issue became so contentious that Jahn eventually refused to speak with Thai officials. The Jahn position was that historical Thai

8.31 *Helmut Jahn came up with a controversial ultramodern design for the Second Bangkok International Airport.* Murphy/Jahn.

character would be reflected in the interior artwork and artifacts sprinkled about the terminal and the floor design, while the character of Thailand in the twenty-first century would be represented by the modern design of its airport terminal. According to architect Bryan O'Connor, the design reflects its global character—"the image of what Thailand will be, not what it was."[151] Moreover, height limitations imposed because of the proximity of the runways to the terminal prohibited building wat-like spires in the traditional architecture of Thailand.[152]

The Helmut Jahn design calls for a terminal which is overlaid by an enormous trellis roof, like an umbrella, for airspace. The glass leading down from the roof changes in shade from opaque to clear at eye level. (See Fig. 8.31.) Sunshine at that latitude, coupled with the humidity of Southeast Asia, creates enormous heat, so an innovative and energy-efficient way was devised to cool the terminal by running water through the floor. The concourses are designed to differ visually from the terminal so as to convey the message that they serve different functions. According to O'Connor, "The best airport is one without a sign, designed so that passengers can tell where to go because it is obvious."[153]

The terminal will be 500,000 m^2 (5.4 million ft^2) large on four or five levels. It will have 46 gates and 25 hardstands.[154] It is billed as larger and more modern than Singapore's Changi International Airport, one of the world's best.[155] The airport's master plan calls for an initial H-shaped terminal to the north, with a future addition of a T-shaped remote terminal, and ultimately a mirror-image H-shaped south terminal.[156] Sufficient land has been reserved for expansion between the terminal and the airfield, which has sufficient space for four parallel runways. But the H-shaped terminal is far less efficient for aircraft movements than a series of parallel I-shaped concourses that have been built an other major airports.

Midstream Design Changes at the World's New Airports

Airport design is a dynamic process. Airport master plans typically must be revised every few years. Technology changes, from air and ground navigation technology to aircraft size and shape. Demand may fail to meet, or may exceed, projections. The anticipated mix of local origin-and-destination versus connecting (transfer), or domestic versus international, passengers may change. Political will may be rallied for or against proposed infrastructure projects. The economic environment may encourage or discourage them.

All this suggests that agility should be a valuable trait of airport planners—the ability to modify the airport's plans to meet its contemporary and future needs, as reasonably as can be anticipated. But, as we shall see, change can be either costly or cost-saving, depending on when decisions are made. Moreover, changing any essential airport feature may require unanticipated and costly downstream changes to accommodate the initial changes. Project management software can assist in bringing the project in closer to its projected timetable and budget. Bechtel has developed an interactive airport simulation model, a "virtual airport." It allows simulation of passenger, concession, aircraft, and roadway activity that drives space planning and design programs. By changing one variable, the model can predict the impact on the rest of the project.[157] Let us examine four airports, and the method by which they dealt with change.

Chicago O'Hare International Airport

A 1946 article in *Fortune* magazine summed up the inadequacy of Midway, then Chicago's only airport, and then the busiest airport in the United States:

> *At almost all hours every telephone booth is filled, with people lined up outside. The dingy airport café is filled, with standees. To rest the thousands there are exactly 28 broken-down leather seats....The traveler consigned to hours of tedious waiting can only clear a spot on the floor and...while oversmoking, drearily contemplate his sins.*[158]

Planning for the new O'Hare Airport began during the closing days of World War II. Retired city engineer Ralph Burke prepared the master plan for Chicago O'Hare Airport in 1952. Though most airports of the day had their buildings along the periphery of the airport, Burke called for the terminal complex to be encircled by concentric taxiways connecting as many as six sets of parallel runways, fanning out from the central complex like a pinwheel.[159] Burke's concept included pier fingers with perhaps the first modern jetway, a light rail system to and from the Loop, and a multilevel passenger system.

But Burke died before completion of the project, so the city hired C.F. Murphy Associates to complete the design in 1957. The introduction of the Boeing 707 and McDonnell-Douglas DC-9 rendered some of Burke's design obsolete. Though the essential elements of Burke's plan were retained, Murphy modified the sizes, proportions, quantities, and construction timetables. Completed in 1963, the terminal was composed of four semiautonomous buildings with fingers protruding onto the airfield. The terminal was accessible from the land side via a dual-level roadway, with the upper level designated for departures (ticketing and check-in) and the lower level designated for arrivals (baggage claim).[160] This dual-roadway land-side access mode has been common with the majority of airports designed since. O'Hare reached maximum capacity by 1968, and has since been expanded with several rounds of additions and remodelings. In 1973, a hotel and garage were added, and in 1981, several projects were begun, including replacement of the old International Terminal.[161] In 1988, United's dynamic new remote terminal, designed by Helmut Jahn, opened. (See Fig. 8.32.) Wood Lockhart described it as follows:

8.32 *Helmut Jahn's United Airlines midfield terminal at Chicago O'Hare International Airport is among aviation's most distinctive modern architectural statements.* Murphy/Jahn.

Quickly dubbed "the diner" by irreverent pilots and air traffic controllers, who sensed in this steel and glass structure an echo of an earlier architecture that predates it at the airport. If it does evoke something of the great railway terminals of the nineteenth century, it is not because of a confusion of purpose, but because, like those earlier structures, it expresses an understanding of the possible relationships between transportation technology and architectural form. User friendly for both pilot and passenger, this terminal, with its remote satellite connected by an underground moving sidewalk, provides easy access to over forty aircraft, which when parked at the gate seem almost to be extensions of the architecture.[162]

O'Hare is an example of an airport that has evolved over time to meet demand well beyond any anticipated at its inauguration. As noted in Chap. 3, though several prominent Illinois politicians, including Republican Congressman Henry Hyde and Democrat Congressman Jesse Jackson, Jr., have teamed up to support a third Chicago metro airport, at this writing, United and American Airlines have exerted sufficient clout to prevent any movement toward construction of a third airport.

Munich's Franz Josef Strauss Airport

Strauss is an example of an airport that probably should have made design changes before construction, but could not because politics and litigation got in the way. The need for a new airport at Munich was recognized as early as 1954, and accelerated after a tragic crash in 1960. In 1963, the state of Bavaria and city of Munich established a commission to select a site for a new airport. Several sites were considered over the ensuing years, with the Freising/Erding-Nord site chosen in 1969.

During the next decade, more than 26,000 complaints were filed by private parties, and 180 statements were submitted by various authorities. After considering them, the government of Bavaria approved planning for the new airport in 1979, hoping to have it open by 1986. Construction began in November 1980, only to be halted by the Bavarian Higher Administrative Court 6 months later on grounds that the airport's land requirements were excessive. The Munich Airport Authority downsized its site from 2050 ha to 1387 ha. After having lain dormant for nearly 5 years, work on the site was allowed to continue in May 1985. The new airport was opened on May 17, 1992.[163]

This downsizing to meet environmental and legal objections resulted in two consequences. First, it constrained the size, and therefore the capacity, of the facility. It reduced terminal, apron, and cargo capacity, and eliminated the possibility of a runway for general aviation. The terminal needed more gates. A four-runway airport became a two-runway airport, hemming in terminal expansion to the north and south. A terminal that should be on two levels (one for arrivals, another for departures), has only one. Second, it meant that even as beneficial design changes were identified, they could not be made because any design change potentially would be subject to additional administrative review and litigation. The original layout of the airport had been adopted in 1968, as an O&D facility. A decade later, it was apparent that the airport would be a major connecting hub. Yet the terminal called for gates on only one side, rather than a satellite connecting aircraft from both sides.[164] However, space exists to the east of the main terminal for a future remote satellite concourse. And though it may consume years of administrative review and litigation, the possibility exists for expansion onto undeveloped farmland as demand warrants.

Denver International Airport

Denver International Airport (DIA) is an example of belated design changes sending costs skyrocketing. From the outset, Denver attempted to persuade the major airlines at Stapleton International Airport to participate in the design of the new airport. Though both United and Continental Airlines, Stapleton's principal tenants, were dead set against the construction of a new airport (both because of its cost and the capacity it might pose for new competitive entry) and refused to sign lease agreements until after the airport was well under construction, the airlines and their associations did participate early on in the design process. The result is one of the best designed airfields of any airport in the world, with enormous space (53 mi^2) for as many as 12 runways, ample runway separation for simultaneous instrument landings, high-speed exit ramps onto taxiways, center field concourse location, and ample space between the concourses for two-way taxiing and backing out without congestion.

But the airlines were late to sign up for gates, and when they did, they exacted a quid pro quo from the city in the form of significant design changes. Of the two major airlines serving Denver, Continental was the most nimble, able to make decisions quickly. United, the world's largest airline, has a bureaucratic corporate culture, less able to act swiftly. In the 1980s, Continental was headed by maverick Frank Lorenzo who, though he had his faults (which ultimately banished him from the airline industry) could make a decision. Delta had been the first carrier to sign up for gates when Atlanta Hartsfield International Airport was reconfigured, and it was able to lock up gates on all of Concourse A (the closest concourse to the main terminal) and half of Concourse B. Sensing a marketing advantage in being closer to the main terminal, in April 1990, Lorenzo cut a deal with Denver to occupy all the domestic gates on Concourse A. In exchange, Lorenzo insisted that all the international gates be moved from the main terminal to Concourse A, and a 600-ft pedestrian bridge be built over the taxiway linking Concourse A with the main terminal—the only bridge over an active aircraft taxiway in the world. These changes required all three concourses to be moved south, closer to the main terminal. As a consequence, DIA's traffic control tower, planned to be on the north side of Concourse B, is on the south side of Concourse C; the high-speed exit taxiways do not line up as neatly with the concourse aprons as originally planned.

But in December 1990, Continental Airlines fell into Chapter 11 bankruptcy, creating uncertainty as to Continental's future, while the city delayed construction at Concourse A. In 1992, Continental signed up for 20 gates on Concourse A—15 for five years, and 5 for ten years. The delay caused significant acceleration costs. In 1993, Continental began to downsize its Denver hub. It renegotiated its leases again, down to 10 gates, of which it would sublease most to other carriers, and occupy only three. Thus, an entire concourse had been built, and the airport redesigned, for a carrier which would ultimately occupy only three gates. Such is the volatility created by airline deregulation.

After Continental initially signed up for Concourse A, United Airlines stepped up to the table and began to negotiate for Concourse B in 1991. The most significant change United wanted was a destination-coded vehicle (DCV) baggage system for its concourse. The city consented, but decided that to prevent United from monopolizing space in the baggage tunnels, the DCV system would be put in all three concourses. But by 1991, much of the main terminal had been completed. Installing the automated baggage system meant adding a new mezzanine level inside the main terminal, adding structural steel to support the weight, ripping out fully or partially complete electrical, plumbing, and other systems, and piercing the levels to allow vertical movement of baggage. The baggage system cost $193 million directly, and another $100 million in related design and construction changes. But the city chose not to extend the October 1993 scheduled opening date of the new airport, even though these belated design changes were of enormous magnitude. As a consequence, the city missed several scheduled opening dates, incurring millions of dollars in finance costs, and leading to still another massive expenditure ($63 million) for a back-up baggage system when the DCV proved incapable of handling the load as opening neared. By 1994, change orders exceeded half a billion dollars. The airport itself did not open until 1995.[165]

Lynne Haraway of Bechtel Corporation described the difficulty encountered when designers of one part of an airport fail to anticipate the downstream impact of their design changes:

> [B]efore coming to Bechtel, I was involved in a terminal development program which called for the creation of a snack bar at the throat of the airport's main concourse. After the area was completed to shell condition, we took a walk through and discovered that the slab to slab distance was less than seven feet. In other words the ceiling was about three feet too low. We demanded that the ceiling be demolished and raised, naturally as a change order, only to discover that the ceiling had been lowered by an engineer as a change order to accommodate the unusually large air mixing box that was resting on the ceiling. We ruled out the problem after significant delays, some very expensive change orders and a few red faces, but this was an instance in which coordination failed, not because of any person's incompetence, but because some of the stakeholders could not visualize the downstream effects of their decisions.[166]

Yet another problem at Denver was the change in personnel. Between DIA's September 28, 1989, groundbreaking, and its opening on February 28, 1995, all the key players had changed. The airport had a new director and new chief engineer, the city had a new mayor and new head of the chamber of commerce, and one of the two principal airline tenants eliminated its Denver hub. This lack of continuity, coupled with belated design changes undertaken well after construction had begun in earnest, as well as a failure to move the opening date to accommodate changes of this magnitude, cost Denver much embarrassment, and took an airport originally projected to cost only $1.5 billion, into the stratosphere, where costs ultimately were $5.3 billion.[167]

Hong Kong International Airport

Hong Kong International Airport is an example of an airport able to adapt its airfield and terminal in a cost-effective way to anticipated changes in aircraft technology. The master plan for the new airport was completed in 1991. The new airport was to open with the second runway and one terminal pier (the northwest concourse) not yet complete. As construction began on the first runway and main terminal, planners at Airport Authority Hong Kong began to meet with airport officials to discern their long-term needs. Megacarrier alliances were beginning to percolate, suggesting more hub-and-spoke operations feeding the alliance networks at strategically located hubs, and the possibility of the development of new very large aircraft ferrying passengers between alliance partners' hubs.

In 1996, it appeared that the new large aircraft (NLA) would be in operation by 2000. The spatial requirements of the next generation of aircraft would be different from the Boeing 747s and 777s by causing a shift in the location of air bridges (jetways). If the piers were not extended, this would result in a net loss of positions by reducing the number of stands. To accommodate the gates at Hong Kong, the northwest concourse would have to be extended to a length 51 m longer than the southwest concourse (they look like a Y from the air). But that would be too close to the north runway. The answer lay in shifting the north runway centerline 15 m north. The island on which the airport lay had originally included a perimeter road for an occasional security vehicle. By eliminating the road, the runway could be moved north to accommodate the extension of the northwest concourse. The extension of the concourse allows passenger loading bridges to accommodate five Code F (as opposed to five 747) positions. If the new-generation aircraft do not materialize, the space is sufficient for 10 narrow-body aircraft positions.[168]

Teething Pains, or the Titanic Syndrome?

Several of the new "green field" airports missed their opening dates or opened with disastrous consequences because essential systems became dysfunctional, and back-up systems either could not handle the problem, or worse, were nonexistent. Some were labeled "teething problems," while others appeared far more serious.

Call it the Titanic syndrome—the collision of ego with reality. The Titanic was designed to be unsinkable. A few lifeboats were added as a precaution against the impossible—that the vessel might actually sink. Hollywood glorified the tale of man's arrogance colliding with an iceberg in the frigid waters of the North Atlantic.

Office towers have been designed with windows that cannot open in the belief that air conditioning systems will not break down. When they do, the buildings become uninhabitable. Even those with functional air conditioning eventually develop "sick building" syndrome, whereby the recirculation of the same air over and over again fosters bacteria, microbes, and viruses. Only a few office plants provide oxygen to those imprisoned therein.

Denver International Airport was the most technologically sophisticated airport ever built. Belatedly in the design process, after construction on the main terminal had begun, United Airlines asked for an automated baggage system for its concourse, Concourse B. Not wanting United to monopolize the baggage tunnels, the city extended the baggage system to all three concourses, on a grand scale never before attempted, and on a timetable that never contemplated it. The system runs 3500 destination-coded vehicles over 19.2 mi of track at speeds up to 19 mi/h, powered by 2400 induction motors and 6400 standard motors, and routed by 17,003 photo-sensors and 56 lasers.[169] A few hundred million dollars in cost overruns and delays later, the city abandoned the automated baggage system in all but Concourse B, where it was used only for outbound luggage. Traditional, tried-and-true tug-and-cart/conveyor belt technology that worked was substituted for a technologically sophisticated state-of-the art automated baggage system that ate bags.

In gathering material for the book *Denver International Airport: Lessons Learned* (McGraw-Hill, 1997), this author had the opportunity to interview many of DIA's planners. I recall one conversation in which a designer and I were standing over a thick stack of blueprints. I pointed to the 6500-foot-long tunnels linking the concourses with the main terminal building. At DIA, Concourse C is 1.2 mi from the main terminal and 900 ft from Concourse B. The only link between Concourses B and C is via the underground train, a 22-car automated transit system capable of moving 7000 people an hour—a Westinghouse-designed system earlier installed at Seattle and Tampa.[170]

"I see two parallel tunnels for the automated trains, flanked by a tunnel for the baggage system," I said. "Where's the pedestrian tunnel?"

"Don't have one," he said.

"But DIA is modeled after Atlanta's new airport, with remote concourses linked by underground train." I noted. "At Atlanta Hartsfield International Airport, if the train breaks down or is too full, a pedestrian can walk on a movable sidewalk to the concourses via a parallel tunnel."

"Don't need one here," came the response. "The automated people mover system we're installing has a 99.9% reliability ratio. You can count on it."

"O.K.," I conceded. "But what happens if it does break down? Have you stocked baby formula for the stranded passengers?"

"Well, we've installed these crossings," he said, pointing to the X-type crossover track tunnels linking the parallel rail tunnels at strategic points. "In the highly unlikely event a train might break down, we can push it on to a siding, or route the other trains around it via these crossover tracks." As I think back on that conversation, I have visions of the lifeboats on the Titanic. The designers of the Titanic assumed that it would not, it could not, sink. Investment in lifeboats was a wholly unnecessary extravagance.

The train did break down, about a year after the airport opened. The airport responded with a contingency plan to deal with the extremely unlikely event it might happen again. The plan called for marching passengers across the bridge linking the main terminal to Concourse A, running them through security, then funneling them aboard buses which would ferry them to the remote concourses.

But when the train became dysfunctional again, 3 years after the airport opened, it was chaos. There was no communication with the sardine-packed passengers stranded aboard the train which had lost a wheel and gnawed on the tunnel's cable. Those who escaped found the catwalk on the wrong side of the tunnel, and all emergency exits locked. They had to hop over an electric rail to reach the sliding concourse doors, and pry them open. Thankfully, no one was killed.

Only four metal detectors had been set up at the Concourse A bridge. The clogged bottlenecks created enormous delays and stranded thousands of passengers. Instead of busing Concourse B and C passengers from the main terminal, the contingency plan ensured that a broken train would disrupt travel for Concourse A passengers as well, though the bridge which had been built over the taxiway between Concourse A and the main terminal provided direct pedestrian access. Something which should never have happened has now happened several times, and the airport was unable to cope. Between the airport's opening in early 1995 and the middle of 1998, the contractual requirement that the trains have no more than 30 delays of more than a minute per month was breached in more than half of the months; in June 1998, DIA's trains experienced 24.5 hours of reduced service. The trains' service availability performance deteriorated from 99.7% in 1997, to only 99.3% in 1998, the worst year on record.[171]

One alternative was to buy some Disneyland-type vehicles to move passengers in the baggage tunnels. Another would be to dig the parallel pedestrian tunnel that should have been installed in the first place.

Denver's "all weather airport" was also shut down in early 1998 when Peña Boulevard (named after the mayor who imagined a great airport) bogged down in snow, trapping all but a busload of Denver Broncos. Why did it take 3 years for the airport to erect adequate snow fences? Shouldn't it have occurred to the people in charge that it sometimes snows in Colorado, and that the wind can create huge drifts on the Plains?

Perhaps no system was built to prevent snowdrifts from blocking the only highway to the airport because no sophisticated technology yet exists to perform that function. But at every turn in the road, DIA's planners opted for stretching the technological envelope, discounting tried-and-true, less sophisticated technology. Yet the more sophisticated the hardware and software, the more it has a tendency to malfunction. That suggests that back-up systems should have been incorporated into the airport's design. For the most part, they were not, for those who planned DIA assumed that these sophisticated systems were fail-safe.

Denver was not alone in having critical systems fail shortly after it opened. Several new airports which have opened in the 1990s have shared a common problem—high-tech collapses that cost millions of dollars, created tremendous public relations disasters, and inconvenienced tens of thousands of people.

At Osaka's Kansai International Airport, the airport was filled with lost bags months after it opened in 1994 because of confusion over the airport's three-letter code, KIX rather than OSA, the code for

Osaka's original airport. When Vancouver's new terminal opened in 1996, a year passed before the system allocating gates communicated properly with the system displaying flight information.[172]

The first problem that emerged when the new Kuala Lumpur International Airport opened on June 30, 1998, was that technicians unfamiliar with the new loading bridges were unable to connect them to aircraft. The opening date had been set, and completion hurried, to meet the opening of the Commonwealth Games in September, 1998, and to upstage the opening of the new airport at Hong Kong.

A total airport management system was installed at Kuala Lumpur International Airport, with airportwide communications and information technology, allowing airport tenants and operators to communicate and use information from shared databases.[173] But soon after opening, the US$168 million central computer network crashed, leaving thousands of travelers stranded. The central computer controlled everything from escalators to flight information monitors.[174] Ticketing had to be done manually. Communications lines broke down, as did mechanical baggage handling systems. Since there was no back-up communications system, the staff had to use cellular phones to communicate.[175] For several days, flights were delayed, check-in counters had long lines, and tons of perishable goods rotted.[176] Another problem was that the new airport was infested with rats. Pest exterminators were hired and hundreds of rats were killed. The fear was that the rats might chew on critical computer cables, as well as frighten the passengers.[177]

At the new Hong Kong International Airport at Chek Lap Kok, what has been described as the "world's most advanced communications galaxy" integrates various airport computer systems while allowing each to stand alone as a separate computer system. Flight display information, security, telephone, baggage handling, public address, building management, fire alarm, mobile radios, time generation and display—all are integrated.[178] The problem is, when it breaks down, it breaks down on a massive scale.

The world's most expensive airport, US$20 billion Hong Kong International, opened on July 6, 1998, to computer bugs, passenger delays, lost luggage, and cargo constipation in Asia's largest airfreight hub. The premature opening, 2 months before the airport was ready, was described as causing "interminable passenger delays, lost baggage, security breaches, near riots, foodless restaurants, drinkless bars, waterless bathrooms and useless telephones."[179] Originally, the opening had been scheduled for April, and though moved back to July, would not be moved again. The opening apparently was hastened by a desire to meet the first anniversary of the handover of Hong Kong to the People's Republic of China. Beijing's top administrative official, Tung Chee-kwa, was informed the airport would not be ready in time and was asked to so inform his superiors in Beijing. Tung replied that Beijing insisted there would be no turning back; the airport had to be open on schedule to accommodate the visit of U.S. President Bill Clinton and Chinese Premier Jiang Zemin.[180] According to Anson Chan, the government's top civil servant, delaying the opening was ruled out because it might damage the government's reputation.[181]

But the opening was marred by a collapse of essential systems. One source observed:

> *[Hong Kong International Airport was] bedeviled by computer bugs, stalled escalators, non-operating telephones, backed-up toilets, frustrating passenger delays and costly disruption to cargo distribution at Asia's busiest air freight hub....Erroneous information typed into a central database triggered a domino effect that sent the new facility into almost comic confusion—flights taking off without luggage, airport officials tracking flights with plastic pieces on magnetic boards and airlines calling confused ground staff on cellular phones to say where even more confused passengers could find their planes.[182]*

The Hong Kong Air Cargo Terminal Ltd. (HACTL), long the industry's standard setter, was unable to cope with an accelerated completion schedule and a computer system that went haywire. According to HACTL managing director Anthony Charter, "We really had not foreseen the impact of building contamination—the dust on the 14,000 sensors and reflectors." The day after the new airport opened, Charter announced, "With immediate effect, all freighter loads will be handled at Terminal 2 at Kai Tak. It appears we have some sort of bug which is deleting records."[183] Charter would later implicitly blame the airport authority, saying, "The whole feeling we got throughout this project was the airport would open on a certain date regardless of whether we were ready or not."[184] Operations were not fully restored at Hong Kong International Airport until August 24, seven weeks after it opened. Air cargo makes up nearly 19% of Hong Kong's US$188 billion in exports, and 24% of its imports. The breakdown in air cargo operations cost the Hong Kong economy 0.35% of its annual GDP.

One Hong Kong legislator said, "Our airport has become the laughing stock of the world. This was meant to be a first class project, but it has turned into a ninth class airport and a disgrace."[185] According to Richard Siegel, Director of Civil Aviation, the problems with flight information systems, and at HACTL, were complete system failures, not teething problems.[186] Another source wrote:

> Analysts generally attribute the botched premieres [at Kuala Lumpur and Hong Kong] to a combination of hubris and blind political pressure that forced a premature rush to start-up operations. Result: A public relations disaster of deeply embarrassing proportions and severe financial losses amid a fiasco of massive technical collapse and human error.
>
> What hurts most, especially in the Asian cultural contest, is incalculable loss of face accompanying the fiascos.[187]

At each of these airports, the opening was dysfunctional because of a collapse of essential systems. The fundamental problem was not just in stretching the technological envelope, for the more sophisticated the system, the more venues deep within the recesses of computer hardware and software for lapses and failures. The problem was also in stretching the envelope without installing simpler, less-expensive back-up systems to kick in when the primary systems fail.

In contrast, prior to opening Munich's Franz Josef Strauss Airport took 7 months to tweak its flight and baggage information display systems (FIDS and BIDS, respectively) prior to opening, and months more to ensure that the baggage system worked properly.[188]

Addressing Design Flaws

As we saw in Chap. 5, "Airport Planning," airport master plans tend to be drafted years ahead of completion of the airport infrastructure they identify as necessary. Between the time the need is identified and designers and engineers begin drafting architectural plans and the project is complete, demand and technology may have so changed to make the infrastructure improvement partially or totally obsolete, like a square peg being fit into a round hole. Wholly new problems may have arisen that were not contemplated by the master plan. Or, the airport designers may simply have got it wrong.

At the outset, it should be emphasized that as the year 2000 approaches, every airport must have satisfactorily resolved the "Y2K" problem with computers prone to meltdown as the last two digits become 00. So much has been written about that, only terse treatment is given the problem here. But suffice it to say, those airports which have not addressed the issue appropriately may face catastrophe.

Second, as the preceding section emphasizes, every essential primary system at an airport should have a back-up system in place, and appropriate contingency planning and preparation to deal with potential problems when they arise.

Third, every airport will have design flaws. Perhaps they weren't flaws when they were designed (or perhaps they were), but virtually every airport will have a part of its property that is lacking in its capability, or is inhibiting the efficiency of some other part of the airport infrastructure because of an oversight or design flaw.

Let's examine Denver International Airport (DIA). One must commend the designers of DIA for a runway and taxiway configuration which moves aircraft efficiently, expeditiously, and safely to and from one of the world's busiest airports. They deserve accolades as well for the midfield terminal design, with remote concourses linked to a land-side main terminal by underground train. That is the concept borrowed from Atlanta Hartsfield International Airport, and it satisfies the needs of hub connecting passengers efficiently. Walking distances between connecting flights are minimized with midfield terminals, for aircraft can crowd around all sides of the remote terminal like pigs at a trough. So, too, we should tip our hats to the decision to reserve 53 mi^2 (a land mass the size of the District of Columbia) for future expansion for up to a dozen runways, five concourses, and a two-thirds expansion of the main terminal building. Foresight marks the spatial planning of DIA.

But there are two overriding design flaws. First, as noted above, DIA's planners embraced cutting-edge technology without providing for back-up systems should primary systems fail. Stretching the technological envelope is always risky, for sophisticated systems do sometimes collapse. Indeed, the more sophisticated the system, the greater likelihood that something in the system will malfunction, for there are more parts of the software and hardware that can collapse. DIA's baggage system and its underground train are two obvious examples. Nor is there an alternative roadway into and out of DIA when four-lane Peña Boulevard (which links DIA with Interstate 70) becomes constipated, though snow fences and abundant snow-removal equipment will help.

The second design handicap involves the sacrifice of hub connection efficiency for origin-and-destination (O&D) local passenger efficiency. Given that hub efficiency was a top priority, the sacrifice of O&D efficiency was inevitable. That meant longer driving distances to downtown, and parking located remotely from departing gates, for example.

Nevertheless, these design problems can be addressed to ameliorate their adverse impacts. Contingency and emergency planning can be implemented to address system failures. When the automated baggage system held up DIA's opening, Denver installed a tug-and-cart back-up baggage system. A bridge links the main terminal to Concourse A. But when the train fails, there is no efficient way to move passengers to Concourses B and C. A pedestrian tunnel parallel to the train (as in Atlanta) would do the trick, but that fix is costly.

DIA has installed some fixes. Communications have been improved, so that a human being (rather than a recording) speaks with stranded passengers and informs them what the situation is and remedies are. The trains, concourses, and main terminal must have a public address system to alert passengers where the problem lies and what to do about it. Emergency supplies (food, drink, diapers, and formula) must be stocked to serve stranded passengers. An onsite hotel is being built.

The O&D experience could be improved in other ways. DIA is the only new airport in the world built without a rail connection. At one point, Denver Mayor Wellington Webb advocated a monorail to serve the airport. More recently, the city has advocated a less-costly (but expensive nonetheless) rail line to link the airport with the lower downtown Union Station. A rail line would reduce travel time, highway congestion, and air pollution, and serve as a lifeline during periods of inclement weather. A rail line would also allow Amtrak trains to be brought to the main terminal, so that airlines could connect passengers seamlessly to trains departing through the Rocky Mountains on the most spectacularly scenic ride in North America. Passenger buses are being allowed to connect at the remote concourses the way commuter aircraft park at concourse gates, further improving the seamlessness of intermodal connections.

Remote parking lots, initially placed miles and miles from DIA's main terminal, are being moved closer. Up-to-date signage is being installed in DIA's massive parking garages to inform drivers where the empty spaces are. At this writing, plans are underway to move the highway toll booths closer to the terminal, to allow that group of drivers only dropping off or picking up passengers to avoid the toll plaza altogether.

All airport improvements must undergo a cost/benefit analysis for prioritization, and some design flaws at DIA are too costly to fix. Installing an alternative to the underground train that links DIA's main terminal with its remote concourses would cost $30–75 million dollars, an enormous expenditure for redundancy to a system that breaks down only rarely (though it creates enormous problems when it does break down). At DIA, baggage claim is on level 5, where only public transport, taxicabs, and car rental vans pick up passengers. Private automobiles pick up passengers on level 4, meaning such passengers must lug their bags down a level via escalator or elevator. Though many airports allow free use of baggage carts, and several have carts which can shimmy up an escalator like a mountain goat, DIA charges $1.50 for a cart which can do nothing but roll horizontally. Departing passengers arriving on levels 4 or 5 must make their way up escalators effectively behind the ticket counters on level 6, a circuitous an unnatural passenger flow. A passenger arriving by automobile to drop a party off for check-in on level 6 must make a 3-mi loop around before parking the car and joining the party at the gate.

Inadequate space produces bottlenecks at the main terminal automated people-mover train departure area, ticketing queues on the west side of the main terminal (a feature insisted on by United Airlines to allow passengers to pass through between ticket counters, though unlike flow at O'Hare, the natural flow of DIA passengers is dead-end into a wall), and queues at the main terminal food court. Aesthetics could be improved by adding glass to the west end of each concourse, to allow waiting passengers to enjoy the fabulous view of the Rocky Mountain Front Range, from Long's Peak down to Pike's Peak.

At this writing, DIA's new aviation director, Bruce Baumgartner, has dedicated himself to four goals: (1) improving internal and external communications, (2) developing contingency and emergency plans, (3) making the organization stronger in terms of customer service, and (4) developing a comprehensive strategic business plan.[189] He met regularly with citizen advisory groups to listen to constructive criticism and respond to it. That process is producing a better airport, one more customer-oriented and user-friendly.

In mending or ameliorating an airport's design flaws, one should remember the words of Albert Einstein: "The significant problems we face today can not be solved at the same level of thinking we were at when we created them."

Airports of the Future

What will our airports of the future offer for the harried passenger? Here is one observer's prediction:

> *Like Japan's just-in-time production systems, railways will deliver passengers to airports 20 minutes before takeoff;*
>
> *Bar-coded flight information will speed travelers through check-in, coded passports through immigration;*
>
> *Travelators will whisk them through their gates, and underground people-movers will carry them under tarmacs to distant terminals;*
>
> *No luggage will be carried up stairs; no endless concourses will be walked; no bad food eaten.*
>
> *Vacationers, as opposed to business travelers, have different needs and more time. They will be able to swim, shower, shop, tinker around on a computer, enjoy a decent cup of coffee or just sit in the terminal, bathed in natural light and soothed by sympathetic color tones, and watch planes land. Comfortable terminal lounges, plush carpets, international chain restaurants, video games and bars all beckon from the next century.[190]*

The airport of the past is a place for processing passengers, then baggage, air freight, and mail. The airport of the future is a shopping center, a business center, indeed a small city, surrounded by an airfield. As we shall see in Chap. 10, such commercial centers will be powerful economic engines for the regions they serve.

Summary and Conclusions

Like all commercial activities, airports should identify their customers and their needs. Historically, airports have viewed the airlines which serve them as their principal customers. This dominant view is changing to recognize that the passenger is the customer whose needs must be satiated.

Airport planners and designers need first to determine what the airport's particular market will be. Is the airport to be primarily a passenger or a cargo facility? Is its primary customer to be an O&D or a connecting passenger? Dallas/Fort Worth and Kansas City International Airports work well as O&D airports, but awkwardly as connecting airports. In contrast, Atlanta Hartsfield and Denver International Airports work well as connecting airports, but relatively less well as O&D airports.

As we shall examine in greater detail in Chap. 9, terminals should be located and designed for efficient and seamless flows of aircraft and ground vehicles to them, and passengers through them. They should be directionally sensible, well signed, with natural passenger flows and without bottlenecks.

An airport's architecture can, and arguably should, be an aesthetic statement of the cultural heritage and natural features of the region. An airport is the first and last impression travelers will receive about the city and nation they are visiting. Airports at Seoul and Kuala Lumpur, for example, attempt to acquaint the visitor with a pleasant first and last glimpse of the cultural and natural beauty and heritage of the nation. The positive impression will likely enhance tourism and other business sectors, as well as giving local residents pride in their history, their culture, and their heritage.

Airport planners and designers would be well advised to adopt proven technology for critical systems or, in other words, "Keep it simple, stupid" (KISS), and to install back-up systems to kick in when critical primary systems fail. Munich learned from Frankfurt's experience with a high-tech sophisticated baggage system, and adopted "tried and true" off-the-shelf technology for critical systems. While it may not have all the bells and whistles of cutting-edge technology, it works fine. The most critical failure of Denver International Airport was the decision to install a highly sophisticated baggage system throughout the airport on a fast track without time to test it. Stretching the technological envelope also led to teething problems and costly and embarrassing failures at Hong Kong International and Kuala Lumpur International Airports. It is important that airport planners set aside ample time to test-run the technology. Munich set aside a sufficient time period to test the airport's systems and fine-tune them in order to "get the bugs out."

A well designed airport is an efficient and expeditious place to do business for all the airport's customers—airlines, airfreight companies, postal services, and passengers. Hopefully, it is also a pleasant environment in which to work or through which to travel.

9

Intermodal Transport Connections

"You can measure the dedication of an airport's planners to intermodal transportation by the size of the airport's parking garage."[1]—GIL CARMICHAEL, CHAIRMAN, UNIVERSITY OF DENVER INTERMODAL TRANSPORTATION INSTITUTE; FORMER ADMINISTRATOR, FEDERAL RAILROAD ADMINISTRATION

"An inclusive multi-modal terminal is the most important ingredient of seamless transportation."[2]—CRAIG LENTZSCH, CEO, GREYHOUND LINES, INC.

In Chap. 9, we examine:

- Principles of airport design which attempt to achieve seamless passenger flows between air and surface modes of transportation
- Case studies reviewing how intermodal issues have been addressed at several of the world's major airports
- The negative effect that earmarked funding has on the development of a comprehensive intermodal network
- Legislative efforts to provide funding for multimodal planning and project development, thereby increasing flexibility and cooperation among federal and state transportation agencies

Introduction

An airport is a transit point—no more, no less. It is that point at which passengers, cargo, and mail are moved from one aircraft to

405

another, or from an aircraft to a surface mode of transport. It is the latter—the intermodal movement—that is the subject of this chapter.

Air travel is a combination of time in the air, time in the airport, and time on the ground getting to or from the airport. Unfortunately, passengers often spend more time in delay on access and circulation roads than they do on the airfield. Time is money. It has been estimated that more than $20 million per year is lost by congestion due to difficult access to New York Kennedy International Airport alone.[3]

Though with hubbing, many passengers connect between flights without leaving the hub airport, very, very few passengers fly to a destination airport without going beyond its doors, and most need some sort of ground transportation to get to their ultimate destination. Airport designers and planners need to be cognizant of the passenger's need for seamless connecting service, not only between connecting flights, but between air and surface modes of transport.

Though more needs to be done, numerous successful examples exist of facilitating seamless connections of passengers from ground to air modes, and substituting surface modes for movements in congested air corridors. In places like Frankfurt and Osaka, short-haul traffic has been diverted to road or high-speed rail networks. At Frankfurt, a passenger can descend escalators to a rail terminal linking the airport with high-speed trains that whisk them away to the business and commercial capitals of Europe. Airports in Frankfurt and Paris are connected to as many as six separate urban, intercity, and long-distance rail and bus lines.[4] Improvement of the road linking Kuala Lumpur to Bangkok has slowed traffic growth in that air corridor. Ferries link Hong Kong to Macau and its airport.[5] But particularly in the United States, airport connections to modes other than the automobile are notoriously poor. Unfortunately, government aviation ministries concern themselves with laying runways, highway ministries concern themselves with building roads, rail and transit ministries concern themselves with laying rail. As in a Tower of Babel, they cannot seem to talk to one another, much less coordinate user-friendly seamless intermodal connections.

Access to the airport will be by car, truck, van, taxi, bus, streetcar, or train. Integration of the terminal with surface transportation connections via rail and road is important if passengers and freight are to flow with minimum inconvenience.

Passenger unloading points should be as close as possible to the processing positions in the terminal building. Ideally, they should be on the same floor as the ticketing and departure level. The distance between the modal connections can discourage their use. The distance between the surface connection (be it automobile, taxi, bus, or rail) and the entrance to the arrivals area of the passenger terminal should be extended only where necessary, and if extended, should be enclosed and offer the passenger the convenience of movable sidewalks and/or trams. Enclosure is essential both to shield the passenger against inclement weather and to provide the passenger with a sense of security. The distance between the modal connections can be 2 or 3 times longer if they are in an enclosed environment.[6]

In planning intermodal transportation facilities, four C's have been identified:

1 *Connections.* The convenient, rapid, efficient, and safe transfers of people and goods among modes that characterize comprehensive and economic transportation service.

2 *Choices.* Opportunities afforded by modal systems that allow transportation users to select their preferred means of conveyance.

3 *Coordination.*

4 *Cooperation.* With item 3, collaborative efforts of planners, users, and transportation providers to resolve travel demands by investing in dependable, high-quality transportation service, either by a single mode or by two or more modes in combination.[7]

Automobile Connections

Curb length is determined by the number and average size of the vehicles serving the airport given an average standing time for unloading of about three minutes. Some airports, such as Newark International and Denver International, reduce automobile dwell time by offering passengers the option of valet parking. Curb lengths will influence the building configuration of the terminal.[8] Curb area can be expanded by adding islands to the roadway for taxi stands, and hotel and car rental vans or buses, limousines, or courtesy vans. Curb space can also be expanded by adding levels to the terminal with highway viaducts serving each level. For example, different levels can serve arrivals and departures, and public and private pick-up.

The curb area should be properly signed to direct passengers to their preferred airline ticketing and baggage claim areas. The curb area should also be sufficiently wide to accommodate passenger flows and curbside baggage check-in. Additional curb space can be created by adding multiple levels to the main terminal to accommodate different functions (e.g., arrivals, departures, public transport). As in all parts of the airport, the design should seek to achieve economy and efficiency of operations, as well as passenger convenience.[9]

An airport should also be concerned about the congestion of the highways leading to and from it. For example, the highway leading from Narita Airport to downtown Tokyo is tremendously congested with local traffic. The highway leading from Don Muang International Airport to downtown Bangkok is worse. In contrast, the palm tree–lined highway leading from Changi to downtown Singapore is not only aesthetically pleasant, but also traffic moves quite well on it. The airport design with the best highway is Dulles International, with a dedicated highway on which local traffic is excluded between the Washington, D.C., interstate beltway out to the airport. Airports should be designed with dedicated highways, from which local traffic is excluded, with ample space for additional lanes and rapid transit in the future. Denver International Airport built a highway with ample space for future lanes or transit, but did not prohibit local ingress and egress. Over time, that highway could come to look like the bumper-to-bumper situation between Narita and Tokyo.

Unfortunately, there is an economic incentive at many airports to favor automobile access over transit and rail, for the airport earns significant revenue from the enormous parking garages which are typically built next to the main terminal, and from automobile rental lots on the airport property. At many airports, a passenger leaving for an extended trip can pay more than US$100 for the privilege of parking at the departure airport. Typically, the airport earns little or no additional revenue from a passenger arriving and departing via train or transit.

Transit Connections

In the United States, an early example of a dedicated rapid transit line serving an airport is Cleveland's "Rapid," which connected Cleveland's Hopkins International Airport to the central business district (CBD). London, Paris, New York, Philadelphia, Chicago, Atlanta, and

Munich extended their existing rapid transit systems to serve their airports. Today, among U.S. airports, Cleveland Hopkins, Philadelphia, Washington National, Atlanta Hartsfield, and Chicago O'Hare have rail transit stations within walking distance of their terminals, while Washington Dulles, Boston Logan, Baltimore/Washington, and Oakland have shuttle bus service to a transit station. Rail use at airports with transit accessibility within walking distance accounts for between 2% (at Philadelphia) and more than 9% (at Washington National and Atlanta Hartsfield) of passenger surface movements. Rail systems have been most successful in the United States serving frequent travelers with little or no baggage, and where rail offers faster or more reliable transport times because of highway congestion or parking inadequacy at the airport.[10]

At Washington, D.C., Dulles International Airport is connected to the Washington, D.C., beltway circling the city by a dedicated limited-access interstate highway segregated from the local road network. The construction of the Washington, D.C., "Metro" subway system took a tortuous path. Metro sought to bring its station to a point directly in front of Washington National Airport's main terminal building. But the site was used for V.I.P. parking (for members of Congress, Supreme Court Justices, and other high-ranking government officials). The Federal Aviation Administration (FAA) had jurisdiction over Washington National Airport, and the Urban Mass Transit Administration had jurisdiction over transit projects. Though both were subsidiaries of the U.S. Department of Transportation, they could not work it out. The result was that the Metro station was built some distance away from the airport terminal building, and shuttle buses were brought in to ferry passengers between them. The design was obviously flawed. So since the DOT couldn't cut through the bureaucratic inertia to build the subway system next to the airport terminal, the new terminal at Washington's Ronald Reagan National Airport was built next to the subway station. Such is the logic of decision making in Washington, D.C.

The Port Authority of New York and New Jersey is also attempting to improve surface transport connections to its airports. At Newark International Airport, the Port Authority has formed a joint venture with New Jersey Transit to build a $415 million monorail line linking the airport to Amtrak and a commuter rail line. To be completed in 2001, the monorail will connect an existing monorail that shuttles between the airport's three terminals to a new station at Waverly

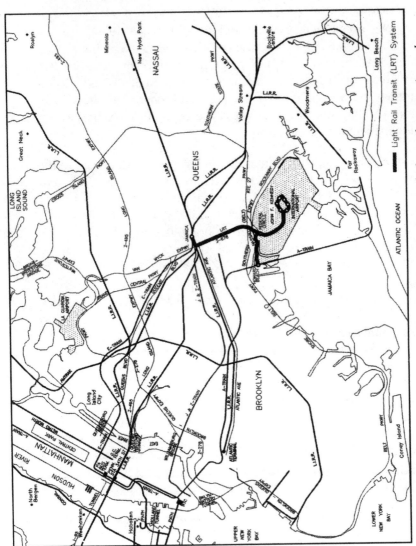

9.1 At New York Kennedy International Airport, an 8.4-mi light-rail system is under construction to link the airport's nine terminals to the surface transportation hubs at Howard Beach and Jamaica, where passengers can transfer to subway, rail, and bus lines. U.S. Transportation Research Board.

9.2 *Most airports are served by transit buses, such as this one at Copenhagen Airport.* Copenhagen Airports.

Yards in southwestern Newark, where passengers can connect to Amtrak or commuter trains bound for Newark or New York City. Already, a $75 million hotel and conference center is planned around the new terminal. A $1.5 billion train line is under construction to link New York John F. Kennedy International Airport to the Borough of Queens. Completion is expected in 2002.[11] (See Fig. 9.1.) Transit bus connections also provide connections between airports and the central business districts of the cities they serve. (See Fig. 9.2.)

Intercity Rail Connections

In Europe, major airports are linked to their nation's intercity railroad grid. Examples include Amsterdam, Barcelona, Birmingham, Brussels, Düsseldorf, Frankfurt, Geneva, Pisa, Rome, Vienna, and Zurich. At Vienna, a train runs every 30 minutes. The Austrian government has committed to reduce time between trains to 15 minutes by 2000, and to reduce the transit time from 31 to 15 minutes.[12] At Düsseldorf, a new station has been built on the main German railway network, with a monorail to shuttle passengers to the airport terminal.[13] In contrast, in the United States, only Baltimore/Washington International Airport is conveniently linked to Amtrak, and it only via a shuttle bus.[14] In 1998, motor coach service was established between San Francisco International Airport and the Amtrak station at Embryville, in the Sacramento Valley of California. Compared to Europe, rail connections at U.S. airports are primitive.

More should be done to link airports to the central business districts they serve via high-speed rail. Such would reduce highway congestion and air pollution. Multiple airports serving a single city should also be linked with high-speed rail, allowing a passenger arriving at one airport to transfer to a departing flight at another airport. This would enable one airport to specialize in long-haul, larger aircraft operations, while allowing a secondary airport to specialize in commuter and general aviation flights.

Investment in new rail technologies will not only allow more economical transit between airports and nearby cities, but will also ease airport congestion. By 1998, 12 countries had trains operating at average speeds above 125 mi/h. With a cruising speed of 185 mi/h, the TGV cut the transit time between Paris and Brussels to 85 minutes. Germany is spending US$5.4 billion on a high-speed train that will cut transit time between Berlin and Hamburg to under an hour. *The Economist* notes:

> *For journeys of between 100 and 400 miles, rail is now usually quicker than either air or road from city centre to city centre, if access times are taken into account. High-speed rail is also more reliable, more energy-efficient, more comfortable and arguably safer than any competing mode of transport.*[15]

Intercity Bus Connections

Intercity buses are the most efficient way of connecting passengers to an airport from cities located within 150 mi of the airport. Yet for bus connections to realize their full potential, several improvements must be made. Schedules, ticketing, and joint fares need to be coordinated with airlines. Only a few airlines code-share or contract with bus companies for through seamless service.[16] In 1998, the following airlines had established intercity feeder services in the United States:

Airtran Airlines: Atlanta–Macon, Georgia; Atlanta, Georgia–Chattanooga, Tennessee

American Airlines: Chicago–Rockford, Illinois

Continental Airlines: Newark, New Jersey–Allentown, Pennsylvania

Frontier Airlines: Denver–Boulder, Colorado

United Airlines: Chicago–Rockford, Illinois; San Francisco–San Jose, California; Denver–Fort Collins, Colorado; Portland–Salem/Albany–Corvallis, Oregon[17]

The other innovation that needs to occur is for more airports to allow buses to connect their passengers from the air side, planing and deplaning passengers at gates the same way commuter aircraft enplane or deplane passengers. Provided security and safety concerns are adequately addressed, sealed buses should be allowed to drive over the tarmac and up to the concourse gates in the way that commuter aircraft coming from small communities taxi up to commuter gates. This concept has been successfully pursued at Newark International and Denver International airports.

But some airports have even refused to allow intercity bus companies ticketing and baggage counters in their terminals. Chattanooga Airport refused to allow Greyhound to lease terminal facilities, forcing Greyhound to set up a facility off the airport property. At Chattanooga, bus passengers are forced to pass through two security screenings.[18] In contrast, Phoenix Sky Harbor Airport allowed Greyhound to build a terminal at it, which brings 120 passengers per week to the airlines from rural Arizona.[19] Again, smooth intermodal connections require the establishment of intermodal transportation facilities which bring together buses, railroads, airlines, vans, and automobiles.

Maritime Connections

Many of the world's major airports are bordered by bays, estuaries, or rivers. Only a few have realized the potential of using the water as a highway for passenger connections. One notable exception is Boston Logan International Airport, which has a ferry leaving from the airport every 15 minutes to downtown Boston, a ride that takes only 7 or 8 minutes.

The Remote Terminal Concept

A few cities have tried the center-city check-in terminal concept, allowing a passenger to be ticketed and check bags at the downtown terminal, then be ferried by bus to the airport. In essence, the remote terminal becomes an extension of the airport for ticketing and baggage-check purposes. New York City was the first to try such a concept, with a downtown terminal near Grand Central Station. In San

Francisco and Dallas, connecting bus lines allow remote check-in of baggage. Remote baggage checking is a great convenience for passengers. There is an old German saying, "Kein mann mit gepach ist frei," or translated, "No man with a suitcase is free."

Brussels and Hong Kong also offer passengers remote baggage check-in, but send the passengers and their luggage to the airport via rail.[20] Consumers are spared the ordeal of dragging bags aboard trains. Sparing them this inconvenience makes the intermodal connection eminently more attractive. The bags are processed by the airline, whisked to the airport via rail, and placed into the belly of the aircraft carrying the passenger.

One missed opportunity was at Denver, where unfortunately, no one considered use of the terminal at Denver Stapleton International Airport, located only 7 mi from downtown, as a remote ticketing and baggage facility, with trains whisking passengers and their bags to the remote concourses at the new Denver International Airport, more than 20 mi from downtown. Instead, Stapleton was closed, and within only a few years, dilapidated from hail damage and bursting pipes and unusable for anything.

The Multimodal Terminal Concept

Miami is establishing a $6.2 billion Miami Intermodal Center (MIC) to allow convenient rail, transit, bus, taxi, car rental, and private automobile access to Miami International Airport. The planned high-speed rail connection will allow passengers to connect to Palm Beach and Miami. Using French technology, trains at speeds of up to 200 mi/h will whisk passengers from MIC to Orlando and Tampa.[21] The MIC will serve as a regional hub for Amtrak, Tri-Rail, and Metrorail. It will serve buses, taxis, automobiles, bicycles, and pedestrians, and provide airport land-side terminal functions, including ticketing and baggage service, while being connected to Miami International Airport via an automated fixed guideway transit system.[22]

Terminal Design and Intermodal Access at the World's Major Airports

Let us examine several case studies of airport intermodal design, beginning with several of Europe's established airports, and moving to many of the world's new airports.

Frankfurt Main Airport

A pioneer in intermodal transportation, Frankfurt Main Airport built an airport railway station in the 1970s. Located beneath Terminal 1, the station is served by the S-Bahn to the Frankfurt Hauptbahnhof (transit time: 12 minutes) and cities throughout Germany, and InterCity trains from cities throughout Europe. By the turn of the century, a second rail station will be operable, handling InterCity Express trains. It will be linked to Frankfurt's air terminals to create an integrated "intermodal travel complex." The airport also has a Sky Line people-mover system which shuttles passengers between Terminals 1 and 2. The 90-ha CargoCity South at the airport is served by a rail line.[23] According to airport Deputy Chairman Manfred Schölch, "Frankfurt is a strong believer in intermodality."[24] He predicts that once nine high-speed trains serve the airport per hour, "Europe will not be a growth market for smaller aircraft," for short-haul trips will be handled by rail.[25]

Paris Charles de Gaulle Airport

In Paris, the high-speed TGV railway makes a great loop around the city, stopping at Euro-Disney as well as Charles de Gaulle and Orly airports, connecting with the Paris-Lyon main line.[26] In the very center of Paris Charles de Gaulle Airport is an intermodal complex which allows passengers arriving or departing by aircraft to connect directly with TGV high-speed rail, suburban trains, and minimetro "people movers" connecting the different terminals. (See Fig. 9.3.) It is one of the few airports in the world with a high-speed rail line in its core. Aeroports de Paris described the intermodal complex in these words:

> *Although it serves many functions and is highly complex, the vast multimodal complex is designed to be reassuring for passengers, who can easily see how the whole thing works, and which route they have to take. Its architecture makes it a strikingly visible feature of the airport, while emphasizing its multiple purposes.*

> *The multimodal complex comprises two curved structures 350 meters long, aligned along the same axis as CDG-2. Perpendicular to them is the belowground railway station, which is covered by huge glass roofs. The whole structure is topped by an oval-shaped hotel. Perched over the railway lines are four levels, each purpose-built for a specific function, plus the hotel, giving a total floor area of 100,000 square meters.[27]*

LE MODULE D'ECHANGES ET LA GARE
"AEROPORT CHARLES DE GAULLE TGV–RER"
MISE EN SERVICE LE 13 NOVEMBRE 1994

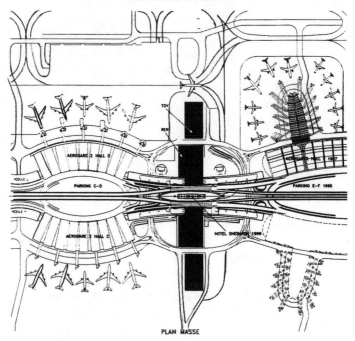

PLAN MASSE

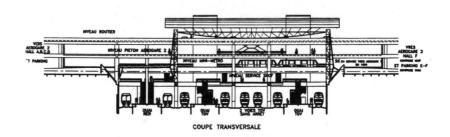

COUPE TRANSVERSALE

9.3 *Paris Charles de Gaulle Airport was one of the first modern airports to incorporate an intercity rail terminal into the heart of the airport.* Aeroports de Paris.

The uppermost of the four levels (level 4, or the "Viaduc et Terrasse") is devoted to road traffic, with drop-off and pickup points for taxis, shuttle buses, and coaches serving the railway station. Below that (level 3, or "Traffic") is a pedestrian level with moving walkways carrying passengers between the air terminals and railway stations. It also contains stores and restaurants, a 1100-m² business center, airport services, and the hotel entrance. Level 2 ("Mezzanine") contains connections between the two minimetros which transport passengers to CDG-1 and CDG-2. Level 1 ("Gare") is the lobby for the railway station, served by the long-distance TGV and the suburban railroads. Level 0 contains eight railway tracks and pedestrian platforms.[28]

Copenhagen Airport

In 1998, Copenhagen Airport completed an Air Rail Terminal below ground, just in front of the airport's International Terminal. The rail terminal sits on either side of the impressive wing-shaped entrance to Terminal 3. Pedestrian walking distance from train to airline check-in counters is only 100 m. A train departs for Copenhagen Central Station every 15 minutes, and takes only 10 minutes of travel time. An extension is being built to provide a fixed link across Oresund, to allow direct train service to Sweden from the airport. Long-distance routes will be served by InterCity Express trains.[29]

Munich's Franz Josef Strauss Airport

In designing Munich's new Franz Josef Strauss Airport, a primary objective was to interconnect different modes of transport as efficiently as possible. "Short distances for everyone" was the theme which dominated, making transfers from the land side to the air side as quick and convenient as possible.[30]

Strauss is located 25.8 km (16 mi) northeast of downtown Munich, beside the Isar River. Munich had the foresight to link the airport to its center city, not only with its ubiquitous autobahn system, but with direct subway routes. Connections to the central city railroad station take 38 minutes, with trains departing every 20 minutes, and the passenger paying only about US$8 one-way.[31] (See Fig. 9.4.) In contrast, the new airport is a US$60 taxi ride from downtown Munich.[32] Planners anticipated that 40% of the airport's passengers would arrive by train, while 60% would take buses, automobiles, or taxis. Train and bus passengers are able to check their bags in the railway station and proceed directly to their gates.[33] The tram makes 13 stops between

9.4 *Rail transit serves most airports in Europe, providing expeditious service to the center of the city. Most link the airport with the central railway station, where passengers can transfer to the major cities of the continent.* Flughafen München.

München Haubtbahnhof (central railway station) and Flughafen München (the airport). An express train would need another track. Unfortunately, local environmental opposition prohibited building a rail corridor to link Strauss to Europe's high-speed intercity rail network, objecting to the noise.[34]

Osaka's Kansai International Airport

The airport island on which Kansai International Airport (KIA) was built is linked to Osaka by a 4-km double-decker six-lane highway/railway truss bridge. KIA has highways, light rail connections to bullet trains, and express ferries.[35] The train runs into a railway terminal immediately in front of the air terminal building, making passenger flows smooth and expeditious. (See Fig. 9.5.) Regularly scheduled trains allow passenger movements from the airport to Osaka city center in 29 minutes. A marine terminal allows high-speed passenger boat service between the airport and Kobe in 28 minutes. Air terminals were established in Kyoto, Osaka, and Kobe to allow passengers to check in before reaching the airport.[36] Kansai serves Osaka, Kobe, and Kyoto, which are western Japan's major distribution center, Japan's largest port, and one of the nation's leading tourist destinations, respectively.[37]

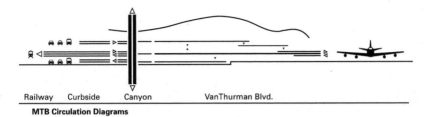

Railway   Curbside   Canyon   VanThurman Blvd.

MTB Circulation Diagrams

9.5 *At Osaka's Kansai International Airport, the train terminal is immediately in front of the air terminal, facilitating seamless intermodal connections directly from land side to air side.* Renzo Piano Building Workshop.

Denver International Airport

Denver International Airport (DIA) reflects America's preoccupation with the automobile, for it is linked to the city it serves only by a circuitous four-lane highway. The 12-mi-long highway, Peña Boulevard, is named after Federico Peña, Denver's mayor who championed the new airport and later served as U.S. Secretary of Transportation. Instead of being designed in a straight line, the shortest distance between two points, the highway was designed in the shape of an upside down and reversed L, which adds several miles to the journey. Moreover, the highway is not dedicated to airport operations, as was the highway at Washington Dulles International Airport, and will likely fill up over time with local traffic, as is the case Tokyo's Narita International Airport. Unfortunately, the bridges on DIA's highway were built out to only two lanes in each direction, a myopic short-term cost savings, but a long-term costly decision. On occasion DIA's highway has become impassable with snowdrifts, and why it didn't occur to its planners that snow fences would be needed defies explanation.

Fortunately, sufficient land was reserved between the parallel lanes on Peña Boulevard for a future rail track. There has been talk of building an "Air Train" to link DIA with Denver's downtown Union Station, but by 1998, no real action had been taken. The Union Pacific Railroad has a line with ample right-of-way within only a few miles of DIA. It is unfortunate that DIA's planners did not have the foresight to build a spur to the site of DIA's main terminal, use the spur to bring in the concrete and steel, and build the terminal around it. Amtrak's Zephyr could then make a scheduled stop on its way from Chicago to San Francisco. Eventually, the city or state might have been prompted to purchase rolling stock to provide rail service to downtown, both to ease highway congestion and reduce automobile pollution.

Originally, the airport's cargo terminals were to be placed on the north side of the property, far away from the main interstate highway that connects it to Denver. Fortunately, this oversight was corrected, and the cargo campus was moved to the south side of the terminal.

Macau International Airport

A new four-lane bridge links the airport island of Taipa to the Macau peninsula, making the new Macau International Airport about a 10-minute drive from downtown Macau.[38] A second interisland bridge is under construction to handle airport traffic.[39] The north side of the airline terminal will include a ferry terminal, for transfer to Hong Kong, about 40 mi away, or mainland China (Macau is 70 mi southeast of Guangzhou).[40] Highway and railway lines are anticipated to link the airport to the Shuhai special economic zone and Guangzhou.[41] The railway station, located west of the air terminal, will block expansion in that direction.[42] Dramatic improvements in rail and road infrastructure in the Pearl River basin are anticipated, which should feed traffic into the Macau airport.[43]

Kuala Lumpur International Airport

The new Kuala Lumpur International Airport is one of the few new "green field" airports to open without a rail link. This leaves passengers at the mercy of traffic-clogged roads for travel to the new airport, some 70 km south of Kuala Lumpur.[44] Construction of a rail line was delayed after currency devaluation, and will not likely open until after 2000.[45] The requirement that passengers check in 3 hours prior to departure at the new airport, plus the time consumed in getting to the airport by road, caused Malaysian Airlines to lose domestic passengers to the automobile. As a consequence, Malaysian Airlines moved some domestic flights back to the old Subang Airport.[46]

Hong Kong International Airport at Chek Lap Kok

Immediately in front of the terminal building at Hong Kong's Chek Lap Kok Airport is a transportation center, containing arrival and departure platforms for the high-speed rail to the city's urban and business districts. It also serves as a bus depot and taxi and rental car facility. A ferry terminal also provides sea access to Hong Kong, Macau, and points along the coast of southern China and the Pearl River estuary.[47] The airport has six transportation links, including a suspension bridge more than 4500 ft long, two railways (one exclusive to the airport), a 1.2-mi six-lane tunnel, and three highways.[48]

The new airport is about 15 mi from central Hong Kong, as the crow flies.[49] It is linked to the city by tunnel, as well as the 1377-m Tsing Ma Bridge, the world's longest suspension bridge carrying both a railroad and highway.[50] It is tied to the city by a 34-km railway, subways, highway tunnels, and container ship terminals.[51] More than 40% of the airport's passengers are expected to arrive and depart by rail.[52] The railway cars are spacious, with ample space for baggage, and consumer-friendly, with television monitors mounted in the seats to provide passengers with information and entertainment. A lighted display reveals how far the train is from one of its three stations—Tsing Yi, Kowloon, or Hong Kong Station. (See Fig. 9.6.) Stations then provide free shuttle bus service to the area's hotels, and connections to the MTR subway network. The rail stations are also effectively extensions of the airport terminal, allowing airline passengers to check their bags, which are then whisked away to the bellies of the aircraft they will board. Trains depart every 8 minutes, and take only 23 minutes from Hong Kong Station at Victoria to the airport. A one-way trip costs less than US$12. Buses depart every 10 to 15 minutes, take 65 to 70 minutes, and cost about half of the price of a train trip. One other attractive feature is the availability of an abundant number of complimentary baggage carts for passengers as they depart the trains. This enhances convenience, and expedites passenger movements.

The road and bridge infrastructure was built by the Mass Transit Railway Corporation; a private sector franchisee will run the Western Harbour Crossing.[53] An Airport Authority spokesman said, "the Government has taken into account the need for a fully functional and integrated land transport system to serve this world-class airport."[54]

Yet there are a couple of design features of the rail terminals that deserve criticism. Passenger flows are not natural, and signage is poor, leading to confusion as to where the rail and connecting bus platforms are located. Ideally, boarding platforms should be behind the airline ticketing counters, but at Kowloon they are well off to the side and down a level. Further, the rail platform is a wall of glass,

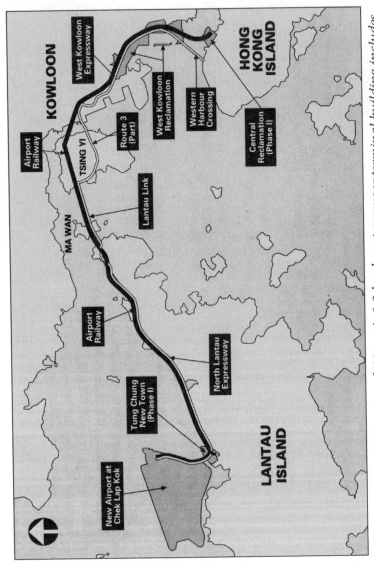

9.6 Hong Kong International Airport's 1.3-km-long passenger terminal building includes a transportation center along its front which will provide high-speed rail and highway linkage to Hong Kong and Kowloon. Airport Authority Hong Kong.

with really no visual differentiation as to which glass panels open to allow access to the train.

Oslo Airport Gardermoen

The new Oslo Airport at Gardermoen incorporated rail into its design, with a high-speed rail line capable of transporting 170 passengers at 210 km/h in about 19 minutes from the central business district to a rail terminal just in front of the main airport terminal building. In peak hours, this capacity can be doubled. An express train departs from Oslo Central Station for the airport every 10 minutes, while trains depart every 20 minutes to the airport from the western portion of suburban Oslo. The airport has set a target of 50% of all passengers arriving via rail.[55] To accommodate automobile traffic, the airport built 8000 parking spaces, of which 4000 are in Norway's largest multistory parking garage.[56] The rail terminal and parking garage are situated on either side of the air traffic control tower, immediately in front of the main airport arrival/departure terminal. (See Fig. 9.7.)

Seoul's Inchon International Airport

Intermodal plans for the new Inchon International Airport near Seoul call for an expressway, helicopter, ferry, and rail service to connect the airport to downtown Seoul.[57] A new 54.5-km expressway, a high-speed railway, and a suspension bridge are being built to link the airport with Seoul's industrial area and its seaport.[58] The upper level of the 4.4-km suspension bridge, to be completed in 2000, will be used for automobiles and trucks, while the lower level will be used for rail traffic.[59] Running between the new Inchon Airport and the Seoul Highspeed Railroad Station and through Yongsan Station, a 61.5-km rail line will link with the Seoul-Pusan Highspeed Railroad, to be completed in 2017 at a projected cost of US$3.5 billion.[60] The rail line will be built in two stages, one connecting Inchon International Airport and Kimpo Airport, to be completed in 2003, and the second connecting Kimpo Airport and the Seoul Yongsan Station.[61] In 1997, the rail project was described as being "behind schedule, over budget and having shoddy workmanship."[62] Seaport facilities will be constructed to allow intermodal cargo movements.[63]

Second Bangkok International Airport

Bangkok's existing airport has extremely poor surface access, consisting of a clogged series of highways perpetually in near-gridlock. (See Fig. 9.8.) As John Meredith observes:

9.7 *At the new Oslo Airport Gardermoen, a high-speed rail terminal is located directly outside the airport's terminal.* Illustration by Ole A. Krogness. Oslo Lufthavn.

9.8 *Bangkok's roads are some of the most severely congested of any major city in the world.* Photo by P. S. Dempsey.

> *In Bangkok heavy road congestion often makes journeys to and from the city longer than the flight time to adjacent Asian centers. Surface access is often one of the most neglected aspects of airport development.*
>
> *For the passenger journey does not begin or end at the airport. At both ends of the journey passengers have to go through an often lengthy and arduous process involving travel to and from the airport. If the passenger's journey is to be rapid, efficient and comfortable, it is essential that ground links to and from airports are improved to match the increasing capacity planned for airways and airports. Failure to make these improvements will simply transfer congestion in the skies, on runways, or in terminals to the road and rail links that serve airports.[64]*

The government is planning a road to link the two airports, for Don Muang will likely become the nation's primary domestic airport, while Second Bangkok International Airport (SBIA) will be the international airport.[65] At Don Muang, bus service to Thai cities is difficult to find, with poor to nonexistent signage. Buses to coastal cities, for example, are caught from Terminal 2, and ticketed under a sign for limousine and helicopter service.

Financing Intermodal Connections

An issue related to revenue diversion is whether airport revenue may be spent on surface transportation infrastructure to link the airport to its community. Airlines have generally opposed the use of airport revenue for such purposes, insisting that highway or transit funds be spent on surface transport connections. Nonetheless, it is unclear how passengers can fully use an airline's services if denied convenient access to the airport it serves.

All modes of transportation (i.e., air, rail, highway, transit, and maritime), and their corresponding federal institutions, tend to jealously guard their independent source of infrastructure financing. The segregation of funding along modal lines inherently creates institutional roadblocks to the facilitation of intermodal connections, as the Federal Aviation Administration seeks to have airport trust funds dedicated to airport infrastructure, the Federal Highway Administration (FHWA) seeks to have highway trust funds dedicated to highway construction, and the Federal Transit Administration (FTA) seeks to build transit. All three agencies are subsidiaries of the U.S. Department of Transportation, which should have the foresight and ability to facilitate seamless transportation between modes, among the fundamental purposes of the institution as set forth in its statutory charter. Though the DOT has set up a special unit within the Office of the Secretary to facilitate intermodal connections, and though Congress in 1991 passed the Intermodal Surface Transportation Efficiency Act (ISTEA) to facilitate intermodal transportation, jurisdictional turf battles and bureaucratic inertia inevitably inhibit seamless connections. Nonetheless, as we shall see, the agencies do sometimes come together and cooperate effectively.

The Federal Aviation Act requires that public airports accepting AIP funding agree that all revenue generated by the airport be used exclusively for the capital or operating costs of the airport, the local airport system, or facilities owned or operated by the airport directly and substantially related to the air transportation of persons or property.[66] The question has arisen whether airport funds can be spent on building or operating transit or rail lines or stations to be owned or operated by the airport, and whether these facilities are directly and substantially related to the air transportation of passengers.

Federal Aviation Administration regulations provide that airport access projects must preserve or enhance the capacity, safety, or security of the national air transportation system, reduce noise, or provide an opportunity for enhanced competition between carriers.[67] Such projects must also be for exclusive use of the airport patrons and employees, be constructed on airport-owned land or rights-of-way, and be connected to the nearest public access of sufficient capacity.[68] The FAA insisted that AIP funds be limited to land-side expenditure, "which encompasses the area from the airport boundary where the general public enters the airport property to the point where the public leaves the terminal building to board the aircraft. Typical eligible landside development items include such things as terminal buildings, entrance roadways and pedestrian walkways."[69] As we shall see, more recent interpretations by the FAA have liberalized this rather constricted view of the types of land-side projects which are appropriate for federal airport funding.

In 1996, the FAA approved the request of the Port Authority of New York and New Jersey to use PFC funds to extend Newark Airport's monorail line 4400 ft to an Amtrak/New Jersey Transit station off airport grounds.[70] At this writing, the largest intermodal project approved by the FAA for PFC funding was in 1998 for a $1.5 billion rail line linking New York's John F. Kennedy International Airport with the Long Island Rail Road and the E, J, and Z subway lines to Manhattan at Jamaica Station, and to Howard Beach. (See Fig. 9.1.) The Port Authority of New York and New Jersey alleged that the line would create "a more efficient vehicular flow at the airport by removing buses, shuttle vans, and private autos currently used by air passengers, airport visitors, and airport employees at JFK...", and that without the line, "ground access congestion would constrain projected O&D passenger growth at JFK and adversely affect the national air transportation system."[71]

The FAA concluded that PFC expenditures on the JFK rail link would satisfy the statutory and regulatory requirements by alleviating ground congestion on airport roadways and terminal frontages, by enhancing the efficient movement of airport employees, by freeing up capacity on the roadways for additional passengers, and by improving the airport's connection to the regional transportation network. It found, "Where ground access is shown to be a limiting factor to an airport's growth, a project to enhance ground access may qualify as preserving or enhancing capacity of the national air transportation system."[72] The FAA found that the rail line would enable an additional 3.35 million passengers to use JFK annually by the year 2013, and "therefore must be construed to have a substantial capacity enhancement effect on JFK, as measured in air passengers accommodated by the airport."[73] The FAA concluded that the rail link would "serve to preserve or enhance the capacity of JFK and the national air transportation system...."[74] The $3 per ticket Passenger Facility Charge would generate about $45–50 million a year, enabling the airport to pay off the cost of the line in 20 years.[75]

Rail lines at Atlanta, Chicago, Cleveland, and Washington, D.C., have been financed by transit systems rather than airports. The ISTEA legislation included a special appropriation for extension of the Bay Area Rapid Transit System (BART) to San Francisco International Airport (SFO). The Federal Transit Administration committed $750 million, or about 64% of the $1.2 billion project. The remaining $417 million will come from state and local funding sources.[76] The FAA approved airport funding for construction of a BART station at SFO.[77] The 8.7-mi extension, the largest since BART was built in the early 1970s, will have four stations. About 68,000 riders a day are expected to use the line when it opens in 2001.[78]

The Federal Transit Administration has also committed to contribute 72% of the construction costs of the $399 million extension of the Saint Louis Metrolink to Mid-America Airport in St. Clair County, Illinois. This light rail system already connects to St. Louis Lambert International Airport.[79]

The ISTEA legislation provided for flexible funding (up to $70 billion of federal highway funds and $10 billion of federal transit funds over 6 years) to support multimodal planning and project development. Though only $6 million was transferred from the highway trust funds to transit in the year preceding promulgation of ISTEA, by 1995, more than $802 million was being transferred annually.[80] Flexible funding allowed the various federal, state, and local transportation units to coordinate development of the Miami Intermodal Center, for example, which seeks to facilitate seamless passenger connections between air, rail, bus, and ferry modes.[81]

The Federal Highway Administration is financing 80% of the $11.6 billion 7.5-mi highway/tunnel extension of the interstate highway link to Boston Logan International Airport.[82] Federal and state highway departments have partnered successfully with airport authorities to connect road networks with airports at many cities, including Las Vegas and Pittsburgh. More than $300 million in PFC funding was approved for building an access road and tunnel at Las Vegas Mc-Carran International Airport, while National Highway System funds were used to construct the highways outside the airport property.[83]

In summary, federal funding of an airport with the surrounding highway, rail, or transit networks can come from the FAA, FHWA, or FTA. ISTEA's effort to foster more cooperation between these agencies has had limited, but significant, success.

Summary and Conclusions

At airports around the world, mass transit and high-speed intercity rail terminals are incorporated into the airport passenger terminal design. Except in the United States, most of the world's major new airports incorporate intercity rail directly into the airport terminal, allowing efficient, high-speed, environmentally sound ground access. America's infatuation with the automobile leads to the design of car-friendly airports with the main terminal building surrounded by (as at Atlanta Hartsfield International), or surrounding (as at Baltimore/Washington International, for example) enormous parking garages. In the long term, the congestion and pollution costs of the automobile must lead the way to rail.

Conventional networks, built up around the individual modes, have created capacity and service problems. In many cities, road and highway congestion have achieved gridlock. Highway fatalities continue to produce enormous human carnage. Inadequate land-side connectivity has devalued the speed and efficiency of commercial aviation. Land use, congestion, and pollution have become chronic problems in many regions. Paradoxically, transportation is both the cause and a potential solution to many of these problems. By integrating the separate transport modes into a single, unified intermodal network, transportation will not only meet the economic and mobility needs of a society, but can alleviate problems of pollution, congestion, safety, and energy consumption.[84] The strengths and weaknesses of each mode should be identified, means must be developed to minimize negative impacts, and an efficient and integrated transportation system should be established that is consonant with the goal of sustainable development.[85]

Airport planners should attempt to develop a comprehensive intermodal and multimodal transportation hub. At Miami, and new airports at Seoul, Hong Kong, Macau, and Osaka, we see efforts to link all modes of transport together, and a key desire to accommodate the rapidly growing cargo sector.

Airports should be designed to maximize efficiency of passenger, cargo, and mail movements. The goal must be to design airports which provide seamless surface transport connection to expedite the movement of passengers, freight, and mail through the airport.

10

Economic Impacts of Airport Development

"The economic role of airports cannot be overestimated, not only as engines for their region but also as vehicles for regional restructuring."[1]—Dr. Wilhelm Bender, President, Flughafen Frankfurt

"We recognize that the development of civil aviation has to keep advancing or else economic development can't keep up."[2]—Maritza Salinas, Honduras Secretary of Public Works and Transportation

In Chap. 10, we examine:

- The direct, indirect, and induced economic impact of commercial aviation upon world commerce
- The economic impact of activity at major airports on the regions served by them

Introduction

As the gateways to an increasingly global market, airports are the arteries through which commerce flows and economic growth is stimulated. Airports are an essential building block of the transportation and economic infrastructure of cities, regions, and nations, as well as the global economy. In terms of their economic contribution, airports stimulate trillions of dollars in trade, commerce, and tourism in the communities which build, maintain, and operate them.

Few industries are as important to the economic and social well-being of a nation as transportation. Transportation is a prime factor in shaping patterns of human social, economic, and cultural existence.[3] By shrinking the planet, aviation is a principal means of intermingling and integrating disparate economies and cultures and stimulating social and cultural cross-fertilization, economic growth, and diversity in an increasingly interdependent global environment. Whole economic sectors (e.g., hotels, automobile rental firms, convention business, and tourist destinations) depend on safe, reliable, efficient, and reasonably priced commercial air transportation.[4] "Just-in-time" inventory has moved to a global scale with the expeditious movement of cargo by air. The economic ripple effect throughout industrial and commercial sectors and geographic regions is profound. Aviation is an integral part of the infrastructure of a global economy. Because it defies national boundaries, air transport is uniquely international in scope.

As a fundamental component of the infrastructure upon which economic growth is built—the veins and arteries of commerce, communications, and national defense[5]—a healthy transportation system offering reasonable prices and ubiquitous service to the public is vitally important to the health of the region it serves. For that reason, governments the world over have promoted and encouraged its development by providing infrastructure, research and development, protective regulation, and occasionally outright ownership of airlines.[6]

The Economic Impact of Commercial Aviation

Commercial aviation facilitates the efficiency of business and government transactions, enabling a larger variety of relationships which, under the law of comparative advantage, stimulate broader economic growth.[7] As President John Kennedy observed, "A rising tide raises all ships." Thomas Petzinger said, "Like bees, airlines pollinate the world's financial system with capital. They create, mobilize, and transport wealth in proportions vastly exceeding the fares paid by the passengers."[8] They also create wealth far in excess of the tariffs paid by freight shippers.

Airlines and airports are essential components of tourism and travel, arguably the world's largest industry, accounting for 5.5% of the world's GNP, 12.9% of consumer spending, 7.2% of worldwide capital investment, and 127 million jobs, employing one in every 15 workers. By 2001, some 1.8 billion passengers (nearly a third of the

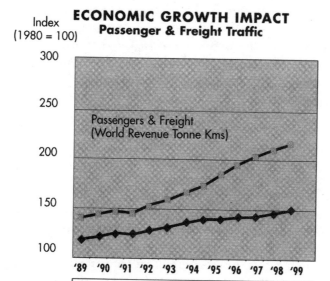

10.1 *The worldwide economic growth impact of air transportation has been on an upward trajectory throughout the past decade.* Dallas/Fort Worth International Airport.

world's population) will travel by air.[9] Within 20 years, the number of passengers is expected to double.[10]

The economic potential this increased passenger flow represents is beyond comprehension. It has been estimated that airports contribute $380 million to the U.S. economy annually.[11] In 1994, the aggregate economic impact of air transportation on the world economy was US$1.14 trillion; by 2010, it is estimated to be US$1.7 trillion. Some US$1.5 trillion worth of goods is flown around the world each year.[12] (See Fig. 10.1.) It is anticipated that by 2020, there will be 1.6 billion international tourist arrivals worldwide, and that those tourists will spend US$2 trillion. This reflects an annual growth rate of 4.3% and 6.7% respectively, far above the anticipated 3% growth in the world's wealth.[13]

Professors Kevin O'Connor and Ann Scott note that an airport is perhaps the most important single piece of infrastructure in the battle between cities and nations for economic growth.[14] Cities and regions invest in transportation infrastructure in order to "put a larger cup in the stream," and realize greater economic growth. Though not without their economic and social costs, the major air transportation hubs of Chicago, London, Frankfurt, Paris, Tokyo, and Hong Kong generate enormous economic and social benefits for their communities. Such cities are key centers for investment and trade in an increasingly global economy.

As *de facto* cities unto themselves, airports are major employers. For example, London Heathrow Airport employs 55,000 people in such professions as air traffic controllers, meteorologists, pilots, cabin crews, cleaners, caterers, ticketing and check-in staff, baggage handlers, mechanics, security personnel, fire fighters, concession sales personnel, and so on. Germany's airports employ 144,000 people. Over the next decade, it is anticipated that 100,000 new jobs will be created at German airports. Indirect employment is responsible for between 700,000 and 1 million people.[15] Every deutschemark earned at an airport produces induced income of DM1.70 in the region.[16] In 1995, the gross value added by German airlines was DM11.5 billion, while the gross value added by German airports was DM5.1 billion.[17] With 53,000 employees, Frankfurt Main airport is Germany's largest employer at a single location, surpassing Volkswagen's largest assembly site at Wolfsburg.[18] More than 15,000 people work at Vienna International Airport, and, with an average annual salary of ATS450,000 (about US$37,500), they have considerable purchasing power.[19] New York's LaGuardia Airport employs 9000 people and contributes $5.7 billion to the New York metropolitan area economy, generating some 63,000 jobs.[20] Saint Louis Lambert International Airport generates more than $184 million annually in direct wages.[21]

Amsterdam's Schiphol airport describes itself as "the engine powering Holland's economy."[22] It handles 30 million passengers a year, 40% of whom are in transit, and 1.1 million tons of cargo, 65% of which is in transit. By 2015, it is estimated Schiphol will handle 44 million passengers, 5 million high-speed rail passengers, 3.3 million tons of cargo, and about 430,000 aircraft movements a year. By 2015, the Schiphol Mainport will employ 60,000 people, and create another 66,000 jobs indirectly, or a total of 126,000 jobs. It will contribute US$18 billion, or 2.8% of Netherlands GNP. Thus, almost 3 cents of every dollar earned in the Netherlands will come from Schiphol.[23]

Aeroports de Paris estimates its activities account for approximately 9% of the region's total economic activity. Direct economic impacts in terms of salaries paid are Ffr38 billion per year, while indirect economic impacts (spending by airline passengers in the greater Paris region on hotels, restaurants, and such) account for 30 billion francs annually. Induced economic impacts (spending by airport employees, local companies, and local authorities) are estimated to be Ffr130 billion annually. The 1000 companies at Paris Orly and Charles de Gaulle Airports employed 76,000 people in 1994. Every million additional passengers creates about 1300 additional jobs.[24]

In 1911, Atlanta Alderman William Hartsfield predicted that his city's future lay in aviation. In the 1920s, he campaigned vigorously to turn a vacant automobile racetrack into an airfield, which the city did in 1925. Today, Atlanta's airport is named for this city leader. Atlanta reconfigured and expanded Hartsfield International Airport in 1980. That development was estimated in the 1980s to have produced some $7 billion in economic spin-off growth, the movement of *Fortune* 500 corporate headquarters to Atlanta (such as transport behemoth United Parcel Service), and the movement to Georgia of more than 1000 foreign firms.[25] In 1998, Hartsfield employed 40,000 people directly or indirectly, generated more than $170 million a year in revenue, and had an overall impact on the Georgia economy of $15 billion.[26]

Early estimates for the then-new Dallas/Fort Worth International Airport (DFW) were for 18 million passengers, 23,000 employees, and $630 million annually in direct and indirect economic activity generated by the airport by 1975.[27] By the mid-1980s, DFW was estimated to have produced $5.5 billion in spin-off growth and the relocation of major *Fortune* 500 corporate headquarters to Dallas and Fort Worth, while some 750 international companies had regional or national headquarters in the Las Colinas development near DFW.[28] Salt Lake City International Airport employs 12,500 people, whose total wages, salaries, and benefits exceed $450 million. The operating expenditures of the airlines, airport tenants, and the airport authority at Salt Lake are $750 million per year, nearly double the level of a decade earlier.[29]

Professor Andrew Goetz has found that populations in cities with higher air passenger/population ratios grew faster in periods subsequent to air travel growth than cities with lower ratios.[30] Other studies have noted the importance of reliable and frequent service in the location decisions of large corporations, and the relationship between service connectivity and professional administrative and auxiliary employment. Cities with abundant air travel appear to grow faster than cities without.

But a word of caution is in order. It is the classic question of which came first—the chicken or the egg. Does air transportation growth stimulate economic growth, or does economic growth (occurring for other economic reasons) increase demand for air travel? Actually, it appears to be a case of mutual causation, or bidirectionality in the relationship. Cities grow from enhanced transport connections, while cities that are growing receive enhanced airline service. Macroeconomic cycles; the metamorphosis in the production system; government policies; and changes in the social, environmental, cultural, demographic, and educational mixes of the population all impact the urbanization process. It is not a "build it, and they will come" phenomenon. Urban population and economic growth is not strictly reliant on air transportation infrastructure investment.[31] The two influence one another, but there is vast uncertainty as to how, and there is certainly no conclusive evidence of a cause-and-effect relationship between airport infrastructure investment and economic growth.

Economic Impacts at the World's New Airports

Munich's Franz Josef Strauss Airport

It is estimated that almost 25,500 people are employed by 300 different companies linked to the new Franz Josef Strauss Airport, making it one of the largest sources of employment in Bavaria. The airport employs 2800 people directly, while Germany's dominant airline, Lufthansa, employs 1600.[32]

Osaka's Kansai International Airport

Kansai is Japan's second most significant economic region (behind only Tokyo), accounting for 2% of world GNP. The Itami Airport handled only 10% of Japan's cargo and 17% of its international passenger traffic. But the new Kansai International Airport (KIA) was expected to increase the total cargo volume to 40% of Japan's total. By the year 2000, the volume of international cargo is anticipated to be 5 times greater than at Itami, and the number of international passengers to be 4 times greater.[33] In its first year of operation, KIA generated 660 billion yen for the Kansai region, consisting of 500 billion yen in commercial activity, 20 billion yen from the flow of goods, and 140 billion yen from airport-related projects.[34] The second-phase expansion is anticipated to generate 2760 billion yen and create 190,000 jobs.[35]

A principal reason for construction of the new airport was a desire to boost Osaka's economic position. The Osaka/Kobe region accounts for about 20% of Japan's gross national product. Dr. Yoshio Takeuchi, president of the Kansai International Airport Co., said, "We want to make the airport the center of economic activity in Kansai. We plan to build a new 21st century city around the airport that will embody a highly flexible electronic world."[36]

Airport officials hope KIA will tap dormant air transport demand in western Japan, which does not fly because of the inconvenience and congestion of Tokyo's Narita Airport.[37] Kansai is expected to stimulate a "revival" of the Kansai region, which is Japan's second-largest economic center. As Tetsuro Kawakami, head of the Kansai Economic Federation, said, "I feel that the effect of Kansai International Airport, together with the development of the Osaka Bay area, will let flourish the region's economic and cultural potential."[38] International flights at KIA increased 93%, from 338 in 1994 (the year it opened) to 653 in 1998.[39]

Macau International Airport

Before construction of the new Macau International Airport, the last scheduled service to Macau was in 1940, involving flying boats.[40] With the Portuguese colony scheduled to revert to China in 1999, later than Hong Kong, much capital has been transferred from Hong Kong, creating an economic boom. Not until now did Macau have the need, or the ability, for the construction of an airport.

The new Macau International Airport was originally planned as a facility for second-tier air carriers, or for primary carriers using Macau as a complement to Hong Kong. But China gave the territory the freedom independently to negotiate bilateral air transport agreements with other countries even after it reverts to China in 1999.[41] Moreover, delays in constructing Hong Kong's new airport, the fact that it will open with only a single runway, as well as its formidable cost, have stimulated interest in Macau's airport. The airport may be able to capitalize on its potential as a regional hub for the Pearl River Delta region, a fast-growing area which includes the industrial city of Guangzhou and eight other cities, with a population of 20 million.[42] (See Fig. 10.2.) The airport's proponents anticipate that the new airport will assist in the transformation of Macau from a gambling haven into a business and tourist gateway to China, similar to neighboring Hong Kong.[43]

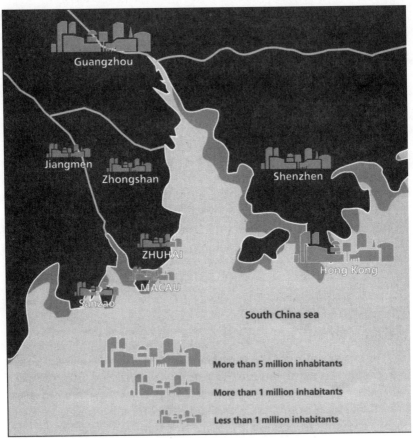

10.2 *The new airport at Macau provides access to major cities in the lower Pearl River Basin of China, which is projected to have 20 million inhabitants by 2000. The mainland provinces near Macau account for 7.1% of the gross national product of the People's Republic of China.* Autoridade de Aviacão Civil de Macau.

Kuala Lumpur International Airport

Malaysian officials perceive that the new Kuala Lumpur International Airport (NKLIA) will be a strong magnet to bolster Kuala Lumpur as an international business and travel hub. But it will compete with Bangkok to the north (which is building a new airport) and Singapore to the south (which will have a third terminal by 2000).[44]

Hong Kong International Airport at Chek Lap Kok

About 20% of Hong Kong's enormous trade, as measured by value, moves by air. Seven million tourists fly to Hong Kong every year, spending about US$7 billion. At the time the new Hong Kong International Airport (HKIA) opened, Hong Kong's existing airport, Kai Tak, had already become the world's fourth busiest international airport, and had reached full capacity.[45]

As yet, it is unclear what impact the enormous cost of HKIA will have on air traffic at Hong Kong. Nevertheless, its proponents insisted the new airport was essential if Hong Kong was to remain a commercial center.[46] The project also calls for development of a new town with a population of more than 200,000. The airport itself generates 45,000 jobs.[47]

Two other new airports will vie for Pearl River Basin regional development. As described above, Macau is building its first airport 23 mi west of Chek Lap Kok, while the new Shenzhen Airport was recently opened some 23 mi to the north. Airport proponents predict that, with such an enormous regional population base, the airports will complement, rather than compete, with each other.[48] As John Mok of the Airport Authority Hong Kong observed, "We believe that the airports planned in Hong Kong and in the Southern China region will, in all likelihood, serve their own specialized markets, ultimately working in harmony rather in competition with each other."[49] The room for economic development is vast. Half the world's population resides within a 6-hour flight of Hong Kong.

Seoul's Inchon International Airport

Between 1962 and 1991, the South Korean economy grew at an astounding 8.6% per year.[50] The Korean Transport Ministry predicts the new Inchon International Airport is "destined to become the hub of Northeast Asian air traffic."[51] At the groundbreaking of the terminal for the new Inchon International Airport, South Korean President Kim Young-sam said, "The 24-hour operating airport will play a role as a hub airport in the Northeast Asian region. It will also emerge as the center of international trade and finance."[52] In 1994, Seoul's existing airport was the fastest-growing airport in the world, with a 19.6% growth rate.[53]

At Inchon, priority has been placed on developing office, hotel, exhibition hall, conference center, and shopping facilities near the new airport. Two deluxe hotels and a large shopping center are planned near the passenger terminals.[54] A full-service tourism district may also be built on four islands near the new airport (i.e., Sin-do, Si-do,

Mo-do, and Jangbong-do) which will include three five-star hotels, an ocean dome, casinos, swimming pools, and a miniature race track.[55] A "Business City," accommodating 100,000 foreign business people and exhibition specialists is also planned.[56]

Second Bangkok International Airport

Thailand adopted a Vision 2020 plan, to grow from the world's twenty-sixth biggest economy to the sixteenth. It plans to spend billions of dollars on its Eastern Seaboard Development Project, an integrated system of industrial parks and rail, road, seaport, and airport infrastructure.[57]

Summary and Conclusions

Economists such as Adam Smith and David Ricardo recognized that, under the law of comparative advantage, specialization in production and increasing the geographic reach of markets enhances the wealth of nations. As catalysts for trade, major airports are responsible for regional economic activity well in excess of their direct costs. They make it possible for manufacture in one part of the globe to reach markets in another. Airports are an essential part of the infrastructure facilitating the free flow of capital and goods, raising the living standards of workers, and enhancing the choices of consumers.

That is not to endorse a "build it and they will come" philosophy. It is to say that if major population centers are to participate in a rising tide that elevates all ships, they must have adequate airport capacity relative to passenger and freight demand.

11

Airlines and Airports

A Symbiotic Relationship

"I never heard a word spoken about hub and spoke in the entire debate leading up to deregulation.... The hub and spoke system served to concentrate power at major airports in the hands of a few players and made it more difficult to have real competition in the marketplace."[1]
—CONGRESSMAN JAMES OBERSTAR, RANKING MINORITY MEMBER, U.S. HOUSE OF REPRESENTATIVES AVIATION SUBCOMMITTEE

"Why pave over some of America's prime farmland?"[2]
—ROBERT CRANDALL, CEO, AMERICAN AIRLINES

In Chap. 11, we examine:

- The volatility of airline profitability created by deregulation, and the dangers this creates for airport managers who may rely on ephemeral commitments of service
- The operational and economic advantages and disadvantages of airline hub-and-spoke versus linear route systems
- The costs to consumers of hub domination by major airlines
- The efforts of major airlines to suppress competition by engaging in predatory practices against smaller competitors
- The public policy dimensions of competition as a means of improving airline service and disciplining airline pricing

Cost versus Operating Efficiency

Airlines and airports are inextricably intertwined in an interdependent, symbiotic relationship. Neither can survive without the other. Airlines cannot take off or land their aircraft without runways, nor can they board or deplane their passengers without terminals. Airports rely on the revenue stream created by their principal tenants (the airlines) and their passengers for their very existence.

Airports are the hearts that pump the circulatory system in which airline routes serve as veins and arteries. In a less metaphorical sense, airlines are the airports' most important customers. Both join forces to provide seamless service to the passenger. Airports are the essential venue for funneling passengers into the air transportation network.

Airports and airlines typically confront several common problems— capacity, safety, efficiency, finance, and environmental concerns (principally noise). Several of these are legal and/or economic in nature. Many require design, planning, technology, and engineering for their solution. It is for that reason that it is prudent for airport planners to consult with and invite input from airline officials at the planning and design phase of airport expansion or new airport development.

But major differences exist between these two institutions. Airports are, by and large, publicly owned and operated; airlines are usually publicly traded private corporations. Airports have a long planning horizon; under deregulation, airline managers tend to think of operations only a few quarters into the future, thereby demanding extreme flexibility.[3] Airports, however, build infrastructure to last decades.

The construction of additional airport capacity is of direct concern to the primary tenants, the airlines. From the airlines' perspective, airport expansion has positive and negative components. On the positive side of the ledger, demand-driven expansion of capacity can reduce congestion and delay, leading to enhanced utilization of aircraft and labor, and reduced consumption of fuel. New infrastructure can enhance carrier efficiency and productivity in serving a growing customer base. In 1995, the U.S. Federal Aviation Administration estimated that delays cost the airline industry about $2.5 billion per year in higher operating expenses. Delta Air Lines estimated that air traffic inefficiencies cost it $300 million per year.[4]

On the negative side of the ledger, while some airport infrastructure costs are borne by passengers, taxpayers, and concessionaires, and the sale and lease of real estate, most of the cost of new and expanded infrastructure must be borne by the airlines (in the form of landing fees, terminal fees, aircraft parking fees, gate and hangar rental, ground handling services, air traffic control charges, and fuel taxes) and their passengers (in the form of passenger facility charges, parking, and tolls). For example, European airports with the highest turnaround costs based on airport charges, ground handling, and fuel are as follows:

1 Frankfurt (FRA)

2 Paris (CDG)

3 Düsseldorf (DUS)

4 Athens (ATH)

5 London (LHR)

6 Brussels (BRU)

7 Amsterdam (AMS)

8 Faro (FAO)

9 Manchester (MAN)

10 Santiago de Compostela (SCQ)

11 Stockholm (ARN)

12 London (LGW)

13 Madrid (MAD)

14 Larnaca (LCA)[5]

From the perspective of the airports, user costs are a relatively modest portion of airline operating expenses—a mere 4.1% of total airline average annual operating costs since 1978.[6] But from the airlines' perspective, whose net profit margins in the United States ranged between 2 and 3% before deregulation, and collapsed to less than 1% thereafter, even a modest economic burden is an onerous one.

During the 1980s, airline user charges constituted between 70 and 90% of airport revenue (although other sources insist that passenger carriers pay only about a quarter of airport costs, about the same as concessions).[7] ICAO predicts an average 9% annual increase in airport landing and associated charges, and an average 12% annual increase in route facility charges, through the end of this decade.[8] What is clear is that in recent years, airport and route charges imposed upon airlines

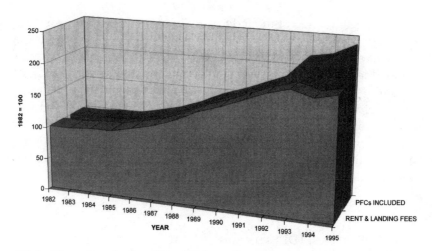

11.1 *Though only 5 to 6% of U.S. airline costs, airport rates and charges (per passenger) have risen significantly in recent years.*
Air Transport Association.

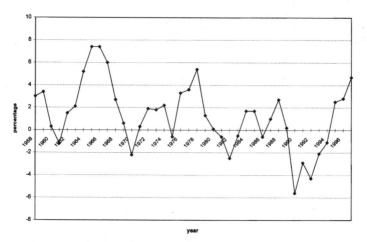

11.2 *U.S. airline profitability has been poorer since deregulation in 1978 than before it. The volatility of profitability has endangered commitments made to airports for infrastructure.*

have grown faster than most other operating expenses, and the ability of airline operating revenue to digest them.[9] (See Fig. 11.1.)

While airport capital equipment needs will total between $250 and $350 billion by the year 2010 (with much of that paid, directly or indirectly, by the airlines), airline capital needs worldwide (mostly for new aircraft) will, by some estimates, total $815 billion by the year 2000.[10]

But the cyclical nature of airline profitability, coupled with the uncertainty created by deregulation and liberalization, imperils airlines' ability to satiate their capital needs, and thereby jeopardizes their ability to fulfill long-term commitments to a wide range of creditors, including airports. (See Fig. 11.2.) Economic recession dampens passenger demand, thereby relieving some pressure on the infrastructure, and squeezing airline profits, making it more difficult for carriers to bear the cost of airport development. It is said of airlines that they order aircraft in good times and take delivery in bad. Of airports, it can be said that construction is sometimes begun in good times, and completed in bad.

Airline Route Structures: Hub-and-Spoke versus Linear Systems

Since deregulation, network economies of scale, density, and scope[11] have motivated most major airlines to increase the number of routes served from a centralized connecting airport—the infamous hub-and-spoke system. On the marketing side of the equation, hubbing allows carriers to offer a geometrically increasing array of city-pair products with every additional spoke. It also allows carriers to satiate consumer desires for frequent flights to that wide array of destinations. Hubs generate higher revenue, and can create barriers to market entry.[12]

Subsequent to deregulation, airlines began consolidating their operations around hubs. Hubs account for 70% of the flights offered by U.S. domestic airlines.[13] The large peaks and valleys in flights that hubbing necessitates impose enormous demands on airport infrastructure.

What are the characteristics of an airport that make it an attractive venue for a hub? A prudent airline seeks these attributes:

1 An interior point geographically situated for flow from several directions, particularly east to west, since that is the routing of most business traffic (the most lucrative share of the market)

2 A large population base to enhance high-yield origin and destination (O&D) traffic, preferably white collar (again, because business travelers pay more for air transportation)

3 Preferably, no nearby hubs or competing airports dominated by another airline[14]

Table 11.1. Largest U.S. Metropolitan Areas (1990) and Dominant Airline[15]

Metropolitan area	Population, millions
1 New York	18.1
2 Los Angeles	14.5
3 Chicago (United, American)	8.1
4 San Francisco (United)	6.3
5 Philadelphia (USAirways)	5.9
6 Detroit (Northwest)	4.7
7 Boston	4.2
8 Washington (United at Dulles)	3.9
9 Dallas (American at DFW, Southwest at Love)	3.9
10 Houston (Continental at Bush, Southwest at Hobby)	3.7
11 Miami (American)	3.2
12 Atlanta (Delta)	2.8
13 Cleveland (Northwest)	2.8
14 Seattle	2.6
15 San Diego	2.5
16 Minneapolis (Northwest)	2.5
17 Saint Louis (TWA)	2.4
18 Baltimore	2.4
19 Pittsburgh (USAirways)	2.2
20 Phoenix (America West)	2.1
21 Tampa	2.0
22 Denver (United)	1.8
23 Cincinnati (Delta)	1.7
24 Milwaukee	1.6
25 Kansas City	1.6
34 Charlotte (USAirways)	1.2
38 Salt Lake City (Delta)	1.1
40 Nashville	1.0
41 Memphis	1.0
44 Dayton	1.0
54 Raleigh-Durham	0.7
? Colorado Springs	0.3

Table 11.2. Ten Largest U.S. Airports (1990)[16]

Airport	Total passengers	Scheduled operations
Chicago O'Hare	58,775,486	775,687
Dallas/Fort Worth	48,915,464	713,958
Atlanta	47,629,438	569,438
Los Angeles	45,530,880	612,428
San Francisco	30,355,338	397,524
New York Kennedy	29,428,400	282,126
Denver	27,383,602	305,660
Miami	25,838,398	281,180
New York LaGuardia	22,789,260	333,512
Newark	22,207,200	356,957

According to the 1990 census, the largest metropolitan area populations of U.S. cities were as shown in Table 11.1.

Note that Nashville, Dayton, Raleigh-Durham, and Colorado Springs had too small a population to create a sufficient O&D traffic base to support sustained hubbing operations by American Airlines, USAirways, and upstart Western Pacific Airlines, respectively. Northwest has also downsized its Memphis hub. Table 11.2 lists the largest airports in the United States.

Chicago dominates U.S. air transportation because of geographic proximity and huge metropolitan population (8 million people, compared to Detroit's 4.7 million, Saint Louis' 2.4 million, or Minneapolis' 2.5 million). Dallas dominates the south central region, and Atlanta the southeast, for the same reasons—population base and geographic proximity. Atlanta, for example, has but one airport serving a metropolitan population of 2.8 million compared to the surrounding southern hubs of Charlotte, Nashville, Raleigh, and Memphis of less than half the people. The three largest U.S. airports are dominated by the three largest U.S. airlines—American, Delta, and United.

Compare these data with the number of passengers and operations at the largest foreign airports in Table 11.3.

Again, the three largest U.S. airlines—American, Delta, and United—have the largest U.S.-flag presence at the two largest foreign airports

Table 11.3. Ten Largest Foreign Airports (1990)[17]

Airport	Total passengers	Commercial operations
London Heathrow	42,647,235	388,289
Frankfurt	29,631,427	324,387
Paris Orly	24,205,570	191,421
Paris Charles de Gaulle	22,094,122	233,000
London Gatwick	21,047,089	203,211
Stockholm	14,822,450	257,606
Copenhagen	12,080,978	190,767
Düsseldorf	11,576,506	139,147
Munich	11,218,119	163,282
Vancouver	9,912,429	279,788

(American and United are the dominant U.S.-flag carriers at Heathrow; Delta is dominant at Frankfurt). And abroad, the national carrier of each nation is dominant. British Airways dominates London Heathrow; Lufthansa dominates Frankfurt and Munich; Air France dominates Paris; and SAS dominates Stockholm and Copenhagen. International carriers employ their gateways as venues for "sixth freedom" connecting traffic. For example, KLM puts enough capacity on the North Atlantic to transport the entire population of the Netherlands to the United States in a single summer. Most of the traffic is funneled through its hub at Amsterdam Schiphol, from or to points beyond.

The dominant megatrend on the U.S. deregulation landscape is the growth of hubs and spokes. Some call them "fortress hubs," where a single airline controls a major share of gates, takeoffs and landings, and passengers. Hubs became the dominant megatrend on the deregulation landscape for nearly all major airlines. Before deregulation, while Atlanta (for Delta) and Pittsburgh (for Allegheny, now USAirways), were moderately concentrated, no airline dominated more than 50% of the market (measured by gates, passengers, or takeoffs and landings) at any major airport in the United States. After deregulation, dominant airlines control more than half the enplanements at more than half of the nation's 50 largest airports, as shown in Table 11.4.

The infrastructure of gates and landing slots at several major airports has been consumed by the megacarriers, leaving little room

Table 11.4. Concentrated Hub Airports[18]

Airport	Carrier	Market share, %
Atlanta	Delta	79
Baltimore	USAirways	50
Charlotte	USAirways	95
Chicago (Midway)	Southwest	50
Cincinnati	Delta	91
Dallas/Fort Worth	American	61
Dallas (Love)	Southwest	100
Denver	United	63
Detroit	Northwest	77
El Paso	Southwest	64
Houston (Hobby)	Southwest	78
Houston (Intercontinental)	Continental	81
Memphis	Northwest	76
Miami	American	62
Minneapolis	Northwest	83
Nashville	American	64
Newark	Continental	54
Philadelphia	USAirways	60
Pittsburgh	USAirways	91
Raleigh/Durham	American	75
St. Louis	TWA	65
Salt Lake City	Delta	72
San Francisco	United	57
San Juan	American	66
Syracuse	USAirways	52
Washington (Dulles)	United	69

for significant new entry.[19] Some upstarts have focused on the remaining, smaller airports. In the early 1980s, America West focused on Phoenix and Las Vegas. In the 1990s, Western Pacific began operations at Colorado Springs, and American Trans Air focused operations on Indianapolis.[20] Several major carriers (i.e., TWA, Braniff, and Eastern) unsuccessfully attempted to establish a hub at Kansas City. In the 1990s, upstart Vanguard Airlines also focused operations at Kansas City.

One source notes:

> *A product of deregulation, the hub system was initially a great success. It enabled more airlines to envelop huge geographical regions like giant spiderwebs, snare passing traffic and expand market share. By replacing linear routes, it multiplied customers' flight options—and customers. American Airlines, for example, [in 1990 had] 455 daily departures from Dallas/Fort Worth International Airport compared with 137 in pre-hub 1979. Hubs also integrated remote cities into a national and international route network.*[21]

Strategically located hubs are designed to allow the carriers to blanket the nation with ubiquitous and frequent service. For example, United established hubs at Chicago, Denver, San Francisco, and Washington, D.C. (Dulles). American Airlines developed hubs at Chicago, Dallas/Fort Worth, Miami, San Juan, San Jose, Nashville, and Raleigh/Durham. Delta built hubs at Atlanta, Dallas/Fort Worth, Salt Lake City, and Cincinnati. Northwest has hubs at Detroit, Minneapolis/St. Paul and Memphis.

In contrast, TWA has a domestic hub only at St. Louis. Before its demise, Pan Am dominated no domestic airport. Among the airlines which have fallen into bankruptcy, only Continental had multiple strategically located hubs—at Houston, Denver, Cleveland, and Newark (the last it acquired from People Express on that airline's death bed).[22]

But this is a dynamic process. Hubs have been dismantled or downsized at Washington Dulles Airport (United), Denver (Continental), Dallas/Fort Worth (Delta), Dayton (USAirways), Kansas City (TWA, Eastern, and Braniff), and San Jose, Nashville, and Raleigh/Durham (American). Many do not have a sufficient O&D traffic base on which to load fixed costs. Downsizing a hub is a painful process, for every spoke eliminated deprives other spokes of traffic feed, causing the synergies of the hub to unravel.[23] Nevertheless, airline management must be sufficiently agile to withdraw from markets which are producing unsatisfactorily, and redeploy resources to more lucrative markets. Of course, airports which lose hub carriers often are left with excessive capacity built for a higher traffic base.

Hubbing is advantageous to airlines for a number of reasons. It allows enhanced marketing opportunities via the geometric proliferation of the number of possible city-pair markets which can be served. The number of passengers undergoes a corresponding exponential growth, while labor staffing increases at a much more moderate rate.[24]

To work, a hub must have a large number of flights ("banks," as they are called) from a large number of origins converging at an airport in close time proximity, so that passengers can readily transfer to flights departing to an equally large number of destinations. This requires a large number of airport runways, gates, and ground personnel and equipment, which are heavily utilized during the hub rotation, then sit idle between rotations. But the peak period drives infrastructure needs. Mel Olsen, who did the initial planning for American Airlines' hub networks, calculated that adding a new city to an existing bank of flights added 73 new passengers to the network, at an average fare of $180. The $13,140 in additional revenue compared favorably with the projected additional cost of only $560.[25] Thus, significant networking economies may be achieved via hubbing.

Both O&D and many connecting passengers pay a yield premium for the frequent service hubbing allows. At the concentrated "fortress hub," consumption of airport infrastructure often translates into higher yields. Yields at concentrated airports are more than 20% higher per mile for passengers who begin or end their trips there than at unconcentrated airports.[26] Hubbing also results in a yield premium for certain connecting traffic, particularly in the large majority of city-pair markets not served nonstop, for city-pairs less than 1500 mi in distance, and for smaller cities without multiple hub connections. Some hub carriers have learned to focus on this high-yield connecting traffic, and avoid the local O&D price wars with Southwest Airlines or low-cost nonunion upstart airlines.[27]

Airlines with more gates, takeoff and landing slots (at capacity-constrained airports), and/or code sharing agreements charge significantly higher prices than those without, according to the U.S. General Accounting Office. In fact, flights at airports where majority-in-interest clauses reduce expansion opportunities result in 3% higher fares; flights at slot-controlled airports result in 7% higher fares; and carriers with code-sharing arrangements charge 8% higher fares.[28]

In 1988, the eight largest airlines controlled 96% of the landing and takeoff slots at the four slot-constrained airports (i.e., Chicago O'Hare, Washington National, and New York's Kennedy and LaGuardia). In

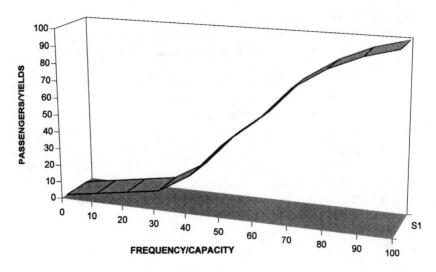

11.3 *The S-curve relationship between frequency and capacity along one axis, and passengers and yields along the other, reveals the incentive for market domination by airlines. An airline enjoys a disproportionate revenue advantage vis-à-vis its competitors if it provides a disproportionately higher number of flight frequencies than they do.*

1985, before the U.S. Department of Transportation decreed these public resources could be bought and sold in the market, the eight largest airlines controlled only 70% of the slots.[29] An airline which doubles the number of its gates enjoys a 3.5% increase in fares.[30]

These yield advantages are achieved because of a broader economic principle, the S curve, which posits that the dominant carrier in terms of frequency and capacity in any market will enjoy a disproportionate share of the traffic in terms of higher load factors and higher yields.[31] (See Fig. 11.3.)

Several sources have criticized hubbing as inefficient for short-haul operations, because of the increase in delay and congestion, which has a debilitating effect on labor and aircraft productivity. They point to Southwest's average of 20.4 minutes of ground time, compared to American's 50.3 minutes.[32] Southwest's half hour less ground time translates into enhanced aircraft utilization, 22% higher than the industry norm.[33] Moreover, the absence of banking flights into congested hub airports also results in more efficient use of ground personnel. Table 11.5 provides comparisons of aircraft utilization of selected major U.S. carriers.

Table 11.5. Major Airlines' Aircraft Utilization Per Day[34]

Carrier	Average stage length, miles	Daily aircraft utilization, hours flown
Southwest	380	10:55
America West	637	10:35
United	826	9:44
USAir	518	9:44
Delta	626	9:35
Continental	793	9:29
American	835	9:25
Northwest	705	9:08
TWA	695	9:01

Despite the growth and profitability of Southwest Airlines and its linear route clones, American Airlines' Chairman Robert Crandall argues, "hubs will continue to be the most efficient way, in most markets, of providing the frequent time-of-day choices travelers like even more than they like nonstop service. In fact, intense competition between multiple carriers offering very frequent service to many destinations via multiple hubs tends to make most nonstop service unfeasible."[35] He continues:

> One of our greatest strengths is a huge and well-integrated domestic and international route system centered around our six hubs. This hub-and-spoke system allows us to serve thousands of markets, thus generating a large network revenue benefit....
>
> While a hub-and-spoke system is admittedly more expensive to operate than a comparably-sized system of point-to-point routes, the system's incremental costs are more than offset by its enormous revenue benefits. For example, we estimate that there are fewer than 500 city pair markets in the United States big enough to adequately support point-to-point jet service. However, our hub-and-spoke system makes it possible for American to effectively serve over 10,000 markets—and realize a large revenue per available seat mile premium relative to point-to-point carriers.[36]

Nevertheless, hubbing sacrifices equipment and labor utilization and consumes more fuel and airport infrastructure, and is less convenient for passengers than a linear route system in city-pair markets sufficiently dense to support nonstop service. Clearly also, the United States is overhubbed by duplicative parallel route networks connecting virtually every conceivable city-pair market. As noted above, to trim costs and reduce capacity, in the 1990s, carriers began to downsize or close selected hubs.

Moreover, the only consistently profitable U.S. major airline, Southwest, embraces a point-to-point linear route system, which allows more productive equipment and labor utilization, and more efficient fuel consumption than does a hubbed operation. Southwest avoids congested airports, focusing instead on secondary airports in many markets, further facilitating a quick turnaround time (15 minutes is the goal). Southwest offers high-frequency point-to-point service in dense short-haul markets. By offering numerous nonstop flights a day between city pairs, it satiates consumers' desires for frequent service. By utilizing its gates and ground services at both end points throughout the day, Southwest enjoys economies of density. Because it shuns connections, and avoids congested airports wherever possible, Southwest is able to enjoy greater productivity in the utilization of its aircraft and labor, and consume relatively less fuel vis-à-vis the network carriers.

Think of an aircraft as a $30 million to $180 million factory that produces consumer goods—in this case, seats flown from here to there. A factory that operates more hours per day produces more seats. Southwest's planes sit on the ground between flights only 15 to 20 minutes. United's sit at its hub airports for 45 to 55 minutes, during which time they produce no product. Southwest also enjoys enhanced asset utilization by using its gates 10 to 12 times a day compared to United's six times a day, or American's seven.

By the early 1990s, several of the megacarriers appeared interested in following Southwest's lead, with Continental inaugurating CALite; United launching U-2, or United Shuttle; Delta creating Delta Express; and USAir developing Operation High Ground. America West tried a different approach to improving daily aircraft utilization— adding banks of discounted "red-eye" trips to Las Vegas (in 1996, these accounted for more than 14% of America West's flights).[37] Nonetheless, one source predicts that hubs will continue to dominate air transportation:

> *While there is increasing demand for point-to-point services and carriers willing to offer them like Southwest, Continental*

Lite, and a number of new entrants, the actual amount of traffic carried on the flights is only about 6 to 7 percent of the total traffic. Most city pairs are too small to justify point-to-point service so the maximum growth in traffic will probably never exceed 20 percent of the total traffic. Thus, at least 80 percent of all passengers are still expected to utilize hub services into the foreseeable future.[38]

While building a ubiquitous network produces marketing and revenue advantages, it imposes significant costs. Hub-and-spoke carriers do enjoy economies of density at the hub airport; the recurring banks of passengers allow enhanced utilization of gate and ground personnel and equipment, at least at the hub, although hubbing requires the leasing of many more gates than does a linear route system. Attempting to land and take off large waves of aircraft at a central point creates congestion, which causes delay (worsened when the weather becomes inclement), resulting in poorer aircraft and labor utilization and increased fuel consumption.

Hubbing has also led airlines to invest in relatively smaller aircraft than was the trend before deregulation. In the early 1980s, enthusiasm for the relatively small Boeing 737s replaced orders for larger aircraft such as the 747, for in hubbing, carriers do not need large aircraft to fly long distances; instead, they need small aircraft to fly relatively short distances. Thus, both the economies of aircraft size and stage length have been significantly sacrificed by hubbing. According to Melvin Brenner:

The deregulation-encouraged emphasis on smaller planes means that the industry will be losing the unit-cost efficiencies of larger aircraft. Many of the costs involved in aircraft operation do not increase proportionately to increased plane size. The result is that larger planes normally provide greater seat-mile cost efficiency.[39]

In sum, that which drives the airline industry to produce excess capacity (the high-frequency overlapping ubiquitous hub-and-spoke networks) has forced an erosion of systemwide efficiency and productivity in the postderegulation period. This has forced airlines to slow the pace of price decreases significantly from prederegulation trends. From 1950 to 1978, productivity improvements (primarily attributable to the economies of scale of larger and larger aircraft flying longer distances, as well as advances in engine technology) allowed real (inflation-adjusted) yields to decline 2.5% per year on average. In contrast, from 1979 to 1993 real yields fell only 1.7% per year.[40]

By the mid-1990s, U.S. domestic commercial aviation seems to have divided itself into two dominant types of service providers: (1) the ubiquitous network hub carriers (e.g., United, American, Delta, Northwest) emphasizing connecting traffic and (2) the short-haul nonstop point-to-point carriers (e.g., Southwest) focusing on O&D traffic. Internationally as well, new long-haul point-to-point carriers have emerged (e.g., Virgin Atlantic) to compete alongside the established international network carriers (e.g., British Airways). As noted in earlier chapters, airports which serve connecting traffic should be designed significantly differently from those whose primary customer is the O&D passenger.

Airline Concentration at Major Airports

Deregulation theory

Prior to deregulation, the consensus among many economists was that removal of governmental barriers to entry and pricing for airlines would result in a healthy competitive environment, one perhaps approaching the model of perfect competition.[41] Destructive competition, whose purported existence gave birth to regulation of the industry in the 1930s, was deemed unlikely to occur.[42] It was predicted that airport concentration was highly unlikely, for new entry (actual or potential) would keep the industry hotly competitive.[43]

But transportation has turned out not to be the ideal model of perfect competition that proponents of deregulation insisted it was. There appear to be significant economies of scale, scope, and density, and economic barriers to entry in the airline industry. Since deregulation, widespread bankruptcies, mergers, and strategic alliances have reduced the number of competitors to the point that major oligopolies now exist. Large airlines dominate the infrastructure of airports and their gates and landing slots, as well as their vertically integrated distribution vehicle, computer reservations systems.[44]

The theory of contestability served as a major intellectual justification for deregulation. The theory posits that if a monopolist or oligopolist begins to earn supracompetitive profits, new competitive entry, or the threat thereof, will restore prices to competitive levels. Should incumbents raise rates above competitive levels, new entrants should be attracted like sharks to the smell of blood to discipline the monopolist.[45] The theory is premised upon the belief that economic barriers to entry and economies of scale in the airline industry are relatively modest. Hypothetically, even the threat of entry should be sufficient to discipline the market and restore the competitive equilibrium.[46]

But the theory collided head-on with reality. Contestability has not proven to be a significant competitive catalyst in the deregulated airline industry. Because of formidable barriers to entry, many airlines are now able to exert market power in the markets which they dominate. With the creation of frequent flyer programs, travel agent commission overrides, and megacarrier dominance in fortress hubs and computer reservations systems, the entry of new carriers, and their growth, has been circumscribed. In a situation where market power exists, prices rise and/or the level of service deteriorates, excessive wealth is transferred from consumers to producers, and society's resources are misallocated, as consumers purchase alternative products or services it costs society more to produce.

Within deregulation's first decade, more than 120 new airlines appeared, although most were small, commuter lines.[47] This flood of entry caused pricing to spiral downward.[48] While a short-term boon for consumers, the pricing competition which emerged from deregulation was financially catastrophic for the airline industry.[49] More than 200 airlines have ceased operations or been acquired in mergers,[50] and a decade after deregulation began, only 74 carriers remained.[51]

In one important sense, the economic characteristics of transportation differ from those of most other sectors of the economy, and make it inherently vulnerable to overcapacity. Because of the revenue premium earned by offering greater frequency, all carriers have incentive to offer lots of frequency in their most important markets. If a manufacturer or retailer suffers a period of slack demand, it can usually store unsold inventory and sell it another day, when demand improves. In contrast, transportation firms sell what is, in essence, an instantly perishable commodity. Once an aircraft taxis down the runway, any unused capacity is lost forever. Empty seats cannot be warehoused and sold another day as could, say, canned beans. It is as if a grocer was faced with spoilage of all his or her canned goods on a daily basis—as if they had the properties of open jars of unrefrigerated mayonnaise. A grocer selling produce having the perishability of airline seats would be forced to have a fire sale every afternoon. Such an economic environment inevitably leads to distress sale pricing during weak demand periods, or when excess capacity abounds. Hence, the vicissitudes of the market cycle are particularly brutal for transportation.

In the short term, the pressure of overcapacity has a downward impact upon profitability as carriers scramble to lower prices to fill empty seats. In order to establish themselves as viable competitors, most carriers find they must carve out a geographic territory (usually a region or cluster of city pairs), lease gates, and provide a number of frequencies sufficient to satisfy business passenger demand. The short-term marginal cost of adding another passenger to a scheduled flight are "peanuts"—printing another ticket, adding another meal, a few drops of additional fuel, for example. Any ticket sold makes some contribution. Hence, strong incentives exist to sell empty seats for whatever will lure a bottom to fill them.[52] Carriers competing head-to-head spiral downward in destructive competition.[53] In such circumstances, while carriers cover short-term marginal costs, fixed costs are necessarily ignored.

Coupled with recession, these characteristics of air transport created distress sale pricing in the early 1980s, and again in the early 1990s. While a bonanza for passengers, the first two decades of deregulation were the darkest financial period in the history of domestic aviation. To survive, carriers had no choice but to slash wages, trim service and maintenance, and defer new aircraft purchases.

Airlines perceive that they need monopoly opportunities to stem the economic brutality of destructive competition, so they merged and developed hub-and-spoke systems, giving them regional and city-pair market power. Hubbing became the dominant megatrend on the deregulation landscape. Hubbing allows airlines to bring a number of flights from spoke cities into a central hub airport, interchange the traffic, and send the flights out to their final destinations several times a day. Hub domination allows airlines to raise prices significantly above competitive levels for O&D passengers, who begin or end their trips at the hub.

Airline industry concentration

More than 50 mergers and acquisitions were consummated after airlines were deregulated in the United States. The six largest airlines increased their passenger share from 73% in 1973, to 84% in 1986.[54] The eight largest airlines accounted for 80% of the domestic market in 1978, and 92% in 1990.[55]

Much criticism was leveled at the Department of Transportation for approving every merger submitted to it after it assumed the Civil Aeronautics Board's [CAB] jurisdiction over mergers, acquisitions and consolidations upon the CAB's "sunset" on December 31, 1984. The Airline Deregulation Act of 1978 insisted that the agency guard against "unfair, deceptive, predatory, or anticompetitive practices" and avoid "unreasonable industry concentration, excessive market domination" and similar occurrences which might enable "carriers unreasonably to increase prices, reduce services, or exclude competition...."[56] But these admonitions fell on deaf ears at DOT, which never met a merger it didn't like.

For example, DOT approved Texas Air's (i.e., Continental and New York Air) acquisition of both People Express (which included Frontier), and Eastern Airlines (which included Braniff's Latin American routes);[57] United's acquisition of Pan Am's transpacific routes; American's acquisition of Air Cal; Delta's acquisition of Western; Northwest's acquisition of Republic (itself a product of the mergers of North Central, Southern, and Hughes Airwest); TWA's acquisition of Ozark; and USAir's acquisition of PSA and Piedmont. The major mergers that have been consummated since deregulation are listed below:[58]

American: American, Air Cal, TWA (London Heathrow), Eastern (Latin America), Reno

United: United, Pam Am (transpacific), Pan Am (Latin America), Pan Am (Heathrow)

Delta: Delta, Western, Pan Am (Europe)

Continental: Continental, Texas International, Frontier, New York Air, Rocky Mountain, Britt, PBA

Pan Am: Pan American World, National, Ransome

Eastern: Eastern, Braniff (Latin America)

Northwest: Northwest, Republic (North Central, Southern, Hughes Airwest)

USAirways: US Air (Allegheny), Piedmont, PSA, Empire, Henson

Southwest: Muse, Morris Air

TWA: Trans World Airlines, Ozark

By the mid-1980s, Alfred Kahn characterized the airline industry as an "uncomfortably tight oligopoly."[59] He was been particularly critical of the Department of Transportation's permissive approach to

Table 11.6. U.S. Airline Market Shares
(in Percentage of Revenue Passenger Miles)

Carrier	1978	1983	1984	1989	1990	1991	1994	1995
United	17.4	16.0	15.5	16.4	16.7	18.5	21.0	21.0
American	12.8	21.7	21.4	17.3	17.0	18.6	19.4	19.3
TWA	11.9	10.1	9.6	8.3	7.5	6.3	4.8	4.7
Eastern	11.1	10.5	9.9	2.7	3.7	0	0	0
Delta	10.3	9.6	9.2	14.0	13.0	15.2	16.8	16.0
Pan Am	9.3	10.7	9.5	6.8	6.8	4.1	0	0
Continental	3.8	3.5	3.7	9.1	8.6	9.4	8.1	7.5
Northwest	3.1	6.6	6.7	10.8	11.3	12.0	11.2	11.7
USAirways	1.8	2.7	2.8	8.0	7.8	7.7	7.5	7.2

airline mergers. Said he, "They have been *permitted* by a totally, and in my view indefensibly, complaisant Department of Transportation. It is absurd to blame deregulation for this abysmal dereliction."[60] National levels of concentration rose to unprecedented levels. (See Table 11.6.)

Airport concentration

As we have seen, before deregulation, no single airline accounted for more than 50% of gates, enplanements, or takeoffs and landings at any major airport. Today, more than 20 major airports are dominated by a single carrier, with more than 60 or 70% and sometimes 100% of landings, takeoffs, gates, and passengers.[61]

Since deregulation, all major airlines but one (i.e., Southwest) have gravitated toward the hub-and-spoke means of distribution. Though hubbing increases costs by lowering aircraft, gate, and labor utilization and increasing fuel consumption, airlines have been attracted by their revenue enhancement potential. According to Lehman Brothers, "Airlines that control a greater percentage of their hubs' gates obtain significant benefits in terms of scheduling flexibility and insulation from new competition."[62]

For passengers who live in a city served by an airport which an airline has transformed into a hub, the hubbing phenomenon is both a curse and a blessing. The blessing is that O&D passengers enjoy frequent nonstop service to a wide array of destinations. The curse is that they pay exorbitant fares for such service, well above competitive levels.

Adding a spoke to an existing hub geometrically increases the number of city-pair markets an airline can sell, and adds incremental connecting passengers to other spokes at the hub, thereby improving load factors. Hub dominance also enables the dominant airline to increase the number of city-pair monopolies radiating from the hub, allowing monopoly fares to be imposed on origin-and-destination passengers. It is this monopoly exploitation that, in a deregulated environment, only competition can remedy. Knowing that, hub dominant airlines use a variety of means (lawful and not) to suppress competition which threatens their monopolies.

Table 11.7 reveals the growth in airline concentration at major U.S. airports. Concentration has increased at 20 of these 22 airports over the past two decades, and at most it has increased significantly.

Table 11.7. Airline Concentration at Major U.S. Airports by Market Share (Percent)[63]

Airport (dominant airline)	1977	1982	1987	1996
Atlanta Hartsfield (Delta)	n.a.	50.3	54.5	78.6
Charlotte Douglas (USAirways)	n.a.	0	1.1	92.8
Chicago O'Hare (United and American duopoly)	n.a.	57.9	77.3	81.4
Cincinnati International (Delta)	35.0	49.4	72.1	93.6
Cleveland Hopkins (Continental)	n.a.	0.4	14.5	47.4
Dallas/Fort Worth International (American)	n.a.	57.3	61.9	65.4
Dallas Love (Southwest)	n.a.	88.6	96.4	100.0
Denver Stapleton/International (United)	n.a.	30.9	43.0	69.2
Detroit Metropolitan (Northwest)	n.a.	13.8	61.2	80.1
Houston Hobby (Southwest)	n.a.	66.3	53.7	80.6
Houston Bush Intercontinental (Continental)	20.4	22.2	73.7	78.1
Memphis International (Northwest)	n.a.	0	84.9	78.5
Miami International (American)	n.a.	2.0	3.7	66.4
Minneapolis/St. Paul Int'l (Northwest)	45.9	42.6	79.5	84.3
Newark International (Continental)	n.a.	2.5	40.0	52.4
Philadelphia International (USAirways)	n.a.	22.8	36.2	60.0
Phoenix Sky Harbor (America West)	n.a.	0	44.8	39.2
Pittsburgh International (USAirways)	43.7	65.5	83.8	88.1
St. Louis Lambert International (TWA)	39.1	52.6	82.3	68.8
Salt Lake City International (Delta)	n.a.	3.0	59.0	74.9
San Francisco International (United)	n.a.	32.6	36.4	60.5
Washington Dulles (United)	n.a.	28.6	35.0	57.4

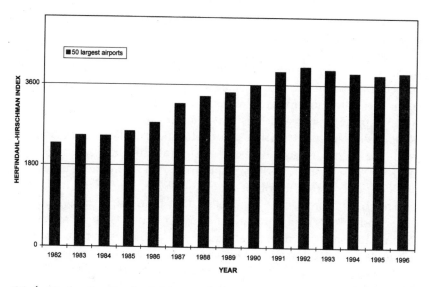

11.4 *Under the Herfindahl-Hirschman (HHI) analysis, employed by the U.S. Department of Justice to evaluate mergers, an HHI of 1800 or above is deemed "highly concentrated." The chart shows the airline concentration at the 50 largest U.S. airports.*

Dr. Julius Maldutis performed several studies of concentration at the nation's 50 largest airports, calculating the Herfindahl-Hirschman Index (HHI) for each. The HHI is the methodology employed by the U.S. Department of Justice for determining acceptable levels of concentration for antitrust review. It provides a measure based on squaring the market share of the dominant firm. For example, a firm with a 100% monopoly would have an HHI of 10,000. Under the Justice Department's analysis, an HHI below 1000 is presumed unconcentrated; an HHI of between 1000 and 1800 is believed moderately concentrated; and an HHI of above 1800 is deemed highly concentrated. (See Fig. 11.4.)

The 50 largest airports in the United States account for 81% of total scheduled passenger enplanements.[64] As Fig. 11.4 reveals, these 50 airports have an HHI well above 1800. They are therefore highly concentrated, under the Justice Department's analysis. The weighted average of concentration for all 50 airports rose from an HHI of 2215 in 1977, to 3513 in 1987, to 3877 in 1996.[65] According to Maldutis, this represents "an unprecedented degree of concentration in the airline industry."[66]

Though clearly alarming, Maldutis' calculations actually *understate* the degree of concentration at these airports, for they fail to aggregate the market share of the dominant carrier with the market share of its code-sharing "alliance" partners. Code-sharing, and the computer reservations system bias associated therewith, allows a dominant carrier to monopolize the connecting traffic at a dominant hub. For example, using Maldutis' data, adding together the 1996 passenger enplanement market share of United in Denver of 69.21%, with the market shares of its code-sharing affiliates, Mesa (3.48%) and Air Wisconsin (2.76%), results in a 75.45% market share over which it exercises control, or an HHI of 5693, one of the highest in the nation.[67] More recent data reveal that by April 1998, United Airlines had a 70.2% market share at Denver, while United Express affiliates had a 5.6% market share. If we add the market share of United's proposed partner, Delta (4.7%), United's Star Alliance will have an 80.6% market share at Denver, or an HHI of 6496.[68] (See Fig. 11.5.)

Hub concentration translates into escalating fares for origin-and-destination passengers. *The New York Times* observed, "Passengers who live in a hub city and begin their flight there end up paying higher fares, in some cases 50 percent more than they would had deregulation not occurred."[69] The General Accounting Office found that, after its merger with Ozark, TWA increased fares 13 to 18% on formerly competitive routes radiating from St. Louis.[70] A similar study compared fares in markets radiating from Minneapolis/St. Paul in

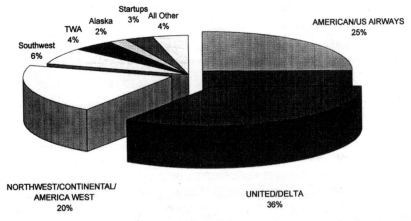

11.5 *The proposed U.S. megacarrier alliances would result in more than 80% of the passenger market (as measured by domestic revenue passenger miles in 1997) controlled by three massive alliances.*

which Northwest and Republic formerly competed, and found that fares rose between 18 and 40%.[71] In 15 of the 18 hubs in which a single carrier controls more than 50% of the market, passengers pay significantly more than the industry norm.[72]

Between June 1996 and June 1997, average air fares increased 9% for origin-and-destination passengers flying to or from concentrated hub airports, 3 times the national average. Full coach fares increased 15% at Newark, 16% at Dallas/Fort Worth, 17% at Saint Louis, and 26% at Denver.[73] City-pair competition is also declining significantly. In 1992, some 12,500 routes enjoyed competitive alternatives. By 1998, only 9400 routes were competitive, a 28% drop.[74]

The most comprehensive studies of the effect of airport concentration upon pricing have been those performed by the U.S. General Accounting Office (GAO). The GAO compared prices at 15 concentrated[75] hub airports and 38 relatively unconcentrated airports. It found that prices were 27% higher in the concentrated hubs.[76] A decade after deregulation, prices per mile charged by dominant airlines at concentrated hubs were 38% higher than those charged at unconcentrated airports.[77]

The U.S. Department of Transportation also studied the impact of concentration on airline pricing, and concluded as follows:

> *The average fare per mile at the eight most concentrated hubs is higher than the national average. Adjusting for the average trip distance and the size of the market served at the eight most concentrated hubs, fares were on average 18.7% higher than similar markets for other airports. This finding supports the conclusion that high hub concentration leads to high fares for passengers traveling to and from such cities. Fares are highest for travel between large cities within 1,000 miles of the hub.*[78]

More recently, the DOT found that, "In the absence of competition, the major carrier is able to charge fares that exceed its fares in non-hub markets of comparable distance and density by upwards of 40 percent."[79]

Kenneth Mead, then director of the GAO's transportation division, found that "no single factor is responsible for higher fares at concentrated airports, but that it is the interaction of a number of barriers that allows carriers at these airports to charge higher fares."[80] The GAO found several factors correlating with higher fares:

1 The larger the share of gates a carrier leased at an airport, especially on a long-term, exclusive-use basis, the higher the fares.

2 Flights at airports where a majority-in-interest clause might reduce expansion opportunities have about 3% higher fares.

3 Flights at airports where entry is limited by slot controls have about 7% higher air fares.

4 Airports with congested runway capacity and limited expansion due to majority-in-interest clauses have about 3% higher fares.

5 Carriers with a code-sharing agreement at one of the airports on a route charge fares almost 8% higher than carriers do on routes on which they do not code-share.[81]

Though deregulation proponents predicted competition would drive airline prices to marginal costs,[82] in fact, hub pricing has put the anticipated cost/price relationship on its head. The least costly of airline operations is nonstop point-to-point service, as the costs of Southwest Airlines confirm. Though Southwest flies a relatively short average stage length, by focusing on point-to-point nonstop flights it enjoys superior aircraft, labor, gate, and fuel utilization compared to

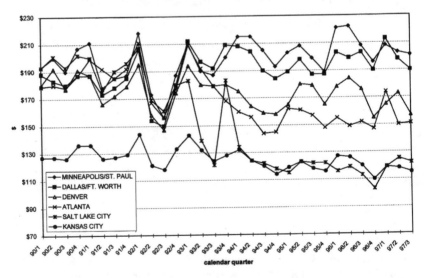

11.6 *Average airline fares are significantly higher at concentrated hubs, such as Minneapolis, than at unconcentrated airports, like Kansas City.*

its competitors. Network connecting service is far less efficient from an aircraft, labor, gate, and fuel utilization perspective. Yet high-cost connecting service typically is priced at a relatively low level vis-à-vis low-cost nonstop service from a dominant hub. Thus, pricing in the airline industry under deregulation appears to be driven more by competition than by costs. Airlines generally face more vigorous competitors for long-haul connecting traffic (for the greater the distance, the more alternative hubs over which a passenger can connect) than for short-haul nonstop service, particularly to and from the hubs they dominate.

Concentration levels often correlate with price levels—higher concentration tends to equate with higher prices in many markets. But the identity of the competitor can also significantly influence pricing. The presence of a low-cost/low-fare competitor (such as Southwest, Frontier, Vanguard, Reno, Spirit, Kiwi, or Pro Air, for example) can result in significant competitive discipline and consumer savings. According to the DOT, fares tend to be $80 higher on average when no low-fare competitor is present on the route.[83] This is reflected in Fig. 11.6, which shows historical average one-way fares at six airports— Minneapolis/St. Paul, Dallas/Fort Worth, Denver, Atlanta, Salt Lake City, and Kansas City.

At various times, Braniff, Eastern, and TWA have attempted to establish a hub at Kansas City. Each failed. The result is that Kansas City remains unconcentrated (Southwest is the largest carrier, with 23% of enplanements), and its consumers enjoy average fares among the lowest of any city its size. Note that average fares at Salt Lake City, Atlanta, and Denver marched in virtual lockstep with fares at Dallas and Minneapolis until 1994. In that year, Southwest acquired Salt Lake City–based Morris Air. Average fares dropped by 50% in Southwest's markets, while traffic tripled. By late 1995, average fares in markets served by Southwest were only one-third the level of fares in other Salt Lake City markets.[84] By 1996, Southwest accounted for 12% of enplanements, and Salt Lake City's average fares were as low as Kansas City's. At Atlanta, ValuJet's entry brought fares down, though it accounted only for 8% of enplanements. At Denver, first MarkAir's, then Frontier's entry brought fares down, though Frontier accounts for only about 4% of enplanements.[85]

Contrast these price declines with the relatively higher prices at Dallas/Fort Worth. Dallas/Fort Worth International Airport is dominated by two megacarriers—American (65%) and Delta (19%). Neither is a vigorous price competitor, though prices are somewhat disciplined in short-haul markets by the presence of Southwest at Dallas Love Field. Minneapolis/St. Paul is by far the worst of the group in terms of exorbitant air fares, because Northwest dominates the hub with an 84% market share, no low-cost/low-fare carrier accounts for even a 1% market share, and there is no secondary airport at Minneapolis.[86] One study revealed that concentration at the Minneapolis/St. Paul hub caused a 72% increase in prices from 1988 to 1995, and that by 1995, its residents were paying $693 million above the national average for tickets.[87] In fact, average round-trip fares at Minneapolis since 1994 of $412 are 72% higher than at Kansas City or Salt Lake City, where the average round-trip fare is $240. If Denver's fares were set at Minneapolis' levels, Denver's 20 million annual origin-and-destination passengers would pay $750 million in higher fares per year. Conversely, if Frontier were to grow to a size comparable to Southwest at Salt Lake City, consumer savings could total an additional $1 billion per year.

Barriers to entry

Barriers to entry have been defined as "any factor that prevents a new firm from competing on a equal footing with existing firms."[88] These factors are numerous in the airline industry, ranging from the consumption by incumbent airlines of airport gates and landing slots to computer reservations systems.

In one sense, barriers to entry appear deceivingly small, and were deemed inconsequential by deregulation's architects. A large used-aircraft leasing market and a large number of skilled workers (individuals who had been laid off by the major airlines or lost their jobs because of major carrier liquidation) were available in the early 1990s. Despite their financial collapse, airlines remain a glamorous industry. Coupled with investor and lender enthusiasm for new airline ventures, this led to the emergence of a number of new airlines. But entering and surviving are two entirely different things.[89] More than a hundred new airlines have emerged since deregulation, and the overwhelming majority have collapsed in bankruptcy. Moreover, even entering a single market where the incumbent enjoys supracompetitive profits is difficult, given that an overwhelming number of nonstop city-pair routes appear able to support only a single airline, and that new entry must manifest itself inflexibly in plane-load lots.[90]

For several reasons, it is difficult for a new entrant to compete successfully with a megacarrier. First, the infrastructure of gates, terminal facilities, and—at four of America's busiest airports (i.e., Chicago O'Hare, Washington National, and New York's LaGuardia and Kennedy)—landing slots, has been consumed. Sixty-eight percent of U.S. airports have no gates to lease to new entrants.[91] Even if an incumbent would be willing to lease a gate to an upstart airline (and at a carrier's hub, few are so willing), the incumbent could exact monopoly rents for the lease. The decision of DOT to allow carriers to buy and sell landing slots means that the deeper-pocket carriers can purchase market share, and thereby enjoy market power to reap monopoly profits.[92] The DOT seems to have recognized this by opening up a few slots to new entrant airlines in the late 1990s. Incumbent airlines also control hub airport expansion through "majority-in-interest" clauses,[93] which gives incumbents effective veto power over new airport expansion. Moreover, restrictions on the type of equipment that carriers can use at several noise-restricted airports also constrain new entry.

Second, the largest airlines today own the largest computer reservations systems (CRSs), from which 90% of tickets are sold.[94] Many critics argue that such vertical integration offers the incumbents the potential to enjoy various forms of system bias (including screen bias, connecting point bias, and database bias).[95] The General Accounting Office has also found that the airline-owned systems are so dominant that they stifle competition in the industry.[96] An airline which owns a CRS has a 13% to 18% greater likelihood of selling its tickets through the system.[97] United and American own the dominant computer reservations systems, which together account for about 75% of passenger bookings.

Moreover, the advantages of being listed in the computer as an "on line" connection with one of the major airlines have led 48 of the 50 small air carriers to affiliate themselves with the megacarriers, renaming their companies (to, for example, United Express, Continental Express, or American Eagle) and repainting their aircraft in megacarrier colors. Ninety percent of the 31.7 million passengers who flew aboard regional airlines in 1987 were carried aboard code-sharing airlines.[98] The small carriers have become, in effect, franchisees of the behemoths of the industry, and are therefore an unlikely source from which new competition will spring. They are also declining in number. The regional airlines, peaking at 246 in

1981, dwindled to 168 by 1987.[99] Sophisticated computers also give airlines the ability to manage yield in a way to adjust the number of seats for which discounts are offered on an hourly basis, depending on passenger demand for seats.[100]

Third, large airlines have more attractive frequent flyer programs, an indirect price rebate, which serve as a lure to business travelers, the most lucrative segment of the market. Once committed to a carrier's frequent flier program and having some investment in accumulated mileage, a business traveler may prefer that carrier over its rivals even when the rivals' flights are cheaper. After all, most business travel is not paid for by the individual flying, but by his or her firm. Frequent flier–induced brand loyalty makes it difficult for a new rival to find a niche, particularly when its frequent flier program offers free travel to decidedly less exotic destinations.[101]

Not only are the frequent flier programs creating passenger loyalty, but commission overrides and promotions are generating travel agent loyalty.[102] Hence, both the passenger and agent often prefer a more expensive, established airline, to a discount carrier. Indeed, the travel agent has been given an incentive to engage in consumer deception. Say the consumer calls and asks whether there is a flight on Carrier A at noon. There is, but the agent is working toward commission overrides on Carrier B this month. How easy it would be for the agent to say, "Sorry, the noon flight is sold out. But I can get you a seat at 1:30 on Carrier B."

Fourth, although new entrants enjoyed significantly lower labor costs in the inaugural years of deregulation, the squeeze on carrier profits unleashed by deregulation has forced management to exact serious concessions in terms of labor wages and work rules. Thus, the margin of labor cost and productivity between a new entrant and an established airline has been narrowed.

Fifth, incumbents have shown that they will not sit idly by while new rivals rob them of market share. When the new entrants offer lower fares, the incumbents almost always match them with below-cost pricing. Further, they may turn up the heat on the new entrant by dumping excess capacity in the market, expanding travel agent commission overrides, and entering into exclusive dealing contracts with corporate purchasers and regional feeder airlines.

In summary, because of the higher brand recognition of incumbent airlines, their vast networks and exclusive alliances with regional feeder carriers, their ability to bias the computer reservations systems they own against competing interline connections, their ability to bribe travel agents with commission overrides to steer business their way, and their frequent flyer programs, they have enormous advantages which can be overcome only if the new entrant can offer a better price.

Predation and monopolization

Consumers at concentrated airports often are starved for price relief, for the average fares they pay are far above competitive levels. New airline entrants typically bring with them lower fares than those prevailing prior to their entry. The new entrants typically can offer a lower price because of their lower cost structure. In a commodity business such as transportation, a company with a lower cost structure should be able to take some reasonable share of the market. If both companies price on the basis of cost, then the new entrant will likely be sustainable. The new entrant may stimulate increased demand among price-elastic travelers, expanding the size of the pie for all competitors.

But new entry cannot be sustained where the incumbent airline is willing to endure significant short-term losses in a below-cost predatory pricing strategy designed to force the new entrant out of the market (or into bankruptcy) so that after the new entrant leaves, the incumbent can resume its monopoly price gouging well above competitive levels.

An incumbent airline can respond to new entry by matching its low fares on frequencies in close proximity to the new entrant's departures, meeting the new competitor's introductory fares and locking them in (i.e., refusing to follow the new price leader's fares up after the promotional period), dumping additional capacity (flights) into the market, or sandwiching the new competitor's frequencies (with a departure within a few minutes on both sides of the new entrant's departure) until the new entrant is financially exhausted and withdraws.[103] As Severin Borenstein notes, airport dominance may intensify the retaliatory threat:

> *Besides the advantage in attracting customers to its flights over a competitor's, airport dominance might also allow an airline to deter entry of competitors. This could be done with a threat of retaliation, possibly made more credible due to*

airport dominance, or by blocking access to scarce gates or landing slots at an airport.[104]

Predatory pricing has been defined as pricing below an appropriate measure of cost for the purpose of eliminating competitors in the short-term and reducing competition in the long-term.[105] Under neo-classical free market theoretical beliefs, predation is irrational, for the dominant firm engaging in the predatory behavior must be able to recover the short-term losses it incurs in the longer term after it has driven the new entrant from the market; since theoretically, it can never hope to recover its short-term losses, it will not likely engage in such predation. Hence, some commentators argue that predatory pricing schemes are rarely attempted, and even more rarely successful.[106]

Despite the theoretical opposition to predation based on its hypothetical irrationality, airline observers have seen numerous examples of predatory behavior in the airline industry attempted since deregulation, with various degrees of success. Evaluating the postderegulation experience, during which he served as CEO of a small airline (New York Air), Michael Levine concluded, "I believe predation is possible and that it occurs....[I]t is possible for an incumbent to impose on prospective entrants nonrecoverable costs by pricing in a way that seeks to ensure that they do not attract a significant share of passengers regardless of the incumbent's own costs."[107] Dr. Alfred Kahn concurred, criticizing Northwest Airlines for its "scorched-earth" policy of substantially undercutting People Express' price while simultaneously increasing the number of flights in the market, saying:

> *If predation means anything, it means deep, pinpointed, discriminatory price cuts by big companies aimed at driving price cutters out of the market, in order then to be able to raise prices back to their previous levels. I have little doubt that is what Northwest was and is trying to do.*[108]

The Economist summarized the problem in these words:

> *Predatory pricing. Deregulation left the previously related companies with advantages and disadvantages.... But the deregulated companies also entered the new era with substantial patches of market in their grasp. They could try to defend them by setting some of their prices below cost, fending off smaller invaders one by one.*

Mr. Kahn reckons, for example, that Northwest Airlines tried to squeeze People Express off its route from Minneapolis/St. Paul to New York last year. When People started to fly the route, Northwest cut its regular fare from $263 to $99 (weekdays) and $79 (weekends and off-peak). That two-thirds cut matched People's fares, and Northwest still ran its traditional high-quality service. Evidently the idea was to drive the upstart from the market, after which Northwest could put its fares back up.[109]

Though some theorists claim that predation is economically irrational and therefore theoretically cannot exist,[110] allegations of predatory behavior have been widespread in the airline industry. Continental Airlines, Northwest Airlines, and America West Airlines—three major carriers—alleged American Airlines had engaged in predatory pricing.[111] UltrAir alleged its demise was caused by predatory pricing by Continental Airlines.[112] Pro Air's Chairman and CEO Kevin Stamper pointed out that since his airline went aloft in July of 1997, it was forced to withdraw from Milwaukee and reduce service to Indianapolis because of predatory pricing. Said Stamper:

Matching prices is a normal competitive behavior. But it can be a powerful predatory tactic if its purpose is simply to drive the new entrant out of business and restore the markets to their prior monopoly status....

Too much government involvement can stifle innovation and competition, but too little can have exactly the same effect. Our concern is that predatory activities on the part of carriers that already possess many competitive advantages may drive small airlines out of the industry and lead to monopolies and market abuse.[113]

Similarly, Mark Kahan, Vice Chairman of Spirit Airlines, testified that after Spirit entered the Detroit-Philadelphia market, Northwest dropped its yields 54% while increasing its capacity 15%. Sharply reduced prices coupled with capacity dumping allegedly forced Spirit to withdraw from the Detroit-Boston and Detroit-Orlando markets as well. Said Kahan:

It is probable that Northwest sacrificed out-of-pocket a lot more than $10 million because of its fare decreases and capacity increases in the Detroit-Boston and Detroit-Philadelphia *markets in the third quarter of 1996 alone. These actions clearly made no sense unless Northwest was confident that Spirit would be obliged to exit the market....*

You will pardon us for believing that Northwest tried to put Spirit out of business in the third quarter of 1996.[114]

Vanguard has also complained to DOT about the anticompetitive practices of Northwest Airlines.[115] In 1993, ValuJet complained to DOT that Delta offered below-cost fares over 40% of its capacity and 300% of ValuJet's capacity, saying, "We believe that Delta is pricing its product well below its costs and is embarking on this course of predatory pricing for the express purpose of driving ValuJet out of business."[116] Chronicling the alleged efforts of Delta Air Lines to drive ValuJet from its Atlanta base, Professor Fred Allvine observed:

Government study after study shows that the airline industry is not perfectly competitive.... These studies show that the major airlines employ many monopolistic practices that contribute to the market power to raise and maintain prices above the competitive level. In imperfect competitive markets it makes perfectly good sense for large firms to use predatory pricing to destroy competition that threatens the monopoly prices charged.[117]

Lewis Jordan, President and COO of ValuJet, testified, "While ValuJet welcomes fair competition and does not seek to be insulated from such competition, we believe that no airline should be subjected to pricing actions designed to force it out of markets with the attendant likely consequence of fares immediately being restored to unnecessarily high levels. We urge a vigorous enforcement of the antitrust laws and reexamination of the predatory pricing doctrine to ensure fair competition for big and small carriers alike."[118] After ValuJet announced it was surrendering the Atlanta-Mobile route to Delta Air Lines, Delta raised its fares from $58 to $404, almost 600%.[119]

Continental Airlines sued United Airlines for predatory conduct, receiving $77 million in an out-of-court settlement.[120] Laker Airways filed a $1.7 billion lawsuit alleging predatory practices by six large airlines drove it out of business.[121] The suit settled 2 years later for $60 million.[122] Gulf Air filed an antitrust suit against Continental, Eastern, and other Texas air carriers alleging they had engaged in predatory business practices.[123] Pacific Express filed suit against United Airlines for alleged predatory activity.[124] Alaska Airlines filed suit against United Airlines for monopoly leveraging and violation of the essential facilities doctrine. TACA International Airlines filed complaints against United Airlines, American Airlines, and Continental Airlines for predatory pricing.[125] Virgin Atlantic brought suit against British Airways for predation. ValuJet filed suit against TWA and Delta for monopolization. USAirways filed suit against British Airways and American Airlines on antitrust grounds.[126] And most recently, Reno Air filed suit against Northwest Airlines for predation following its entry into the Reno-Minneapolis market. If there is no fire, what is causing all the smoke?

The U.S. Department of Transportation has observed, "there have been instances in which a new, small carrier has offered low price service between a major carrier's hub and a spoke city, only to find the major carrier cutting its own airfares and increasing the number of seats—or even airplanes—on that route and sacrificing short term profits with only one goal in mind: to drive the new entrant out of the market and then raise its own fares to their original level or higher, and cut back its service."[127]

An established carrier which finds its spokes assaulted by a new entrant typically will cut prices to meet the competition. Both will lose money, but large carriers have the ability to cover short-term revenue losses from profits derived from less competitive markets.[128] Typically, the major airline offers the low fare only on local O&D traffic on flights in close time proximity to the new entrant's, extracting additional revenue from passengers connecting to the assaulted spokes. This revenue advantage may neutralize the new entrant's cost advantage and will deleteriously impact its staying power.[129] Levine notes, "The ability of an incumbent to respond rapidly and cheaply to the prices and output of new entrants contradicts perhaps the most critical assumption of contestability theory."[130]

If contestability theory applied to the airline industry as deregulation proponents had anticipated, the need for actual entry to protect the consumer interest would be nil. The theory of contestability posits that the absence of barriers to entry and economies of scale will allow potential entrants to prohibit monopolists from earning supracompetitive profits. While many proponents of deregulation originally embraced this theory, most have since rejected it on the basis of the empirical behavior of airlines in the postderegulation period, and the existence of economies of scale, scope, and density, as well as the significant costs of and barriers to successful entry.[131]

Most empirical studies have demonstrated that deregulated airline markets are not perfectly contestable,[132] and that there is a positive relationship between concentration and fares.[133] Ticket prices in city-pair markets with two competitors were about 8% lower than in monopoly markets, and markets with three competitors were another 8% less still. A potential competitor has one-tenth to one-third the competitive impact of an actual competitor.[134] The exit of a competitor results in a 10% average price increase for the remaining incumbents.[135] Other studies reveal that the number of competitors is not nearly as significant as their identity (e.g., Southwest's presence in a market creates deeper pricing competition than, say, Delta's).[136] Some have insisted that the airline industry is "imperfectly contestable."[137] Without doubt, "imperfect" is an appropriate adjective to describe airline economics.

Economies of scale, scope, and density also appear to exist in the airline industry, although the fact that new entrant airlines have lower ASM costs than established major airlines might suggest the contrary to those who do not look more deeply. Significant informational economies are associated with incumbency—a small carrier must invest in relatively higher advertising, marketing, and ramp-up costs in introducing its service to a city-pair market, while a large established carrier adding that city-pair to its existing hub network has relatively lower start-up costs. Economies of scope are achieved as a carrier increases frequency in a market (spreading more customers over its station costs, for example). Enhanced frequency has a disproportionate impact on demand for its product (the carrier with more frequency enjoying a disproportionately larger share of passengers and higher-yield revenue). Hub carriers also enjoy network economies by adding a spoke to an existing hub network, offering a vast increase in the number of city-pair products it can offer. According to Levine, "We have seen the creation of a large number of hub monopolies because of the economies of scale and scope at the hubs."[138] Kahn has insisted, "We advocates of deregulation were misled by the apparent lack of evidence of economies of scale."[139]

Add to network economies the vast increase in product lines that are added when large networks are joined together in code-sharing relationships, relationships from which new entrants are generally excluded. In fact, some carriers refuse to enter into joint-fare and code-sharing relationships with independent regional jet airlines.

Then there are the "induced" scale and scope effects, including frequent flyer programs (for which larger network carriers have a manifest advantage vis-à-vis their smaller competitors) which attract higher-yield business travelers, and travel agent commission overrides, which essentially bribe agents to steer business toward the carrier which offers them. These have been described in the literature as the "principal-agent" problem.[140] As Levine has noted, "by constructing incentive commission programs and by inventing frequent flyer programs, big airlines learned to create economies of scope and scale that are not present in the basic technology."[141]

Levine cataloged the multitude of developments not anticipated by the pro-deregulation economists—mergers and consolidations, vertical integration, hub-and-spoke systems, complicated fare structures, frequent flyer programs, travel agent commission overrides, computer reservations systems (CRSs), slot and gate monopolies, predation, and the high mortality rate among new entrants. From these developments, he concluded:

> [T]hese unanticipated effects of deregulation seem to stem from the economics of information and from related economies of scope and scale, and from production indivisibilities (such as the problems of providing frequent and convenient service in city-pair markets with small traffic flows).... Frequent flyer programs, the importance of travel agents and travel agent incentive programs, computer reservations systems, and hub and spoke systems all are techniques of utilizing economies of scale and scope to take best advantage...of the costs of communicating a complex web of service and service attributes to consumers.... The information and transaction costs are real....[142]

On the distribution side of the equation, the largest airlines, which own the computer reservations systems, through which the vast majority of flights are sold, have incentive to display their competitors' flights more poorly (for example, CRSs add the equivalent of 24 hours in time to non-code-sharing connections so as to push them off the first page of the screen, where 85% of seats are sold), and earn significant revenue from their competitors' CRS bookings and sales. Some newer entrants have responded by attempting to sell their products directly to consumers through 800 numbers and heavy advertising.

The three most prominent architects of airline deregulation—Alfred Kahn, Elizabeth Bailey, and Michael Levine—all have jettisoned the notion that contestability theory explains market behavior in the airline industry. For them, market reality trumped economic theory. According to Kahn, "I bear some responsibility for promulgating the notion, before deregulation, that because the industry's capital equipment is physically mobile, neither destructive competition nor significant monopoly power were likely to emerge; but I had long since ceased to do so after it became very clear that contestability is in fact very imperfect, at best."[143] Although an early proponent of the application of contestability theory to the airline industry, Bailey has concluded that airline "...markets are not perfectly contestable, so that carriers in concentrated markets are able to charge somewhat higher fares than carriers in less concentrated markets."[144]

Levine said it even more strongly: "Unfortunately, those theories turned out to be wrong as they applied to the airline industry..."[145] and "airline markets cannot be modeled by any reasonably pure version of contestability theory."[146] Levine acknowledged that "economists committed to a high degree of airline market contestability have historically maintained that predation is doomed to failure and is therefore unlikely because the capital assets involved in airline production are mobile....This contestability analysis is unfortunately inconsistent with much observed behavior since deregulation...large holdover incumbents are not easily susceptible to predation, but smaller new entrants are."[147] Levine concluded that new industrial organization theory better describes the airline industry than the perfect competition/contestability model.[148] Assistant Attorney General Charles Rule concluded, "Most airline markets do not appear to be contestable, if they ever were....[D]ifficulties of entry, particularly on city pairs involving hub cities, mean that hit-and-run entry is a theory that does not comport with current reality."[149] DOT Assistant Secretary Pat Murphy has declared contestability theory "deader than dead" in its ability to describe the market reality of commercial aviation.[150]

The consensus among economists (outside the Washington, D.C., *laissez faire* think tanks) today is that the airline industry does not reflect theoretical notions of perfect competition or contestability. The high degree of pricing discrimination between consumers and markets suggests that the industry may better reflect economist Joan Robinson's theory of "imperfect competition"[151] or Edward Chamberlin's theory of "monopolistic competition."[152] The literature lends strong support to the conclusion that, if airline industry is to be competitive, the competition laws must be applied with full force to it.

Finally, with more than 150 airlines having failed since 1978, many having been pushed into the abyss of bankruptcy by the predatory behavior of their larger rivals, investor confidence in new airline ventures has been severely diluted.[153] Hence, new entry is far more difficult in the deregulated airline industry than the proponents of deregulation perceived it to be.[154] The dominance by incumbent carriers of airport gates, terminal space, landing and takeoff slots, computer reservations systems, and the most attractive frequent flier programs makes it unlikely that major new entrants will emerge to challenge the megacarriers.[155]

The theory of contestable markets, which provided an intellectual justification for deregulation, has been refuted by an overwhelming body of empirical evidence.[156] The theory was premised upon the false assumption that transportation was inherently competitive, and that the only barriers to entry were governmental requirements that carriers obtain certificates of public convenience and necessity before being allowed to compete. Here again, deregulation's proponents overestimated the competitive nature of the industry. As Alfred Kahn noted, "Certainly one of the assumptions behind airline deregulation was that entry would be relatively easy."[157] Former DOT Assistant Secretary Matthew Scocozza confessed, "To be very honest, in 1978 we envisioned that there would be a hundred airlines flying to every major hub."[158]

The foundation upon which the theory rested has been shattered by a decade of evidence that proves that economic barriers to entry, significant advantages in terms of traffic density, and economies of scale and scope *do* exist in the airline industry, and are of some significance.

But even if new entry is unlikely, why should we be concerned with the high level of concentration which has emerged in the airline industry under deregulation? After all, even though Coke and Pepsi dominate the soft drink industry, don't we still have pricing competition between them? Although other industries are dominated by huge firms, transportation is different in the way it impacts the economy. Melvin Brenner said it best:

> *Other industries, even when comprised of only a few large firms, do not usually end up with a one-supplier monopoly in specific local markets. But this can happen in air transportation.*
>
> *Moreover, because of the nature of transportation, a local monopoly can do greater harm to a community than could*

a local monopoly in some other industry. This is because transportation is a basic part of the economic/social/cultural infrastructure, which affects the efficiency of all other business activities in a community and the quality of life of its residents. The ability of a city to retain existing industries, and attract new ones, is uniquely dependent upon the adequacy, convenience, and reasonable pricing of its airline service.[159]

Airline Manipulation of Airport Development: A Case Study

Airlines often find themselves at odds with airports on issues of infrastructure expansion, and use their political and economic power to their advantage. Let us take a look at some of the major conflicts that arose over the construction of a new airport at Denver.

Former United Airlines' CEO Stephen Wolf said, "I never fought anything so hard in my life" as the new Denver International Airport (DIA).[160] Opposition was predicated on cost, and (though never said by United publicly), the possibility that a large new airport might have sufficient capacity to attract new competition. But once Denver's second hub carrier, Continental Airlines, jumped on the DIA bandwagon, United had little choice but to jump aboard too.

Among the last things Frank Lorenzo did as CEO of Continental was to sign a lease with the city of Denver in the summer of 1990 for 20 gates at the new Denver International Airport. Because Continental became DIA's first hub carrier, it was able to reserve the closest (and therefore most desirable) concourse (A) to the main terminal building, and the city agreed to build a pedestrian bridge linking the terminal directly to that concourse. When United subsequently signed up for Concourse B (which had no pedestrian bridge to the main terminal), it insisted the city make the glass on Continental's Concourse A bridge opaque, so that no passenger could see the splendid view of the Colorado Rocky Mountains from it, for United believed the bridge offered Continental a competitive advantage. Mercifully, DIA engineer Ginger Evans refused, and the city breached its contractual agreement.[161] United chose not to press its case, undoubtedly fearing the public relations fallout once the new airport opened.

United took several actions to pressure Continental to depart. For example, United refused to allow DIA to open until its automated baggage system could deliver 225 bags a minute. Ginger Evans contends DIA could have opened early in 1994 with the baggage system operating at 40 bags per minute, adequate to allow United to meet its connect times.[162] "Virtually every design and construction professional [who] was involved directly or as a consultant...believed at that time the project, including the BAE automated baggage system, could have been completed by October 21, 1993 [the originally scheduled opening date]."[163] If an allegedly malfunctioning baggage system was not the fundamental cause of the delay, what was? One plausible and widely accepted explanation has been proffered by former Denver airport director George Doughty in testimony before the U.S. Congress:

> *United Airlines did not want to go to DIA. United could have cooperated with the City to work out options for manual bag handling, but they did not.... As to exactly what United's rationale [was] one can only speculate, but a few things are clear. United had no incentive to move in 1994. They had just increased their operations at Denver in order to capture an even greater market share that would eventually force Continental to dismantle its hub. It was to their advantage not to move until that was assured....*[164]

In 1992, Continental reached its last high water mark of 285 flights per day (including Continental Express) at Denver. With 38% of the market, Continental was still second to United's 40%. United designated Denver its "major domestic initiative," and increased its capacity by 30% over the next several years in what appeared to be a deliberate move to oust Continental once and for all.[165] UAL's 1992 Annual Report spoke of "an *aggressive plan* for expansion" at Denver: "At Denver, United phased in a dramatic increase in departures during the year, moving from 180 flights last spring to 217 during the summer and, by March of this year, to 247 departures [plus 105 by United Express carriers]."[166] Its 1993 Annual Report stated, "An *aggressive buildup* has made a significant contribution to revenue improvement at [the Denver] hub. United ended 1993 with 257 daily departures in Denver, up from 212 a year earlier....Already the number one carrier in the Mile High City, United had increased its capacity over the last two years by nearly 30 percent."[167] United's aggressive behavior at Denver was clearly targeted at Continental.

United pulled away and never looked back as it steadily increased market share, leading to Continental's decision to pull out.

Toward the end, Continental was losing $10 million a month at its Denver hub, and could not face the prospect of continuing hub operations at the more expensive new airport. Denver airport director Jim DeLong described it as a "fierce battle for dominance of the Denver market."[168] Continental's Annual Report revealed that the company had lost $130 million at Denver in 1993, and lost $500 million at Denver from 1990 to 1993.[169] According to Continental's CEO Robert Ferguson, "Continental's losses are at unacceptably high levels in Denver, even with our reduced flying."[170] Continental's Annual Report said, "Although the new facilities [at DIA] will be greatly superior to those presently serving Continental's Denver passengers, they also will be much more expensive."[171] It was estimated that increased landing fees required to pay off bonds issued to finance DIA would have added another $50 million to Continental's costs.[172]

The advantages Continental gained from signing the first lease at DIA were never fully realized because Continental ultimately was forced to cry "uncle," in March 1994, and announced its decision to abandon its Denver hub operations and relinquish the market to United. Continental had already downsized its presence in Denver from a high of 285 flights a day in February 1992, to 165 flights in August 1993, to 148 flights in January 1994, to 107 flights in March 1994. Continental's Denver operations dropped to 86 flights in July 1994, 59 in September 1994, and 19 in March 1995.[173] Meanwhile, United increased its flights to 280.[174] Following Continental's announcement that it was scaling back its Denver hub beginning in the fall of 1993, the withdrawal proceeded throughout the following year. Continental dropped 26 routes through August 1994, and an additional 23 on October 31, 1994. By the time DIA opened, Continental was down to 13 flights a day to three cities.

Passenger traffic, which was increasing since 1990, began to decline after Continental's pullout. In 1995, only 31 million total passengers were flown to, from, or through Denver, down from 33 million in 1994. A contributing factor in this decline was the fare increase strategy that United enacted as it realized more monopoly opportunities.

United's goal of becoming the dominant carrier in Denver has been fully realized, as United and its code-sharing affiliates now control nearly 80% of the total passenger market at Denver. By September

1996, United flew to 55 cities from Denver and Colorado Springs.[175] In 1994, United's CEO, Gerald Greenwald, confessed that United's strategy to dominate the Denver market had paid off in increased market share and profitability. Outgoing United CEO Stephen Wolf called Denver the "major domestic initiative" for the airline over the preceding 2 years. According to Greenwald, "United has done a fantastic job of building strength in Denver and we'd like to take advantage of that, if anything, and build it stronger."[176]

As noted above, United may have purposely delayed opening of DIA in order to encourage Continental to abandon its Denver hub. Along the way, United also cut several additional deals with the city to disadvantage its competitors. United insisted it be allowed to take over the Concourse C automated baggage system loop, so that only United would have a high-speed baggage system. Ironically, the Concourse C automated system was the only one operating well before United occupied it. Other airlines were relegated to traditional tug-and-cart technology.

Further, United insisted that carriers using Concourse A pay a disproportionate share of the costs of that concourse's automated baggage system, though the system has never been functional. (On a per-passenger basis, Continental Airlines pays the highest costs of any carrier at DIA; it absorbs a portion of the additional costs of its two sublessees on Concourse A; but because of this agreement, half of Concourse A's domestic gates remain empty.) United's insistence assures that the high cost of Concourse A's gates will dissuade new carrier entry on that concourse for years to come, despite its superior location. Unfortunately, DIA's costs have dissuaded many low-fare airlines (including Southwest Airlines) from entering the market, and driven low-cost airlines (e.g., Midway and Morris Air) from it. Finally, United insisted the city build it a hangar directly north of Concourse C, on land a future concourse is supposed to occupy. No such future concourse can be built without tearing down United's hangar.

In summary, though unable to stop construction of a new airport at Denver, United Airlines was able to ensure its fortress hub dominance of DIA by driving up its costs, by withholding permission to open DIA until Continental Airlines had succumbed to United's below-cost pricing and fully committed itself to eliminating its Denver hub, and by ensuring Concourse C carriers would be deprived of use of an automated baggage system and that Concourse A carriers would pay an exorbitant price for its automated baggage system (though a back-up baggage system was welded on top of it, thereby denying Concourse A carriers the automated system whose costs they must pay).

Summary and Conclusions

The competition unleashed by airline deregulation has been beneficial to large segments of the consuming public, particularly discretionary travelers in competitive markets. But competition should be further advanced so that a larger universe of Americans can enjoy the benefits of airline deregulation. In the Airline Deregulation Act of 1978, the U.S. Congress explicitly affirmed its commitment to preventing unfair, deceptive, predatory, or anticompetitive practices in air transportation, avoiding unreasonable industry concentration, excessive market domination, monopoly power, and other conditions that would allow a carrier unreasonably to increase prices, reduce service, or exclude competition in air transportation.

But as deregulation enters its third decade, certain trends appear ominous. Real air fares, though below 1978 levels, are falling at a significantly slower rate than they did in the prederegulation era. Compare the 18 years preceding deregulation with the 18 years for

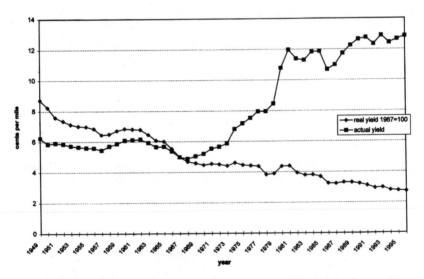

11.7 *Though the fact is counterintuitive, real (inflation adjusted) airline yields were falling at a significantly faster rate before deregulation in 1978 than subsequent to it.* ESG Airline Monitor, Nov. 1994, p. 4; Sept. 1997, p. 11.

which we have data since deregulation. From 1961 to 1978, before deregulation, real (inflation adjusted) yields fell at an average annual rate of 2.77% per year. From 1978 to 1996, since deregulation, real yields have fallen at a rate of only 2.06% per year.[177] (See Fig. 11.7.) Anecdotal evidence for 1997 and 1998 suggests that price declines are slowing more still. The price of business travel increased 17% in 1997 alone.[178] *Business Week* reported in 1998 that the air fare portion of the Consumer Price Index was increasing 37% on an annual basis. Moreover, Alfred Kahn has found that unrestricted fares have increased 73% under deregulation.[179] According to Kahn, "I do not see how anybody can object in principle to calls for reregulation when you clearly have a good deal of monopoly power. Hubbing seems to have contributed to that, and the carriers are ever more exploiting it at the expense of nondiscretionary travelers."[180] Thus, if deregulation is to continue to be perceived as a consumer success, competition must be enhanced.

Prices are not fully competitive for travel to or from airports dominated by a single airline, and for travel to and from small communities. In the first decade of deregulation, 61% of the 514 nonhub communities receiving air service in 1978 suffered declines in flight frequency, 28% lost all service; only 6% enjoyed the inauguration of new service. By 1995, 33% had lost all service, while only 5% had gained new service.[181]

The U.S. General Accounting Office found that passengers flying from small-city airports to major airports paid 34% more if the major airport was concentrated and 42% more if both the small-city and major airports were concentrated. The DOT found that passengers pay an average of $54 less per flight in markets with low-cost carrier service. Market-dominant airlines are using their domination of public resources in a manner to suppress competition. Because airports and the airways are public resources, they should be used for public benefit, and airline competition produces major public benefits.

Despite record profits earned by the major airlines, the number of low-cost low-fare airlines is declining. Since U.S. deregulation in 1978, the 43 prederegulation airlines have dwindled to 15, while two-thirds of the 226 postderegulation entrants have disappeared. Most of the surviving new carriers are cargo, charter, or megacarrier feeders, not independent scheduled passenger airlines. Despite record industry profitability, in recent years the number of new entrant airlines collapsing has exceeded the number of new airlines emerging.

The arsenal of anticompetitive activities used by major airlines against new entrants is not new, though these tactical weapons have been used with increasingly better precision and effectiveness over time by the major airlines. Many low-cost/low-fare airlines have found themselves in the crosshairs of the major airlines. It is also apparent that a failure of government agencies to impose sanctions against such practices has led to a widespread belief among the major airlines that our nation's competition laws do not apply to them. The U.S. Department of Justice and the U.S. Department of Transportation should take such enforcement action against a major hub-dominant airline for blatantly anticompetitive activities such as those described herein as is necessary to preserve competition, while there is still competition to preserve.

New airline ventures fail for a number of reasons. A spike in fuel costs, recession, the failure to find a market niche, or managerial ineptitude can destroy a new airline. So too, can the predatory practices of the major airlines. The homicidal cycle is well established:

1 Major airline establishes monopoly in a market, and raises prices to confiscatory levels.

2 New low-cost airline enters the market, offering low fares.

3 Major airline responds by matching fares (even if below cost), sometimes adding aircraft capacity and frequency. Major airline rebates a portion of the ticket price in the form of frequent flier travel, and bribes travel agents with commission overrides to steer business its way.

4 After suffering severe economic losses, new entrant airline withdraws from the market.

5 Major airline reduces service and raises prices to confiscatory levels, often higher than those prevailing before the new entrant emerged.

The cycle repeats itself, though less often, as the pile of bankrupt airlines has grown higher, and chilled investment enthusiasm for new airline ventures.

What is wrong with this scenario is the economic impact route monopolization has on the local and regional economy, for air transportation is an essential part of the infrastructure of economic development. Business executives need frequent and reasonably priced air transportation, and will locate offices, factories, and warehouses only in those communities which have it.

Deregulation's principal architect, Alfred Kahn, has long decried the U.S. government's "abysmal dereliction" in failing to enforce the competition laws on airlines. In applauding DOT's initiative to address the efforts of incumbent airlines to suppress competition, Kahn told Congress, "The most grievous governmental failure in recent years has...been the failure to prosecute a single case against what appears to have been flagrant cases of predatory competition by incumbent major airlines against new competitors."

Ron Chernow, author of a biography of John D. Rockefeller, *Titan,* observed, "Free markets do not exist in a state of nature. Free markets are things that have to be defined by custom and law."[182] Congress has commanded that the market of air transportation be free of unfair and deceptive practices and unfair methods of competition. As *The New York Times* opined, "Although it can be hard to draw the line between predation and clean competition, the Federal Government needs to try. A flying public upset at the high fares on many routes deserves assurance that pricing practices are fair."[183]

Governmental institutions need not protect an individual competitor from the rigors of the marketplace. They should, however, protect competition. Monopolistic exploitation cannot long be tolerated in infrastructure industries upon which the rest of the economy depends. Better to have a competitor discipline a monopolist than have the government do it. Without application of the competition laws, predation runs riot, and competition is jeopardized. Because of the profound economic externalities airlines impose upon communities and businesses across America, monopolization in the airline industry cannot be tolerated. As Senator John McCain has observed, "The only thing worse than a regulated monopoly is an unregulated one." After all, the people own the airways and the airports, which were built with their tax dollars. These are public resources to be used in the public interest. It is reasonable to insist that air carriers serve the public interest. Monopolistic exploitation of consumers, and anticompetitive conduct designed to achieve it, are antithetical to that duty.

Notes

Chapter 1

1 Melvin Brenner, "Airline Deregulation: A Case Study in Public Policy Failure," 16 *Transportation Law Journal* 179, 189 (1988).

2 Oris Dunham, "Infrastructure Constraints: Deeds Not Words," 7th IATA High-Level Aviation Symposium, Cairo, Egypt, 1993, p. 109.

3 Eldon Griffiths, "U.S. Steams Ahead in Euro-Freight, While Europe Takes Off in World Airports," *World Trade,* June 1, 1998, p. 40.

4 Stephen Dunphy, "Hub Dreams Abound: Asia Airports about to Take Off," *Seattle Times,* Jan. 19, 1997, p. F1.

5 "Singapore to Build Billion-Dollar New Terminal for Super Jumbos," Agence France-Presse, Dec. 2, 1996 (1996 WL 12191726).

6 "Demand for air transportation in the Asia-Pacific region grew 12.1% per annum between 1985 and 1990." *Asia/Pacific Air Traffic Growth and Constraints,* International Air Transport Association, 1994, p. 3.

7 Carl Schwartz, "Airports in Asia Play Catch-Up as Economies Explode," *Milwaukee Journal Sentinel,* July 14, 1996, p. A11.

8 "IATA predicts that the Asia-Pacific region will grow by 7.5% between 1995 and 2000, and 7% between 2000 and 2010." Paul Dempsey and Kevin O'Connor, "Air Traffic Congestion and Infrastructure Development in the Pacific Asia Region," *Asia Pacific Air Transport,* Institute of Southeast Asian Studies, 1997.

9 For a comprehensive summary of airport projects in Asia, see Michael Mecham, "Growth Outpaces Asian Airports," *Aviation Week and Space Technology,* Aug. 29, 1994, p. 54; and Edward Neilan, "Turbulence Ahead for Asia's New Designer Airports," *Tokyo Business,* Aug. 1994, p. 30.

10 Moon-Sik Park, "Korea: Airport Development Market," *Financial Times Asia Intelligence Wire,* May 9, 1997.

11 *Asia/Pacific Air Traffic Growth and Constraints,* International Air Transport Association, 1993.

12 "Malaysia Seeks More 'Open Skies' Pacts," Agence France-Presse, May 5, 1997.

13 Chua Lee Hoong, "Can KL's $5b Total Airport Pip Changi International?" *Singapore Straits Times,* April 26, 1997.

14 Terence Hardeman, "The Necessity for New Airports in Asia-Pacific," paper presented to the IBC International Conference on Airport Development and Expansion, Hong Kong, Oct. 28, 1993.

15 Mark Landler, "Asian Crisis Takes Shine off Glittering Airports' Unveilings," *Seattle Post-Intelligencer,* July 6, 1998.

16 Nasid Hajari, "A Walk In The Clouds," *Time,* June 22, 1998.

Chapter 1 (*Cont.*)

17 Schwartz, op. cit.

18 Dunphy, op. cit.

19 John Meredith, "Room to Boom," *Airline Business,* Jan. 1995, p. 38.

20 "TDA Study Sees up to $7.24 Billion in U.S. Exports for Asian Airports," *Airports,* April 15, 1997, p. 145.

21 U.S. Trade and Development Agency, *Aviation Project Opportunities in Asia,* pp. 16–17, 1997.

22 "Expanding from Regional to International Status," *Jane's Airport Review,* April 1, 1997, p. 15.

23 "Cambodia: Joint Venture Plans for Proposed $1,300,000,000 Build-Operate-Transfer (BOT) Redevelopment Project," *ESP Report on Engineering Construction and Operations in the Developing World,* March 1, 1996 (1996 WL 9173801).

24 "China's Commercial Aviation Market to Open to Foreign Investors," *Air Cargo Report,* April 14, 1994 (1994 WL 3229623).

25 "Airlines Clear Path for Foreign Investment," *China Daily,* May 27, 1994.

26 "Guangzhou to Start Building New Airport Next Year," Agence France-Presse, Aug. 18, 1997 (1997 WL 13378737).

27 "Getting There Is Getting Easier," Successful Meetings, ABI/Inform, March 1997.

28 "Study Outlines Problems and Prospects for Air Transport in China," *World Airport Week,* Sept. 15, 1998.

29 "New Guangzhou Airport Tops China's Near-Term Airport Plans," *Airports,* Sept. 1, 1998, p. 356; Angelica Cheung, "Airport Crisis," *Orient Aviation,* Nov. 1998, p. 42.

30 "New Guangzhou Airport Tops China's Near-Term Airport Plans," *Airports,* Sept. 1, 1998, p. 356.

31 Joel Fischl, *China: Air Traffic Control Equipment Market,* Industry Sector Analysis prepared by the American Embassy in Beijing, March 4, 1997 (1997 WL 9850420).

32 Peter Trautmann, "The Need for New Airport Infrastructure," paper delivered before International Conference on Aviation and Airport Infrastructure, Denver, Dec. 5, 1993.

33 "China Opens New Airport in Nanjing," Agence France-Presse, June 12, 1997 (1997 WL 2132506).

34 Renee Schoof, "China Booming with New Airport Development," *Seattle Times,* April 8, 1995, p. D1.

35 Fischl, op. cit.

36 Ibid.

37 Ibid.

38 Angelica Cheung, "Airport Crisis," *Orient Aviation,* Nov. 1998, p. 42.

39 "Beijing's Billion-Dollar Airport Expansion Picks Up Steam," Agence France-Presse, July 15, 1997 (1997 WL 2152857).

40 Madelyn C. Ross, "Bursting at the Seams, China's Airports are Struggling to Keep Up with Demand for Air Travel," *China Business Review,* July 1, 1995 (1995 WL 9770806).

41 "Beijing's Billion-Dollar Airport Expansion Picks up Steam," Agence France-Presse, July 15, 1997 (1997 WL 2152857).

42 "SDB Loans to Beijing Capital Airport," Xinhua News Agency, Nov. 15, 1996 (1996 WL 12533723).

43 "Business Conference and Management Reports," *Asian Infrastructure Monthly,* March 1, 1996 (1996 WL 9522531).

44 "Aviation Sector Prepared for Direct Air Service across Strait," Xinhua English Newswire, March 4, 1997 (1997 WL 3748169).

Chapter 1 (*Cont.*)

45 "China Opens New Airport in Nanjing," Agence France-Presse, June 12, 1997 (1997 WL 2132506).

46 "China to Open New Airport in Fujian Province," The Associated Press, June 2, 1997 (1997 WL 4868802).

47 "Hume in RM810m Chinese Airport Project," *New Strait Times,* April 26, 1994 (1994 WL 14010055).

48 "China to Open New Airport in Fujian Province," The Associated Press, June 2, 1997 (1997 WL 4868802).

49 "China Opens New Airport in Nanjing," Agence France-Presse, June 12, 1997 (1997 WL 2132506).

50 "Xiamen Gaoqi Airport under Expansion," Xinhua News Agency, Nov. 4, 1994 (1994 WL 9102344).

51 "Airport Development Projects in Southern China Fujian Province: Xiamen Gaoqi International Airport," *Airports,* Nov. 17, 1992 (1992 WL 2290051).

52 Michael Mecham, "Air Transport Cathay Breaks the Ice, Invests in Chinese Airport," *Aviation Week and Space Technology,* Aug. 22, 1994 (1994 WL 2603595).

53 "Xiamen International Airport Upgraded," Xinhua English Newswire, March 15, 1997 (1997 WL 3751032).

54 "Guangzhou to Start Building New Airport Next Year," Agence France-Presse, Aug. 18, 1997 (1997 WL 13378737).

55 Schwartz, op. cit.

56 "Guangzhou to Start Building New Airport Next Year," Agence France-Presse, Aug. 18, 1997 (1997 WL 13378737).

57 Dunphy, op. cit.

58 "Other Airport News," *Air Cargo Report,* Aug. 4, 1994 (1994 WL 3229396).

59 Ibid.

60 John Ridding, "Survey—Macao: Economy," *The Financial Times,* May 13, 1997 (1997 WL 3793355).

61 "China Opens New Airport in Nanjing," Agence France-Presse, June 12, 1997 (1997 WL 2132506).

62 "Fuyang Attracts Foreign Capital," AsiaInfo Daily News Service, Dec. 22, 1997 (1997 WL 12772148).

63 "China Opens New Airport in Nanjing," Agence France-Presse, Jun. 12, 1997 (1997 WL 2132506).

64 "Hangzhou International Airport to be Located in Xiaoshan," Xinhua News Agency, March 11, 1994 (1994 WL 9099369).

65 "New Airport Opens to Flights in Nanjing," Xinhua News Agency, June 30, 1997 (1997 WL 8997207).

66 "China Opens New Airport in Nanjing," Agence France-Presse, June 12, 1997 (1997 WL 2132506); "New International Airport Opens in Nanjing, Fuzhou," Agence France-Presse, July 1, 1997 (1997 WL 2144302).

67 "Airlines Settled in Nanjing's New Airport," Xinhua News Agency, Oct. 31, 1997 (1997 WL 13732440).

68 Chen Shenzhang, "China: Jiangsu: New Transport Links Boost Economy," *Beijing Review,* Feb. 17, 1997 (1997 WL 10062542).

69 "Big Airport Completed in Nanjing," AsiaInfo Daily News Service, July 4, 1997 (1997 WL 7226525).

70 Chen Shenzhang, op. cit.

Chapter 1 (*Cont.*)

71 "Shanghai's Pudong Airport Gets 361 Million Dollar Loan: Report," Agence France-Presse, July 8, 1997 (1997 WL 21485078).

72 Ibid.

73 "Shanghai Builds China's Longest Runway," Xinhua News Agency, July 19, 1998.

74 "Guanyin Airport Begins Operation in Eastern China," Xinhua News Agency, Nov. 10, 1997 (1997 WL 13732781).

75 "China Opens New Airport in Nanjing," Agence France-Presse, June 12, 1997 (1997 WL 2132506).

76 "New Airport Plan off the Ground in Jiangsu," *China Daily,* Jan. 13, 1996 (1996 WL 11205104).

77 "China Opens New Airport in Nanjing," Agence France-Presse, June 12, 1997 (1997 WL 2132506).

78 "Farmer Owned Airport Operational," Xinhua News Agency, Nov. 9, 1993 (1993 WL 12195912).

79 Dunphy, op. cit.

80 "Guangzhou to Start Building New Airport Next Year," Agence France-Presse, Aug. 18, 1997 (1997 WL 13378737).

81 Raymond Lamont-Brown, "Japan Looks to Her Asian Neighbors," *Contemporary Review,* June 1, 1996 (1996 WL 13113775).

82 "Guangzhou to Start Building New Airport Next Year," Agence France-Presse, Aug. 18, 1997 (1997 WL 13378737).

83 Philippe Ries, "Adds IATA, Airport Authority Statements," Agence France-Presse, March 13, 1997 (1997 WL 2076303).

84 "China's Commercial Aviation Market to Open to Foreign Investors," *Air Cargo Report,* April 14, 1994 (1994 WL 3229623).

85 "India: Fresh Study on Bangalore Airport Project," Business Line (*The Hindu*), Feb. 22, 1997 (1997 WL 8206153).

86 "KSIIDC, Tata Consortium Sign MOU for Airport Project," *Financial Express,* Dec. 30, 1995 (1995 WL 14238937).

87 "India: Joint Venture Construction Start-up on $400,000,000 Build-Own-Operate Airport is Scheduled for Second Half of 1996," *ESP Report on Engineering Construction and Operations in the Developing World,* March 1, 1996 (1996 WL 9173830).

88 "India: Fresh Study on Bangalore Airport Project," Business Line (*The Hindu*), Feb. 22, 1997 (1997 WL 8206153).

89 Shirish Nadkarni, "Second Airport on the Cards for Battling Bombay," *South China Morning Post,* July 27, 1994 (1994 WL 8780386).

90 "India: Maharashtra Pact with Mumbai Airport Alliance for International Airport," Business Line (*The Hindu*), May 9, 1997 (1997 WL 10896219).

91 "India: 55 Projects on Offer at Advantage Maharashtra Meet," Business Line (*The Hindu*), Feb. 19, 1997 (1997 WL 8205858).

92 "Private Sector May Develop Navi Mumbai Airport," *Financial Express,* Nov. 27, 1996 (1996 WL 14479988).

93 "India: 55 Projects on offer at Advantage Maharashtra Meet," Business Line (*The Hindu*), Feb. 19, 1997 (1997 WL 8205858).

94 "Cargo Airport at Nashik Soon," *Financial Express,* March 2, 1996 (1996 WL 8365029).

95 U.S. Trade and Development Agency, *Aviation Project Opportunities in Asia,* 1997, p. 34.

96 Schwartz, op. cit.

97 Ibid.

Chapter 1 *(Cont.)*

98 "Singaporeans Must Realize How Tough the Competition Has Become," *Straits Times,* May 2, 1997, p. 46.

99 U.S. Trade and Development Agency, op cit., p. 57.

100 Ibid., p. 69.

101 Ibid., p. 73.

102 Catherine Wheeler, "Special Report: Indonesia," *The Financial Post,* Aug. 16, 1996 (1996 WL 5738219).

103 Horace Greeley, "Taiwan Investments in China and Southeast Asia," *Asian Survey,* May 1, 1996 (1996 WL 13007360).

104 "German Group Applies to Manage 3 Indonesian Airports," *Asia Pulse,* Nov. 21, 1996 (1996 WL 16263666); Thomas Fuller, "Malaysian Airport on Course Project Escapes Delay Amid Nation's Crisis," *International Herald Tribune,* Sept. 13, 1997 (1997 WL 4493614).

105 "Indonesian Pulp & Paper Co. to Build US$20 Mln Airport," *Asia Pulse,* Dec. 2, 1996 (1996 WL 16357432).

106 Schwartz, op. cit.

107 "Japan: Expo 2005 in Aichi: Opportunities," *International Market Insight Trade Opportunities Inquiries,* Report from the American Consulate in Nagoya, July 7, 1997 (1997 WL 12319681); Miki Tanikawa, "Projects for 2005 Expo Fuel a Rally on Nagoya," *International Herald Tribune,* Sept. 6, 1997 (1997 WL 4493500).

108 "Japan: Expo 2005 in Aichi: Opportunities," *International Market Insight Trade Opportunities Inquiries,* Report from the American Consulate in Nagoya, July 7, 1997 (1997 WL 12319681).

109 Michiyo Nakamoto, "A Model for the Next Century," *The Financial Times,* July 15, 1997 (1997 WL 11041533).

110 Ibid.

111 Tom Ballantyne, "Under Siege!," *Orient Aviation,* Nov. 1998, p. 58.

112 Haruhiro Fukui and Shigeko N. Fukai, "Pork Barrel Politics, Networks, and Local Economic Development in Contemporary Japan," *Asian Survey,* March 1, 1997 (1997 WL 13007329).

113 "Japan's Area Import Promotion Program Approved," *Asia Pulse,* Oct. 30, 1996 (1996 WL 16353824).

114 Shozo Okuno, "Japan: Airport Projects in Kansai Market," Industry Sector Analysis prepared by the American Consulate in Osaka Kobe, June 13, 1997 (1997 WL 9850556).

115 "Other Airport News," *Air Cargo Report,* Aug. 4, 1994 (1994 WL 3229396).

116 "China's Commercial Aviation Market to Open to Foreign Investors," *Air Cargo Report,* April 14, 1994 (1994 WL 3229623).

117 "Singapore Airport Expands to Meet Passenger and Cargo Traffic," Agence France-Presse, April 23, 1997 (1997 WL 2101447).

118 Michael Mecham, "Seoul Turns to Inchon to Solve Airport Woes," *Aviation Week and Space Technology,* Oct. 14, 1996, p. 50.

119 "Profile—Japan's Air Transport Industry," *Asia Pulse,* July 1998.

120 Brent Hannon, "Gateways to Growth," *Asia, Inc.,* Nov. 1995, p. 62.

121 Dunphy, op. cit.

122 Schwartz, op. cit.

123 Matthew Chance, "Malaysia's Slump Saves the Forest," *The Independent—London,* Sept. 11, 1997 (1997 WL 12345682); Andrew Taylor, "Blue Circle Stays on Course," *Financial Times,* Sept. 9, 1997 (1997 WL 11052291).

124 "Give Priority to Bintulu Airport," *Borneo Bulletin,* Sept. 18, 1997 (1997 WL 12222987).

125 "Plan for New Airport in Kuching," *Borneo Bulletin,* Sept. 6, 1997 (1997 WL 12222750).

Chapter 1 (*Cont.*)

126 Azman Ujang, "Malaysia Achieves Big Savings in New Airport Construction," *Asia Pulse,* March 25, 1997 (1997 WL 10227999).

127 Ibid.

128 Thomas Fuller, "Malaysian Airport on Course Project Escapes Delay Amid Nation's Crisis," *International Herald Tribune,* Sept. 13, 1997 (1997 WL 4493614).

129 Ibid.

130 Ujang, op. cit.

131 Fuller, op. cit.

132 Ujang, op cit.

133 "Mahathir Says Bakun Dam Project, Others Delayed Due to Currency Weakness," *AFX News,* Sept. 5, 1997 (Westlaw database: EURONEWS).

134 "TDA Study Sees Up to $7.24 Billion in U.S. Exports for Asian Airports," *Airports,* April 15, 1997, p. 145.

135 "Malaysian Officials Claim Kuala Lumpur Charges Will Be Reasonable," *Asian Aviation News,* April 4, 1997.

136 Peter Comparelli, "Profile: Ting's Towering Tasks: Model of Plaza Rakyat," *Asia, Inc.,* April 1, 1997 (1997 WL 10060063).

137 "Rangoon Airport Extension Proceeding without Japan Aid," Agence France-Presse, March 8, 1997 (1997 WL 2073186).

138 Ibid.

139 Ted Bardacke, "Ital-Thai Invests in Southeast Asia to Grow Contract Business," *The Financial Post,* Aug. 9, 1996 (1996 WL 5737712).

140 "Singapore Firm in Burmese Airport Deal," Agence France-Presse, June 6, 1995 (1995 WL 7812698).

141 "Transport Briefs: Thailand Approves Loan to Burma for Airport," *The Journal of Commerce,* March 22, 1996 (1996 WL 8121797).

142 "Rangoon Airport Extension Proceeding without Japan Aid," Agence France-Presse, March 8, 1997 (1997 WL 2073186).

143 "Japan Moves to Loan Burma Airport Money," *Airline Industry Information,* March 16, 1998 (1998 WL 4675320).

144 "Pakistan: Joint Venture Negotiations for $220,000,000 Airport Construction Contract Proceed Ahead," *ESP Report on Engineering Construction and Operations in the Developing World,* June 1, 1996 (1996 WL 9174299); "Pakistan: Joint Venture Contract Award for $250,000,000 Airport Expansion Project," *ESP Report on Engineering Construction and Operations in the Developing World,* Oct. 1, 1996 (1996 WL 13073665).

145 "China's Commercial Aviation Market to Open to Foreign Investors," *Air Cargo Report,* April 14, 1994 (1994 WL 3229623).

146 U.S. Trade and Development Agency, op cit., p. 136.

147 Schwartz, op. cit.

148 "Getting There Is Getting Easier," *ABI/Inform,* March 1997, p. 3.

149 Ibid., pp. 150–157.

150 "ADB Approves 93-Million-Dollar Loan to Improve Philippines," Agence France-Presse, Sept. 16, 1997 (1997 WL 13395554).

151 Ibid.

152 Philippines Department of Transportation and Communications, MIADP 2000, 1996.

153 "Starting in Cebu," *Travel Agent,* Dec. 23, 1996 (1996 WL 14227872).

154 Sharon Desker Shaw, "ATF Site Ready Soon," *Travel Trade Gazette Asia,* May 9, 1997 (1997 WL 8942774).

Chapter 1 *(Cont.)*

155 Interview with Romeo Bersonda, Mactan-Cebu International Airport Authority Operations Manager, Sept. 3, 1998.

156 "The Philippines in 1996: A House Finally in Order?" *Asian Survey,* Feb. 1, 1997 (1997 WL 11023362).

157 "Philippines: Joint Venture Construction Start-up on $160,000,000 Airport Terminal 1995," *ESP Report on Engineering Construction and Operations in the Developing World,* Dec. 1, 1995 (1995 WL 11679161).

158 "Lockheed Martin Eyes Ninoy Aquino International Airport 3, Cebu, Gensan Airport Operations," *BusinessWorld (Manila),* Dec. 18, 1996 (1996 WL 14370723).

159 "Philippines: Joint Venture Construction Plans for Proposed $520,000,000 Airport Terminal Construction Project," *ESP Report on Engineering Construction and Operations in the Developing World,* Sept. 1, 1997 (1997 WL 12986459).

160 E. P. Patanne, "DHL's Asia Hub Thrives near Manila International Airport," *The Wall Street Journal,* Sept. 23, 1996.

161 "Subic, Philippines," Agence France-Presse, Nov. 3, 1996 (1996 WL 12171041).

162 U.S. Trade and Development Agency, op. cit., pp. 143, 144.

163 "Singapore Airport Expands to Meet Passenger and Cargo Traffic Boom," Agence France-Presse, April 23, 1997.

164 Chua Lee Hoong, op. cit.

165 "Getting There Is Getting Easier," *ABI/Inform,* March 1997, p. 3.

166 Chua Lee Hoong, op. cit.

167 "Singapore Airport Expands to Meet Passenger and Cargo Traffic Boom," Agence France-Presse, April 23, 1997.

168 Ibid.

169 Thar Pavilions, "Limousine with Liveried Jeeves," *India Today Plus,* July 1, 1997 (1997 WL 11279269).

170 "Singapore to Build Billion-Dollar New Terminal for Super Jumbos," Agence France-Presse, Dec. 2, 1996 (1996 WL 12191726); "Singapore to Build a Third Airport Terminal for Super Jumbos," Agence France-Presse, Dec. 2, 1996 (1996 WL 12191583).

171 "Singapore to Build Billion-Dollar New Terminal for Super Jumbos," Agence France-Presse, Dec. 2, 1996 (1996 WL 12191726); "Singapore to Build a Third Airport Terminal for Super Jumbos," Agence France-Presse, Dec. 2, 1996 (1996 WL 12191583).

172 "Singapore to Build Billion-Dollar New Terminal for Super Jumbos," Agence France-Presse, Dec. 2, 1996 (1996 WL 12191726).

173 "Singapore Airport Expands to Meet Passenger and Cargo Traffic," Agence France-Presse, April 23, 1997 (1997 WL 2101447).

174 "Singapore Airport Expands to Meet Passenger and Cargo Traffic," Agence France-Presse, April 23, 1997 (1997 WL 2101447).

175 "South Korea: Review 1997," *Asia and Pacific Review,* May 1997.

176 "TDA Study Sees up to $7.24 Billion in U.S. Exports for Asian Airports," *Airports,* April 15, 1997, p. 145.

177 U.S. Trade and Development Agency, op. cit., pp. 103, 104.

178 Interview with I. M. Razeek, Manager of Colombo's Bandaranaike International Airport, Sept. 8, 1998.

179 Schwartz, op. cit.

180 Hannon, op. cit.

181 CKS International Airport Annual Report 1996.

182 "Asian Airport Oversupply," *Jane's Airport Review,* Oct. 1, 1998, p. 1.

Chapter 1 (*Cont.*)

183 "Thailand in 1996: Economic Slowdown Clouds Year," *Asian Survey,* Feb. 1, 1997 (1997 WL 11023355).

184 "South Korea: Kim Softens Stance," *Far Eastern Economic Review,* Jan. 30, 1997 (1997 WL-FEER 2009779).

185 "Thai Cabinet Approves Reduced Budget for Airport Expansion," Agence France-Presse, May 27, 1997 (1997 WL 2121926).

186 "Thai Cabinet Approves Reduced Budget for Airport Expansion," Agence France-Presse, May 27, 1997 (1997 WL 2121926).

187 U.S. Trade and Development Agency, op. cit., p. 213.

188 "Turkey: $81 Million Airport Contract," *ESP Report on Engineering Construction and Operations,* Jan. 1, 1993 (1993 WL 2663675).

189 "Uzbekistan-EBRD Tashkent Airport Rehabilitation Project: Opportunities," International Market Insight Trade Opportunities Inquiries from the American Embassy in London, June 17, 1997 (1997 WL 10398623).

190 Tom Ichniowski and Paul Rosta, "Asia Projects Taking Off," *Engineering News-Record,* April 28, 1997, p. 16.

191 www.tradeport.org/ts/countries/vietnam, April 10, 1998.

192 Schwartz, op. cit.

193 www.tradeport.org/ts/countries/vietnam, April 10, 1998.

194 Ichniowski and Rosta, op. cit.

195 "TDA Study Sees up to $7.24 Billion in U.S. Exports for Asian Airports," *Airports,* April 15, 1997, p. 145.

196 U.S. Trade and Development Agency, op. cit., p. 247.

197 "Vietnam: Business Briefs," *Saigon Times,* Sept. 20, 1996 (1996 WL 11758722).

198 "Vietnam: Construction Plans for Airport Upgrade Project, Civil Aviation Administration (Vietnam)," *ESP Report on Engineering Construction and Operations in the Developing World,* Aug. 1, 1995 (1995 WL 2367556).

199 Ibid.

200 "Vietnam: Plans For $1,800,000,000 Airport Upgrading Project, Ministry of Transport, Civil Aviation Department & Tan Son Nhat Airport," *ESP Report on Engineering Construction and Operations,* Aug. 1, 1994 (1994 WL 2477224).

201 "Vietnam: Stop Press," *Saigon Times,* Nov. 17, 1996 (1996 WL 11758830).

202 Cong Thanh, "Call for $1.8b Tan Son Nhat Facelift," *Vietnam Investment Review,* May 16, 1994 (1994 WL 14153324).

203 "Vietnam Needs $US15 Billion for Transport Development," *Asia Pulse,* July 31, 1996 (1996 WL 16346901).

204 Cong Thanh, op. cit.

205 "Vietnam: Aviation and Aircraft Overview," International Market Insight Reports, Telegraphic Report from the American Embassy in Hanoi, April 23, 1997 (1997 WL 10397322).

206 "New Cargo Terminal Opens in Tan Son Nhat Airport in Vietnam," *Asia Pulse,* Jan. 28, 1997 (1997 WL 10599287).

207 Russell Barling, "VN Cargo to Increase," *CNA,* Sept. 8, 1997.

208 Nguyen Manh Hung, "Vietnamese Will Build Tan Son Nhat," *Vietnam Investment Review,* Oct. 28, 1996 (1996 WL 8391871).

209 "Vietnam: Aviation and Aircraft Overview," International Market Insight Reports, Telegraphic Report from the American Embassy in Hanoi, April 23, 1997 (1997 WL 10397322).

210 Ibid.

211 Anne Paylor, "Gulf Airports Rally to New Challenges," *The Middle East,* May 1, 1995 (1995 WL 14950144).

Chapter 1 (*Cont.*)

212 Philippa Nugent, "Looking to the Future," *The Middle East*, April 1, 1997 (1997 WL 9741896).

213 "No. 1 in Middle East," http://www.bahrainairport.com/bia/news2.htm, June 18, 1998.

214 Nadia Hussein, "Gulf Airports," *The Middle East*, Nov. 1, 1997 (1997 WL 9742021).

215 Anton LaGuardia, "International: Palestine Airport Still Awaits Israeli Clearance for Take-off," *The Daily Telegraph—London*, June 6, 1997 (1997 WL 2315316); "Arafat Intends to Use Gaza Airport for First Time," Agence France-Presse, Sept. 18, 1997 (1997 WL 13397116); Philippa Nugent, "The Flying Squads," *The Middle East*, Jan. 1, 1997 (1997 WL 9741859).

216 Lee Hockstader, "Palestinian Airport Cheered," *The Denver Post*, Nov. 25, 1998, p. 20A.

217 "Iran: Procurement Budget Estimated to Reach $50,000,000 for Proposed Airport Terminal Construction Project," *ESP Business Opportunities in Africa and the Middle East*, Jan. 1, 1996 (1996 WL 9172208).

218 Andrew Album, "Paving the Road to Prosperity," *The Middle East*, June 1, 1995 (1995 WL 14950159).

219 "Jordan: Feasibility Study for Proposed $36,000,000 International Airport Is Scheduled for Completion by March 1996," *ESP Report on Engineering Construction and Operations in the Developing World*, Dec. 1, 1995 (1995 WL 11679116); "Rift Valley Project Shows Few Signs of Concrete Development," *The Middle East*, July 1, 1995 (1995 WL 14950174).

220 "Updates," Agence France-Presse, Aug. 13, 1997 (1997 WL 13386817).

221 "Lebanon: Joint Venture's Low Bid of $490,000,000 for Airport Expansion Project Is Likely to Lead to Final Contract Award," *ESP Report on Engineering Construction and Operations*, June 1, 1994 (1994 WL 2477289); Nugent, "The Flying Squads," op. cit.

222 John McLaughlin, "Special Report on Lebanon: New Airport Nears Take-off," *Lloyds List*, June 29, 1995 (1995 WL 11773707).

223 Brian Walters, "East-West Hub Revitalized," *Jane's Airport Review*, Sept. 1, 1998, p. 62.

224 Daniel Pearl, "Rocky Takeoff: Lebanon Scrambles to Rebuild Airport Battered in Civil War," *The Wall Street Journal Europe*, April 7, 1998 (1998 WL-WSJE 3514374).

225 "Beruit to Receive $66 Million Upgrade Courtesy of EIB," *World Airport Week*, Jan. 3, 1995 (1995 WL 8286452).

226 McLaughlin, op. cit.

227 "Lebanon: Airports & Free Zone BOT Contracts," International Market Insight Trade Opportunities Inquiries, Telegraphic Report from the American Embassy in Beruit, July 17, 1997 (1997 WL 12319864).

228 Paylor, op. cit.

229 "Oman: Letter of Intent for $21 Million Airport Expansion Project," *ESP Report on Engineering Construction and Operations*, Jan. 1, 1993 (1993 WL 2663662).

230 Paylor, op. cit.

231 "Qatar: Invitation to Bid on Construction of $105,000,000 Airport Upgrade Project Is Tentatively Scheduled to be Issued in Late 1996 or Early 1997," *ESP Business Opportunities in Africa and the Middle East*, Aug. 1, 1996 (1996 WL 9172395); Nugent, "The Flying Squads," op. cit.

232 "Air Travel," http://www.saudi.net/profile/transcom/transcom_air.html, Nov. 14, 1997.

233 Ibid.

234 "Saudi Arabia: Airport Expansion Plans," International Market Insight Trade Opportunities Inquiries, Telegraphic Report from the American Embassy in Jeddah, June 3, 1997 (1997 WL 10398468).

235 "Gulf Airport Expansion Takes-Off," http://www.arabia.com/Gulfbusiness/cover/airports.htm, Nov. 14, 1997; Nugent, "The Flying Squads," op. cit.

236 "Second Satellite for Abu Dhabi," *Jane's Airport Review*, March 1, 1998 (1998 WL 10233079).

Chapter 1 (*Cont.*)

237 Maeve Fatheldin, "U.A.E.: Airport Development Market," Industry Sector Analysis Prepared by the American Embassy in Abu Dhabi, Jan. 31, 1997 (1997 WL 9850404).

238 "ADP To Install Radars at Abu Dhabi Airports," *Aviation Europe,* Oct. 31, 1996 (1996 WL 14116354).

239 "Gulf Airport Expansion Takes-Off," http://www.arabia.com/Gulfbusiness/cover/airports.htm, Nov. 14, 1997.

240 David Woolley, "New UAE Airport Attracts a Wealth of CIS Involvement," *Air Cargo World,* May 1, 1994 (1994 WL 13078817).

241 Fatheldin, op. cit.

242 "U.A.E.: Invitation to Bid on $300,000,000 Airport Expansion Works Scheduled for March 1996," *ESP Report on Engineering Construction and Operations in the Developing World,* Dec. 1, 1995 (1995 WL 11679219).

243 Nadim Kawash, "Dubai Gearing Airport Up for 21st Century," Agence France-Presse, March 2, 1997 (1997 WL 2068867).

244 Roger Bray, "Dubai Terminal Set to Open," *Financial Times,* March 23, 1998 (1998 WL 3541883).

245 "U.A.E.: Invitation to Bid on $300,000,000 Airport Expansion Works Scheduled for March 1996," *ESP Report on Engineering Construction and Operations in the Developing World,* Dec. 1, 1995 (1995 WL 11679219).

246 "Investment in the Gulf Adopts New Technology," *Jane's Airport Review,* March 1, 1998 (1998 WL 10233077).

247 Rod Pascoe, "Seizing Fresh Funds," *Jane's Airport Review,* Jan. 1, 1998 (1998 WL 10232985).

248 "Thousands Object to Airport Master Plan," *Australasian Business Intelligence,* June 1, 1998 (1998 WL 9699145).

249 "$300m Airport Expansion Trebles Jobs, Passengers," *Australasian Business Intelligence,* March 18, 1998 (1998 WL 9658611).

250 Pavilions, op. cit.

251 Nikki Tait, "Survey—Australia: Not All for Going for Gold," *The Financial Times,* Nov. 14, 1996 (1996 WL 13948474); "Australia: Environmental Auditor for Airport EIS," *SciTech Newsletter,* Oct. 1, 1996 (1996 WL 11756742).

252 John Meredith, "Room to Boom," *Airline Business,* Jan. 1995, p. 38.

253 Pascoe, op. cit.

254 Ibid.

255 Ibid.

256 Ibid.

257 "Roundup: Airport Renovation Under Way in NZ Cities," Xinhua English Newswire, Oct. 30, 1997 (1997 WL 11207638).

258 "Airport Expansion to Be Opened Soon," *Christchurch Press,* Sept. 2, 1998 (1998 WL 9491230).

259 Kevin Taylor, "New Airport Terminal among Best," *Christchurch Press,* Sept. 16, 1998, p. 4.

260 Pascoe, op. cit.

261 "Roundup: Airport Renovation Under Way in NZ Cities," *Xinhua English Newswire,* Oct. 30, 1997 (1997 WL 11207638).

262 "Vanuatu: Contract to Upgrade 2 Airports Is Currently Being Negotiated—Thiess Contractors," *ESP Report on Engineering Construction and Operations in the Developing World,* Nov. 1, 1994 (1994 WL 2721320).

Chapter 1 (*Cont.*)

263 David Feldman, "Commercial Magnetism," *Airline Business,* Dec. 1996, p. 36.

264 Barry D. Wood, "Letter From Tirana," *Europe,* May 1, 1996 (1996 WL 13158617).

265 "Armenia: Construction Contract Award for $28,000,000 Airport Cargo Terminal, Bill Harbert International Construction Inc.," *ESP Business Opportunities in Eastern Europe and the CIS,* Dec. 1, 1996 (1996 WL 15500479).

266 Vienna International Airport, "Where West Goes East," 1998.

267 "Azerbaijan Airlines Development Strategy," International Market Insight Reports from the American Embassy in Baku, May 8, 1997 (1997 WL 10397627).

268 "Equipment—News," *Jane's Airport Review,* Oct. 1, 1998, p. 31.

269 Brussels Airport Authority, "Runway to the Future," 1997.

270 Tony Carding, "Look at the Future Plans Europeans Have for Trans-Atlantic Trade!," *Air Cargo World,* June 1, 1996 (1996 WL 9777903); "TNT Pours $66 Million into New European Air Hub," *Air Cargo Report,* March 14, 1996 (1996 WL 8435171).

271 "TNT Pours $66 Million into New European Air Hub," *Air Cargo Report,* March 14, 1996 (1996 WL 8435171).

272 "Bulgaria Tenders," *Privatisation International,* Sept. 1, 1997 (1997 WL 8935933).

273 "Spotlight on Croatia," *Commuter/Regional Airline News,* Aug. 24, 1998.

274 "Airport Management Co. Commissions New Check-in Terminal Called JIH 2 in Prague-Ruzyne," *Ekonomicke Zpravodajstvi,* Sept. 3, 1997 (1997 WL 14052699).

275 "Czech Prague Airport Expansion to Open June 3," *CTK, Dow Jones International News,* April 4, 1997 (Westlaw database: EURONEWS).

276 Ibid.

277 "Denmark: Airport Security Equipment Market Overview," International Market Insight Report from the American Embassy in Copenhagen, April 14, 1997 (1997 WL 10397122).

278 Copenhagen Airports, "Building for the Future," July 1998.

279 Bruce Barnard, "Europe's Entrepreneurs: Creating Jobs Globally," *Europe,* Sept. 1, 1996 (1996 WL 13158684).

280 "Denmark: Airport Security Equipment Market Overview," International Market Insight Report from the American Embassy in Copenhagen, April 14, 1997 (1997 WL 10397122).

281 "Estonia-EBRD Tallinn Airport Passenger Terminal Project: Opportunities," International Market Insight Trade Opportunities Inquiries, report from American Embassy in London, May 21, 1997 (1997 WL 10398412).

282 Sara-Anna Yrjola, Harri Makinen, "Finland: Airport & Ground Support Equipment Market," Industry Sector Analysis prepared by the American Embassy in Helsinki, March 28, 1997 (1997 WL 9850441).

283 Ibid.

284 Ibid.

285 Ibid.

286 "Canada: Airport Commercialization Opportunities," *International Market Insight Trade Opportunities Inquiries,* June 2, 1997 (1997 WL 10398449).

287 Ibid.

288 Ester Laushway, "Marseille: An Up and Coming Mediterranean Hub," *Europe,* April 1, 1997 (1997 WL 10572262).

289 "Aéroports de Paris prévoit un 't chaud' et cinq années difficiles," *Tribune Desfosses,* March 13, 1997 (1997 WL 8315665).

290 Aeroports de Paris, Orly, "Mastering the Future," 1997.

Chapter 1 (*Cont.*)

291 "Aéroports de Paris prévoit un 't chaud' et cinq années difficiles," *Tribune Desfosses,* March 13, 1997 (1997 WL 8315665).

292 "The Expansion of Roissy Will Be Announced Soon," *Les Echos,* Sept. 16, 1997 (1997 WL 13354132).

293 Eric Drosin, Charles Goldsmith, Lucia Piana, "A Special Roundup of Business, Industry and Financial Developments," *The Wall Street Journal Europe,* March 5 1997 (1997 WL-WSJE 3807531).

294 Ibid.

295 "Private Funds for Germany," *Jane's Airport Review,* Dec. 1997, p. 13; Frederick Studeman, "Three Link for Airport Bid," *The Financial Times,* March 5, 1997 (1997 WL 3777760).

296 Michael A. Taverna, "Berlin OKs Airport Privatization, New International Facility," *Aviation Week and Space Technology,* Oct. 6, 1997 (1997 WL 14817512).

297 Adeline Ong, "Berlin Braces for Air Traffic Growth," *Business Times,* Oct. 15, 1997 (1997 WL 13786713).

298 Taverna, op. cit.

299 Ibid.

300 Ibid.

301 Frederick Studemann, "States Agree Sell-off Details," *Financial Times,* Sept. 10, 1997 (1997 WL 11052627).

302 Gunter Endres, "Bidding for Berlin," *Jane's Airport Review,* July 1, 1998 (1998 WL 10233263).

303 "Bidders for Berlin Airport Can Hope for Assistance," *Handelsblatt,* July 6, 1998 (1998 WL 7552190).

304 "Germany: Cologne and Dortmund Airport Expansion Plans," *International Market Insight Trade Opportunities Inquiries,* April 2, 1997 (1997 WL 10397922).

305 Ibid.

306 Ibid.

307 Brendan McGrath, "Aer Rianta on Short-list for German Airport Stake," *The Irish Times,* July 1, 1997 (1997 WL 12012019).

308 "Germany: Cologne and Dortmund Airport Expansion Plans," *International Market Insight Trade Opportunities Inquiries,* April 2, 1997 (1997 WL 10397922).

309 Barbara Shea, "On the Go," *Newsday,* Nov. 10, 1996 (1996 WL 2543893).

310 "Pen-Pusher's City Gets New Signature," *Financial Times,* Sept. 7, 1998, p. 16.

311 Ian Mader, "Frankfurt Airport Gets Makeover," *The Denver Post,* Nov. 9, 1997, p. 8J.

312 Carding, op. cit.

313 Ibid.

314 Ibid.

315 "Private Funds for Germany," *Jane's Airport Reivew,* Dec. 1997, p. 13.

316 Ibid.

317 Ibid.

318 Leonard Hill, "Countdown on Munich II," *Air Transport World,* June 1, 1991, p. 70.

319 Richard Rowe, "Host with the Most," *Airports International,* May 1, 1997, p. 18 (1997 WL 9706715).

320 Jenny Paris, "Heard in Athens: Athens Shares Go for the Gold," *The Wall Street Journal Europe,* Sept. 9, 1997 (1997 WL-WSJE 12211415).

321 Pavilions, op. cit.

Chapter 1 (*Cont.*)

322 "Canada: Airport Commercialization Opportunities," *International Market Insight Trade Opportunities Inquiries,* June 2, 1997 (1997 WL 10398449).

323 Ibid.

324 "Budapest to Be Centre," *The Budapest Sun,* May 29, 1997 (1997 WL 8262622).

325 Brendan McGrath, "Irish Aviation Authority to Report Profits Up 10%," *The Irish Times,* June 9, 1997 (1997 WL 12010997).

326 Ibid.

327 Robert J. Guttman, "John Hume: Northern Ireland Peacemaker," *Europe,* July 1, 1995 (1995 WL 15052626).

328 Robert J. Guttman, "Northern Ireland: You Will Never Know Unless You Go," *Europe,* July 1, 1995 (1995 WL 15052631).

329 Peter Trautmann, "The Need for New Airport Infrastructure," International Conference on Aviation and Airport Infrastructure, Denver, Dec. 5, 1993.

330 Martin du Bois, "Floating an Idea: Holland is All at Sea over the Future of Its International Airport," *The Wall Street Journal Europe,* June 16, 1997 (1997 WL-WSJE 3812967).

331 Aeroporti di Milano, "For You," 1997.

332 "New Air Terminal at Fiumicino," *Il Sole 24 Ore,* Aug. 1, 1997 (1997 WL 13232024).

333 "Italy: Airport Development Equipment/Services Market," *Industry Sector Analysis,* June 29, 1998 (1998 WL 11163462).

334 "Business News Abstracts: Baltics," *East/West Commersant,* March 15, 1997 (1997 WL 9404698).

335 "Latvia: Riga Airport Set for Expansion: Opportunity," International Market Insight Trade Opportunities from the American Embassy in Riga, April 2, 1997 (1997 WL 10397929).

336 Thomas Orszag Land, "Berlin: Capital on the Move," *Contemporary Review,* Feb. 1, 1997 (1997 WL 10530010).

337 Amsterdam Airport Schiphol, Fact Sheet 3: Mainport Study, 1997.

338 Pavilions, op. cit.

339 Bruce Barnard, "Gateway Europe (Netherlands)," *Europe,* March 1, 1995 (1995 WL 15052566); Roel Janssen, "The Dutch Job Machine," *Europe,* Sept. 1, 1996 (1996 WL 131589691); Gordon Cramb, "Dutch May Reverse Airport Noise Curbs," *Financial Times,* Sept. 16, 1997 (1997 WL 11053837).

340 Amsterdam Airport Schiphol, "Schiphol's Masterplan," 1997.

341 Amsterdam Airport Schiphol, Fact Sheet 3: Mainport Study, 1997.

342 "Norway: Overview of Economic Trends," *International Market Insight Reports,* May 6, 1997 (1997 WL 10397587).

343 Leonard Hill, "Nordic Newcomer," *Air Transport World,* Feb. 1, 1995, p. 80.

344 Habeeb Salloum, "Portugal Prepares for Tourist Onslaught," *Contemporary Review,* Feb. 1, 1996 (1996 WL 13113741).

345 "Portugal Govt. Forms Group to Study New Airport Project," *Dow Jones International News,* July 4, 1997 (Westlaw database: EURONEWS).

346 Ibid.

347 "Structural Funds: Portugal's Controversial Alqueva Project Approved," *European Report,* July 30, 1997 (1997 WL 13046127).

348 "Overview: Romania," *Privatisation International,* Sept. 1, 1997 (1997 WL 8935944).

349 "Romania: Airport Security, Safety Equipment Market Overview," International Market Insight Reports from the American Embassy in Bucharest, April 7, 1997 (1997 WL 10396999).

Chapter 1 (*Cont.*)

350 "Russia: Sheremetevo Airport Completed Construction of New Runway," *Kommersant,* July 20, 1997 (1997 WL 9687236).

351 "Russia: Transit Cargo Terminal Will Be Built in Pulkovo by Lufthansa Cargo (Germany)," *Delovoi Mir,* Aug. 5, 1997 (1997 WL 14053785).

352 "Russia: Plans to Improve, Expand Samara Airport: Opportunities," *International Market Insight Trade Opportunities Inquiries,* April 29, 1997 (1997 WL 10398247).

353 Ibid.

354 "Russia: Reconstruction of Koltsovo Airport Near Ekaterinburg Will Be Supported by Tissen (Germany)," *Delovoi Mir,* Aug. 5, 1997 (1997 WL 14053786).

355 "Spanish Airport Authority Aena Will Invest 95 Billion Pesetas in 30 Airports This Year," *Expansion,* April 6, 1998 (1998 WL 7547584).

356 Carlta Vitzthum, "Travel: In Bilbao, There's a Work of Art on the Horizon," *The Wall Street Journal Europe,* June 27, 1997 (1997 WL-WSJE 3813633).

357 "Spain: Balearic Islands Commercial Profile," *International Market Insight Reports,* May 20, 1997 (1997 WL 11135608).

358 "EIB Financing for Four Airports in the Canary Islands," *European Report,* June 14, 1997 (1997 WL 8517543).

359 Michael Knipe, "Catalonia Drives Towards the Millennium," *The Times of London,* April 23, 1997 (1997 WL 9200555).

360 "Expansion (Spain)," Information Access Company, May 24, 1997 (1997 WL 8317015).

361 "Barajas to Take Off in a Year and a Half with a Third Runway, the Largest in Europe," *Expansion,* July 24, 1997 (1997 WL 13349279).

362 Ibid.

363 Ibid.

364 Rowe, op. cit.

365 "New SF100 Mil Terminal Finger to be Built at Geneva-Cointrin Airport as Part of SF400 Mil Expansion Program for 1996–2005," *Journal de Genève & Gazette de Lausanne,* July 12, 1997 (1997 WL 9814847).

366 Pavilions, op. cit.

367 Rose Rouse, "Travel: Navel Manoeuvres," *The Independent—London,* June 14, 1997 (1997 WL 10473022).

368 "Turkey Court Clears BOT Model for Istanbul Airport," *Dow Jones International News,* Sept. 19, 1997 (Westlaw database: EURONEWS).

369 Ibid.

370 "Birmingham, U.K., Airport Begins Next Phase of Terminal Expansion," *Airports,* Aug. 25, 1998, p. 344.

371 Rachel Thomas Garrett Nagle, "Terminal Gain? (Pros and Cons of New Heathrow Airport Terminal)," *Geographical Magazine,* March 1, 1996 (1996 WL 13219665); du Bois, op. cit.

372 Richard Halstead, "Terminal 5 Waits on Inquiry," *The Independent—London,* April 6, 1997 (1997 WL 10458745); "BAA/FY-3: Passenger Traffic Up over 5% This Year," *Dow Jones International News,* June 9, 1997 (Westlaw database: EURONEWS).

373 Halstead, op. cit.

374 "Report BAA to Invest 250M GBP in Heathrow Baggage System," *Dow Jones International News,* Aug. 31, 1997 (Westlaw database: EURONEWS).

375 Nagle, op. cit.

376 Carding, op. cit.

Chapter 1 (*Cont.*)

377 Nick Small, "Hubbing the Airways," *Geographical Magazine,* Feb. 1, 1995 (1995 WL 15097568).

378 Pavilions, op. cit.

379 Richard Wolffe, "High Hopes for Phoenix Estate," *The Financial Times,* April 3, 1997 (1997 WL 3784370).

380 Matt Benson, "Airport Expansion," *The Times of London,* June 11, 1997 (1997 WL 9208791); Nicci Gerrard, "Child of the Revolution: Tunnel of Love for Denise and Grand-pappy, the Earth Moved Under Manchester Airport," *The Observer,* June 15, 1997 (1997 WL 10816368).

381 Richard Adams, "Facing Up to the Problems," *The Financial Times,* Nov. 11, 1996 (1996 WL 13947355).

382 Paul Dempsey, "Airlines in Turbulence: Strategies for Survival," 23 *Transporation Law Journal* 15, 81 (1995).

383 Paul Stephen Dempsey, *Airline Deregulation and Laissez Faire Mythology,* Quorum Books, 1992, pp. 272–274.

384 Robert Moskal, "Canada: Cargo Handling Equipment Market," *U.S. and Foreign Commercial Service: Industry Sector Analysis,* Aug. 20, 1997 (1997 WL 9850758).

385 "Canada: Calgary Airport Expansion: Opportunities," *International Market Insight Trade Opportunities Inquiries,* April 2, 1997 (1997 WL 10397920).

386 Jac MacDonald, "Construction Predicts Surge; Trade Shortages Foreseen with Heavy Industrial Projects on Horizon; Area's Biggest Projects for '97," *The Edmonton Journal,* Dec. 17, 1996 (1996 WL 5161484).

387 Mairi MacLean, "Airport Hires Reno Manager," *The Edmonton Journal,* Oct. 17, 1996 (1996 WL 5152261); Trish Worron, "In (Sky) High Dudgeon; Airport Braces for Turbulence from Travellers as New Tax Takes Off," *The Edmonton Journal,* April 7, 1997 (1997 WL 6769876).

388 Darcy Henton, "Airport Tax Grab Angers Fliers: User Fee Pays for Improvements to Terminals," *The Toronto Star,* April 13, 1997 (1997 WL 3831578).

389 Anne Sutherland, "Traffic Plan Aims to Cut Snarls Near Dorval Airport," *Montreal Gazette,* Nov. 7, 1996 (1996 WL 4212127).

390 Lynn Moore, "Airport Move Grounded," *Montreal Gazette,* Feb. 13, 1997 (1997 WL 4610394).

391 Joanne Laucius, "Airport Deal Builds Framework for Transfer to Local Authority," *The Ottawa Citizen,* Sept. 6, 1996 (1996 WL 3616223).

392 Habeeb Salloum, "Canada Prepares for Ski Championships," *Contemporary Review,* Jan. 1, 1995 (1995 WL 15010060).

393 Moskal, op. cit.

394 David Israelson, "Airport Overhaul Gets Kick-Start," *The Toronto Star,* March 26, 1997 (1997 WL 3828518).

395 Greater Toronto Airports Authority, "Terminal Development Master Plan," 1997.

396 Vancouver International Airport Authority, "Expansion '96," 1995.

397 Richard Rowe, "Vibrant Vancouver," *Airports International,* July 1, 1996 (1996 WL 8980187); Richard Rowe, "Canadian Club," *Airports International,* July 1, 1997 (1997 WL 9706745).

398 "The Changing Face of Business Travel," *Canadian Business,* May 1, 1997 (1997 WL 10685265).

399 Alan Daniels, "Airport Sprints on to New Runway," *Vancouver Sun,* Nov. 1, 1996 (1996 WL 5028314).

Chapter 1 (*Cont.*)

400 Richard Rowe, "Canadian Club," op. cit.

401 Bruce Constantineau, "Tough Times Mean Shippers' Boom," *Vancouver Sun,* March 10, 1997 (1997 WL 6106721).

402 "Yellowknife Company to Build Addition to 2nd Flight Hangar and Renovate RCMP Detachment Building," Canada NewsWire, Sept. 5, 1996.

403 Ian Cruickshank, "Golf Comes to Costa Rica," *The Financial Post,* April 25, 1997 (1997 WL 4093152).

404 "New Crops Bring Mixed Blessings," *Latin America Weekly Report,* Sept. 7, 1995 (WL database: PAPERSLAT).

405 "Crushed Stone from Canada Gets Rocky Reception in Jamaica," Montreal Gazette, Jan. 6, 1997 (1997 WL 4604258); W. Lee Whittingham, "Jamaica," Knight-Ridder Info. (WL database: PAPERSLAT).

406 "Weekly Report," Knight-Ridder Info., March 21, 1996 (WL database: PAPERSLAT).

407 Jeff A. Wright, "Aeroexpo to Launch Industry," WorldSources Online, Inc.: Emerging Markets Datafile, Jan. 23, 1997 (1997 WL 8251766).

408 Canute James, "Governor Gives Pledge," *Financial Times,* Sept. 19, 1997 (1997 WL 11054617).

409 "Panama: A Country of Services," Dec. 1994 (WL database: PAPERSLAT).

410 "Region: Debt & Lending Round-up, Weekly Report," Knight-Ridder, Jan. 28, 1997 (WL database: PAPERSLAT).

411 Alex Finer, "Travel: To Sleep, Perchance Tobago Twenty Years On," *The Sunday Telegraph London,* March 23, 1997 (1997 WL 2297200).

412 National Civil Aviation Review Commission, "Avoiding Aviation Gridlock & Reducing the Accident Rate," 1997, p. B-7.

413 Ibid., p. B-8.

414 Nasid Hajari, "A Walk in the Clouds," *Time,* June 22, 1998.

415 U.S. General Accounting Office, "Airport Privatization," Feb. 29, 1996, p. 2.

416 Gail Butler and Martin Keller, "Airports and Airlines: Analysis of a Symbiotic, Love-Hate Relationship," in *Handbook of Airline Economics,* D. Jenkins, ed., McGraw-Hill, 1995, pp. 87, 90.

417 National Civil Aviation Review Commission, "Avoiding Aviation Gridlock & Reducing the Accident Rate," 1997, p. II-9.

418 "USA Today's Delayed Takeoff Index," *USA Today,* July 27, 1998, p. 3B.

419 "Passenger Survey Ranks DIA 5th of 36 Airports," *Rocky Mountain News,* Feb. 17, 1998, p. 16A.

420 "Albany County Airport Authority Selects Nortwl," Canada NewsWire, Aug. 5, 1998.

421 Federal Aviation Administration, "Airport Capacity Enhancement Plan," 1996, App. B, p. 6.

422 Federal Aviation Administration, "Airport Capacity Enhancement Plan," 1996, App. B, p. 13.

423 Paul Stephen Dempsey, Andrew R. Goetz, and Joseph S. Szyliowicz, *Denver International Airport: Lessons Learned,* McGraw-Hill, 1997, p. 503.

424 Massport, "Logan Modernization: Works in Progress," 1998.

425 Federal Aviation Administration, "Airport Capacity Enhancement Plan," 1996, App. B, p. 20.

426 Gary Wisby, "U.S. Airport Boom," *Chicago Sun-Times,* July 26, 1998, p. 31.

427 Ibid.

428 Letter from Rep. Henry Hyde to DOT Secretary Rodney Slater, July 2, 1998.

Chapter 1 *(Cont.)*

429 Julie Johnsson, "Giant O'Hare Overhaul Plan Takes Flight," *Crain's Chicago Business,* Oct. 13, 1997, p. 1.

430 Federal Aviation Administration, "Airport Capacity Enhancement Plan," 1996, App. B, p. 19.

431 Federal Aviation Administration, "Airport Capacity Enhancement Plan," 1996, App. B, p. 28.

432 Margaret Allen, "Hansel Phelps Lands Contract for DFW Job," *Denver Business Journal,* Oct. 9, 1998, p. 18A.

433 "DFW Expands to Meet Predicted Traffic Increase," *DFW Blue Skies,* Jan. 1998, p. 1.

434 "Could Love Field Be a 'Field of Dreams' for Legend Airlines?" *World Airport Week,* Sept. 29, 1998.

435 Federal Aviation Administration, "Airport Capacity Enhancement Plan," 1996, App. B, p. 27.

436 Federal Aviation Administration, "Airport Capacity Enhancement Plan," 1996, App. B, p. 30.

437 "Host Your Next Meeting in the Motor City," *Association Management,* Sept. 1998.

438 "NWA to Move to New Spot at Detroit," *Travel Weekly,* Jan. 6, 1997 (1997 WL 8946435).

439 "Detroit Expands Global Terminal," *Star-Tribune,* Feb. 26, 1998 (1998 WL 6343147).

440 "NWA To Move To New Spot At Detroit," *Travel Weekly,* Jan. 6, 1997 (1997 WL 8946435).

441 Federal Aviation Administration, "Airport Capacity Enhancement Plan," 1996, App. B, p. 33.

442 Robin Stark, "Welcome to Las Vegas," *International Airport Review,* 1998.

443 "McCarran International Airport Construction Facts," 1998.

444 Shelby Grad and Lorenza Munoz, "California and the West Southland Airport Planners Face Frustrating Paradox," *Los Angeles Times,* March 1, 1998 (1998 WL 2403623).

445 Jim Newton, Jodi Wilgoren, and Lorenza Munoz, "Battle Over LAX Expansion Leaves Behind Turbulence Growth," *Los Angeles Times,* May 29, 1998 (1998 WL 2432011).

446 Laura Mecoy, "Air Traffic Boom: L.A. Turbulence," *The Sacramento Bee,* Aug. 2, 1998 (1998 WL 8834194).

447 Grad and Munoz, op. cit.

448 Ibid.

449 Mecoy, op. cit.

450 Jeff Gottlieb and Shelby Grad, "Where Will Orange County Land in 2020?" *Los Angeles Times,* May 26, 1998 (1998 WL 2430967).

451 Grad and Munoz, op. cit.

452 Federal Aviation Administration, "Airport Capacity Enhancement Plan," 1996, App. B, p. 60.

453 Dempsey, Goetz, and Szyliowicz, op. cit.

454 Federal Aviation Administration, "Airport Capacity Enhancement Plan," 1996, App. B, p. 61.

455 Address of Sunil Harman, Chief of Planning of Miami International Airport Before the University of Denver North American Intermodal Transportation Summit, Oct. 16, 1997.

456 Federal Aviation Administration, "Airport Capacity Enhancement Plan," 1996, App. B, p. 63.

Chapter 1 (*Cont.*)

457 "Commission Votes to Build New Charter Air Terminal," *Minneapolis Star-Tribune,* Aug. 23, 1998, p. 3B.

458 "Germans Eye Grand Strand for Cargo and Tourism Opportunities," *Air Cargo Report,* Aug. 1, 1996 (1996 WL 8435284).

459 Metropolitan Nashville Airport Briefing Paper, March 1998.

460 Dempsey, Goetz, and Szyliowicz, op. cit., p. 503.

461 "New Terminal Due At JFK Airport," *Rocky Mountain News,* Nov. 16, 1997, p. 3T.

462 "A Facelift at 50," *Air Transport World,* Sept. 1998, p. 35.

463 "Cargo Terminals Open at JFK," *Crain's New York Business,* Sept. 14, 1998, p. 1.

464 Port authority of New York and New Jersey, *LaGuardia,* Winter 1997–1998.

465 Ibid.

466 "USA Today's Delayed Takeoff Index," *USA Today,* July 27, 1998, p. 3B.

467 Philadelphia International Airport, "Runway 8-26 Development Project," 1998.

468 Dempsey, Goetz, and Szyliowicz, op. cit., pp. 503, 504.

469 Dempsey, Goetz, and Szyliowicz, op. cit., p. 505.

470 Philadelphia International Airport, "Yearbook & Directory," 1997–98, p. 20.

471 "S&P Rates Phoenix," *Business Wire,* July 27, 1998.

472 Dempsey, Goetz, and Szyliowicz, op. cit., p. 503.

473 Ron Lietzke, "Pittsburgh Prepares for Airport to Take Off," *Columbus Dispatch,* Sept. 30, 1992 (1992 WL 3228456).

474 Sandra Livingston, "Flying into the Future," *The Plain Dealer,* Nov. 29, 1992 (1992 WL 4307948).

475 "Panetta Announces Pittsburgh Airport Grants," U.S. Department of Transportation, Sept. 20, 1996 [1996 WL 532843 (D.O.T.)].

476 "Airports—News," *Jane's Airport Review,* Oct. 1, 1998, p. 3.

477 "Airports—News," *Jane's Airport Review,* Sept. 1, 1998, p. 3.

478 Emily Bazar, "Roomier Airports Nearing Reality," *Sacramento Bee,* Sept. 23, 1998, p. B1.

479 Mei-Ling Hopgood and Mark Schlinkmann, "New Runway May Cut Flight Delays in Half," *St. Louis Post-Dispatch,* Dec. 23, 1997, p. A7.

480 Mei-Ling Hopgood and Mark Schlinkmann, "FAA Endorses Lambert Plan," *St. Louis Post-Dispatch,* Dec. 23, 1997, p. A1.

481 Robert Goodrich, "New Airport Says NBC Criticism Is Attracting Potential Customers," *St. Louis Post-Dispatch,* Jan. 13, 1998, p. A1.

482 Salt Lake City International Airport, "Airport Master Plan," 1996.

483 Max Knudson, "Planes Taxi from Gate—Then Wait," *Deseret News,* July 29, 1998, p. A1.

484 "USA Today's Delayed Takeoff Index," *USA Today,* July 27, 1998, p. 3B.

485 Dempsey, Goetz, and Szyliowicz, op. cit., p. 503.

486 "USA Today's Delayed Takeoff Index," *USA Today,* July 27, 1998, p. 3B

487 Gary Wisby, "U.S. Airport Boom," *Chicago Sun-Times,* July 26, 1998, p. 31.

488 San Francisco International Airport, "Building an Airport for a New Century," 1998.

489 "$1 Billion Plan for San Jose Airport under Federal Review," *World Airport Week,* Sept. 8, 1998.

490 Federal Aviation Administration, "Airport Capacity Enhancement," 1996, App. B, p. 88.

491 P&D Aviation, "Master Plan Update Final Report 5-1," Jan. 1996.

492 Dempsey, Goetz, and Szyliowicz, op. cit., p. 503.

493 "National Airport on Final Approach to Opening New Terminal, *World Airport Week,* July 15, 1997, p. 4.

Chapter 1 (*Cont.*)

494 Dempsey, Goetz, and Szyliowicz, op. cit., p. 505.

495 Metropolitan Washington Airports Authority, "Washington Dulles International Airport," 1997.

496 "Schematics Almost Ready for Bradley Field's Makeover," *World Airport Week,* Sept. 1, 1998.

497 "Adaption Key to Obtaining Privatization Business in Latin America," *World Airport Week,* July 22, 1997, p. 1.

498 "Argentina Awakening," *Privatisation International,* Aug. 1, 1997 (1997 WL 8935789).

499 "New Airport Proposed for Buenos Aires," *Airline Industry Information,* April 2, 1998 (1998 WL 4675426).

500 "Bolivia Keeps It Simple," *Infrastructure Finance,* June 1, 1997 (1997 WL 9629326).

501 Sally Bowen, "Bolivian Private Airports," *Privatisation International,* April 1, 1997 (1997 WL 8935337).

502 "Bolivia Keeps It Simple," *Infrastructure Finance,* June 1, 1997 (1997 WL 9629326).

503 Edvaldo Pereira Lima, "International Influence: Airports Are Being Expanded and Upgraded in Brazil, to Cope with Explosive Traffic Growth," *Air Transport World,* Nov. 1, 1997 (1997 WL 9286899).

504 Gunter Endres, "Brazilian Airports Join the Space Race," *Jane's Airport Review,* Oct. 1, 1997 (1997 WL 9501216).

505 Ibid.

506 Pereira Lima, op cit.

507 Ibid.

508 Endres, op. cit.

509 Ibid.

510 Pereira Lima, op. cit.

511 "Construction Ahead," *Aviation Week and Space Technology,* April 13, 1998 (1998 WL 8143343).

512 Endres, op. cit.

513 Ibid.

514 Ibid.

515 Ibid.

516 Ibid.

517 Ibid.

518 Ibid.

519 "Santiago Airport Expansion Bids," *Privatisation International,* July 1, 1997 (1997 WL 8935668).

520 "Chile Lines Up Santiago Airport for Privatization before Year End," *World Airport Week,* July 29, 1997 (1997 WL 8131628).

521 "Chile: Agunsa to Begin Work on Airport in May," *South American Business Information,* Nov. 22, 1997 (1997 WL 16414594).

522 "Chile Lines Up Santiago Airport for Privatization before Year End," *World Airport Week,* July 29, 1997 (1997 WL 8131628).

523 "Chile to Invite Private Investment in Concepcion's Airport," *Airports,* Oct. 21, 1997 (1997 WL 8541153).

524 "Latin American Airports, Plotting a Flight Path," *Project and Trade Finance,* Aug. 1, 1997 (1997 WL 9853768).

525 Beth McGoldrick, "The 10 Most Creative Deals of 1996," Part 4, *Infrastructure Finance,* Feb. 1, 1997 (1997 WL 9629223).

Chapter 1 (*Cont.*)

526 "Latin American Airports, Plotting a Flight Path," *Project and Trade Finance*, Aug. 1, 1997 (1997 WL 9853768).

527 "Privatization Projects Weigh Pros and Cons of Equity and Debt Financing," *World Airport Week*, July 22, 1997 (1997 WL 8131621).

528 "Ecuador: Airport Construction Projects: Opportunities," *International Market Insight Trade Opportunities Inquiries*, Dec. 2, 1997 (1997 WL 15020963).

529 "Ecuador Looks to Privatization to Prime Nation's Economic Engine," *World Airport Week*, May 27, 1997 (1997 WL 8131560).

530 Ibid.

531 "More Business Draws More Delegates at Latin American Privatization Conference," *World Airport Week*, July 22, 1997 (1997 WL 8131619).

532 "Airports—News," *Jane's Airport Review*, Oct. 1, 1997 (1997 WL 9500926).

533 "Peru to Take Bids in 1998 for Lima Airport Expansion," *Seattle Post-Intelligencer*, Sept. 8, 1997 (1997 WL 3206716).

534 "Trinidad and Tobago Is Ripe with Temptations," *Travel Agent*, Feb. 17, 1997 (1997 WL 8973639).

535 "Tobago Acting to Improve Air Service," *Tour and Travel News*, Feb. 3, 1997 (1997 WL 8587294).

536 "Uruguay Regional Update: Firms Bid for Airport Job," *World Airport Week*, May 1, 1998 (1998 WL 10366858).

537 "Uruguay Initiates Prequalification Round for Carrasco," *World Airport Week*, Jan. 13, 1998 (1998 WL 7874139).

538 "Uruguay Selects Firms to Assist in Montevideo Airport Privatization," *Airports*, Aug. 19, 1997 (1997 WL 8540667).

539 "Airports in Venezuela: An Overview," *International Market Insight Reports*, Aug. 11, 1997 (1997 WL 12319462).

540 Ibid.

541 Ibid.

542 "Quo vadis Venezuela?" *Jane's Airport Review*, April 1, 1998 (1998 WL 10233146).

543 "SNC-Lavalin Gets Algeria Job," *The Wall Street Journal*, Sept. 12, 1997 (1997 WL-WSJ 14166054).

544 Joseph Takougang, "The 1992 Multiparty Elections in Cameroon: Prospects for Democracy and Democratization," *Journal of Asian and African Studies*, June 1, 1996 (1996 WL 12843409).

545 "Civil Aviation Authority of Egypt Adopts Upgrading Plan at Cairo International Airport," Press Release—Egypt's Minister of Tourism, May 14, 1997, http://touregypt.net/press9.htm.

546 "Egypt: Upcoming Invitation to Bid on Construction of $380,000,000 Airport," *ESP Report on Engineering Construction and Operations in the Developing World*, July 1, 1995 (1997 WL 2367525).

547 "Improvements of Airport Facilities in Egypt," Press Release—Egypt's Minister of Tourism, April 29, 1997, http://touregypt.net/press4.htm.

548 "Upgrades and Expansions in Luxor Airport," Press Release—Egypt's Minister of Tourism, May 14, 1997, http://touregypt.net/press10.htm.

549 "Egypt: Contract Award for Planned $35,000,000 Build-Operate (BO) Airport Project," *ESP Business Opportunities in Africa and the Middle East*, Sept. 1, 1997 (1997 WL 9762033).

550 "Egypt: Negotiations for Planned $35,000,000 Build-Own-Operate-Transfer (BOOT) Airport Project and Associated Property Development Scheme," *ESP Business Opportunities in Africa and the Middle East*, Feb. 1, 1997 (1997 WL 9761859).

Chapter 1 *(Cont.)*

551 "The Gambia," *Geographical Magazine,* Nov. 1,1996 (1996 WL 13219707).

552 "New Kenyan International Airport to Open on Wednesday," Agence France-Presse, Sept. 16, 1997.

553 Ibid.

554 Ibid.

555 Ibid.

556 Ibid.

557 "Kenya: Joint Venture to Undertake $82,000,000 Airport Upgrading Project," *ESP Report on Engineering Construct and Operations,* Nov. 1, 1993 (1993 WL 2660350).

558 "Consortium Looking at South African Projects: PM Goh Leaves for 10-day visit to South Africa," *Singapore Straits Times,* Feb. 24, 1997.

559 Ibid.

560 Ibid.

561 Ibid.

562 Hennie Tojaard, "Cape Town International Airport—Something on the Table for Everyone," 2 *International Airport Review* 13 (1998).

563 "South Africa: Project Feasibility Study for Proposed $60,000,000 Airport Track Transit System is Tentatively Scheduled for Completion by Mid-December 1997," *Business Opportunities in Africa and the Middle East,* Oct. 1, 1997.

564 Ibid.

565 Ibid.

566 "New Airport Plan for Kruger Park South Africa: Government Plans to Build an International Airport Near the Kruger National Park," *Travel Trade Gazette UK and Ireland,* May 28, 1997, p. 64.

567 Ibid.

568 Ibid.

569 Ibid.

570 "SNC-Lavalin—Airport Program -2: To Relocate Airport," Dow Jones News Service, March 17, 1997 (WL database: CANADANEWS).

571 "Swaziland Airport to be Upgraded," Agence France-Presse, May 28, 1997 (1997 WL 2123216).

572 "Zimbabwe—An Emerging Tourism Nation," Part 2, Xinhua English Newswire, May 20, 1996.

573 Andrew Meldrum, "Zimbabwe's Cabinet 'Loots' Pensions," African News Service, April 25, 1997.

574 Lewis Machipisa, "Zimbabwe Economy: Business Is Not the State's Forte," Inter Press Service, March 18, 1997.

575 "Zimbabwe: Consultancy for $178 Million Airport Terminal Upgrade Project Overview," *ESP Report on Engineering Construction and Operations,* Feb. 1, 1993 (1993 WL 2663607).

576 Paul Stephen Dempsey and Laurence E. Gessell, *Airline Management: Strategies for the 21st Century,* Coast Aire, 1997.

577 "Malaysia Seeks More 'Open Skies' Pacts," Agence France-Presse, May 5, 1997.

578 Michael Mecham, "Growth Outpaces Asian Airports," *Aviation Week and Space Technology,* Aug. 29, 1994, p. 54; Edward Neilan, "Turbulence Ahead for Asia's New Designer Airports," *Tokyo Business,* Aug. 1994, p. 30.

579 Renee Schoof, "China Booming with New Airport Development," *Seattle Times,* April 8, 1995, p. D1.

Chapter 2

1 FAA Forecast Conference, March 5, 1996.

2 *The Quotable Traveler,* Running Press, 1994, p. 109.

3 Joel Millman, "Mexican Airport Auction Reignites Asset Sales," *The Wall Street Journal,* Dec. 9, 1998, p. A14.

4 Stanley Scott and Levi Davis, *A Giant in Texas: A History of the Dallas-Fort Worth Regional Airport Controversy 1911–1974,* Nortex, 1974, p. 76.

5 Brent Hannon, "Gateways to Growth," *Asia, Inc.,* Nov. 1995, p. 62.

6 Leonard Hill, "Countdown on Munich 11," *Air Transport World,* June 1, 1991, p. 70.

7 Michael Mecham, "Germany Struggles to Meet Airport Needs," *Aviation Week and Space Technology,* Aug. 26, 1991, p. 38.

8 Patrick O'Driscoll, "Munich's Facility Reflects Future of DIA," *The Denver Post,* July 31, 1994, p. 24A.

9 Peter Trautmann, "The Need for New Airport Infrastructure," International Conference on Aviation and Airport Infrastructure, Denver, Dec. 5, 1993.

10 O'Driscoll, op. cit.

11 Hill, op. cit.

12 "Munich Selects German Firm to Design New Terminal," *Airports,* Aug. 11, 1998, p. 327.

13 "Kansai Int'l Airport Inaugurated," *Kyodo News International,* Sept. 5, 1994.

14 "Japanese Carriers Launch New Service," *Aviation Week and Space Technology,* Sept. 5, 1994, p. 33.

15 "Kansai Airport to Introduce New Flight Paths over Land," *Daily Yomiuri,* July 27, 1998.

16 Eric Lassiter, "Japan to Open Much-Delayed Kansai Airport," *Travel Weekly,* Aug. 29, 1994, p. 4.

17 Alexandra Black, "Trade and Tourist Boom from New Airport," Inter Press Service Global Information Network, Sept. 9, 1994.

18 "Growth Outpaces Asian Airports," *Aviation Week and Space Technology,* Aug. 29, 1994, p. 57.

19 Black, op. cit.

20 "Kansai International," *Aviation Week and Space Technology,* Sept. 12, 1994, p. 19.

21 Lassiter, op. cit.

22 "Growth Outpaces Asian Airports," *Aviation Week and Space Technology,* Aug. 29, 1994, p. 57.

23 "Top 50 Airports 1995," *Airline Business,* Dec. 1996, p. 42.

24 "Kansai International Airport," International Flights at KIX, July 31, 1998.

25 Zenjiro Ogawa, "Kansai International Airport Projects," International Conference on Aviation and Airport Infrastructure, Denver, Dec. 8, 1993.

26 Francis Fiorino, "Airline Outlook Yen for Growth," *Aviation Week and Space Technology,* Sept. 5, 1994, p. 31.

27 Stephen Leonard and Thomas Noel, *Denver Mining Camp to Metropolis,* University of Colorado Press, 1992, p. 435.

28 Paul Stephen Dempsey, Andrew R. Goetz, and Joseph S. Szyliowicz, *Denver International Airport: Lessons Learned,* McGraw-Hill, 1997, Chap. 9.

29 "Macau's Presence in Pearl River Delta Will Be as Reliever Airport," *World Airport Week,* July 18, 1995.

30 Autoridade de Aviacão Civil de Macau, "Macau International Airport Master Plan Summary," June 1994, p. 12.

Chapter 2 (*Cont.*)

31 "Far East Facility Slates December Opening," *Travel Weekly,* July 3, 1995, p. 29.

32 "Macau Air Officials See Bulk of First Year Traffic Coming from Kai Tak," *World Airport Week,* May 23, 1995.

33 Stephen Dunphy, "Hub Dreams Abound—Asia Airports about to Take Off," *Seattle Times,* Jan. 19, 1997, p. F1.

34 J. A. Donoghue, "The Pearl-Y Gateways," *Air Transport World,* Feb. 1, 1995, p. 75.

35 "U.S. Funds Feasibility Study for Second Stage of Macau Airport," *Aviation Week and Space Technology,* June 6, 1994, p. 36.

36 "April Is Start of Preparation for Macau International Cargo Facility," *Air Cargo Report,* March 30, 1995.

37 Donoghue, op. cit.

38 "U.S. Funds Feasibility Study for Second Stage of Macau Airport," *Aviation Week and Space Technology,* June 6, 1994, p. 36.

39 "Macau Airport Sees Summer Traffic Surge," *China Daily,* Sept. 3, 1998, p. 3.

40 Chua Lee Hoong, "Can KL's \$5b Total Airport Pip Changi International?" *The Singapore Straits Times,* April 26, 1997.

41 K.L. International Airport, "The Future Is Here," 1997.

42 Chua Lee Hoong, op. cit.

43 Leonard Hill, "Asia's Newest 'Dragon,'" *Air Transport World,* Sept. 1, 1993, p. 66.

44 John Stackhouse, "Crisis or Not, Sepang and Chek Lap Kok to Open This Year," *Interavia Business and Technology,* Feb. 1, 1998 (1998 WL 11048663).

45 "Malaysia's New Airport Begins Commercial Operations," Agence France-Presse, June 30, 1998 (1998 WL 2312931).

46 Rapael Pura, "Delays Hit Mahathir-Endorsed Projects," *The Asian Wall Street Journal,* Sept. 8, 1997 (1997 WL-WSJA 11015757).

47 Steven Mufson, "Accord Boosts Hong Kong Airport," *The Washington Post,* Nov. 5, 1994, p. A19.

48 Thomas Darmody, "The Design and Development of World Class Airports," IBC International Conference on Airport Development and Expansion, Hong Kong, Oct. 28, 1993.

49 "Fasten Your Seat Belts," *The Economist,* Nov. 12, 1994, p. 42.

50 "Hong Kong Airport Construction Site Strictly High-Tech," *The Vancouver Sun,* June 3, 1995, p. A13.

51 Darmody, op. cit.

52 John Mok, "The Development of Hong Kong's New International Airport," International Conference on Aviation and Airport Infrastructure, Denver, Dec. 8, 1993.

53 Ibid.

54 Mark Lander, "Asian Crisis Takes Shine off Glittering Airports' Unveilings," *Seattle Post-Intelligencer,* July 6, 1998.

55 Chris Donnolley, "The New Hong Kong International Airport," 2 *International Airport Review* 31 (1998). At this writing, the pier is still under construction.

56 Airport Authority Hong Kong, "Hong Kong International Airport," 1998.

57 "Growth Outpaces Asian Airports," *Aviation Week and Space Technology,* Aug. 29, 1994, p. 58.

58 Darmody, op. cit.

59 Airport Authority Hong Kong, "News," Jan. 1998.

60 Interview with George Brewer, Denver, Dec. 12, 1998.

61 Oslo Airport, "Europe's Safest and Most Efficient Airport," 1997.

Chapter 2 (*Cont.*)

62 "Ample Capacity Creates Great Interest," *Airport Business Communiqué*, 1997–1998, p. 17.

63 "Gardermoebanen Running on Time," *Airport Business Communiqué*, 1997–1998, p. 10.

64 "Shanghai's Pudong Airport Gets 361 Million Dollar Loan: Report," Agence France-Presse, July 8, 1997 (1997 WL 21485078).

65 Jonathan Lemberg, Xiaohu Ma, and James Zimmerman, "Financing and Investing In China's Airport Development," *China*, 1998.

66 "Shanghai's Pudong Airport Gets 361 Million Dollar Loan: Report," Agence France-Presse, July 8, 1997 (1997 WL 21485078).

67 "Shanghai Build's China's Longest Runway," Xinhua News Agency, July 19, 1998.

68 "China's Major Airport Projects," *Orient Aviation*, Nov. 1998, p. 43.

69 "Korea Inchon Forward," *Air Cargo World*, Oct. 1, 1997 (1997 WL 10135155).

70 Jong-Heui Shin, "Airport Developments in Korea," International Conference on Aviation and Airport Infrastructure, Denver, Dec. 8, 1993.

71 "Air Traffic Growth Forces Expansion of Seoul Airport Construction," *World Airport Week*, May 30, 1995.

72 Shin, op. cit.

73 James Ruggia, "Matter of Direction," *Travel Agent*, March 10, 1997, p. 106.

74 Ibid.

75 Michael Mecham, "Seoul Turns to Inchon to Solve Airport Woes," *Aviation Week and Space Technology*, Oct. 14, 1996, p. 50.

76 Moon-Sik Park, "Korea: Airport Development Market," Financial Times Asia Intelligence Wire, May 9, 1997.

77 "South Korea: Review 1997," *Asia and Pacific World of Information*, May 1997.

78 "Korean Economy Powers Airlines," *Asian Aviation News*, Oct. 18, 1996.

79 Curtis Fentress, "Revitalizing the Excitement of Travel," *Passenger Terminal '95*, p. 28.

80 Roger Bray, "N. Asian Gateway," *The Financial Times*, Dec. 16, 1996, p. 16.

81 "Air Traffic Growth Forces Expansion of Seoul Airport Construction," *World Airport Week*, May 30, 1995.

82 Ibid.

83 "NSIA Considers Opening with Extra Runway," *World Airport Week*, July 1, 1995.

84 Ibid.

85 Mecham, "Seoul Turns to Inchon to Solve Airport Woes," op. cit.

86 "Private Money Permitted in S. Korea Center Construction," *Asia Pulse*, April 10, 1997.

87 "NSIA Privatization Opportunities," *World Airport Week*, June 20, 1995.

88 Shin, op. cit.

89 "Korean Economy Powers Airlines," *Asian Aviation News*, Oct. 18, 1996.

90 Galina Slepneva, "Olympic Flame Sparks Athens Construction Boom," *The European*, May 4, 1998 (1998 WL 8699822).

91 Kerin Hope, "Greek Airports," *The Financial Times*, June 18, 1998 (1998 WL 12246072).

92 Ibid.

93 "Greece's New International Airport to Be Open in 2001," Xinhua English Newswire, June 10, 1998 (1998 WL 12162997).

94 Hope, op. cit.

95 Ibid.

Chapter 2 *(Cont.)*

96 "Greece's New International Airport to Be Open in 2001," Xinhua English Newswire, June 10, 1998 (1998 WL 12162997).

97 Slepneva, op. cit.

98 Ibid.

99 Donoghue, op. cit.

100 Craig Smith, "Zhuhzi Airport Opens to Uncertain Future," *The Asian Wall Street Journal,* May 31, 1995, p. 6.

101 Donoghue, op. cit.

102 "China Seeks Investors for Airports," *The Asian Wall Street Journal,* July 19, 1995.

103 Helen Brower, "Mainland Adds, Upgrades Airports," *Travel Weekly,* Aug. 17, 1995, p. 24; "China's Guangzhou to Build New International Airport," AAP Newsfeed, Jan. 15, 1998.

104 "Guangzhou to Build New International Airport," Xinhua News Agency, Jan. 13, 1998; "China: Guangzhou to Build New Airport," *China Daily,* Jan. 13, 1998, p. 5.

105 "China: Guangzhou to Build New Airport," *China Daily,* Jan. 13, 1998, p. 5.

106 Donoghue, op. cit.

107 Michael Mecham, "Seoul Ranks No. 1 in Airport Growth," *Aviation Week and Space Technology,* April 10, 1995, p. 34.

108 Airports Authority of Thailand, "Second Bangkok International Airport Implementation Program," 1994.

109 "NSIA Considers Opening with Extra Runway," *World Airport Week,* July 1, 1995.

110 Ibid.

111 Airports Authority of Thailand, op. cit.

112 Tyler Davidson, "Thailand Reroutes Money to Don Muang Facility," *Travel Weekly,* March 17, 1997, p. 63.

113 "Thailand: Bangkok's Airport Woes," *PATA Travel News Asia Pacific,* March 1, 1997.

114 Davidson, op. cit.

115 "Bangkok to Add Gateway," *Travel Trade Gazette Europa,* April 8, 1993, p. 20.

116 "Fewer Arrivals So Airport Faces Scale-Down," *Aerospace* (Singapore), Sept. 1993, p. 26.

117 "Thailand to Revive Second International Airport Scheme," Agence France-Presse, Nov. 19, 1997 (1997 WL 13437145).

118 "New Airport Budget to Be Set at 101bn Baht," Financial Times Asia Intelligence Wire, April 12, 1997.

119 Ibid.

120 "Expanding from Regional to International Status," *Jane's Airport Review,* April 1, 1997, p. 15.

121 "Thailand: Bangkok Int'l Airports Update," Financial Times Asia Intelligence Wire, April 1, 1997.

122 "Nong Ngu Hao: Company Board Endorses Adjusted Construction Plan," *Bangkok Post,* March 27, 1997, p. 4.

123 "Thailand: Bangkok's Airport Woes," *PATA Travel News Asia Pacific,* March 1, 1997.

124 "Thailand: International Airports Update," *International Market Insight Trade Opportunities Inquiries,* March 12, 1998 (1998 WL 8988221).

125 Ibid.

126 Ibid

127 "Thailand to Raise Airport Departure Tax," Agence France-Presse, March 17, 1998 (1998 WL 2243310).

Chapter 2 (*Cont.*)

128 Ian Thomas, "Airline Solidarity on Second Airport," *Australian Financial Review,* April 8, 1998 (1998 WL 9425641).

129 Ibid.

130 "Government Wants to Build Second Sydney Airport at Badgerys Creek," *Daily Commercial News,* April 9, 1998, p. 9 (1998 WL 9666645).

131 Ian Thomas, "Badgerys on Back Burner," *Australian Financial Review,* Aug. 3, 1998 (1998 WL 12570522).

132 Nick Hordern, "Trains and Votes and Planes," *Australian Financial Review,* Aug. 5, 1998 (1998 WL 12570666).

133 "No Second Airport until 2006," *Sydney Morning Herald,* Feb. 28, 1998, p. 1 (1998 WL 9641202).

134 Ibid.

135 Airport facilities often have few alternative uses. Denver Stapleton built a beautiful new $53 million Concourse E in 1990, and was closed in 1994. By 1998, it could find no tenants for that concourse or the other portions of the main terminal building, and decided to raze the building. Ann Imse, "Time to Scrap Stapleton?" *Rocky Mountain News,* Jan. 19, 1998, p. 5A.

136 Keith Mordoff, "Air Transport Munich's New International Airport Expected to Begin Operations in 1991," *Aviation Week and Space Technology,* Feb. 22, 1988, p. 92.

137 Lassiter, op. cit.

138 Hill, "Asia's Newest 'Dragon,'" op. cit.

Chapter 3

1 *The New York Times,* Oct. 27, 1957.

2 Paul Stephen Dempsey, Andrew R. Goetz, and Joseph S. Szyliowicz, *Denver International Airport: Lessons Learned,* McGraw-Hill, 1997, p. 83.

3 Federal Aviation Administration, "Airport Master Plans," 1985, p. 52.

4 Thomas Darmody, "The Design and Development of World Class Airports," IBC International Conference on Airport Development and Expansion, Hong Kong, Oct. 28, 1993.

5 Michael Buckley, Alan Samagalski, Robert Storey, Chris Taylor, and Clem Lindenmayer, *China,* Lonely Planet, 1994, pp. 24, 25.

6 Steven Mufson, "Accord Boosts Hong Kong Airport," *The Washington Post,* Nov. 5, 1994, p. A19.

7 "Fasten Your Seat Belts," *The Economist,* Nov. 12, 1994, p. 42.

8 Peter Stein, "Hong Kong's Tycoons Now Bet Their Future on Hong Kong's Big Market," *The Wall Street Journal,* July 1, 1997, p. A1.

9 Stuart Becker, "Airports: Chinese Opera," *Far Eastern Economic Review,* April 6, 1995, p. 54.

10 Carl Schwartz, "Airports in Asia Play Catch-Up as Economies Explode," *The Milwaukee Journal Sentinel,* July 14, 1996, p. A11.

11 Darmody, op. cit.

12 "Hong Kong Builds $20.3 Billion Airport," *The Orlando Sentinel,* April 9, 1995, p. A20.

13 Michael Mecham, "Airport Fees Stable for Asian Carriers," *Aviation Week and Space Technology,* Aug. 28, 1995, p. 51.

14 Becker, op. cit.

15 Mufson, op. cit.

16 Ibid.; "Tung's Tiananmen Headache," *The Economist,* June 7, 1997, p. 36.

Chapter 3 *(Cont.)*

17 Becker, op. cit.

18 Becker, op. cit.

19 "Growth Outpaces Asian Airports," *Aviation Week and Space Technology,* Aug. 29, 1994, p. 58.

20 Darmody, op. cit.

21 "Hong Kong Passes Bill to Set Airport Body," *The Asian Wall Street Journal,* July 20, 1995, p. 3.

22 Stein, op. cit.; Nicholas Kristof, "Hong Kong Wags the Chinese Dog," *Rocky Mountain News,* July 1, 1997, p. 24A.

23 "The Thai Experience," *Sawasdee,* June 1997, p. 60.

24 "Thailand: Bangkok's Airport Woes," Financial Times Asia Intelligence Wire, March 1, 1997.

25 Satya Sivaraman and Jurawan Ngamman, "Political Tug-of-War over Airport Project," *Asia Times,* Jan. 24, 1997, p. 1.

26 "New Life for Don Muang," *The South China Morning Post,* Jan. 24, 1997, p. 6.

27 New Bangkok International Airport Co., Annual Report 1996–1997.

28 Somkian Wanthana, "In Memory of Black May," *The Bangkok Post,* May 17, 1997, p. 8.

29 New Bangkok International Airport Co., Annual Report 1996–1997.

30 Airports Authority of Thailand, "6th Anniversary," July 1, 1985, p. 37.

31 Ibid.

32 Wanthana, op. cit.

33 New Bangkok International Airport Co., Annual Report 1996–1997.

34 Airports Authority of Thailand, "6th Anniversary," July 1, 1985, pp. 37, 38.

35 Airports Authority of Thailand, "6th Anniversary," July 1, 1985.

36 "Bangkok to Add Gateway," *Travel Trade Gazette Europa,* April 8, 1993, p. 20.

37 "Thailand: Bangkok's Airport Woes," Financial Times Asia Intelligence Wire, March 1, 1997.

38 "A New Airport Has Villagers Raging," *Business Week,* July 24, 1995, p. 5.

39 "NSIA Considers Opening with Extra Runway," *World Airport Week,* July 1, 1995.

40 "Thailand: Bangkok's Airport Woes," Financial Times Asia Intelligence Wire, March 1, 1997.

41 Interview with George Emery of TAMS, Bangkok, Thailand, June 16, 1997.

42 Sivaraman and Ngamman, op. cit.

43 Tom Ballantyne, "Bangkok Hub Plan Snubbed," *Airline Business,* March 1997, p. 20.

44 New Bangkok International Airport Co., Annual Report 1996–1997.

45 "Expanding from Regional to International Status," *Jane's Airport Review,* April 1, 1997, p. 15.

46 "Thailand: Bangkok Int'l Airports Update," Financial Times Asia Intelligence Wire, April 1, 1997.

47 "Expanding from Regional to International Status," *Jane's Airport Review,* April 1, 1997, p. 15.

48 "Thailand: Bangkok Int'l Airports Update," Financial Times Asia Intelligence Wire, April 1, 1997.

49 Ibid.

50 Sivaraman and Ngamman, op. cit.

51 "New Life for Don Muang," *The South China Morning Post,* Jan. 24, 1997, p. 6.

Chapter 3 (*Cont.*)

52 "Thailand: Bangkok Int'l Airports Update," Financial Times Asia Intelligence Wire, April 1, 1997.

53 "New Airport Budget to Be Set at 101Bn Baht," Financial Times Asia Intelligence Wire, April 12, 1997; "Nong You See It, Now You Don't," *Airline Business,* April 1997, p. 37.

54 "Thailand: Bangkok Int'l Airports Update," Financial Times Asia Intelligence Wire, April 1, 1997.

55 "Thailand: Bangkok's Airport Woes," Financial Times Asia Intelligence Wire, March 1, 1997.

56 "Thailand: Bangkok Int'l Airports Update," Financial Times Asia Intelligence Wire, April 1, 1997.

57 Tom Ballantyne, "Bangkok Hub Plan Snubbed," *Airline Business,* March 1997, p. 20.

58 "Thailand Opts to Postpone Bangkok's 2nd International Airport," Deutsche Presse-Agentur, Feb. 11, 1997.

59 "New Life for Don Muang," *The South China Morning Post,* Jan. 24, 1997, p. 6.

60 "Expanding from Regional to International Status," *Jane's Airport Review,* April 1, 1997, p. 15.

61 Ballantyne, op. cit.

62 Somchai Meesane, "Bangkok's Citizenry Growing Old Waiting for Nong Ngu Hao Airport," *The Bangkok Post,* Feb. 1, 1997 (1997 WL 7475456).

63 "Airport: Singapore Irked by Montri's Claims: Island State Says It Is a Rival and a Friend," *Bangkok Post,* Jan 31, 1997 (1997 WL 7475384).

64 Meesane, op. cit.

65 Ibid.

66 "Nong You See It, Now You Don't," *Airline Business,* April, 1997, p. 37; "Thailand: Bangkok Int'l Airports Update," Financial Times Asia Intelligence Wire, April 1, 1997.

67 Tyler Davidson, "Thailand Reroutes Money to Don Muang Facility," *Travel Weekly,* March 17, 1997, p. 63.

68 "Thailand: Bangkok Int'l Airports Update," Financial Times Asia Intelligence Wire, April 1, 1997.

69 Ibid.

70 "Asia Faces the Double Burden of Capacity Squeeze and Entrenched Military Interests to Meet Demand Growth," *Jane's Airport Review,* April 1, 1997, p. 15.

71 "Thailand Tracking Delay of New Airport Plan," Xinhua News Agency, Sept. 21, 1998.

72 Amornrat Mahitirook, "Nong Ngu Hao: Company Board Endorses Adjusted Construction Plan," *The Bangkok Post,* March 27, 1997, p. 4.

73 Joshua Elliot, *Malaysia and Singapore Handbook,* 1996, p. 86.

74 Ibid., p. 538.

75 Ibid., p. 93.

76 "Malaysia to Build Major Kuala Lumpur Airport," *Airports,* Sept. 10, 1991, p. 371.

77 Daniel Moss and Christopher Donville, Financial Times Asia Intelligence Wire, April 29, 1997.

78 "Singapore Airport Expands to Meet Passenger and Cargo Traffic Boom," Agence France-Presse, April 28, 1997.

79 Leonard Hill, "Asia's Newest 'Dragon,'" *Air Transport World,* Sept. 1, 1993, p. 66.

80 Elliot, op. cit., p. 87.

81 "Pressure Tactics Not the Way, Says PM," *Business Times,* April 12, 1998 (1998 WL 2967706).

Chapter 3 (*Cont.*)

82 Ibid.

83 Azman Ujang, "Malaysia Achieves Big Savings in New Airport Construction," *Asia Pulse,* March 25, 1997.

84 "Government Won't Counter British Press Allegations," *Bernama,* May 5, 1998 (1998 WL 6597583).

85 Ibid.

86 "Government Has No Control over Media Reporting, Says British Minister," *Bernama,* May 5, 1998 (1998 WL 6597600).

87 "Government Won't Counter British Press Allegations," *Bernama,* May 5, 1998 (1998 WL 6597583).

88 "Government Has no Control Over Media Reporting, Says British Minister," *Bernama,* May 5, 1998 (1998 WL 6597600).

89 Elliot, op. cit., p. 91.

90 Neil Winton, "Malaysia's Mahathir Lauds High-Tech Zone's Freedom," *Reuters European Business Report,* May 21, 1997; Syed Azman, "Malaysia to Launch Futuristic 'Cybercity,'" Reuters North American Wire, May 16, 1997.

91 Elliot, op. cit., pp. 83, 89.

92 Elliot, op. cit., pp. 544, 547.

93 "S. Korea's Inchon Airport behind Other Top Airports in Asia," *Asia Pulse,* March 7, 1997.

94 Laxmi Nakarmi, "The Slush Fund That's Shaking up Seoul," *Business Week,* Nov. 13, 1995, p. 60.

95 J. H. Yoon, "Groundbreaking Ceremony for Inchon International Airport," *Business Korea,* June 1996, p. 54.

96 "South Korea: Review 1997," *Asia and Pacific Review,* May 1997.

97 *Keyes v. School District No. 1,* 413 U.S. 189 (1973).

98 Robert Kowalski, "Turbulence Marks DIA History," *The Denver Post,* March 1994, p. 22.

99 Ibid.

100 Many have suggested that the real reason for his defeat was the Blizzard of December 1982 which dumped more than 20 in on the Colorado Front Range, and left much of Denver impassable. The people may not expect much from their elected leaders, but they do expect them to keep the roads open. Peña himself would face a similar blizzard only a few years later. His response was to roll garbage trucks up and down Denver's side streets, which created deep, frozen ruts making much of Denver impassable for weeks. Mercifully for Peña, however, it was not an election year.

101 These included building a convention center, planning a large new dam (Two Forks), attracting a baseball team with a new baseball stadium, and building a new library and various other infrastructure improvements.

102 After a decade and a half of unprecedented losses following deregulation, Wall Street had downgraded most airline debt to "junk" status. The airlines could ill afford a gold-plated airport they did not need.

103 Fred Brown, "Romer Says Reign Humbling," *The Denver Post,* Dec. 27, 1998, p. 23A.

104 Dempsey, Goetz, and Szyliowicz, op. cit.

105 Ibid.

106 The city's public relations fiasco was not helped one morning when Mayor Webb overslept and missed a scheduled live appearance on ABC's *Good Morning America.* The absent mayor was symbolic of a city whose leader appeared asleep at the wheel. Webb was to make opening remarks at an international conference showcasing DIA at the Denver

Chapter 3 *(Cont.)*

Convention Center attended by 700 delegates from 23 nations around the world, but at the last minute sent city public works director Mike Musgrave with the excuse that Webb was out of town and his flight had been delayed. Musgrave was then informed that Webb had been seen at the convention center only an hour and a half earlier, at another function. Musgrave got on the telephone and a few minutes later came back with another excuse—Webb had a hard day and needed to rest. Webb seemed to have no conception of how the rest of the world was beginning to view Denver's leadership.

107 Don Phillips, "Delay-Plagued Airport Becoming Excessive Baggage in Denver," *The Washington Post,* May 16, 1994, p. A6; Louis Sahagun, "Denver Airport to Bypass Balky Baggage Mover," *The Los Angeles Times,* Aug. 5, 1994, p. A3. Some American Indians said that the letters stand for Dead Indian Airport, and that somewhere on the 53-mi^2 site are ancestral Indian bones that need honoring. Said a member of the Cheyenne tribe, "Listen to what the spirits are trying to tell you. Try to do something to appease the spirits. When the spirits calm down, then maybe your troubles will be over." Dan Meyers, "Airport's Wing and a Prayer," *The Philadelphia Inquirer,* Feb. 1, 1995, p. A2. Nefarious gremlins seemed to be sabotaging the software and hardware of the highly sophisticated automated baggage system. The city commissioned an Indian ritual exorcism, and things began to settle down.

108 Denver agreed to pay BAE $37 million beyond the $195 million it had already committed to complete the job, and pay Rapistan Demag $63 million to install a "back-up" baggage system (actually the primary baggage system for Concourse C carriers). The city quietly allowed United to take over the BAE system on Concourse C, and surreptitiously relieved United of its commitment to bring more than a thousand reservations jobs to Denver. Any potential lawsuits the city might have against BAE would also be dropped.

109 The baggage system wasn't the only reason the airport failed to be completed on schedule. Shortly after Webb became mayor, he signed an agreement with United Airlines to occupy 45 gates on Concourse B. Part of the deal was installation of the automated baggage system, which United wanted on its concourse, but the city decided should be in all three concourses—a 23-mi long system. Moreover, United wanted significant design changes in an airport that was already well under construction. Those changes were monumental, and required putting a full basement under the terminal and moving Concourse C farther north. United also wanted the city to build it a sprawling maintenance hangar, located coincidentally where a future Concourse might otherwise one day be built. The scope of the airport had changed radically, but Webb stuck with the existing opening date, not wanting to take the political heat in announcing a delay. Had Webb had the courage to take the heat and announce that United's changes required a delay in the opening, he would have spared himself much embarrassment later on, and saved the city millions of dollars in accelerated completion costs.

110 The cost disproportionately would be borne by local origin-and-destination (O&D) passengers. East-to-west connecting traffic could just as easily be routed via a competing carrier's hub (e.g., Dallas, St. Louis, Chicago, Houston, Salt Lake City, or Phoenix), and was therefore immune from significantly higher fares to cover the cost of DIA. And United, now the dominant megacarrier in the market, would raise fares on traffic some $40 per ticket over prices at Stapleton, saddling local O&D passengers with a half a billion dollars a year in ticket hub monopoly rents.

111 Dan Meyers, "Taking the Baggage Test," *The Philadelphia Inquirer,* March 1, 1995, p. A1.

112 Paul Stephen Dempsey, "A Close Look at 'Federico's Folly,'" *Journal of Commerce,* Feb. 28, 1995, p. 8A.

113 *Aterand Contractors, Inc., v. Pena,* 515 U.S. 200 (1995).

114 Clifford May, "Webb v. DeGroot," *Rocky Mountain News,* May 28, 1995, p. 104A.

115 "The Mayoral Match-Up," *The Denver Post,* May 28, 1995, p. 12A.

116 Ibid. Under Webb's leadership, the airport had failed to meet scheduled opening dates of October 29, 1993; December 19, 1993; March 9, 1994; and May 13, 1994. The airport did

Chapter 3 (*Cont.*)

not actually open until February 28, 1995, a few months before the mayoral election, after the city agreed to pay BAE another $40 million to fix the primary baggage system (and drop all claims against the baggage handling manufacturer), and another $63 million to Rapistan Demag to build a "back-up" baggage system. The 16 months of delay cost $100 million in overruns and $361 million in financing costs.

117 Examples of Webb's cronyism were abundant. Webb gave city jobs to his brother and sister-in-law, and ethical questions surfaced when it became known that his chief-of-staff, Venita Vinson, had not filed income tax returns in years. As *The Denver Post* observed, "The issue of cronyism became glued to Webb when friends of his were among those awarded concession contracts at DIA." DeGroot and other city council members alleged that the city's contracting procedures were circumvented so that Webb's friends would enjoy preferential treatment. Mark Eddy, "Cronyism 'Already an Issue for the Public,' Says Pollster," *The Denver Post,* May 22, 1995, p. 9A. Webb's sister-in-law, Marilyn Webb, was given a $48,000 a year job in the Mayor's office of Employment and Training. Katie Kerwin, "DeGroot Blasts 'Cronyism,'" *Rocky Mountain News,* April 13, 1995, p. 21A.

118 Mark Eddy, "Webb's Cronyism Woes Began Early," *The Denver Post,* May 22, 1995, p. 1; Brian Weber, "Webb," *Rocky Mountain News,* April 16, 1995, p. 28A.

119 Eddy, ibid., p. 9A.

120 Patrick O'Driscoll, "Webb Allegedly Tried to Bend DIA Hiring Rules,"

121 Eddy, "Cronyism 'Already an Issue for the Public,' Says Pollster," op. cit.

122 Brian Weber, "Webb, DeGroot Swap Claims of Bias, Dirty Politics," *The Denver Post,* May 23, 1995, p. 4A.

123 In the October 1991 bond issuance, the firm earned $146 million from the $600 million sale, compared with the $162,476 it had earned during the Peña Administration. Kevin Flynn, "You Be the Judge," *Rocky Mountain News,* May 28, 1995, p. 101A.

124 Adriel Bettelheim, "Potential Influence Peddling, Shoddy Work Probed," *The Denver Post,* Feb. 26, 1995, p. 19A.

125 SEC MSRB Rule G-37 (Apr. 7, 1994); Charles Garparino, "'Pay to Play' Getting New SEC Review," *The Wall Street Journal,* Dec. 17, 1998, p. C1.

126 *Blount v. SEC,* 61 F.3d 938, 944 (D.C. Cir. 1994), cert. denied, 116 S.Ct. 1351 (1996).

127 Burt Hubbard, "22% of Webb Funds Have Airport Link," *Rocky Mountain News,* April 10, 1995, p. 4A.

128 Katie Kerwin, "Crider Says Webb Sells City Contracts," *Rocky Mountain News,* April 11, 1995, p. 4A.

129 Bettelheim, op. cit.

130 George Doughty, Denver's director of aviation (the city's top airport official), believed the mayor's office was slowing completion of DIA, "often to maneuver political influences into contracts." Kevin Flynn, "Airport Proves Too Much for Webb," *Rocky Mountain News,* Feb. 19, 1995, p. 4A. Specifically, Doughty believed that Webb refused to adopt a process for selecting concessionaires on the basis of merit and quality because he could not figure out how to give such business opportunities to his supporters. Webb's chief of staff, Venita Vinson, was determined to give an existing concessionaire $8 million in rent relief for a deal under which it would turn over some business to selected minority businesses. DIA bond issues were also moved to a selected minority firm. Patrick O'Driscoll, "Mayoral Cronyism at DIA Alleged," *The Denver Post,* May 26, 1995, p. 18A. Like several senior airport officials, Doughty too would become disenchanted with the Webb administration, resigning to become director of the Allentown, Pa., airport.

131 Richard Boulware, head of public relations at Stapleton International Airport from 1984 to 1993, claimed that Webb's city attorney Mike Dino directed him to give a minority-owned

Chapter 3 (*Cont.*)

public relations firm a subcontract for airport publicity. Kevin Flynn, "Ex-Airport Spokesman: Webb Aide Lied," *Rocky Mountain News,* March 2, 1995, p. 17A. Claiming that political meddling by the mayor's office led to his demotion, Boulware would subsequently win an Equal Employment Opportunity Commission (EEOC) complaint against the city alleging age and gender bias.

132 Aviation director Trommeter testified at a city personnel hearing that Webb's office had accused him of bigotry and failing to maximize DIA contracts for "disadvantaged business enterprises" (firms owned by minorities and women), an allegation Webb denied. Patrick O'Driscoll, "Mayoral Cronyism at DIA Alleged," *The Denver Post,* May 26, 1995, p. 18A.

133 As an example, the city gave the operating and maintenance contract for the error-prone baggage system to a firm other than the one that built it, largely because of doubts that BAE would hire enough minority workers (BAE pledged to hire 36% local and minority workers, but apparently this wasn't enough). Chief airport engineer Ginger Evans described this as the "worst management decision" possible. Mark Eddy, "BAE Operating Exclusion Blasted," *The Denver Post,* May 1, 1994.

134 Kevin Flynn, "Denver Program Gets a Grilling," *Rocky Mountain News,* May 23, 1995, p. 10A.

135 Ibid., p. 12A.

136 Lou Kilzer, Robert Kowalski, and Steven Wilmsen, "Minority Firm Carves out an Empire," *The Denver Post,* April 7, 1995, pp. 1A, 10A.

137 Ibid., p. 10A.

138 Kevin Flynn, "'Disadvantaged' Building Firm Racks Up Millions," *Rocky Mountain News,* March 26, 1995, p. 4A.

139 Webb promised to reform the minority contracting program by imposing harsh penalties for firms that lie about their eligibility, ensuring companies no longer eligible are not given contracts, and generally tightening overall management of the program. Mark Eddy, "Webb Faces List of Campaign Vows," *The Denver Post,* June 11, 1995, p. 7C.

140 Brian Weber, "Mayor Candidates Question Minority Contracts," *Rocky Mountain News,* March 8, 1995, p. 23A.

141 "A Second Term for Webb," *Rocky Mountain News,* June 7, 1995, p. 39A.

142 "DeGroot Would Provide Fresh Direction at City Hall," *The Denver Post,* April 16, 1995, p. 2D.

143 "Many Leaders Share Blame for DIA's Slow Takeoff," *The Denver Post,* Feb. 27, 1995, p. 7B.

144 Kevin Flynn, "Airport Proves Too Much for Webb," *Rocky Mountain News,* Feb. 19, 1995, p. 4A.

145 Ibid.

146 "DIA Cost Comparisons Can Be Very Misleading," *The Denver Post,* March 2, 1995, p. 6B.

147 Lou Kilzer, Robert Kowalski, and Steven Wilmsen, "DIA: Dream to Disappointment," *The Denver Post,* Feb. 12, 1995, p. A1; "The Untold DIA Story," *Rocky Mountain News,* Feb. 5, 1995, p. 1A.

148 "Peña's Folly," *The Wall Street Journal,* Feb. 23, 1995, p. A12.

149 Brian Weber and Katie Kerwin, "Webb's Image Soars with Airport's Takeoff," *Rocky Mountain News,* Mar. 5, 1995, p. 4A; "Airborne...At Last," *The Denver Post,* Feb. 28, 1995, p. 1A; "At Last," *Rocky Mountain News,* Feb. 28, 1995, p. 1A.

150 Gil Spencer, "An Election Not to Be Forgotten," *The Denver Post,* June 11, 1995, p. 1E; Mark Eddy, "Will Webb Keep Word?" *The Denver Post,* June 11, 1995, p. 1C. Webb promised to demand the resignations of all 57 political appointees in city government (except for the police chief, who had come out swinging against DeGroot in the campaign),

Chapter 3 (*Cont.*)

and to make some serious staff changes. Kevin Flynn, "Most Webb Appointees Will Likely Stay Put," *Rocky Mountain News,* June 11, 1995, p. 6A. After the election, many predicted Webb would make few changes at city hall. Ibid.

151 See Kevin Flynn and Burt Hubbard, "Webb Berates United for $40 Fare Increase," *Rocky Mountain News,* Jan. 29, 1995, p. 4A.

152 Jeffrey Leib, "United Toys with Its Rivals," *The Denver Post,* May 21, 1995, p. 1G.

153 DeGroot had been something of a loner on the city council. She had introduced little legislation, and built no significant coalitions. The 17th Street business community was unfamiliar with her, and it was difficult for voters to understand what kind of mayor she would be. The people were favorably impressed with the new airport, and DIA was flooded with tourists coming to gaze and gawk. And it was difficult for DeGroot to make DIA much of an issue—if the airport had such serious problems, why didn't she persuade the city council to correct them? Webb was able to put together a political machine of 1000 volunteers, composed of political appointees and Democratic Party regulars, to walk the neighborhoods and drive voters to the polling places. Brian Weber, "Incumbency, Philosophy, Image Beat DeGroot, Consultant Says," *Rocky Mountain News,* June 8, 1995, p. 10A; Fred Brown, "Political Machine Drove Webb Home," *The Denver Post,* June 8, 1995, p. 1A.

154 Katie Kerwin, "How DeGroot, Webb Compare," *Rocky Mountain News,* June 4, 1995, p. 21A.

155 See "A Grubby First for Denver," *Rocky Mountain News,* June 6, 1995, p. 32A: "The fact that a moderate to liberal Democrat with no verbal or professional record that even hints at bias can be so casually smeared suggests that Denverites are in for a great many nasty contests in years to come."

156 Brian Weber, "Good from the Bad," *Rocky Mountain News,* June 11, 1995, p. 93A.

157 Chuck Green, "The Mayor: Decency in Triumph," *The Denver Post,* June 11, 1995, p. E1.

158 "Webb says it is time to heal the wounds and bring the city together. That will be tough after he alleged DeGroot is a racist. The Anglo population continues to shrink. The white flight to the suburbs by young families is a direct result of the poor quality of Denver Public Schools' education." Gene Amole, "DeGroot Couldn't Beat the Machine," *Rocky Mountain News,* June 8, 1995, p. 5A.

159 Michael Booth, Mark Eddy, and Christopher Lopez, "It's Still 'Mr. Mayor,'" *The Denver Post,* June 7, 1995, p. 1A.

160 Weber, "Good from the Bad," op. cit.

161 While DIA is a splendid connecting airport, local passengers are inconvenienced with a 24-mi drive from downtown, parking that is expensive and difficult to find, with only public transportation available on the baggage claim level, and again, the monopoly pricing of United Airlines. But then, many of DIA's O&D passengers reside in Denver's suburbs, and are therefore precluded from voting in Denver's mayoral election.

162 In retrospect, had the airport come in on time and on budget, probably not a fifth of the investigations would ever have been launched.

163 See, for example, "Probe Zeros in on Peña," *The Denver Post,* March 12, 1995, p. 1A.

164 Mark Eddy, "Grand Jury to Probe Funding of Delegation's Trip to Africa," *The Denver Post,* Feb. 1, 1995, p. 4A; "Agent to Plead Guilty in DIA Case," *The Denver Post,* Oct. 19, 1995, p. 3B; Kevin Flynn, "Witness Will Have Little on DIA, Acquaintances Say," *Rocky Mountain News,* Oct. 19, 1995, p. 8A.

165 Michelle Johnston, "Ex-DIA Spokesman Joins Critics of Ritter's Probe," *The Denver Post,* Sept. 12, 1996, p. B4.

166 Penelope Purdy, "Silence Deafening on DIA Funds," *The Denver Post,* June 9, 1996, p. D3.

167 "Fed Judge Approves Denver Airport Litigation Settlement," *The Bond Buyer,* Dec. 9, 1997, p. 1.

Chapter 3 (*Cont.*)

168 Mike McPhee, "Airport Contractor Is Sued," *The Denver Post,* Nov. 26, 1998, p. B1.

169 Terry Burns, "Studies Find Travelers Would Benefit from Third Chicago Airport," *The State Journal-Register,* Jan. 25, 1997 (1997 WL 6984327).

170 "IDOT Says 3rd Chicago Airport Feasible, Vital," *The Pantagraph,* Jan. 25, 1997 (1997 WL 2461242).

171 Burns, op. cit.

172 Ibid.

173 Ibid.

174 Ibid.

175 "IDOT Says 3rd Chicago Airport Feasible, Vital," *The Pantagraph,* Jan. 25, 1997 (1997 WL 2461242).

176 Burns, op. cit.

177 Jon Hilkevitch, "City Solicits Help in Bid to Ground Peotone Airport," *The Chicago Tribune,* March 23, 1998 (1998 WL 2837841).

178 "IDOT Not Giving Up on 3rd Airport," *The Chicago Tribune,* Oct. 11, 1997 (1997 WL3597992).

179 Ibid.

180 "Relieving O'Hare," *The Economist,* Jan. 10, 1998, p. 22.

181 Jon Hilkevitch, "Airlines Study Chicago's Fight against 3rd Airport," *The Chicago Tribune,* March 27, 1998 (1998 WL 2839570).

182 Ibid.

183 John Schmeltzer, "Flight, not Fight, Southwest Vows," *The Chicago Tribune,* April 8, 1998 (1998 WL 2843569).

184 Ibid.

185 Yvette Shields, "Chicago Officials Seek Bonds to Fix Midway Airport," *The Bond Buyer,* June 11, 1998 (1998 WL 13142199).

186 Fran Spielman, "Bonds for Midway Overhaul OKd—Airlines off the Hook if Peotone Built," *The Chicago Sun-Times,* July 8, 1998 (1998 WL 5588549).

187 Ibid.

188 Debbie Howlett, "Congress Jumps into Chicago's Flight Fight," *USA Today,* July 9, 1998 (1998 WL 5729881).

189 Ibid.

190 "Relieving O'Hare," *The Economist,* Jan. 10, 1998, p. 22.

191 Michael Phillips, "Cameroon Is Seen as Most Corrupt in Study of Nations," *The Wall Street Journal,* Sept. 23, 1998, p. B17.

192 "Getting a Reputation," *The Wall Street Journal,* Sept. 23, 1998, p. B17; Michael Phillips, op. cit.

193 Donald R. Cruver, *Complying with the Foreign Corrupt Practices Act,* ABA Publishing, 1994, p. 2.

194 Ibid., p. 4.

195 Ibid., pp. 4, 5.

196 Ibid., p. 6, note 8. The FCPA contains two parts: antibribery provisions and accounting requirements. The two provisions were designed to work together, with the accounting requirements meant to improve corporate accountability and prevent off-the-record payments, while the antibribery provision was to deter corporate bribery with the threat of criminal prosecution. Under the Act, the SEC would be responsible for conducting investigations and bringing civil injunctive actions, while referring cases to the Department of Justice for criminal prosecution. Ibid., p. 8.

Chapter 3 (*Cont.*)

197 Ibid, p. 9.

198 15 U.S.C. §78m(b)(2)(A).

199 15 U.S.C. §78m(b)(2)(B).

200 15 U.S.C. §78m(b)(7).

201 15 U.S.C. §78m(b)(4).

202 15 U.S.C. §78ff(a). However, an issuer who holds 50% or less of the voting power with respect to a domestic or foreign firm is required only to make a good faith effort to persuade the firm to comply with the accounting requirements in this provision. Demonstration of such good faith efforts shall provide a conclusive presumption of compliance with the accounting provisions. 15 U.S.C. §78m(b)(6).

203 The two sections are essentially identical, with the only difference being the entity addressed. Sec. 78dd-1 addresses prohibited practices by "issuers," which are defined as an entity "which has a class of securities registered pursuant to section 781 of title 15, or any officer, director, employee, agent, or stockholder acting on behalf of the issuer." 15 U.S.C. §78dd-1(a) (issuers), 78dd-2(a) (domestic concerns). Sec. 78dd-2 relates to "domestic concerns," which are not issuers subject to Sec. 78dd-1. This includes citizens, nationals, or residents of the United States, as well as any business entity with its principal place of business in the United States, or which is organized under the laws of the United States or its territories. 15 U.S.C. §78dd-2(h)(1)(A)(B).

204 15 U.S.C. §78dd-1(a) (issuers), 78dd-2(a) (domestic concerns).

205 15 U.S.C. §78dd-1(f)(2)(A)(B) (issuers), 78dd-2(h)(3)(A)(B) (domestic concerns).

206 15 U.S.C. §78dd-1(f)(3)(A) (issuers), 78dd-2(h)(4)(A) (domestic concerns).

207 15 U.S.C. §78dd-1(b) (issuers), 78dd-2(b) (domestic concerns).

208 Marianne Lavelle, "Nations Try to Match U.S. on Biz Bribe Law, But Lawyers Still Argue Over the Meaning of the FCPA," *National Law Journal,* Jan. 20, 1997 (NLJ B1, col. 4).

209 15 U.S.C. §78dd-1(c)(2)(A), (B) (issuers), 78dd-2(c)(2)(A), (B) (domestic concerns).

210 Cruver, op. cit., p. 51.

211 15 U.S.C. §78ff(c) (issuers), 78dd-2(g) (domestic concerns).

212 15 U.S.C. §78 (b).

213 15 U.S.C. §78(a).

214 15 U.S.C. §78ff(c) and 78dd-2(g).

215 15 U.S.C. §78ff(c)(1) and 78dd-2(g)(1).

216 15 U.S.C. §78ff(c)(2) and 78dd-2(g)(2).

217 15 U.S.C. §78ff(c)(3) and 78dd-2(g)(3).

218 Wendy C. Schmidt and Jonny J. Frank, "FCPA Demands Due Diligence in Global Dealings," *National Law Journal,* March 3, 1997 (NLJ B16, col. 1).

219 Ibid.

220 Ibid.

221 A recent example involves the IBM Corporation. Auditors for the Argentine tax agency reported that a subcontractor, Capacitacion y Computacion Rural S.A. (CCR), was paid $21 million by IBM, which received no services in return. The money was then paid over to a phantom contractor by CCR. Ibid.

222 "A Blow against Bribery," *Financial Times,* Feb. 23, 1998 (1998 WL 3534986).

223 Finlay Lewis, "34 Nations Sign Treaty Outlawing Business Bribes to Foreign Officials," *The San Diego Union-Tribune,* Jan. 18, 1998 (1998 WL 3987135).

224 Stanley S. Arkin, "Bribery of Foreign Officials: Leveling the Playing Field," *New York Law Journal,* Feb. 19, 1998 (NYLJ 3, col. 1).

225 Ibid.

Chapter 3 (*Cont.*)

226 C. Gopinath, "Corruption and Crony Capitalism," *Business Line,* June 16, 1997 (1997 WL 10899367).

227 Alex Brummer, "Corruption Is a Blot on World Markets," *Airline Business,* Aug. 1, 1997 (1997 WL 9065688).

228 Lavelle, op. cit.

229 OECD Convention on Combating Bribery of Foreign Public Officials, visited July 22, 1998.

230 Ibid.

231 Ibid.

232 Michael Gillard, "Time Is Running out for Global Anti-Bribery Treaty," *The Observer,* Aug. 9, 1998 (1998 WL 13205738).

233 Ibid.

234 Ibid.

235 "Banning Bribes, Finally," *The Washington Post,* Nov. 28, 1997 (1997 WL 16221003).

236 Lewis, op. cit.

237 "Sri Lankan Opposition Party Investigates Air Lanka Sale," *Airline Industry Information,* May 28, 1998 (1998 WL 4675729).

238 "Indonesia to Investigate Refinery Project for Corruption," *Asia Pulse,* June 10, 1998 (1998 WL 12617046).

239 "Thai Transport Ministry Investigating Railway Corruption," Dow Jones International News, Feb. 21, 1998 (Westlaw, TRANNEWS database).

240 "Rail Builders Are Arrested in Hong Kong," *The Asian Wall Street Journal,* July 15, 1998 (1998 WL-WSJA 12982327).

241 "Cargo Delay at Hong Kong Airport to Continue Until End of August," *The Asian Wall Street Journal,* July 16, 1998 (1998 WL-WSJA 12982415).

242 "Controversial New Harare Airport Project to Start at Last," Deutsche Presse-Agentur, Jan. 30, 1997 (DCHPA 07:18:00).

243 David Brodherson, "An Airport in Every City," in *Building for Air Travel,* John Zukowsky, ed., 1995, p. 85.

244 Brodherson, op. cit., p. 93.

245 David Pendered, "Battle Lines Are Drawn on Set-Asides," *The Atlanta Journal-Constitution,* Sept. 21, 1997, p. 8C.

246 "NBIA Denies Corruption in Landfill Deal," *The Bangkok Post,* Nov. 26, 1997, p. 4.

247 "Trinidad & Tobago: Airport Project Continuing," *IAC Newsletter,* Aug. 1, 1997.

248 Lewis Machipisa, "Zimbabwe-Human Rights: Rights Lobby Wants Action on Corruption," Inter Press Service, July 14, 1997.

249 Charles Walston and Darryl Fears, "Ethical Dilemma: Voter Trust Eroding," *The Atlanta Journal-Constitution,* Nov. 1, 1997, p. 8D; Jonathan Ringel and Emily Heller, "Arrington's Heavyweight Parade," *Fulton County Daily Report,* Sept. 21, 1997.

250 Consella Lee, "Empire Builds around Airport," *The Baltimore Sun,* Aug. 22, 1997, p. 1B.

Chapter 4

1 Louis Turpen, adddress before the Annual Meeting of the Canadian Transport Lawyers' Association, Montreal, Dec. 4, 1998.

2 Interview, *Jane's Airport Review,* Dec. 1997, p. 64.

3 See generally International Civil Aviation Organization, *Airport Planning Manual,* 1987, pp. I-23 to I-25.

Chapter 4 (*Cont.*)

4 International Civil Aviation Organization, *Airport Economics Manual,* 1991, p. 76.

5 See generally International Civil Aviation Organization, *Airport Planning Manual,* 1987, pp. I-25, I-26.

6 International Civil Aviation Organization, *Airport Economics Manual,* 1991, p. 68.

7 International Civil Aviation Organization, *Airport Planning Manual,* 1987, p. I-26.

8 Ibid.

9 Charles Barclay, *America's Future in Airport Infrastructure,* 1997.

10 U.S. House Committee on Public Works and Transportation, *Aviation Infrastructure Act of 1993,* H.R. No. 103-240, 103d Cong., 1st Sess. 8 (1993).

11 U.S. Senate Committee on Commerce, Science and Transportation, *Hearings on S. 2642, The Aviation Noise Improvement and Capacity Act of 1992,* 102d Cong., 2d Sess. 49 (May 5, 1992). The higher estimate is by the American Association of Airport Executives. Charles Barclay, op. cit.

12 Robert Hazel, "Airport Economics," in *Handbook of Airline Economics,* McGraw-Hill, 1995, pp. 113, 116.

13 Barclay, op. cit.

14 Henry Hyde and Jesse Jackson, Jr., "The Partnership for Metropolitan Chicago's Airport Future," Oct. 1997, p. 30.

15 U.S. General Accounting Office, "Airport Improvement Program: Military Airport Program and Reliever Set-Aside Update," March 13, 1996, p. 3.

16 U.S. General Accounting Office, "Airport Privatization," Feb. 29, 1996, p. 8.

17 U.S. General Accounting Office, "Airport Financing: Comparing Funding Sources with Planned Development," March 19, 1998, p. 1.

18 Pub. L. No. 97-248, 96 Stat. 671. This act replaced the Airport and Airway Development Act of 1970, Pub. L. No. 91-258, 84 Stat. 219.

19 For a complete list of the taxes charged, see Paul Stephen Dempsey and Laurence Gesell, *Airline Management: Strategies for the 21st Century,* Coast Aire, 1997, pp. 402–413, and Paul Dempsey, Robert Hardaway, and William Thoms, 1 *Aviation Law and Regulation* § 2.26 (1993).

20 National Civil Aviation Review Commission, "Avoiding Aviation Gridlock and Reducing the Accident Rate," Dec. 1997, p. II-26.

21 Hyde and Jackson, op. cit.

22 National Civil Aviation Review Commission, "Avoiding Aviation Gridlock and Reducing the Accident Rate," Dec. 1997, p. II-9. See Gail Butler and Martin Keller, "Airports and Airlines: Analysis of a Symbiotic, Love-Hate Relationship," in *Handbook of Airline Economics,* D. Jenkins, ed., McGraw-Hill, 1995, pp. 87, 91.

23 National Civil Aviation Review Commission, "Avoiding Aviation Gridlock and Reducing the Accident Rate," Dec. 1997, p. II-26.

24 Ibid., pp. II-9, II-41.

25 U.S. General Accounting Office, "Airport Improvement Program: Military Airport Program and Reliever Set-Aside Update," March 13, 1996, p. 2.

26 U.S. General Accounting Office, "Airport Financing: Funding Sources for Airport Development," March 1998, p. 32.

27 U.S. General Accounting Office, "Airport Financing: Comparing Funding Sources with Planned Development," March 19, 1998, p. 1.

28 U.S. House Committee on Public Works and Transportation, *Aviation Infrastructure Act of 1993,* H.R. No. 103-240, 103rd Cong., 1st Sess. 14 (1993).

29 Ibid., p. 27.

30 U.S. General Accounting Office, "Airport Privatization," Feb. 29, 1996, p. 9.

Chapter 4 *(Cont.)*

31 National Civil Aviation Review Commission, "Avoiding Aviation Gridlock and Reducing the Accident Rate," Dec. 1997, p. II-43.

32 See Hyde and Jackson, op. cit.

33 Julie Johnsson, "Giant O'Hare Overhaul Plan Takes Flight," *Crain's Chicago Business,* Oct. 13, 1997, p. 1.

34 49 U.S.C. §40177(a)(3) (1995). U.S. House Committee on Public Works and Transportation, *Federal Aviation Administration Authorization Act of 1994,* 103rd Cong., 2d Sess. 60-61 (1994).

35 Pub. L. 93-44; 49 U.S.C. §40116(e)(2) (1995).

36 Pub. L. 97-248; 49 U.S.C. §47101(a)(12)(13) (1995).

37 49 U.S.C. §2210(a)(1). See James Gesualdi, "Gonna Fly Now: All the Noise about the Airport Access Problem," 16 *Hofstra Law Review* 213, 232–33 (1987).

38 114 S.Ct. 855, 864 (1994).

39 Ibid. For a comprehensive review of this decision, see Rise Peters, "Case Comment," 22 *Transportation Law Journal* 291 (1994).

40 49 U.S.C. §47129.

41 49 U.S.C. §47107.

42 49 U.S.C. §47129.

43 49 U.S.C. §47129(c)(2); see *Trans World Airlines v. City and County of Denver,* DOT Order 95-7-27 (1995).

44 14 CFR Part 302.

45 60 Fed. Reg. 6,906 (Feb. 3, 1995).

46 Robert Span, "Procedures for Determining the Reasonableness of Airport Fees," address before the Transportation Law Institute, San Francisco, Oct. 31, 1995.

47 See, for example, "Second Los Angeles International Airport Rates Proceeding," DOT Order 96-1-18, 1996.

48 *Aviation Daily,* Nov. 27, 1995, p. 307.

49 Darrel Preston, "Trends in the Region," *The Bond Buyer,* Jan. 20, 1998, p. 1.

50 Mary Schiavo, *Flying Blind, Flying Safe,* Avon, 1997, pp. 125–127.

51 Nasid Hajari, "A Walk in the Clouds," *Time,* June 22, 1998; George Raine, "What's Up at SFO?" *The San Francisco Examiner,* Feb. 17, 1998, p. A-1.

52 U.S. General Accounting Office, "Airport Financing: Funding Sources for Airport Development," March 1998, p. 40.

53 Ibid., p. 43.

54 Keith Mordoff, "Air Transport Munich's New International Airport Expected to Begin Operations in 1991," *Aviation Week and Space Technology,* Feb. 22, 1988, p. 92.

55 "Germany Struggles to Meet Airport Needs," *Aviation Week and Space Technology,* Aug. 26, 1991, p. 41.

56 Ibid.

57 Leonard Hill, "Beyond Expectations," *Air Transport World,* June 1, 1993, p. 182.

58 "Munich Airport Still in the Red," *Aviation Europe,* Feb. 2, 1995, p. 6.

59 Robert Moorman, "Osaka to Me," *Air Transport World,* Oct. 1, 1994, p. 62.

60 David Brown, "Japanese Building International Offshore Airport to Serve Osaka," *Aviation Week and Space Technology,* July 13, 1987, p. 38.

61 James Ott, "Kansai Sets Highest Fees," *Aviation Week and Space Technology,* Oct. 25, 1993, p. 21.

62 Brown, op. cit.

Chapter 4 *(Cont.)*

63 "Growth Outpaces Asian Airports," *Aviation Week and Space Technology*, Aug. 29, 1994, p. 57.

64 Moorman, op. cit.

65 Ibid.

66 "Kansai Losses," *Aviation Week and Space Technology*, July 17, 1995, p. 19.

67 Moorman, op. cit.

68 Neil Gross, "Japan Wanted an Airport, It Got a Real Mess," *Business Week*, June 6, 1994, p. 50.

69 Eric Lassiter, "Japan to Open Much-Delayed Kansai Airport," *Travel Weekly*, Aug. 29, 1994, p. 4.

70 "Growth Outpaces Asian Airports," *Aviation Week and Space Technology*, Aug. 29, 1994, p. 57.

71 Moorman, op. cit.

72 "Japan Opens Airport on Man-Made Island," *The San Diego Union-Tribune*, Sept. 5, 1994, p. A18.

73 Alexandra Black, "Trade and Tourist Boom from New Airport," Inter Press Service Global Information Network, Sept. 9, 1994.

74 "Growth Outpaces Asian Airports," *Aviation Week and Space Technology*, Aug. 29, 1994, p. 57.

75 Moorman, op. cit.

76 Gross, op. cit.

77 Brown, op. cit.

78 Kansai International Airport Co., "KIAC Update," July 1998, p. 2.

79 Kansai International Airport Co., "KIAC Update," Jan. 1998, p. 1.

80 The sale of Stapleton's land has been sluggish, and by 1998 had produced only a few million dollars.

81 Stanley Scott and Levi Davis, *A Giant in Texas: A History of the Dallas-Fort Worth Regional Airport Controversy 1911–1974*, Nortex, 1974, p. 68.

82 Paul Stephen Dempsey, Andrew R. Goetz, and Joseph S. Szyliowicz, *Denver International Airport: Lessons Learned*, McGraw-Hill, 1997.

83 Ibid.

84 "New Macau Airport Set to Open July 1995," Phillips Business Information, Jan. 28, 1994.

85 Ibid.

86 "Macau's First Airport to Open Nov. 9," *Aviation Daily*, July 28, 1995, p. 149.

87 J. A. Donoghue, "The Pearl-Y Gateways," *Air Transport World*, Feb. 1, 1995, p. 75.

88 "Airport to Be Macao's Gate to Region," *The Asian Wall Street Journal*, Oct. 31, 1994, p. 3.

89 Ibid.

90 Peter Jansen, "Legal Issues in Airport Construction," IBC Conference on Asia Pacific Airports '94, Singapore, July 25, 1994.

91 "NSIA Considers Opening with Extra Runway," *World Airport Week*, July 1, 1995.

92 John Mok, "The Development of Hong Kong's New International Airport," International Conference on Aviation and Airport Infrastructure, Denver, Dec. 8, 1993.

93 Jansen, op. cit.

94 Mok, op. cit.

95 "Hong Kong Passes Bill to Set Airport Body," *The Asian Wall Street Journal*, July 20, 1995, p. 3.

Chapter 4 (*Cont.*)

96 "Hong Kong Builds $20.3 Billion Airport," *The Orlando Sentinel,* April 9, 1995, p. A20.

97 Stuart Becker, "Airports: Chinese Opera," *Far Eastern Economic Review,* April 6, 1995, p. 54.

98 Steven Mufson, "Accord Boosts Hong Kong Airport," *The Washington Post,* Nov. 5, 1994, p. A19.

99 Mufson, op. cit.

100 "U.K., China Reach Agreement on Hong Kong Airport Financing," *Airports,* July 4, 1995, p. 261.

101 "Growth Outpaces Asian Airports," *Aviation Week and Space Technology,* Aug. 29, 1994, p. 58.

102 Airport Authority Hong Kong, "Newsletter XII," Feb. 1998.

103 "Fasten Your Seat Belts," *The Economist,* Nov. 12, 1994, p. 42.

104 Michael Mecham, "Airport Fees Stable for Asian Carriers," *Aviation Week and Space Technology,* Aug. 28, 1995, p. 51.

105 Peter Negline, "Hong Kong's New International Airport," Salomon Bros., March 9, 1998, p. 7.

106 Oslo Airport, "Europe's Safest and Most Efficient Airport," 1997, p. 10; Oslo Airport, "Commercial Activities at Oslo Airport," 1997, p. 2.

107 Stella Kim, "Rebuilding South Korea," *The International Herald Tribune,* Oct. 29, 1996.

108 John Burton, "Seoul to Raise Dollars 25bn for Transport Hub," *The Financial Times,* July 17, 1996, p. 6.

109 Jong-Heui Shin, "Airport Developments in Korea," International Conference on Aviation and Airport Infrastructure, Denver, Dec. 8, 1993.

110 Ibid.

111 "Korea: Airport Development Market," Financial Times Asia Intelligence Wire, May 9, 1997.

112 Erik Helland, "Not Quite Ready, Yet," *Infrastructure Finance,* Sept. 1996, p. 30.

113 Kim, op. cit.

114 Tom Ballantyne, "Inchon Seeks More Cash," *Orient Aviation,* Sept. 1998, p. 58.

115 Dempsey, Goetz, and Szyliowicz, op. cit., p. 186.

116 U.S. General Accounting Office, "Airport Financing: Funding Sources for Airport Development," March 1998, p. 38.

117 Michael Bell, "Airport Financing," in *Airport Regulation, Law and Public Policy,* R. Hardaway, ed., 1991, pp. 93–95.

118 Alexander Wells, *Airport Planning and Management,* 1992, p. 181.

119 Paul Dempsey and Andrew Goetz, *Airline Deregulation and Laissez Faire Mythology,* 1992, pp. 221–238.

120 Kenneth Mead, "Airline Competition," testimony on passenger facility charges before the House Subcommittee on Aviation, June 19, 1990.

121 Ibid, p. 5.

122 International Civil Aviation Organization, *Airport Economics Manual,* 1991, p. 69.

123 U.S. General Accounting Office, "Airport Financing: Funding Sources for Airport Development," March 1998, p. 35.

124 U.S. General Accounting Office, "Airport Privatization," Feb. 29, 1996, pp. 4, 5.

125 Dempsey, Goetz, and Szyliowicz, op. cit., pp. 187, 190.

126 For example, in Denver, the underwriters who floated Denver International Airport's bonds contributed about 10% of Mayor Webb's contributions to his two mayoral campaigns, and made significant donations to key city council members as well. Dempsey, Goetz, and Szyliowicz, op. cit., pp. 214–220.

Chapter 4 (*Cont.*)

127 U.S. General Accounting Office, "Airport Privatization," Feb. 29, 1996, p. 1.

128 Dempsey, Goetz, and Szyliowicz, op. cit., p. 190.

129 National Civil Aviation Review Commission, "Avoiding Aviation Gridlock & Reducing the Accident Rate," Dec. 1997, p. II-44.

130 Dempsey, Goetz, and Szyliowicz, op. cit., p. 204.

131 International Civil Aviation Organization, *Airport Economics Manual*, 1991, p. 66.

132 U.S. General Accounting Office, "Airport Privatization," Feb. 29, 1996, pp. 1, 3.

133 Turpen, op. cit.

134 Eliot Lees, "Airport Privatization: Latest Trends from around the World," in *ACI World Economic Specialty Conference Proceedings,* Airport Economics in a Technological Age, Denver, April 6–9, 1997, p. 148; Norman Ashford and Clifton Moore, *Airport Finance,* Van Nostrand Reinhold, 1992.

135 U.S. General Accounting Office, "Airport Privatization," Feb. 29, 1996, pp. 1, 5.

136 See Paul Stephen Dempsey, *The Social and Economic Consequences of Deregulation,* Quorum Books, 1989.

137 International Civil Aviation Organization, *Airport Economics Manual*, 1991, p. 8.

138 Paul Stephen Dempsey and Kevin O'Connor, "Air Traffic Congestion and Infrastructure Development in the Pacific Asia Region," *Asia Pacific Air Transport: Challenges and Policy Reforms,* Institute of Southeast Asia Studies, 1997, pp. 23, 36.

139 Dempsey, Goetz, and Szyliowicz, op. cit.

140 Paul Dempsey, "Market Failure and Regulatory Failure as Catalysts for Political Change: The Choices between Imperfect Regulation and Imperfect Competition," 46 *Washington and Lee Law Review* 1 1989.

141 U.S. General Accounting Office, "Airport Privatization," Feb. 29, 1996, pp. 5, 6.

142 U.S. General Accounting Office, "Airport Privatization," Feb. 29, 1996, p. 5.

143 Joel Millman, "Mexican Airport Auction Reignites Asset Sales," *The Wall Street Journal,* Dec. 9, 1998, p. A14.

144 "Private Funds for Germany," *Jane's Airport Review,* Dec. 1997, p. 13.

145 "Dutch Think about Selling Schiphol Interest," *World Airport Week,* July 15, 1997, p. 3.

146 Lees, op. cit.

147 61 Fed. Reg. 31,994 (June 21, 1996).

148 Jonathan Lemberg, Xiaohu Ma, and James Zimmerman, "Financing and Investing In China's Airport Development," *China,* 1998.

149 Executive Order No. 12803 (April 30, 1992). This was affirmed by Executive Order No. 12893 (Jan. 27, 1994), which directed federal agencies to support additional private sector participation infrastructure investment and management.

150 U.S. General Accounting Office, "Airport Privatization," Feb. 29, 1996, p. 7.

151 U.S. General Accounting Office, "Airport Financing: Funding Sources for Airport Development," March 1998.

152 U.S. General Accounting Office, "Airport Privatization," Feb. 29, 1996, p. 4.

153 Butler and Keller, op. cit., pp. 87, 92, 93.

154 U.S. General Accounting Office, "Airport Improvement Program: Military Airport Program and Reliever Set-Aside Update," March 13, 1996, p. 4.

155 U.S. Senate Committee on Commerce, Science and Transportation, Hearings on S. 2642, *The Aviation Noise Improvement and Capacity Act of 1992,* 102d Cong., 2d Sess. 82 (May 5, 1992).

156 U.S. General Accounting Office, "Airport Improvement Program: Military Airport Program and Reliever Set-Aside Update," March 13, 1996, p. 1.

157 Emily Bazar, "New Airport Terminal," *Sacramento Bee,* March 26, 1998, p. B2.

Chapter 4 (*Cont.*)

158 *Convention on International Civil Aviation,* done December 7, 1944, 61 Stat. 1180, T.I.A.S. No. 1591, U.N.T.S. 295 [hereinafter cited as *Chicago Convention*], Art. 44.

159 *Chicago Convention,* Art. 15.

160 International Civil Aviation Organization, *Airport Economics Manual,* 1991, pp. 1, 2.

161 *Chicago Convention,* Art. 15.

162 National Civil Aviation Review Commission, "Avoiding Aviation Gridlock and Reducing the Accident Rate," Dec. 1997, p. II-29.

163 International Civil Aviation Organization, *Airport Economics Manual,* 1991, p. 3.

164 For an explanation of these principles, see Paul Stephen Dempsey, *Law and Foreign Policy in International Aviation,* Transnational Publishers, 1987, pp. 286–288.

165 Paul Stephen Dempsey and Laurence Gesell, *Air Transportation: Foundations for the 21st Century,* Coast Aire, 1997, pp. 450, 451.

166 U.S. General Accounting Office, "Airport Financing: Funding Sources for Airport Development," March 1998, p. 41.

167 International Civil Aviation Organization, *Airport Planning Manual,* 1987, p. I-27.

168 See generally, International Civil Aviation Organization, *Airport Planning Manual,* 1987, p. I-27.

169 U.S. General Accounting Office, "Airport Privatization," Feb. 29, 1996, p. 4.

170 "Rise before the Fall," *Asia Airports,* Nov. 1998, p. 55.

171 International Civil Aviation Organization, "Development of Non-Aeronautical Revenues at Airports," 1979, pp. 7–10.

172 Ibid.

173 International Civil Aviation Organization, *Airport Economics Manual,* 1991, p. 55.

174 Robert Bruegmann, "Airport City," in *Building for Air Travel,* John Zukowsky, ed., 1996, p. 205.

175 Ibid.

176 John Zukowsky, "Introduction," in *Building for Air Travel,* John Zukowsky, ed., 1996, p. 16.

177 Hajari, op. cit.

178 Bruegmann, op. cit.

179 "Variety Spices up Concession Operations at Cleveland," *World Airport Week,* July 15, 1997, p. 3.

180 International Civil Aviation Organization, *Airport Economics Manual,* 1991, p. 13.

181 International Civil Aviation Organization, *Airport Planning Manual,* 1987, p. I-28. Airports and airlines employ similar accounting practices. See Dempsey and Gesell, op cit., pp. 130–138.

182 International Civil Aviation Organization, *Airport Planning Manual,* 1987, p. I-33.

183 International Civil Aviation Organization, *Airport Economics Manual,* 1991, p. 13.

184 International Civil Aviation Organization, *Airport Planning Manual,* 1987, p. I-44.

Chapter 5

1 Stanley Scott and Levi Davis, *A Giant in Texas: A History of the Dallas-Fort Worth Regional Airport Controversy 1911–1974,* Nortex, 1974, p. 68.

2 Interview with Denver International Airport Aviation Director Bruce Baumgartner, July 16, 1998.

3 Peter Trautmann, "The Need for New Airport Infrastructure," International Conference on Aviation and Airport Infrastructure, Denver, Dec. 5, 1993.

Chapter 5 (*Cont.*)

4 Ibid.

5 Louis Turpen, address before the Annual Meeting of the Canadian Transport Lawyers' Association, Montreal, Dec. 4, 1998.

6 James Spensley, "Airport Planning," in *Airport Regulation, Law and Public Policy,* R. Hardaway, ed., 1991, pp. 63, 64.

7 *Aviation Daily,* Dec. 14, 1995, p. 413.

8 U.S. General Accounting Office, "Federal Aviation Administration: Issues Related to FAA Reform," Aug. 2, 1995.

9 *Aviation Daily,* Feb. 22, 1996, p. 288.

10 In the 1980s, several FAA Administrators were retired military officers.

11 Paul Stephen Dempsey and Kevin O'Connor, "Air Traffic Congestion and Infrastructure Development in the Pacific Asia Region," *Asia Pacific Air Transport: Challenges and Policy Reforms,* Institute of Southeast Asia Studies, 1997, pp. 23–28.

12 Ibid., pp. 23, 28.

13 Ibid., pp. 23, 34.

14 Federal Aviation Administration, "Airport Master Plans," 1985, p. 19.

15 Ibid., p. 9.

16 See generally Spensley, op. cit., p. 63.

17 International Civil Aviation Organization, *Airport Planning Manual,* 1987, p. I-1.

18 Federal Aviation Administration, "Airport Master Plans," 1985, p. 5.

19 International Civil Aviation Organization, *Airport Planning Manual,* 1987, p. I-7.

20 Robert Horonjeff and Francis McKelvey, *Planning and Design of Airports,* 4th ed., Mc-Graw-Hill, 1994, p. 184.

21 International Civil Aviation Organization, *Airport Planning Manual,* 1987. pp. I-3, I-5.

22 Federal Aviation Administration, "Airport Master Plans," 1985, p. 13.

23 International Civil Aviation Organization, *Airport Planning Manual,* 1987, p. I-9.

24 Ibid.

25 Federal Aviation Administration, "Airport Master Plans," 1985, pp. 14, 15.

26 Pub. L. 97-248.

27 Spensley, op. cit., pp. 69–71.

28 Federal Aviation Administration, "Airport Master Plans," 1985, pp. 23, 29.

29 Federal Aviation Administration, "Airport Master Plans," 1985, p. 17.

30 Horonjeff and McKelvey, op. cit.; Paul Stephen Dempsey, Andrew R. Goetz, and Joseph S. Szyliowicz, *Denver International Airport: Lessons Learned,* McGraw-Hill, 1997, p. 34.

31 Paul Dempsey and Laurence Gesell, *Airline Management: Strategies for the 21st Century,* Coast Aire, 1997; Paul Dempsey and Andrew Goetz, *Airline Deregulation and Laissez Faire Mythology,* Quorum, 1992.

32 Spensley, op. cit., pp. 63–64.

33 Horonjeff and McKelvey, op. cit., p. 215.

34 Norman Ashford and Paul Wright, *Airport Engineering.* 3d ed., Wiley, 1992.

35 Federal Aviation Administration, "Airport Master Plans," 1985, pp. 21, 22.

36 International Civil Aviation Organization, *Airport Planning Manual,* 1987, p. I-17.

37 Federal Aviation Administration, "Airport Master Plans," 1985, p. 23. For busier airports, the FAA recommends calculation of the average peak hour demand of 10% of the days of the year.

Chapter 5 (*Cont.*)

38 International Civil Aviation Organization, *Airport Planning Manual,* 1987, p. I-13.

39 International Civil Aviation Organization, *Airport Planning Manual,* 1987, p. I-15, supplemented with criteria developed in Dempsey, Goetz, and Szyliowicz, op. cit., p. 34; Horonjeff and McKelvey, op. cit., pp. 216, 217; and Federal Aviation Administration, "Airport Master Plans," 1985, pp. 22, 23.

40 Horonjeff and McKelvey, op. cit.

41 Dempsey, Goetz, and Szyliowicz, op. cit., p. 35.

42 Ashford and Wright, op. cit.

43 Federal Aviation Administration, "Aviation Forecasts—Fiscal Years 1995–2006," 1995.

44 Ashford and Wright, op. cit.

45 See, for example, Dempsey and Goetz, op. cit., and Paul Dempsey, *Law and Foreign Policy in International Aviation,* Transnational Publishers, 1987.

46 Dempsey, Goetz, and Szyliowicz, op. cit., pp. 59, 61.

47 Ann Carnahan, "DIA Traffic Aims toward Record," *The Denver Rocky Mountain News,* Dec. 24, 1998, p. 4A.

48 Dempsey and Gesell, op. cit., pp. 31–59.

49 International Civil Aviation Organization, *Airport Planning Manual,* 1987, pp. I-18, I-19.

50 Federal Aviation Administration, "Airport Master Plans," 1985, pp. 23, 24.

51 Dempsey, Goetz, and Szyliowicz, op. cit., p. 36.

52 "World Airline Traffic Increases 5.0 Percent in 1997," Agence France-Presse, April 2, 1998 (1998 WL 2254080).

53 "The World's Airports in 1997 (Preliminary): Airport Ranking by Total Passengers" (visited Aug. 17, 1998), http://www.airports.org/pax97.html.

54 "The World's Airports in 1997 (Preliminary): Airport Ranking by Total Cargo" (visited Aug. 17, 1998), http://www.airports.org/cargo97.html.

55 "The World's Airports in 1997 (Preliminary): Airport Ranking by Total Aircraft Movements" (visited Aug. 17, 1998), http://www.airports.org/move97.html.

56 International Civil Aviation Organization, *Airport Planning Manual,* 1987, p. I-9.

57 Federal Aviation Administration, "Airport Master Plans," 1985, p. 17.

58 Ashford and Wright, op. cit.

59 Dempsey, Goetz, and Szyliowicz, op. cit., p. 36.

60 Federal Aviation Administration, "Airport Master Plans," 1985, p. 18; Horonjeff and McKelvey, op. cit., p. 192.

61 Federal Aviation Administration, "Airport Master Plans," 1985, pp., 18, 19.

62 See Spensley, op. cit., pp. 63, 72.

63 "Weber Demands New Runway at Frankfurt," *Flight International,* Dec. 2, 1997, p. 8.

64 43 U.S.C. § 4321; Federal Aviation Administration, "Airport Master Plans," 1985, p. 2.

65 Federal Aviation Administration, "Airport Master Plans," 1985, p. 19.

66 Spensley, op. cit., pp. 63, 73; Federal Aviation Administration, "Airport Master Plans," 1985, pp. 57, 58.

67 Dempsey, Goetz, and Szyliowicz, op. cit., p. 288.

68 International Civil Aviation Organization, *Airport Economics Manual,* 1991, p. 61.

69 International Civil Aviation Organization, *Airport Planning Manual,* 1987, p. I-13.

70 Federal Aviation Administration, "Airport Master Plans," 1985, p. 20.

71 International Civil Aviation Organization, *Airport Economics Manual,* 1991, p. 62.

72 International Civil Aviation Organization, *Airport Planning Manual,* 1987, pp. I-11, I-12.

Chapter 5 (*Cont.*)

73 Dempsey and Gesell, op. cit., pp. 179–187.

74 International Civil Aviation Organization, *Airport Planning Manual*, 1987, p. I-14.

75 Ibid., p. 1-9.

76 Ibid., p. 1-7.

77 Ibid., p. 1-11.

78 Laurence Gesell, *The Administration of Public Airports*, 3d ed., Coast Aire, 1992, pp. 144–151.

79 Federal Aviation Administration, "Airport Master Plans," 1985, p. 11.

80 Ibid., pp. 2, 5.

81 Ibid., p. 30.

82 International Civil Aviation Organization, *Airport Planning Manual*, 1987, p. I-2.

83 Ibid., p. 1-7.

84 Federal Aviation Administration, "Airport Master Plans," 1985, pp. 10, 11; Horonjeff and McKelvey, op. cit., pp. 187, 188.

85 Salt Lake City International Airport, "Airport Master Plan." 1996, p. 1.

86 Federal Aviation Administration, "Airport Master Plans," 1985, p. 18.

87 Ibid., p. 42.

88 Ibid., p. 30.

89 Ibid., p. 33.

90 International Civil Aviation Organization, *Airport Economics Manual*, 1991, p. 54. See also Salt Lake City International Airport, "Airport Master Plan," 1996, Chap. 3.

91 Interview with Denver International Airport Aviation Director Bruce Baumgartner at Denver, Oct. 15, 1998. For an analysis of the strategic planning process more generally, see Dempsey and Gesell, op. cit., pp. 179–187.

92 For a comprehensive review of these requirements, see Jalal Haidar, "Operations and Certification" in *Airport Regulation, Law and Public Policy*, R. Hardaway, ed., 1991, p. 107.

93 Ibid., pp. 107, 108.

94 Lynne Haraway, "Technology Improving Airport Economics," *ACI World Economic Specialty Conference Proceedings*, Airport Economics in a Technological Age, Denver, April 6–9, 1997, p. 35.

95 Zenjiro Ogawa, "Kansai International Airport Projects," International Conference on Aviation and Airport Infrastructure, Denver, Dec. 8, 1993.

96 Jong-Heui Shin, "Airport Developments in Korea," International Conference on Aviation and Airport Infrastructure, Denver, Dec. 8, 1993.

97 Moon-Sik Park, "Korea: Airport Development Market," Financial Times Asia Intelligence Wire, May 9, 1997.

98 Shin, op. cit.

99 "NSIA Privatization Opportunities," *World Airport Week*, June 20, 1995.

100 U.S. Trade and Development Agency, "Airport Project Opportunities in Asia," 1997, pp. 68, 69.

101 Dempsey and O'Connor, op. cit., pp. 23, 25.

102 61 Fed. Reg. 32,021 (June 21, 1996).

103 Massport, "Logan Modernization: Works in Progress," 1998, p. 4.

104 Dempsey and O'Connor, op. cit., pp. 23, 34.

105 Dempsey, Goetz, and Szyliowicz, op. cit., p. 291.

106 International Civil Aviation Organization, *Airport Planning Manual*, 1987, p. I-1.

107 Ibid., p. 1-3.

108 Dempsey, Goetz, and Szyliowicz, op. cit., pp. 475, 484, 486.

Chapter 6

1 Quoted in James Gesualdi, "Gonna Fly Now: All the Noise about the Airport Access Problem," 16 *Hofstra Law Review,* 213 (1987).

2 Nasid Hajari, "A Walk in the Clouds," *Time,* June 22, 1998.

3 See generally Robert Horonjeff and Francis McKelvey, *Planning and Design of Airports,* 4th ed., McGraw-Hill, 1994, pp. 181–183.

4 Jennifer Stenzel and Jonathan Trutt, "Flying Off Course: Environmental Impacts of America's Airports," Natural Resources Defense Council, 1996, p. 4.

5 Federal Aviation Administration, "Airport Master Plans," 1985, p. 41.

6 Ibid., p. 33.

7 Mark Bouman, "Cities of the Plane," in *Building for Air Travel,* John Zukowsky, ed., 1996, p. 180.

8 Koos Bosma, "European Airports, 1945–1995," in *Building for Air Travel,* John Zukowsky, ed., 1996, p. 63.

9 "Dutch Think about Selling Schiphol Interest," *World Airport Week,* July 15, 1997, p. 3.

10 Federal Aviation Administration, "Airport Master Plans," 1985, p. 43.

11 Federal Aviation Administration, "Airport Master Plans," 1985, p. 32.

12 James Spensley, "Airport Planning," in *Airport Regulation, Law and Public Policy,* R. Hardaway, ed., 1991, p. 72.

13 Horonjeff and McKelvey, op. cit., pp. 199, 200.

14 See Paul Stephen Dempsey, Andrew R. Goetz, and Joseph S. Szyliowicz, *Denver International Airport: Lessons Learned,* McGraw-Hill, 1997, p. 229.

15 Spensley, op. cit., p. 72; Horonjeff and McKelvey, op. cit., p. 193.

16 Federal Aviation Administration, "Airport Master Plans," 1985, pp. 42, 44. Specifically, the following criteria should be considered: (1) operational capability, (2) capacity potential, (3) ground access, (4) development costs, (5) environmental consequences, (6) socioeconomic implications, and (7) consistency with areawide planning.

17 International Civil Aviation Organization, *Airport Planning Manual,* 2d ed., 1987, p. I-42.

18 Federal Aviation Administration, "Airport Master Plans," 1985, p. 31.

19 David Beard, Ken Kaye, and E. Q. Torriero, "Flying the Dangerous Skies," *The Denver Post,* Jan. 17, 1998, p. 17A.

20 International Civil Aviation Organization, *Airport Planning Manual,* 2d ed., 1987, p. I-40.

21 Horonjeff and McKelvey, op. cit., p. 194.

22 John Zukowsky, "Introduction," in *Building for Air Travel,* John Zukowsky, ed., 1996, p. 14.

23 International Civil Aviation Organization, *Airport Planning Manual,* 2d ed., 1987, pp. I-41, I-43.

24 Horonjeff and McKelvey, op. cit., p. 199.

25 Ibid., pp. 193, 194, 206.

26 Robert Bruegmann, "Airport City," in *Building for Air Travel,* John Zukowsky, ed., 1996, p. 198.

27 Bouman, op. cit., p. 181.

28 Ibid., p. 183.

29 Ibid., p. 183.

30 Aeroporti di Milano, "For You," 1997.

31 Flughafen Munich, "Environmental Protection at Munich Airport," 1996, p. 49.

32 Jose Egurbide, "Stop Biting the Hand That Feeds Us," 22 *Pepperdine Law Review* 1089 (1995).

Chapter 6 (*Cont.*)

33 OECD, "Pollution Prevention and Control: Environmental Criteria for Sustainable Transport," 1996, p. 10.

34 NASA Environmental Compatibility Research Workshop, "Where Do We Go from Kyoto?" May 19, 1998.

35 OECD, op. cit., p. 7.

36 National Round Table on the Environment and the Economy, "Draft Sustainable Transportation Principles," Feb. 21, 1996.

37 Federal Aviation Administration, "Airport Master Plans," 1985, p. 47.

38 See Laurence Gesell, *The Administration of Public Airports*, 3d ed., Coast Aire, 1991.

39 "Germany Tests Hydrogen Fuel on APU," *Flight international*, May 24, 1995.

40 Andrzej Jeziorski, "Exhausting Issues," *Flight International*, June 7, 1995.

41 Transport Canada, "The Greening of Aviation," 1996, p. 45.

42 International Civil Aviation Organization, *Airport Planning Manual*, 2d ed., 1987, p. I-41.

43 Horonjeff and McKelvey, op. cit., pp. 719–763.

44 Stenzel and Trutt, op. cit., p. 4.

45 Ibid., p. 5.

46 Amsterdam Airport Schiphol, "Fact Sheet: Introduction to the World of Amsterdam Airport Schiphol," 1997; Amsterdam Airport Schiphol, "Balancing Environment and Economics," 1997.

47 Aeroports de Paris, "Orly, Mastering the Future," 1997, p. 20.

48 Aeroports de Paris, "Charles de Gaulle Airport: Europe's Foremost Transport Hub," 1997, p. 25.

49 Aeroports de Paris, "Environmental Resource Center," 1997.

50 Environment Centre, "The Greenhouse Effect," 1996.

51 Esther Tan, "Transport Sector Tops List of Polluters," *The New Straits Times,* Nov. 21, 1995, p. 15.

52 F. Kaid Benfield, "Running on Empty: The Case for a Sustainable National Transportation System," 25 *Environmental Law* 651 (1995).

53 U.S. Department of Transportation, "Transportation Statistics Annual Report," 1995, p. 65.

54 Benfield, op. cit.

55 Michael MacCracken, statement before the U.S. House Committee on Science, Hearing on U.S. Global Change Research Programs, Mar. 6, 1996.

56 Environment Centre, "The Greenhouse Effect," 1996.

57 MacCracken, op. cit.

58 IPCC, Second Assessment Report, 1996.

59 Jeziorski, op. cit.

60 Transport Canada, "The Greening of Aviation," 1996, p. 24.

61 See Martin Hindley, "Emission Control," *Flight International,* Jan. 31, 1996.

62 Jeziorski, op. cit.

63 U.S. Department of Transportation, "Transportation Statistics Annual Report," 1995, p. 66.

64 "Aviation Growth May Enhance Global Warming and Ozone Depletion," Information Access Company, July 25, 1994.

65 Environment Centre, "The Greenhouse Effect," 1996.

66 OECD, op. cit., p. 56.

67 P. O. Wennberg et al., "Hydrogen Radicals, Nitrogen Radicals, and the Production of O_3 in the Upper Troposphere," *Science,* Jan. 2, 1998, p. 49; "Study Indicates Greater Contribution by Aircraft, Fires to Greenhouse Effect," *BNA National Environment Daily,* Jan. 5, 1998.

Chapter 6 (*Cont.*)

68 Hindley, op. cit.

69 William Stevens, "Global Mercury Rises," *The Denver Post,* Dec. 18, 1998, p. A2.

70 MacCracken, op. cit.

71 Stevens, op. cit.

72 Environment Centre, "The Greenhouse Effect," 1996.

73 MacCracken, op. cit.

74 "Ozone Hole Biggest Ever," *The Denver Post,* Nov. 2, 1996, p. 10A.

75 Egurbide, op. cit., pp. 1089, 1095.

76 MacCracken, op. cit.

77 U.S. Department of Transportation, "National Transportation Statistics," 1995, p. 170.

78 Transport Canada, "The Greening of Aviation," 1996, p. 26.

79 Ibid.

80 Paul Stolpman, "Environmental Impacts of Aviation Emissions," ABA Forum on Air and Space Law, San Francisco, July 10, 1998.

81 Irwin Arieff, "Global Warming Worries OECD Environment Ministers," Reuters, Feb. 20, 1996.

82 Transport Canada, "The Greening of Aviation," 1996, pp. 24, 25, 28.

83 Benfield, op. cit.

84 John Kennedy testimony before the U.S. House Subcommittee on Energy, Committee on Science, Space and Technology, July 14, 1994.

85 Henry Hyde and Jesse Jackson, Jr., "The Partnership for Metropolitan Chicago's Airport Future: A Call for Regional Leadership," Oct. 1997, pp. 21, 40.

86 Stenzel and Trutt, op. cit., pp. 7, 8.

87 Ibid., p. 10.

88 Scott Kafker, "The Build-Up and Clean-Up of Logan Airport: A Massachusetts Environmental Experience," ABA Forum on Air and Space Law, San Francisco, July 10, 1998.

89 OECD, op. cit., p. 21.

90 Transport Canada, "The Greening of Aviation," 1996, p. 11.

91 Ibid., p. 12.

92 "MDC Forecasts 5.7% Growth and Demand for 13,272 Aircraft," *Aircraft Value News,* Sept. 4, 1995.

93 Hindley, op. cit., p. 27.

94 OECD, op. cit., pp. 69, 70.

95 Hindley, op. cit.

96 Jeziorski, op. cit.

97 Ibid.

98 International Civil Aviation Organization, *Airport Planning Manual,* 2d ed., 1987, p. I-43.

99 Spensley, op. cit., p. 79.

100 49 U.S.C. §§1301–1355.

101 49 U.S.C. §4321.

102 Federal Aviation Administration, "Airport Master Plans," 1985, pp. 49, 50.

103 Spensley, op. cit., p. 76.

104 49 U.S.C. §4332(c).

105 See *Stryckers Bay Neighborhood Council v. Karlen,* 444 U.S. 223 (1980).

106 Bouman, op. cit., p. 189.

107 Salt Lake City International Airport, "Airport Development," 1998.

Chapter 6 (*Cont.*)

108 Pub. L. 90-411.

109 14 CFR Parts 21 and 36.

110 49 U.S.C. §4901-4918. See *City of Burbank v. Lockheed Air Terminal,* 411 U.S. 624 (1973).

111 Gesualdi, op. cit., pp. 213, 237.

112 Pub. L. No. 95-609, 92 Stat. 3079.

113 Pub. L. No. 96-193; 94 Stat. 50.

114 49 U.S.C. §2101–2124.

115 14 C.F.R. §150.

116 14 C.F.R. §36.

117 John Jenkins, Jr., "The Airport Noise and Capacity Act of 1990: Has Congress Finally Solved the Aircraft Noise Problem?" 59 *Journal of Air Law and Commerce* 1023 (1994).

118 Cutler & Stanfield, "Airport Noise: A Guide to the FAA Regulations under the Airport Noise and Capacity Act," 1992.

119 49 U.S.C. §2157.

120 Stolpman, op. cit.

121 Barbara Lichman, "From Confrontation to Collaboration: Opportunities for 'Hushing' Airport Noise," Conference on Aviation and Airport Infrastructure, Denver, Dec. 11, 1993.

122 Bouman, op. cit., p. 189.

123 16 U.S.C. §470.

124 16 U.S.C. §471; 36 CFR §800.

125 Gesualdi, op. cit., pp. 213, 246.

126 Paul Stephen Dempsey, Robert Hardaway, and William Thoms, 1 *Aviation Law and Regulation* §§8.03–8.14 (Butterworth 1993).

127 *U.S. v. Causby,* 328 U.S. 256 (1946).

128 See generally J. Scott Hamilton, *Allocation of Airspace As a National Resource,* 22 *Transportation Law Journal* 251 (1994).

129 See *U.S. v. Causby,* 328 U.S. 256 (1946).

130 *Griggs v. Allegheny County,* 369 U.S. 84 (1962).

131 328 U.S. 256 (1946).

132 W. Page Keeton, *Prosser and Keeton on the Law of Torts,* 5th ed., 1984, p. 81.

133 Gesualdi, op. cit., pp. 213, 221.

134 *City of Burbank v. Lockheed Air Terminal,* 411 U.S. 624 (1973). See also *National Aviation v. City of Hayward,* 418 F. Supp. 417 (N.D. Cal. 1976) (city airport proprietor prohibition of night operation of aircraft at noise above a specified level deemed only an incidental burden on interstate commerce); *Arrow Air, Inc. v. Port Authority,* 602 F. Supp. 314 (S.D.N.Y. 1985) (airport proprietors may establish "fair, even-handed and nondiscriminatory regulations" to limit noise).

135 See, for example, *Western Air Lines v. Port Authority,* 817 F.2d 222 (2d Cir. 1987).

136 See, for example, *United States v. New York,* 552 F. Supp. 255 (N.D.N.Y. 1982), aff'd 708 F.2d 92 (2d Cir. 1983), cert. denied 466 U.S. 936 (1984) (curfew banning all flights from 11:00 P.M. to 7:00 A.M. held overbroad because it banned "all aircraft regardless of the degree of accompanying emitted noise"); *United States v. County of Westchester,* 571 F. Supp. 786 (S.D.N.Y. 1983) (similar curfew preempted as "unreasonable, arbitrary, discriminatory and overbroad").

137 Gesualdi, op. cit., pp. 213, 256.

138 Edward Ziegler, *Rathkopf's The Law of Zoning and Planning,* 4th ed., 1999, §60.01.

Chapter 6 (*Cont.*)

139 Scott Hamilton, "Planning for Noise Compatibility," *Airport Regulation, Law and Public Policy,* R. Hardaway, ed., 1991, pp. 85, 86.

140 Koos Bosma, "European Airports, 1945–1995," *Building for Air Travel,* John Zukowsky, ed., 1996, p. 63.

141 "Weber Demands New Runway at Frankfurt," *Flight International,* Dec. 2, 1997, p. 8.

142 Robert Roenqvist and Haakan Jonforsen, "Addition of a Runway at a Major Airport Can Take Several Years to Clear All Hurdles," *ICAO Journal,* Sept. 1997, p. 5.

143 Ibid., pp. 5, 6.

144 Michael Mecham, "Munich Turns out Lights on Riem," *Aviation Week and Space Technology,* May 25, 1992, p. 20.

145 Keith Mordoff, "Air Transport Munich's New International Airport Expected to Begin Operations in 1991," *Aviation Week and Space Technology,* Feb. 22, 1988, p. 92.

146 "Germany Struggles to Meet Airport Needs," *Aviation Week and Space Technology,* Aug. 26, 1991, p. 41.

147 Mordoff, op. cit.

148 "Germany Struggles to Meet Airport Needs," *Aviation Week and Space Technology,* Aug. 26, 1991, p. 41.

149 Leonard Hill, "Countdown on Munich II," *Air Transport World,* June 1, 1991, p. 70.

150 Ibid.

151 Patrick O'Driscoll, "Munich's Facility Reflects Future of DIA," *The Denver Post,* July 31, 1994, p. 24A.

152 Willi Hermsen, "Analyzing Munich's Move" (address 1996).

153 Patrick O'Driscoll, "It Could Be Worse (Really!)," *The Denver Post,* July 31, 1994, p. 25A.

154 Kim Upton, "New Munich Airport Stresses Ease of Travel," *Los Angeles Times,* May 17, 1992, p. 4.

155 Hill, op. cit.

156 Hermsen, op. cit.

157 David Brown, "Japanese Building International Offshore Airport to Serve Osaka," *Aviation Week and Space Technology,* July 13, 1987, p. 38.

158 Robert Moorman, "Osaka to Me," *Air Transport World,* Oct. 1, 1994, p. 62.

159 Ibid.; "How to Avoid that Sinking Feeling," *The Economist,* Feb. 4, 1995, p. 73.

160 Pierre-Antoine Donnet, "Saga of Troubled Narita Airport Continues 20 Years on," Agence France-Presse, May 24, 1998.

161 "On Narita, Taking a Kinder, Gentler Approach," *Daily Yomiuri,* July 23, 1998.

162 Tom Ballantyne, "Under Siege!" *Orient Aviation,* Nov. 1998, p. 58.

163 Airport Authority Hong Kong, "Hong Kong International Airport," 1998.

164 "The Project," www.osl.no/e/develop/project.htm, May 1998.

165 Michael Mecham, "Seoul Turns to Inchon to Solve Airport Woes," *Aviation Week and Space Technology,* Oct. 14, 1996, p. 145.

166 "Inchon City Cleans Up to Achieve Model Goal," CNN, June 2, 1996.

167 International Civil Aviation Organization, *Airport Planning Manual,* 2d ed., 1987, p. I-44.

168 Andrew Goetz, Joseph Szyliowicz, and Paul Dempsey, "Transportation and International Sustainable Development: A Preliminary Conceptualization and Application," *OECD Proceedings,* Toward Sustainable Development, 1996, p. 118.

169 Quoted in David Gergen, "Surplus Agonistes," *U.S. News and World Report,* Jan. 19, 1998, p. 72. Bonhoeffer was hanged by the Nazis in 1945.

Chapter 7

1 Robert P. Olislagers, *Fields of Flying in the Southwest,* Heritage Media Corp., 1996.

2 Airport Authority Hong Kong, "Hong Kong International Airport," 1998.

3 International Civil Aviation Organization, *Airport Planning Manual,* 2d ed., 1987, p. I-139.

4 Ibid., p. I-73.

5 Amsterdam Airport Schiphol, "Balancing Environment and Economics," 1997.

6 Paul Stephen Dempsey and Kevin O'Connor, "Air Traffic Congestion and Infrastructure Development in the Pacific Asia Region," *Asia Pacific Air Transport: Challenges and Policy Reforms,* Institute of Southeast Asian Studies, 1997, pp. 23, 29.

7 Aeroports de Paris, "Orly, Mastering the Future," 1997.

8 John Hamlin, *The Stansted Experience,* 1997, p. 7.

9 Wolfgang Voigt, "From the Hippodrome to the Aerodrome, from the Air Station to the Terminal," in *Building for Air Travel,* John Zukowsky, ed., 1996, pp. 31, 32.

10 Nasid Hajari, "A Walk in the Clouds," *Time,* June 22, 1998.

11 Voigt, op. cit., p. 34.

12 John Zukowsky, "Introduction," in *Building for Air Travel,* John Zukowsky, ed., 1996, p. 14.

13 Ibid., p. 44.

14 Koos Bosma, "European Airports, 1945–1995," in *Building for Air Travel,* John Zukowsky, ed., 1996, p. 54.

15 International Civil Aviation Organization, *Airport Planning Manual,* 2d ed., 1987, P. I-83.

16 David Brodherson, "An Airport in Every City," in *Building for Air Travel,* John Zukowsky, ed., 1996, p. 81.

17 Bosma, op. cit., pp. 55, 57.

18 Brodherson, op. cit., pp. 86, 87.

19 Ibid., p. 81.

20 Mark Bouman, "Cities of the Plane," in *Building for Air Travel,* John Zukowsky, ed., 1996, p. 186.

21 Brodherson, op. cit., p. 93.

22 Robert Horonjeff and Francis McKelvey, *Planning and Design of Airports,* 4th ed., McGraw-Hill, 1994, p. 249.

23 Wood Lockhart, "A Pilot's Perspective on Airport Design," in *Building for Air Travel,* John Zukowsky, ed., 1996, pp. 215, 216, 222, 224.

24 Horonjeff and McKelvey, op. cit., p. 89.

25 International Civil Aviation Organization, *Airport Planning Manual,* 2d ed., 1987, P. I-47.

26 Ibid., p. I-46.

27 Horonjeff and McKelvey, op. cit., pp. 249, 256.

28 See James Spensley, "Airport Planning," in *Airport Regulation, Law and Public Policy,* R. Hardaway, ed., 1991, p. 73.

29 Ibid., p. 72.

30 Brent Hannon, "Gateways to Growth," *Asia, Inc.,* Nov. 1995, p. 62.

31 International Civil Aviation Organization, *Airport Planning Manual,* 2d ed., 1987, pp. I-47, I-48.

32 Ibid., pp. I-49, I-50.

Chapter 7 (*Cont.*)

33 Horonjeff and McKelvey, op. cit., p. 249.

34 International Civil Aviation Organization, *Airport Planning Manual,* 2d ed., 1987, p. I-59; see Laurence Gesell, *The Administration of Public Airports,* 3d ed., 1991, pp. 65–84.

35 Federal Aviation Administration, "Airport Master Plans," 1985, p. 34.

36 Wood Lockhart, address before the National Aerospace Conference, Dayton, Ohio, Oct. 2, 1998.

37 International Civil Aviation Organization, *Airport Planning Manual,* 2d ed., 1987, pp. I-51, I-53; see also Horonjeff and McKelvey, op. cit.

38 Horonjeff and McKelvey, op. cit., p. 124.

39 International Civil Aviation Organization, *Airport Planning Manual,* 2d ed., 1987, p. I-35.

40 Horonjeff and McKelvey, op. cit., p. 259.

41 Ibid., pp. 265, 266.

42 International Civil Aviation Organization, *Airport Planning Manual,* 2d ed., 1987, p. I-39.

43 Horonjeff and McKelvey, op. cit., p. 109.

44 "Birds Force Israel to Close Airport," *The Denver Post,* Jan. 17, 1998, p. 9A.

45 Horonjeff and McKelvey, op. cit., pp. 250–255.

46 Wood Lockhart, "A Pilot's Perspective on Airport Design," in *Building for Air Travel,* John Zukowsky, ed., 1996, p. 221.

47 Horonjeff and McKelvey, op. cit., p. 250.

48 International Civil Aviation Organization, *Airport Planning Manual,* 2d ed., 1987, p. I-59.

49 Ibid., p. I-59.

50 Ibid., p. I-64.

51 Ibid., p. I-61.

52 Ibid., pp. I-66, I-67.

53 Brussels Airport Authority, "Runway to the Future," 1997.

54 Interview, *Jane's Airport Review,* Dec. 1997, p. 64.

55 U.S. General Accounting Office, "Airfield Pavement: Keeping Nation's Runways in Good Condition Could Require Substantially Higher Spending," July 1998, p. 3.

56 Osaka Regional Civil Aviation Bureau, "Kansai International Airport Office," 1998, pp. 18, 19.

57 Aeroports de Paris, "Charles de Gaulle Airport: Europe's Foremost Transport Hub," 1997, p. 20.

58 Ken Leiser, "Lambert Unveils High-Speed Radar System," *The St. Louis Post-Dispatch,* Nov. 20, 1998, p. C1.

59 Munich Airport, "Flight Operation," 1996, p. 9.

60 International Civil Aviation Organization, *Airport Planning Manual,* 2d ed., 1987, p. I-69.

61 Ibid., p. I-69; Osaka Regional Civil Aviation Bureau, "Kansai International Airport Office," 1998, pp. 22, 23.

62 International Civil Aviation Organization, *Airport Planning Manual,* 2d ed., 1987, pp. I-71, I-72.

63 Federal Aviation Administration, "Airport Master Plans," 1985, p. 33.

64 International Civil Aviation Organization, *Airport Planning Manual,* 2d ed., 1987, pp. I-131, I-132.

65 Ibid., pp. I-129, I-130.

66 Federal Aviation Administration, "Airport Master Plans," 1985, p. 33.

67 Frankfurt Main, News Release, Nov. 11, 1996.

68 Source: Airport Council International, 1996.

Chapter 7 (*Cont.*)

69 www.hkairport.com/news/news9711.htm, Nov. 1997.

70 Moon-Sik Park, "Korea: Airport Development Market," Financial Times Asia Intelligence Wire, May 9, 1997.

71 "KIX Highlights Construction Project Bidding Ills," *Mainichi Daily News,* June 24, 1998.

72 Bouman, op. cit., p. 188.

73 Keith Mordoff, "Air Transport Munich's New International Airport Expected to Begin Operations in 1991," *Aviation Week and Space Technology,* Feb. 22, 1988, p. 92.

74 Michael Mecham, "Germany Struggles to Meet Airport Needs," *Aviation Week and Space Technology,* Aug. 26, 1991, p. 38.

75 Leonard Hill, "Countdown on Munich II," *Air Transport World,* June 1, 1991, p. 70.

76 Mecham, op. cit., p. 41.

77 Leonard Hill, "Beyond Expectations," *Air Transport World,* June 1, 1993, p. 182.

78 Mecham, op. cit., p. 38.

79 "Europeans Upgrade ATC System," *Aviation Week and Space Technology,* May 4, 1992, p. 63.

80 Flughafen München, "Munich International Airport," 1996.

81 Peter Trautman, "High Tech" in *Total Management: A Case Study of a World Class Airport—The New Munich Airport* (unpublished address before the IBC Conference on Asia-Pacific Airports '94, Singapore, July 25–27, 1994).

82 Hill, "Beyond Expectations," op. cit.

83 Mathew Brown, "Osaka's New Airport Is a Site to Behold," *Minneapolis Star-Tribune,* June 7, 1998.

84 Zenjiro Ogawa, "Kansai International Airport Projects," International Conference on Aviation and Airport Infrastructure, Denver, Dec. 8, 1993.

85 Kansai International Airport Co., "Kansai International Airport," 1998, p. 13.

86 Brown, op. cit.

87 "Kansai Int'l Airport Inaugurated," *Kyoto News International,* Sept. 5, 1994.

88 "How to Avoid That Sinking Feeling," *The Economist,* Feb. 4, 1995, p. 73.

89 Eric Lassiter, "Japan to Open Much-Delayed Kansai Airport," *Travel Weekly,* Aug. 29, 1994, p. 4.

90 "Japan Opens Airport on Man-Made Island," *The San Diego Union-Tribune,* Sept. 5, 1994, p. A18.

91 Brown, op. cit.

92 Kansai International Airport Co., "Passenger Terminal Building," 1998, p. 1.

93 Ibid., p. 13.

94 Robert Moorman, "Osaka to Me," *Air Transport World,* Oct. 1, 1994, p. 62; "How to Avoid That Sinking Feeling," *The Economist,* Feb. 4, 1995, p. 73.

95 Kansai International Airport Co., "Passenger Terminal Building." 1998, p. 3.

96 Alexandra Black, "Trade and Tourist Boom from New Airport," Inter Press Service Network, Sept. 9, 1994.

97 "Kansai Int'l Airport Inaugurated," *Kyodo News International,* Sept. 5, 1994.

98 Brown, op. cit.

99 "KIX Highlights Construction Project Bidding Ills," *Mainichi Daily News,* June 24, 1998.

100 J. A. Donoghue, "The Pearl-Y Gateways," *Air Transport World,* Feb. 1, 1995, p. 75.

101 "Far East Facility Slates December Opening," *Travel Weekly,* July 3, 1995, p. 29.

102 "U.S. Funds Feasibility Study for Second Stage of Macau Airport," *Aviation Week and Space Technology,* June 6, 1994, p. 36.

Chapter 7 (*Cont.*)

103 Donoghue, op. cit.

104 "Airport to Be Macao's Gate to Region," *The Asian Wall Street Journal,* Oct. 31, 1994, p. 3.

105 Donoghue, op. cit.

106 Ibid.

107 "Far East Facility Slates December Opening," *Travel Weekly,* July 3, 1995, p. 29.

108 "Macau Finance for Airport Terminal," *Privatization International,* April 1, 1994.

109 "New Macau Airport Set to Open July 1995," *Phillips Business Information,* Jan. 28, 1994.

110 "Macau Finance for Airport Terminal," *Privatization International,* April 1, 1994.

111 Hajari, op. cit.

112 S. Jayasankaran, "Thanks Anyway," *Far Eastern Economic Review,* Feb. 8, 1995, p. 61.

113 "Hong Kong Builds $20.3 Billion Airport," *The Orlando Sentinel,* April 9, 1995, p. A20.

114 Steven Mufson, "Accord Boosts Hong Kong Airport," *The Washington Post,* Nov. 5, 1994, p. A19.

115 "Growth Outpaces Asian Airports," *Aviation Week and Space Technology,* Aug. 29, 1994, p. 58.

116 Stuart Becker, "Airports: Chinese Opera," *Far Eastern Economic Review,* April 6, 1995, p. 54.

117 Thomas Darmody, "The Design and Development of World Class Airports," IBC International Conference on Airport Development and Expansion, Hong Kong, Oct. 28, 1993.

118 "Hong Kong Completes Site Preparation for Chek Lap Kok Airport," *Airports,* June 20, 1995, p. 245.

119 Hajari, op. cit.

120 Peter Negline, "Hong Kong's New International Airport," Salomon Smith Barney, March 9, 1998, p. 4.

121 "How to Avoid That Sinking Feeling," *The Economist,* Feb. 4, 1995, p. 73.

122 Michael Mecham, "Seoul Turns to Inchon to Solve Airport Woes," *Aviation Week and Space Technology,* Oct. 14, 1996, p. 50.

123 "Building of Main Facilities at Korean Airport to Start," *Asia Pulse,* Feb. 20, 1997.

124 "Inchon Poised to Become Hub to Southeast Asia," *World Airport Week,* June 16, 1998.

125 Mecham, "Seoul Turns to Inchon to Solve Airport Woes," op. cit., p. 50.

126 Ibid.

127 Interview with Curt Fentress, May 30, 1997, in Denver.

Chapter 8

1 Aeroports de Paris, "Orly, Mastering the Future," 1997, p. 8.

2 Nasid Hajari, "A Walk in the Clouds," *Time,* June 22, 1998.

3 M. Gordon Brown, "A Flying Leap into the Future," *Metropolis,* July/August 1995, p. 78.

4 John Zukowsky, "Introduction," in *Building for Air Travel,* John Zukowsky, ed., 1996, p. 13.

5 Wolfgang Voigt, "From the Hippodrome to the Aerodrome, from the Air Station to the Terminal," in *Building for Air Travel,* John Zukowsky, ed., 1996, pp. 27, 36.

6 International Civil Aviation Organization, *Airport Planning Manual,* 2d ed., 1987, p. I-74.

Chapter 8 (*Cont.*)

7 See Laurence Gesell, *The Administration of Public Airports,* 3d ed., Coast Aire, 1991, pp. 106–116.

8 International Civil Aviation Organization, *Airport Planning Manual,* 2d ed., 1987, p. I-74

9 Ibid., p. I-75.

10 See Federal Aviation Administration, "Airport Master Plans," 1985, p. 32.

11 Prianka Seneviratne and Nathalie Martel, "Criteria for Evaluating Quality of Service in Air Terminals," *Transportation Research Record,* Dec. 1994, p. 24.

12 James Spensley, "Airport Planning," in *Airport Regulation, Law and Public Policy,* R. Hardaway, ed., 1991, p. 73.

13 George Raine, "What's Up at SFO?" *San Francisco Examiner,* Feb. 17, 1998, p. A-1.

14 International Civil Aviation Organization, *Airport Planning Manual,* 2d ed., 1987, p. I-75.

15 Ibid., pp. I-80 to I-82; see also Paul Stephen Dempsey and Laurence Gesell, *Airline Management: Strategies for the 21st Century,* Coast Aire, 1997, pp. 167–278.

16 Koos Bosma, "European Airports, 1945–1995," in *Building for Air Travel,* John Zukowsky, ed., 1996, p. 51.

17 Ibid., p. 52.

18 International Civil Aviation Organization, *Airport Planning Manual,* 2d ed., 1987; Federal Aviation Administration, "Airport Master Plans," 1985, pp. 32, 33.

19 Seneviratne and Martel, op. cit., p. 25.

20 International Civil Aviation Organization, *Airport Planning Manual,* 2d ed., 1987, p. I-83.

21 Carl Robart, "Surface Access to International Airports," *Transportation Research Circular,* May 1995, pp. 51, 52.

22 International Civil Aviation Organization, *Airport Planning Manual,* 2d ed., 1987, p. I-116.

23 Ibid., p. I-92.

24 Ibid., p. I-91.

25 Ibid., pp. I-91, I-92.

26 Ibid., pp. I-94 to I-98.

27 Frankfurt Main Flughafen, news release, Nov. 11, 1996.

28 Frankfurt Main Flughafen, "Terminal 2: FRA Lifts off for the Future," 1997, p. 8.

29 Patrick O'Driscoll, "Bag System No. 1 Priority for Germans," *The Denver Post,* July 31, 1994, p. 25A.

30 Patrick O'Driscoll, "Munich's Airport: A Case Study for DIA," *The Denver Post,* July 31, 1994, p. A1.

31 Interview with Peter Trautman of Flughafen München in Munich, May 27, 1996.

32 Paul Stephen Dempsey, Andrew R. Goetz, and Joseph S. Szyliowicz, *Denver International Airport: Lessons Learned,* McGraw-Hill, 1997, pp. 308–314, 434–437.

33 Adele Schwartz, "Denver Upgrades," *Airport Equipment and Technology,* Winter 1998, p. 11.

34 Kevin Flynn, "United Strokes Baggage System," *Rocky Mountain News,* Dec. 22, 1998, p. 4A.

35 Brent Hannon, "Gateways to Growth," *Asia, Inc.,* Nov. 1995, p. 62.

36 Interview with Bryan O'Connor of Murphy/Jahn in Chicago, July 27, 1997.

37 Kansai International Airport, "Baggage Handling System," 1998.

38 Eric Lassiter, "Japan to Open Much-Delayed Kansai Airport," *Travel Weekly,* Aug. 29, 1994, p. 4.

39 David Parish, "British Airways Trials at Heathrow of RFID Baggage Tags," 2 *International Airport Review* 29 (1998).

Chapter 8 (*Cont.*)

40 Geraldine Poor and Burr Stewart, "Beyond the Airport Terminal: People-Mover Technologies at Seattle-Tacoma International Airport," *Transportation Research Record*, Oct. 1993, p. 47.

41 Washington Airports Authority, "The Dulles Story," 1997.

42 Judy Jacobs, "Korea Plans Gateway Airport," *Business Travel News*, April 22, 1996, p. 24.

43 Dempsey, Goetz, and Szyliowicz, op. cit.

44 American Association of Airport Executives, "Guidelines for Airport Signing and Graphics," 1996.

45 Airport Authority Hong Kong, "News," Jan. 1998.

46 Joanne Lee-Young, "Turbulence: No Way to Escape It," *The Asian Wall Street Journal*, July 6, 1998.

47 American Association of Airport Executives, "Guidelines for Airport Signing and Graphics," 1996.

48 Bosma, op. cit., p. 57.

49 See Paul Dempsey, *Law and Foreign Policy in International Aviation*, 1987, pp. 349–382.

50 49 U.S.C. §44931 *et seq.*

51 49 U.S.C. §§44931–32.

52 49 U.S.C. §44933.

53 49 U.S.C. §§44901–15.

54 49 U.S.C. §46502.

55 49 U.S.C. §46504. See *U.S. v. Jenny*, 7 F.3d 953 (10th Cir. 1993) (intoxicated passenger physically and sexually assaulted flight attendant and passengers).

56 18 U.S.C. §2244(b), 49 U.S.C.App. §1472(k)(1).

57 49 U.S.C. §46505.

58 49 U.S.C. §46507.

59 U.S. General Accounting Office, "Aviation Security: Implementation of Recommendations Is under Way, but Completion Will Take Several Years," April 1998, p. 17.

60 Eric Neiderman, "People Are the Key to Effective 'Higher Tech' Security," 2 *International Airport Review* 52 (1998).

61 Matthew Wals, "FAA's New Bomb Detectors Idle," *The Denver Post*, Sept. 9, 1998, p. 11A.

62 Lynne Haraway, "Technology Improving Airport Economics," *ACI World Economic Specialty Conference Proceedings,* Airport Economics in a Technological Age, Denver, April 6–9, 1997, p. 30.

63 Interview with Colombo Bandaranaike International Airport Manager I. M. Razeek, Sept. 8, 1998.

64 International Civil Aviation Organization, *Airport Planning Manual*, 2d ed., 1987, p. I-98.

65 Ibid., p. I-102.

66 Hajari, op. cit.

67 Rowan Moore, "Serene Symbol of a Volatile City," *The Evening Standard,* London, June 23, 1998.

68 Hajari, op. cit.

69 International Civil Aviation Organization, *Airport Planning Manual*, 2d ed., 1987, p. I-99.

70 Dempsey and Gesell, op. cit., p. 267.

71 International Civil Aviation Organization, *Airport Planning Manual*, 2d ed., 1987, p. I-104.

72 Ibid., p. I-105.

73 Ibid., p. I-112.

Chapter 8 (*Cont.*)

74 29 U.S.C. §794(a).

75 49 CFR §27.5.

76 49 CFR §27.71(a)(2)(i).

77 Paul Stephen Dempsey, "The Civil Rights of the Handicapped in Transportation: The Americans with Disabilities Act and Related Legislation," 19 *Transportation Law Journal* 309, 320–21 (1991).

78 49 U.S.C. §1374(c)(1).

79 These rules are discussed in detail in Paul Stephen Dempsey, Robert Hardaway, and William Thoms, *Aviation Law and Regulations,* Butterworth, 1993, Vol. 1, §4.29. For a more succinct summary see Dempsey and Gesell, op. cit., pp. 269, 270.

80 Stefano Pavarini, "Trentaquattro Cime," *L'Arca,* Nov. 1994, p. 20.

81 Bosma, op. cit., p. 58.

82 See David Brodherson, "An Airport in Every City," in *Building for Air Travel,* John Zukowsky, ed., 1996, p. 88.

83 David Brodherson, "Plates," in *Building for Air Travel,* John Zukowsky, ed., 1996, p. 137.

84 Washington Airports Authority, "The Dulles Story," 1997.

85 Aeroports de Paris, "Charles de Gaulle Airport: Europe's Foremost Transport Hub," 1997.

86 Flughafen Frankfurt Main, "Terminal 2: FRA Lifts off for the Future," 1997, p. 4.

87 Flughafen Frankfurt Main, news release Nov. 11, 1996.

88 Leonard Rau, "Deregulation and Design," in *Building for Air Travel,* John Zukowsky, ed., 1996, p. 227.

89 Letter from Charles Ansbacher (founding director of the DIA art program) to Paul Stephen Dempsey, Aug. 29, 1997.

90 Dempsey, Goetz, and Szyliowicz, op. cit., pp. 393–395.

91 Bosma, op. cit., p. 61.

92 See Jonathan Bousfield and Rob Humphreys, *Austria,* 1998, p. 69.

93 "Germany Struggles to Meet Airport Needs," *Aviation Week and Space Technology,* Aug. 26, 1991, p. 41.

94 Patrick O'Driscoll, "Munich's Airport: A Case Study for DIA," *The Denver Post,* July 31, 1994, p. A1.

95 Leonard Hill, "Beyond Expectations," *Air Transport World,* June 1, 1993, p. 182.

96 Michael Mecham, "Munich Turns Out Lights on Riem," *Aviation Week and Space Technology,* May 25, 1992, p. 20.

97 O'Driscoll, "Munich's Facility Reflects Future of DIA," op. cit., p. 24A.

98 Felicity Munn, "Airports of the Future Are Here and They're User Friendly," *The Montreal Gazette,* May 8, 1994, p. H4.

99 Mecham, op. cit.

100 Keith Mordoff, "Air Transport Munich's New International Airport Expected to Begin Operations in 1991," *Aviation Week and Space Technology,* Feb. 22, 1988, p. 92.

101 "Munich Airport Expansion Plan Includes Second Terminal by 2005," *Aviation Europe,* July 27, 1995, p. 5.

102 John Pierson, "The Europeans Invade U.S. Airport Lounges," Associated Press, Dec. 9, 1994.

103 Patrick O'Driscoll, "Munich's Overnight Move Had Few Hitches," *The Denver Post,* July 31, 1994, p. 25A.

Chapter 8 (*Cont.*)

104 Flughafen München, "Airport and Art," 1996.

105 Keith Sonnier, *Lightway*, 1993.

106 Willi Hermsen, "Analyzing Munich's Move," (address 1996).

107 David Brown, "Japanese Building International Offshore Airport to Serve Osaka," *Aviation Week and Space Technology*, July 13, 1987, p. 38.

108 Mark Branch, "Piano Wins Osaka Competition," *Progressive Architecture*, March 1, 1989, p. 33.

109 Renzo Piano, *The Renzo Piano Logbook*, 1997, p. 150.

110 "Japan Opens Airport on Man-Made Island," *San Diego Union-Tribune*, Sept. 5, 1994, p. A18.

111 "A Brash Builder," *Time Australia*, May 11, 1998, p. 60.

112 "Japan Opens Airport on Man-Made Island," *San Diego Union-Tribune*, Sept. 5, 1994, p. A18.

113 Robert Moorman, "Osaka to Me," *Air Transport World*, Oct. 1, 1994, p. 62.

114 Branch, op. cit.

115 M. Gordon Brown, op. cit., p. 77.

116 "Kansai International," *Aviation Week and Space Technology*, Sept. 12, 1994, p. 19.

117 Richard Rush, "Buy Now, Fly Later," *Progressive Architecture*, April 1, 1995, p. 70.

118 Moorman, op. cit.

119 Zenjiro Ogawa, "Kansai International Airport Projects," International Conference on Aviation and Airport Infrastructure, Denver, Dec. 8, 1993.

120 "Growth Outpaces Asian Airports," *Aviation Week and Space Technology*, Aug. 29, 1994, p. 57.

121 M. Gordon Brown, op. cit., p. 77.

122 Pavarini, op. cit.

123 Letter from Stephen Pacetti to the editor, *The Denver Post*, Feb. 25, 1996, p. D4.

124 Dempsey, Goetz, and Szyliowicz, op. cit., pp. 359–373.

125 Dempsey, Goetz, and Szyliowicz, op. cit., pp. 374–389.

126 "U.S. Funds Feasibility Study for Second Stage of Macau Airport," *Aviation Week and Space Technology*, June 6, 1994, p. 36.

127 J. A. Donoghue, "The Pearl-Y Gateways," *Air Transport World*, Feb. 1, 1995, p. 75.

128 "Far East Facility Slates December Opening," *Travel Weekly*, July 3, 1995, p. 29.

129 "Macau's Presence in Pearl River Delta Will Be As Reliever Airport," *World Airport Week*, July 18, 1995; "Far East Facility Slates December Opening," *Travel Weekly*, July 3, 1995, p. 29.

130 Leonard Hill, "Asia's Newest 'Dragon,'" *Air Transport World*, Sept. 1, 1993, p. 66.

131 Interview with Ambrin Buang of Kuala Lumpur International Airport Berhad in Kuala Lumpur, Malaysia, June 23, 1997.

132 Hajari, op. cit.

133 Interview with Hank Cheriex in Kuala Lumpur, Malaysia, June 23, 1997.

134 Hannon, op. cit.

135 "Malaysia Gives Pacts to Group for Design, Building of Airport," *The Wall Street Journal*, March 27, 1995, p. A11.

136 Hajari, op. cit.

137 "KLIAS to Install Glass-Wall Passenger Loading Bridges," *World Airport Week*, July 18, 1995.

138 John Mok, "The Development of Hong Kong's New International Airport," International Conference on Aviation and Airport Infrastructure, Denver, Dec. 8, 1993.

Chapter 8 (*Cont.*)

139 Moore, op. cit.

140 Hajari, op. cit.

141 Barry Grindrod, "No Dragons, But Many Demons," *Orient Aviation,* Sept. 1998, p. 55.

142 "Growth Outpaces Asian Airports," *Aviation Week and Space Technology,* Aug. 29, 1994, p. 58.

143 Oslo Airport, "Architecture and Design," 1997.

144 "7,200 Meters of Shops and Restaurants," *OSL Monitor,* Jan. 1, 1998, p. 2.

145 Oslo Airport, "Europe's Safest and Most Efficient Airport," 1997, p. 7.

146 "Concept and Sense of Values Behind the Airport's Design," www.osl.no/e/company/visual.htm. Elsewhere, the airport authority wrote: "Individual buildings are designed primarily for functionality and efficient operation. They communicate architectonic values of Norwegian building traditions within a framework of modern expression. Buildings are designed of materials that highlight the constructive make-up of each building. Surface treatments reveal the natural features and characteristics of the materials." Oslo Airport, "Architecture and Design," 1997.

147 "NSIA Privatization Opportunities," *World Airport Week,* June 20, 1995.

148 Jong-Heui Shin, "Airport Developments in Korea," International Conference on Aviation and Airport Infrastructure, Denver, Dec. 8, 1993.

149 Curtis Fentress, "Revitalizing the Excitement of Travel," *Passenger Terminal '95,* pp. 29–31.

150 "Politicians, Squatters Slow New Bangkok International Airport," *Airports,* March 7, 1995, p. 93.

151 Interview with Bryan O'Connor of Murphy/Jahn Inc. in Chicago, July 21, 1997.

152 Interview with George Emery of TAMS Consultants, Inc., in Bangkok, June 11, 1997.

153 Interview with Bryan O'Connor of Murphy/Jahn Inc. in Chicago, July 21, 1997.

154 "NSIA Considers Opening with Extra Runway," *World Airport Week,* July 1, 1995.

155 "Thailand Plans New Bangkok Airport," *Interavia Air Letter,* March 17, 1993, p. 3.

156 Interview with George Emery of TAMS Consultants, Inc., in Bangkok, June 16, 1997.

157 Haraway, op. cit.

158 Hajari, op. cit.

159 Wood Lockhart, "A Pilot's Perspective on Airport Design," in *Building for Air Travel,* John Zukowsky, ed., 1996, p. 213.

160 David Brodherson, "An Airport in Every City," in *Building for Air Travel,* John Zukowsky, ed., 1996, p. 84.

161 Robert Breugmann, "Plates," in *Building for Air Travel,* John Zukowsky, ed., 1996, p. 136.

162 Lockhart, op. cit.

163 Flughafen München, "A History of the New Munich Airport," press release, Nov. 5, 1992.

164 Interviews with Willi Hermsen, Alexander Hoffman, Hans Klos, and Peter Trautman of Flughafen München at Munich, May 26–28, 1996.

165 Dempsey, Goetz, and Szyliowicz, op cit., pp. 421–441.

166 Haraway, op. cit., p. 36.

167 Dempsey, Goetz, and Szyliowicz, op. cit., pp. 441–463.

168 Interview with John Mok of Airport Authority Hong Kong in Hong Kong, Sept. 15, 1998.

169 Schwartz, op. cit.

170 Ibid.

171 Ricky Young, "DIA Trains Snafus Common," *The Denver Post,* Aug. 27, 1998, p. B1.

172 Lee-Young, op. cit.

173 "Harris Wins Contract for Kuala Lumpur Airport," *Aviation Daily,* June 28, 1995, p. 514.

Chapter 8 (*Cont.*)

174 Lee-Young, op. cit.

175 Badrul Hisham Mahzan et al., "Dr. M. Briefed on Airport's Problems," *The New Straits Times,* July 1, 1998.

176 Leonard Hill, "Turbulent Takeoffs," *Air Transport World,* Sept. 1, 1998, p. 75.

177 "Road and Rat Problems for Sepang," *Orient Aviation,* Sept. 1998, p. 55.

178 Frances Fiorino, "Airport Galaxy," *Aviation Week and Space Technology,* June 12, 1995, p. 33.

179 Simon Winchester, "Chinese Lose Face at Airport Far from Ready for Take-Off," *The Sunday Telegraph,* London, July 12, 1998, p. 36.

180 "Airport Heads on the Block," *Orient Aviation,* Nov. 1998, p. 55.

181 "Hong Kong Knew of Airport Problems," *The Asian Wall Street Journal,* Sept. 22, 1998.

182 Leonard Hill, "Turbulent Takeoffs," *Air Transport World,* Sept. 1, 1998, p. 75.

183 "CLK Cargo Chaos," *Orient Aviation,* Sept. 1998, p. 56.

184 "Airport Heads on the Block," *Orient Aviation,* Nov. 1998, p. 55.

185 John Ridding, "Airport Crisis Grounds Chinese Ambitions," *International Herald Tribune,* July 11, 12, 1998.

186 "Hong Kong Cargo Handler Blames Rushed Opening for Systems Breakdown," Agence France-Presse, Sept. 8, 1998.

187 Hill, "Turbulent Takeoffs," op. cit., p. 75.

188 The FIDS and BIDS systems use television display monitors and giant railroad-style arrival and departure boards. Patrick O'Driscoll, "Munich's Airport, a Case Study for DIA," *The Denver Post,* July 31, 1994, p. A1.

189 Interview with DIA Aviation Director Bruce Baumgartner at Denver, July 16, 1998.

190 Hannon, op. cit.

Chapter 9

1 Interview with the author, June 4, 1998.

2 Address before the University of Denver North American Intermodal Transportation Summit, Oct. 16, 1998.

3 U.S. Department of Transportation, "Intermodal Ground Access to Airports: A Planning Guide," Dec. 1996, pp. 2, 15.

4 Jennifer Lee, "It's a Plane!" *The Wall Street Journal,* July 18, 1997, pp. B1, B6.

5 Paul Stephen Dempsey and Kevin O'Connor, "Air Traffic Congestion and Infrastructure Development in the Pacific Asia Region," in *Asia Pacific Air Transport: Challenges and Policy Reforms,* Institute of Southeast Asia Studies, 1997, pp. 23, 24.

6 Interview with Greyhound CEO Craig Lentzsch, June 4, 1998, in Denver.

7 U.S. Department of Transportation, "Intermodal Ground Access to Airports: A Planning Guide," Dec. 1996, p. 167.

8 International Civil Aviation Organization, *Airport Planning Manual,* 1987, pp. I-86, I-87.

9 Ibid.

10 U.S. Department of Transportation, "Intermodal Ground Access to Airports: A Planning Guide," Dec. 1996, pp. 162–166.

11 Dean Starkman, "Up, up and away on Newark's Monorail," *The Wall Street Journal,* Sept. 30, 1998, p. B10.

12 Vienna International Airport, "Facts and Figures," 1996, p. 5.

13 "Pen-Pushers' City Gets New Signature," *The Financial Times,* Sept. 7, 1998, p. 16.

Chapter 9 (*Cont.*)

14 Mark Bouman, "Cities of the Plane," in *Building for Air Travel,* John Zukowsky, ed., 1996, pp. 184, 185.

15 "A Better Way to Fly," *The Economist,* Feb. 21, 1998, p. 21.

16 Betsy Wade, "Intermodal Travel: When the Plane Really Is Just a Bus," *The Denver Post,* Dec. 14, 1997, p. 4T.

17 Comments of Greyhound Lines, Inc., in DOT Docket OST-97-2881, Dec. 10, 1997.

18 Lee, op. cit., p. B1.

19 Craig Lentzsch, "Introduction to Panel II," 25 *Transportation Law Journal* 290, 292 (1998).

20 Bouman, op. cit., p. 184.

21 U.S. Department of Transportation, "Intermodal Surface Transportation Efficiency Act: Flexible Funding Opportunities for Transportation Investments," 1991, p. 13.

22 Sunil Harman, "Intermodal Transportation Issues: Airports," 25 *Transportation Law Journal* 307, 309, 311 (1998).

23 Flughafen Frankfurt news release, Nov. 11, 1996.

24 Interview, *Jane's Airport Review,* Dec. 1997, p. 64.

25 Ibid.

26 Koos Bosma, "European Airports, 1945–1995," in *Building for Air Travel,* John Zukowsky, ed., 1996, p. 63.

27 Aeroports de Paris, "Charles de Gaulle Airport: Europe's Foremost Transport Hub," 1997, pp. 14, 15.

28 Ibid., pp. 15, 16.

29 Copenhagen Airport, "Station and Air Rail Terminal at Copenhagen Airport," 1997.

30 Willi Hermsen, "Analyzing Munich's Move," address, 1996.

31 Kim Upton, "New Munich Airport Stresses Ease of Travel," *Los Angeles Times,* May 17, 1992, p. 4.

32 Michael Mecham, "Munich Turns out Lights on Riem," *Aviation Week and Space Technology,* May 25, 1992, p. 20.

33 Keith Mordoff, "Air Transport Munich's New International Airport Expected to Begin Operations in 1991," *Aviation Week and Space Technology,* Feb. 22, 1988, p. 92.

34 Interview with Alexander Hoffman and Peter Trautman of Flughafen München in Munich, May 28, 1996.

35 "Kansai Int'l Airport Inaugurated," *Kyodo News International,* Sept. 5, 1994.

36 Zenjiro Ogawa, "Kansai International Airport Projects," International Conference on Aviation and Airport Infrastructure, Denver, Dec. 8, 1993.

37 "Growth Outpaces Asian Airports," *Aviation Week and Space Technology,* Aug. 29, 1994, p. 57.

38 "Far East Facility Slates December Opening," *Travel Weekly,* July 3, 1995, p. 29.

39 "Airport to Be Macao's Gate to Region," *The Asian Wall Street Journal,* Oct. 31, 1994, p. 3.

40 J. A. Donoghue, "The Pearl-Y Gateways," *Air Transport World,* Feb. 1, 1995, p. 75.

41 Ed Paisley, "On a Wing and a Prayer," *Far Eastern Economic Review,* May 5, 1995, p. 76.

42 "Macau's Presence in Pearl River Delta Will Be as Reliever Airport," *World Airport Week,* July 18, 1995.

43 Ibid.

44 Mark Landler, "Asian Crisis Takes Shine off Glittering Airports' Unveilings," *Seattle Post-Intelligencer,* July 6, 1998.

45 Nasid Hajari, "A Walk in the Clouds," *Time,* June 22, 1998.

Chapter 9 (*Cont.*)

46 "Rat and Road Problems for Sepang," *Orient Aviation,* Sept. 1998, p. 55.

47 John Mok, "The Development of Hong Kong's New International Airport," International Conference on Aviation and Airport Infrastructure, Denver, Dec. 8, 1993.

48 Lynne Haraway, "Technology Improving Airport Economics," in *ACI World Economic Specialty Conference Proceedings,* Airport Economics in a Technological Age, Denver, April 6–9, 1997, p. 52.

49 "Growth Outpaces Asian Airports," *Aviation Week and Space Technology,* Aug. 29, 1994, p. 58.

50 "Hong Kong Builds $20.3 Billion Airport," *The Orlando Sentinel,* April 9, 1995, p. A20.

51 Steven Mufson, "Accord Boosts Hong Kong Airport," *The Washington Post,* Nov. 5, 1994, p. A19.

52 Thomas Darmody, "The Design and Development of World Class Airports," IBC International Conference on Airport Development and Expansion, Hong Kong, Oct. 28, 1993.

53 Mok, op. cit.

54 Airport Authority Hong Kong, "News," Jan. 1998.

55 "Gardermobanen Running on Time," *Airport Business Communiqué,* 1997–1998, p. 10.

56 Oslo Airport, "Europe's Safest and Most Efficient Airport," 1997, p. 9.

57 "Getting There Is Another Matter," *Asia Now,* March 1997, p. 3.

58 "Air Traffic Growth Forces Expansion of Seoul Airport Construction," *World Airport Week,* May 30, 1995.

59 Jong-Heui Shin, "Airport Developments in Korea," International Conference on Aviation and Airport Infrastructure, Denver, Dec. 8, 1993.

60 "Air Traffic Growth Forces Expansion of Seoul Airport Construction," *World Airport Week,* May 30, 1995; "Railroad to Link Airport to Seoul-Pusan Rail Line," *Asia Pulse,* Feb. 27, 1997.

61 "Major Korean Builders to Form Consortium for Rail Work," *Asia Pulse,* March 14, 1997; "Inchon Poised to Become Hub to Southeast Asia," *World Airport Week,* June 16, 1998.

62 Moon-Sik Park, "Korea: Airport Development Market," Financial Times Asia Intelligence Wire, May 9, 1997.

63 "Air Traffic Growth Forces Expansion of Seoul Airport Construction," *World Airport Week,* May 30, 1995.

64 John Meredith, "Room to Boom," *Airline Business,* Jan. 1995, p. 38.

65 "Bangkok to Add Gateway," *Travel Trade Gazette Europa,* April 8, 1993, p. 20.

66 49 U.S.C. § 47107(b).

67 14 C.F.R. Part 158.

68 FAA Order 5100.3A, para. 553(a), *AIP Handbook,* Oct. 24, 1989.

69 Quoted in U.S. Department of Transportation, "Intermodal Ground Access to Airports: A Planning Guide," Dec. 1996, pp. 16, 202.

70 "Stalled Train to Kennedy Airport," *The New York Times,* Jan. 30, 1998, p. A20; letter from FAA Associate Administrator Susan Kurland to Port Authority Executive Director George Marlin, Nov. 6, 1996.

71 Letter from FAA Associate Administrator Susan Kurland to Port Authority Executive Director Robert Boyle, Feb. 9, 1998, p. 20.

72 Ibid., p. 21.

73 Ibid., p. 24.

74 Ibid.

75 Matthew Wald, "U.S. Approves Plan for Rail Link to Kennedy Airport," *The New York Times,* Feb. 19, 1998.

Chapter 9 (*Cont.*)

76 U.S. General Accounting Office, "Surface Infrastructure: Costs, Financing and Schedules for Large-Dollar Transportation Projects," Feb. 1998, p. 18.

77 Letter from FAA Associate Administrator Susan Kurland to SFO Airport Director John Martin, Oct. 18, 1996.

78 Benjamin Pimentel, "BART's 4-Year Trip to SFO Starts Today," *San Francisco Examiner,* Nov. 3, 1997, p. 1.

79 U.S. General Accounting Office, "Surface Infrastructure: Costs, Financing and Schedules for Large-Dollar Transportation Projects," Feb. 1998, p. 40.

80 U.S. Department of Transportation, "Intermodal Surface Transportation Efficiency Act: Flexible Funding Opportunities for Transportation Investments," 1996, p. 4.

81 Ibid., p. 13.

82 U.S. General Accounting Office, "Surface Infrastructure: Costs, Financing and Schedules for Large-Dollar Transportation Projects," Feb. 1998, p. 57.

83 U.S. Department of Transportation, "Intermodal Ground Access to Airports: A Planning Guide," Dec. 1996, pp. 16, 203.

84 Joseph Szyliowicz, Andrew Goetz, and Paul Dempsey, "The Vision, the Trends, and the Issues," 25 *Transportation Law Journal* 255, 256 (1998).

85 Andrew Goetz, Joseph Szyliowicz, and Paul Dempsey, "Transportation and International Sustainable Development: A Preliminary Conceptualization and Application," *OECD Proceedings,* Toward Sustainable Development, 1996, p. 118.

Chapter 10

1 "Frankfurt Faces the Future," *Airport World,* Aug. 1997, p. 12.

2 Quoted in *The Denver Post,* Jan. 17, 1998, p. 12A.

3 Paul Stephen Dempsey, Andrew R. Goetz, and Joseph S. Szyliowicz, *Denver International Airport: Lessons Learned,* McGraw-Hill, 1997.

4 George James, *Airline Economics,* 1982.

5 For a review of the national defense importance of air transportation, see Richard Kane and Allan Vose, *Air Transportation,* 7th ed., 1979, pp. 1-21 to 1-24.

6 Paul Dempsey and Laurence Gesell, *Airline Management: Strategies for the 21st Century,* Coast Aire, 1997, pp. 1, 2.

7 "Transportation is a fundamental component of economic growth. It is the infrastructure foundation upon which the rest of the economy is built." Paul Dempsey, *The Social and Economic Consequences of Deregulation,* 1989, p. 5. "[T]ransportation has had a profound effect upon the collective economic growth and intellectual development of man." Paul Dempsey and William Thoms, *Law and Economic Regulation in Transportation,* 1986, p. 1. "Aviation is among the most profound of man's technological accomplishments. Like no other invention, it collapses the time/space continuum. Aviation shrinks the planet, intermingling the world's cultures and economies. It is an integral part of the infrastructure essential to commerce, and national defense. Aviation is mobility for the human race, facilitating travel and tourism, arguably the world's largest single industry." Paul Dempsey, Robert Hardaway, and William Thoms, 1 *Aviation Law* sec. 1.01 (1993), citing Paul Dempsey, *Law and Foreign Policy in International Aviation,* 1987.

8 Thomas Petzinger, Jr., *Hard Landing,* 1995, p. 341.

9 Dempsey and Gesell, op. cit.

10 National Civil Aviation Review Commission, "Avoiding Aviation Gridlock and Reducing the Accident Rate," Dec. 1997, p. II-5.

11 Nasid Hajari, "A Walk in the Clouds," *Time,* June 22, 1998.

Chapter 10 (*Cont.*)

12 National Civil Aviation Review Commission, "Avoiding Aviation Gridlock and Reducing the Accident Rate," Dec. 1997, p. II-5.

13 World Tourism Organization, "Tourism 2020 Vision 3," 1997.

14 Brent Hannon, "Gateways to Growth," *Asia, Inc.,* Nov. 1995, p. 62.

15 Ingomar Joerss, "Airports and Air Transport in Germany," 2 *International Airport Review* 25 (1998).

16 "Frankfurt Faces the Future," *Airport World,* Aug. 1997, p. 12.

17 Joerss, op. cit.

18 Flughafen Frankfurt, News Release, Nov. 11, 1996. See also Koos Bosma, "European Airports, 1945–1995," in *Building for Air Travel,* John Zukowsky, ed., 1996, p. 64.

19 Vienna International Airport, "Facts and Figures," 1996.

20 LaGuardia Airport, "LaGuardia," Winter 1997–1998.

21 Andrew Baur, "As Lambert Goes, So Goes the Region," *The St. Louis Post-Dispatch,* July 26, 1990, p. 3C.

22 Amsterdam Airport Schiphol, "Balancing Environment and Economics," 1997.

23 Ibid.

24 Aeroports de Paris, "Charles de Gaulle Airport: Europe's Foremost Transport Hub," 1997, p. 28.

25 Dempsey, Goetz, and Szyliowicz, op. cit., p. 159.

26 "Inter Airport Atlanta '98," 2 *International Airport Review* 43 (1998).

27 Stanley Scott and Levi Davis, *A Giant in Texas: A History of the Dallas-Fort Worth Regional Airport Controversy 1911–1974,* Nortex, 1974, p. 70.

28 Dempsey, Goetz, and Szyliowicz, op. cit., p. 159.

29 Salt Lake City International Airport, "Airport Master Plan," 1996, p. 2.1.

30 Andrew Goetz, "Air Passenger Transportation and Growth in the U.S. Urban System," 23 *Growth and Change* 217–238 (1992).

31 Dempsey, Goetz, and Szyliowicz, op. cit.

32 "Munich Airport Large Source of Jobs," *Aviation Europe,* June 15, 1995, p. 6.

33 Zenjiro Ogawa, "Kansai International Airport Projects," International Conference on Aviation and Airport Infrastructure, Denver, Dec. 8, 1993.

34 Kansai International Airport, "KIAC Update," Jan./Feb. 1996.

35 Ibid., Jan. 1998.

36 David Brown, "Japanese Building International Offshore Airport to Serve Osaka," *Aviation Week and Space Technology,* July 13, 1987, p. 38.

37 Alexandra Black, "Trade and Tourist Boom from New Airport," Inter Press Service Network, Sept. 9, 1994.

38 "Kansai Int'l Airport Inaugurated," *Kyodo News International,* Sept. 5, 1994.

39 Kansai International Airport, "International Flights at KIX," July 31, 1998.

40 J. A. Donoghue, "The Pearl-Y Gateways," *Air Transport World,* Feb. 1, 1995, p. 75.

41 "New Macau Airport Set to Open July 1995," *Phillips Business Information,* Jan. 28, 1994.

42 "U.S. Funds Feasibility Study for Second Stage of Macau Airport," *Aviation Week and Space Technology,* June 6, 1994, p. 36.

43 Ed Paisley, "On a Wing and a Prayer," *Far Eastern Economic Review,* May 5, 1995, p. 76.

44 Leonard Hill, "Asia's Newest 'Dragon,'" *Air Transport World,* Sept. 1, 1993, p. 66.

45 John Mok, "The Development of Hong Kong's New International Airport," International Conference on Aviation and Airport Infrastructure, Denver, Dec. 8, 1993.

46 Steven Mufson, "Accord Boosts Hong Kong Airport," *The Washington Post,* Nov. 5, 1994, p. A19.

Chapter 10 (*Cont.*)

47 Airport Authority Hong Kong, "Hong Kong International Airport," 1998.

48 Thomas Darmody, "The Design and Development of World Class Airports," IBC International Conference on Airport Development and Expansion, Hong Kong, Oct. 28, 1993.

49 Mok, op. cit.

50 "South Korea: Review 1997," *Asia and Pacific Review,* May 1997.

51 "Korea Seeks Further U.S. Participation in New Airport," *Airports,* June 20, 1995, p. 245.

52 "New Int'l Airport Breaks Ground for Passenger Terminal," *Korea Economic Daily,* June 4, 1996.

53 Michael Mecham, "Seoul Ranks No. 1 in Airport Growth," *Aviation Week and Space Technology,* April 10, 1995, p. 34.

54 "Private Money Permitted in S. Korea Center Construction," *Asia Pulse,* April 10, 1997.

55 "Tourism District to Be Located Next to the New Int'l Airport," Comline Daily News from Korea, Nov. 23, 1996.

56 "Construction of Tourist, Business Centers Planned for Yongjong Airport," *Korea Economic Daily,* Aug. 19, 1995.

57 B. G. Lee, "Singaporeans Must Realize How Tough the Competition Has Become," *The Straits Times,* May 2, 1997, p. 46.

Chapter 11

1 Quoted in *Jane's Airport Review,* Dec. 1997, p. 20.

2 "Relieving O'Hare," *The Economist,* Jan. 10, 1998, p. 22. Crandall was commenting on the effort to build a third Chicago airport.

3 Gail Butler and Martin Keller, "Airports and Airlines: Analysis of a Symbiotic, Love-Hate Relationship," in *Handbook of Airline Economics,* D. Jenkins, ed., 1995, pp. 87, 88.

4 National Civil Aviation Review Commission, "Avoiding Aviation Gridlock and Reducing the Accident Rate," Dec. 1997, pp. II-6, II-7. The FAA also predicts that, without infrastructure expansion, serious delays at more than 30 of the nation's largest airports will cause $1.1 billion in additional airline costs by the year 2001. Federal Aviation Administration, "Aviation System Capacity Annual Report," 1993, p. 5.

5 "Study on European Aeronautical Charges Finds Wide Divergence," *World Airport Week,* July 1, 1997, p. 4.

6 Oris Dunham, "Infrastructure Constraints: Deeds Not Words," *Proceedings,* 7th IATA High-Level Aviation Symposium, Cairo, Egypt, 1993, p. 109.

7 Robert Tompkins, "Infrastructure Capacity Financing through User Charges," unpublished address before IBC Conference at Hong Kong, Oct. 28, 1993.

8 Ibid.

9 Ibid.

10 See John Meredith, "Room to Boom," Airline Business, Jan. 1995, pp. 38, 41.

11 Economies of scale are realized when increases in total production simultaneously decrease unit costs; long-run average cost decreases as output increases. As the scale of production grows, the enterprise becomes more efficient. For example, a large capital-intensive piece of equipment operating at full capacity (such as a Boeing 747) can allow significantly lower ASM costs vis-à-vis a smaller aircraft (such as a Boeing 727). A related concept is economies of scope. The unit cost of producing one more item may be diminished when the scope of activity broadens. For example, advertising costs per unit of serving a particular city-pair market are lower the more city-pairs served, for the same ad can offer several city-pair product lines. Similarly, combination carrier airlines can offer "belly" cargo service in their passenger markets. Economies of scale are realized when a firm increases its total production while simultaneously decreasing its cost to produce each

Chapter 11 (*Cont.*)

unit. As the scale of production grows, the enterprise becomes more efficient. The classic example of the phenomenon of economies of scale is the enormous cost savings experienced from producing automobiles on an assembly line rather than one car at a time. The cost savings resulting from economies of scale can be attributed to: (1) indivisibility, a large capital-intensive piece of equipment operates most efficiently at full capacity; and (2) division and specialization of labor; highly specialized labor is more productive labor. Hub-and-spoke operations in the airline industry are largely successful because of significant economies of scale.

In the related concept of economies of scope, a firm achieves economies by combining one or more activities into a single operation. Thus, the additional cost of producing one more item (marginal cost) is diminished when the scope of activity broadens. Thus, scheduled service can easily gobble up charter service. See Joe Bain, *Barriers to New Competition,* 1956; Robert Heilboner and Lester Thurow, *Economics Explained,* 1987; and William Shepard, *Economics of Industrial Organization,* 1979.

Another related concept is *economies of density.* By combining passengers and groups of passengers, an airline can carry the aggregation of passengers more cheaply than if it carried those passengers separately. Through careful scheduling of flights, consolidating operations and routing passengers through its hub, an airline streamlines its system, making it more dense and thus more efficient. The hub-and-spoke scheme employed by all of the major airlines is testimony to this phenomenon. For example, an airline which carries 100 passengers in a single plane to a destination as opposed to carrying 50 passengers in two aircraft to that same destination is making use of economies of density. See Ann Friedlaender and Richard Spady, *Freight Transportation Regulation: Equity, Efficiency and Competition in the Rail and Trucking Industries,* 1980; A. Lamond, *Competition in the General-Freight Motor Carrier Industry,* 1980.

12 Philip Baggaley, "Assessing an Airline's Credit Quality," in *Handbook of Airline Economics,* D. Jenkins, ed., McGraw-Hill, 1995.

13 "American-Sponsored Study Blasts Criticism of Hubs," *Aviation Daily,* July 31, 1990, p. 197.

14 Paul S. Dempsey, Robert Hardaway, and William Thoms, 1 *Aviation Law and Regulation* §2.12 (1993). See also J.P. Morgan Securities, "The U.S. Airline Industry," 1993.

15 *U.S. Statistics,* 1991, p. 797.

16 "U.S. Large Airport Traffic," *Aviation Daily,* Aug. 15, 1990, p. 309. Enplaned passenger figures have been doubled to approximate total passengers, the standard used in the chart for foreign airports. However, the reader should beware that a doubling of enplaned passengers may not be precisely the total number of passengers flown through the airport.

17 "Worldwide Airport Traffic," *Aviation Daily,* Aug. 15, 1991, p. 308.

18 Julius Maldutis, "Airline Competition at the 50 Largest U.S. Airports," May 11, 1994.

19 By 1990, 88% of the gates at the 66 largest airports in the United States were leased to airlines, and 85% of the leases were for exclusive use. "Intelligence," *Aviation Daily,* Aug. 20, 1990, p. 323.

20 *Aviation Daily,* Jan. 30, 1996, p. 150.

21 James Hirsch, "Big Airlines Scale Back Hub-Airport System to Curb Rising Costs," *The Wall Street Journal,* Jan. 12, 1993, p. 1.

22 Continental no longer maintains a hub in Denver.

23 Hirsch, op. cit., pp. A1, A6.

24 Dan Reed, *American Eagle,* 1992, p. 157.

25 Thomas Petzinger Jr., *Hard Landing,* 1995, pp. 137, 138.

26 U.S. General Accounting Office, "Airline Competition: Higher Fares and Reduced Competition at Concentrated Airports," 1990.

Chapter 11 (*Cont.*)

27 Address by Maurice Myers before the Salomon Brothers Transportation Conference, New York, Nov. 17, 1994.

28 Dempsey, Hardaway, and Thoms, op cit., §5.05. General Accounting Office, "Testimony of Kenneth Mead before the Aviation Subcommittee of the U.S. Senate Commerce Committee," April 5, 1990, p. 6.

29 U.S. General Accounting Office, "Airline Competition: Industry Operating and Marketing Practices Limit Market Entry," 1990, p. 4.

30 Ibid., p. 6.

31 Barbara Beyer, "The Curse and Blessing of Hubs," International Conference on Aviation and Airport Infrastructure, Denver, Dec. 5–9, 1993, p. 3.

32 SH&E, "The Facts about American vs. Southwest," unpublished study prepared on behalf of APA, Sept. 13, 1993, pp. 47, 49. Southwest's average stage length is 380 mi, compared to American's 807.

33 Ibid., pp. 48, 49.

34 Mead Jennings, "Staying at the Top," *Airline Business,* March 1994, pp. 28, 31; see also SH&E, op. cit., p. 49.

35 Robert Crandall, "The Hub Debate," *American Way Magazine.*

36 AMR Corporation, 1993 Third Quarter Report, 1993, pp. 2, 3.

37 "Fast Growth Hurts America West," *The Wall Street Journal,* Sept. 10, 1996, p. B4.

38 Beyer, op. cit.

39 Melvin Brenner, James Leet, and Elihu Schott, *Airline Deregulation,* 1985, p. 95.

40 ESG Aviation Services, 7 *Airline Monitor* 1 (Nov. 1994).

41 See Michael Levine, "The Legacy of Airline Deregulation: Public Benefits, But New Problems," *Aviation Week and Space Technology,* Nov. 9, 1987, p. 161. In an environment of perfect competition, no single producer has market power and consumers purchase goods and services closely approximating their marginal costs of production. In such a market, there is no input waste, excess capacity, or "monopoly" profits. In theory, the most efficient producers provide the commodity or service, and the public enjoys an efficient allocation of resources. Paul Dempsey, "Antitrust Law and Policy in Transportation: Monopoly I$ the Name of the Game," 21 *Georgia Law Review* 505, 535 n.182 (1987).

42 A 1978 Senate Committee report on federal regulation provided a fairly typical summary of those attributes of destructive competition deemed not likely to surface in a deregulated air and motor carrier industry: "A justification sometimes offered for regulation is that in the absence of regulation competition would be destructive. In other words, without regulation, an industry might operate at a loss for long periods....When there is excess capacity in a competitive industry...prices can fall far below average cost. This is because individual producers minimize their losses by continuing to produce so long as their variable (avoidable) costs are covered, since they would incur their fixed (overhead) costs whether they produced or not....Similarly, if resources are mobile [as they are in the trucking and airline industries] depressed conditions in an industry or a region would result in the shift of resources to other employments....

"What is 'destructive' about large and long-lasting losses? Some economists have suggested that they would result in long periods of inadequate investment and slow technical progress which in turn might lead to poor service and periodic shortages....

"Another scenario that has sometimes been suggested is that periods of large losses will result in wholesale bankruptcies and the shakeout of many small producers with the result that the industry in question becomes highly concentrated in a few large firms....

"A third and related notion is the possibility that powerful firms might engage in predation....

Chapter 11 (*Cont.*)

"'Destructive competition' seems…unlikely in the cases of airlines and trucks." Study on Federal Regulation, "Report of the Senate Committee on Government Affairs," 96th Congress, 1st Sess. 1978, pp. 13–15.

43 Although almost every airline opposed deregulation in 1978, the nation's largest carrier, United Airlines, was a vigorous proponent. Alfred Kahn has admitted that, in advocating deregulation, he had underestimated the advantages of the large firms in the airline industry. "Hearings before the California Public Utilities Commission," Jan. 31, 1989, pp. 6190, 6223 (testimony of Alfred E. Kahn). As Kahn said, "we underestimated the importance of economies of scale and scope." Ibid., p. 6201. Elsewhere, Kahn has conceded, "We advocates of deregulation were misled by the apparent lack of evidence of economies of scale…." "Surprises from Deregulation," 78 *AEA Papers and Proceedings* 316, 318 (1988).

A decade earlier, Kahn dismissed fears that the industry would become highly concentrated. Testifying before a House Subcommittee in 1978, Congressman Harsha posed the following question: "[Y]ou are going to invite into the area of new entry the severest competition between airlines servicing that particular market and ultimately the big will eat the little, and those who are able to withstand the severe competition and the reduced fares—even below operating expenses—will prevail. Then the airlines that cannot prevail, of course, will have to go out of business or do something else. After that transition period then you are going to see the air fares go back up again and the big will control the airline industry." "Aviation Regulatory Reform: Hearings on H.R. 11145 before the Subcommittee on Aviation of the House Committee on Public Works and Transportation," 95th Congress, 2d Sess., 1978, p. 178. Kahn dismissed these fears as unfounded: "First, *the assumption that you are going to get really intense, severe, cut throat competition just seems to me unrealistic* when you are talking about a relatively small number of carriers who meet one another in one market after another. We don't find in American industry generally when you have a few relatively large carriers competing with one another that they engage in bitter and extended price wars.

"But number two, *the fear that the big will eat the little, that is one that I would really like to nail.* If you look, as I did last week, at the stock market prices of the securities of the big airlines today you will find that while the average certificated carriers' in the United States stock is selling at about two-thirds of book value…three of the five biggest carriers'…stock is selling at 33 to 37 percent of book value….

"That means to me the investors do not believe that prediction." Ibid., pp. 178, 179 [emphasis supplied]. Similarly, in 1977 hearings before the a House Subcommittee, Kahn said, "I do not honestly believe that the big airlines are going to be able to wipe out the smaller airlines, if only because every study we have ever made seems to show that there are not economies of scale." "Aviation Regulatory Reform: Hearings on H.R. 1145 before the Subcommittee on Aviation of the Committee on Public Works and Transportation," 95th Congress, 1st Sess., 1977, p. 1137.

New entry in the airline industry has become extremely difficult, and mergers and bankruptcies have thinned the ranks of competitors, and created high levels of concentration. Of the 121 small airlines which entered after 1978, fewer than 50 were still operating a decade later. "Focus: A Decade of Deregulation," *Traffic World*, Dec. 5, 1988, Supp. A. Kahn characterized the failure rate of new airlines as "frightening." "All That," *Economic Development*, 1987, pp. 91, 93.

In a 1988 article in the *Transportation Law Journal*, Alfred Kahn admitted that prices are likely to rise, saying, "I have little doubt that…the disappearance of most of the price-cutting new entrants and the marked reconcentration of the industry—will produce higher fares…." Alfred Kahn, "Airline Deregulation—A Mixed Bag, but a Clear Success Nevertheless," 16 *Transportation Law Journal* 229, 236 (1988). Similarly, in testimony before the Senate Commerce Committee in 1987, Kahn said, "the industry has become more concentrated at the national level because of mergers and airline failures, and that means in my judgment that price competition may well become less severe in the years ahead." "Safety and Re-Regulation of the Airline Industry: Hearings before the Senate Committee on Commerce, Science, and Transportation," 100th Congress, 1st Sess., 1987, p. 155.

Chapter 11 (*Cont.*)

44 See generally, Paul Stephen Dempsey, *The Social and Economic Consequences of Deregulation,* 1989; Paul Stephen Dempsey, *Flying Blind: The Failure of Airline Deregulation,* 1990.

45 Assumptions about ease of entry and the importance of potential entry as a means of keeping markets competitive served as an essential foundation for airline deregulation. But even assuming a high level of concentration, deregulation theory had a solution. In the late 1970s, Kahn proclaimed, "almost all of this industry's markets can support only a single carrier or a few: their natural structure, therefore, is monopolistic or oligopolistic. This kind of structure could still be conducive to highly effective competition if only the government would get out of the way; the *ease of potential entry into those individual markets, and the constant threat of its materializing,* could well suffice to prevent monopolistic exploitation." Alfred Kahn, talk to the New York Society of Security Analysts, Feb. 2, 1978, p. 24 [emphasis supplied]. Kahn saw few economies of scale in the industry; hence entry, or the threat of potential entry would keep monopolists from extracting monopoly profits. Ibid., p. 26. Again, this was the essence of contestability theory. In 1977, Kahn testified before a House Subcommittee on the importance of the automatic entry provisions of a pending airline deregulation bill, saying: "[A] realistic threat of entry by new and existing carriers on the initiation of management alone is the essential element of competition. It is only this threat that makes it possible to leave to managements a wider measure of discretion in pricing. It is the threat of entry that will hold excessive price increases in check." "Aviation Regulatory Reform, Hearings before the Subcommittee on Aviation of the Committee on Public Works and Transportation," 95th Congress, 1st Sess. 1977, p. 1111 [emphasis supplied]. Kahn advanced the theory on many occasions as Chairman of the Civil Aeronautics Board. Before another House Committee in 1977, Kahn testified, "Were it not for Government restrictions, entry would be relatively easy,..." "Economic Aspects of Federal Regulation in the Transportation Industry, Hearings before a Task Force of the House Committee on the Budget," 95th Congress, 1st Sess., 1977, p. 19.

In an interview, Alfred Kahn noted, "Certainly one of the assumptions behind airline deregulation was that entry would be relatively easy." Interview with Alfred E. Kahn, *Antitrust,* Fall 1988, pp. 4, 6. But in another, Kahn admitted, "We didn't dream of the way airlines could manipulate fares with such great sophistication....We were a little naive about what 'freedom of entry' meant in the airline business." Robert Kuttner, "Why Air Fares Aren't Falling," *The Washington Post,* Sept. 18, 1988, p. C7.

On the one hand, Kahn appeared to embrace the theory of contestability. In the 1988 reprinting of his book, he said, "all travelers continue to have the protection—admittedly in varying degrees of effectiveness—of the relatively high degree of contestability of airline markets, and of the ability of many travelers to larger towns, where a greater variety of fares is typically available." Alfred Kahn, *The Economics of Regulation,* MIT Press, 1988, p. xix. In his article in the *Transportation Law Journal,* Kahn said: "I have no recollection that in expressing the expectation that the possibility of entry would prevent grossly monopolistic exploitation, the advocates of deregulation clearly distinguished the roles they expected would be played, respectively, by totally new entrants and by existing carriers invading one another's markets. Manifestly, however, it is irrational to conclude, from the unlikelihood of the former, that the anticipated effectiveness of contestability has therefore been disproved." Alfred Kahn, "Airline Deregulation—A Mixed Bag, but a Clear Success Nevertheless," 16 *Transportation Law Journal* 223 (1988). But in testimony delivered in 1987 before a Subcommittee of the Senate Judiciary Committee, Kahn was far less enthusiastic about the potential benefits of contestability: "I attack the easy assumption of the ideologues of laissez-faire that contestability takes care of everything; that private parties cannot monopolize airline markets because the minute they raise their price two bits, there will be a rush of competitors into the market.

"I know of seven studies now of airline pricing since deregulation. They all conclude that while, yes, airline markets are relatively easy to enter, the potential entry of competitors is no substitute for competitors already there...."

Chapter 11 (*Cont.*)

"Now, the view that contestability of airline markets makes antitrust enforcement unnecessary is very close to the position that DOT is taking [in the airline merger cases].

"Contestability is not a sufficient protection, in my opinion, and anybody who looks at the airline industry certainly knows that the likelihood and opportunity of entry, particularly by new carriers—low-cost, price-cutting carriers—has greatly diminished in recent years and is likely to remain much lower than before." "Airline Deregulation, Hearing before the Subcommittee on Antitrust of the Senate Judiciary Committee," 100th Congress, 1st Sess., 1987, p. 64. Similarly, in an article entitled "Deregulatory Schizophrenia," Kahn was quite critical of the theory of contestability: "To return to my schizophrenia thesis, life is much simpler for the economists and lawyers who believe that the mere incantation of 'contestability' holds the answer to all possible concerns about the viability of competition in deregulated industries. According to this view—seldom articulated as baldly as I will here—the comparative ease of entry that helped recommend deregulation of the airline and trucking industries also makes antitrust not very important....In short, if contestability were perfect, there would be no need for antitrust laws at all....

"In my opinion, the *contestability of airline markets does not afford travelers sufficient protection* in those circumstances....

"It seems to me absolutely incontestable that the likelihood of entry into any industry is itself powerfully affected by the previous practices of the incumbent firms. The recent history of the airline industry provides ample documentation of that proposition. Entry, particularly by genuinely new firms, has clearly become much more difficult...." Alfred Kahn, "Deregulatory Schizophrenia," 75 *California Law Review* 1059, 1063 (1987).

46 See generally, William Baumol, J. Panzar, and R. Willig, *Contestable Markets and the Theory of Industry Structure,* 1982.

47 Christopher Power, "The Frenzied Skies," *Business Week,* Dec. 19, 1988, pp. 70, 72.

48 William Stockton, "When Eight Carriers Call the Shots," *The New York Times,* Nov. 20, 1988, p. 3-1. See Paul Stephen Dempsey, "Transportation Deregulation: On a Collision Course," 13 *Transportation Law Journal* 329, 342–352 (1984).

49 James Ott, "Industry Officials Praise Deregulation, but Cite Flaws," *Aviation Week and Space Technology,* Oct. 31, 1988, p. 88. In 1988, the industry's profit margin shrunk to 1.3%, compared to 2% a decade earlier. Stockton, op. cit. Alfred Kahn maintains that the airline industry's profit margin "fell to a puny 0.10 in the 1979–87 period." Alfred Kahn, "Surprises of Airline Deregulation," 78 *AEA Papers and Proceedings* 316 n.1 (1988). As one careful observer of the airline industry, Melvin Brenner, noted: "The eight years of deregulation comprise the worst financial period in airline history. The cumulative industry operations in those eight years generated a loss of over $7 billion, when interest payments are included with operating expenses....The deregulation era is the first time that the industry as a whole has recorded a cumulative loss over an eight-year period....The principal cause of the poor financial results has been the tendency of airlines to engage in *destructive* competition in the absence of regulation—a tendency evident particularly in excess capacity and fare wars....By failing to cover fixed costs, marginal cost reliance jeopardizes the industry's long term viability." Melvin Brenner, "Airline Deregulation—A Case Study in Public Policy Failure," 16 *Transportation Law Journal* 179, 200–201 (1988) [emphasis in original]. Ten years after he implemented airline deregulation as President Carter's Chairman of the Civil Aeronautics Board, Alfred Kahn admitted, "There is no denying that the profit record of the industry since 1978 has been dismal, that deregulation bears substantial responsibility, and that the proponents of deregulation did not anticipate such financial distress—either so intense or so long-continued." Alfred Kahn, "Airline Deregulation—A Mixed Bag, but a Clear Success Nevertheless," 16 *Transportation Law Journal* 229, 248 (1988). After 5 years of deregulation, with carriers going "belly up" in bankruptcy, in an interview published in *The Wall Street Journal,* Alfred Kahn noted that "There's a lot of turmoil, but that's what we intended." Bill Richards, "CAB's Ex-Chairman, Alfred Kahn, Looks at Airline Industry He Helped to Deregulate," *The Wall Street Journal,* Oct. 4, 1983, p. 35. He then acknowledged

Chapter 11 (*Cont.*)

that "the turmoil is more intense and lasting longer than most of us anticipated." He was also willing to concede that the new entrants would likely never account "for more than 5% of the total travel." Ibid. But Kahn then expressed surprise with the magnitude of the turmoil, saying in 1983, "I've been dismayed because the airlines suffered more pain than I envisioned." Stroller, "Al Kahn Has Few Regrets," *Airport Press,* Aug. 1983, p. 1. In 1987 hearings before the Senate Judiciary Committee, Kahn acknowledged that deregulation "caused a great deal of turmoil—more, I think, than most of us would have predicted. Turmoil has, however, socially positive as well as painful aspects: a release of creativity, entrepreneurship and innovation, along with a painful readjustments, bankruptcies and unprecedented financial losses." "Airline Deregulation: Hearing before the Antitrust Subcommittee of the Senate Judiciary Committee," 100th Congress, 1st Sess., 1987, p. 67. Kahn said, "I found it distressing in the middle of this. I hated to be responsible for the industry suffering so. I wanted to be sure that it would always be financially healthy and able to attract capital." "Testimony of Alfred Kahn before the California Public Utilities Commission," Jan. 31, 1989, pp. 6247, 6248. He acknowledged that the low fares consumers had enjoyed were a short-term phenomenon, saying, "I have little doubt that...the disappearance of most of the price-cutting new entrants and the marked reconcentration of the industry—will produce higher fares." Ibid., p. 236.

50 Jeff Pelline, "Bumpy Ride under Deregulation," *San Francisco Chronicle,* Oct. 28, 1988, p. 21. One source estimated that 214 airlines disappeared from the market in the first decade of deregulation. Martha Hamilton, "Is the Airline Industry on the Verge of Going Global?" *The Washington Post,* Dec. 11, 1988, p. K1.

51 Pelline, op. cit.

52 The difficulty airlines face is in managing yield in a way which lures only passengers not otherwise likely to fly. Hence, Saturday stay-over requirements, which are unappealing to business travelers.

53 Paul Stephen Dempsey and Laurence Gesell, *Airline Management: Strategies for the 21st Century,* Coast Aire, 1997, pp. 31–93.

54 Ibid. In the short term, competition unleashed by deregulation reduced the dominance by the largest airlines. Thus, in January 1986, the five largest airlines accounted for 54.3% of the domestic passenger market. But by June of 1987, after a series of unprecedented mergers, their share had soared to 72.2%. Paul Dempsey, "Antitrust Law and Policy: Monopoly I$ the Name of the Game," 21 *Georgia Law Review* 543. Fifteen independent airlines operating at the beginning of 1986 had been merged into six megacarriers by the end of 1987. The mergers, approved by DOT, consolidated about 70% of the nation's aircraft capacity. See Melvin Brenner, "Airline Deregulation—A Case Study in Public Policy Failure," 16 *Transportation Law Journal* 180.

55 "Focus: A Decade of Deregulation," *Traffic World,* Dec. 5, 1988, Supp. D, updated by "Happiness Is a Cheap Seat," *The Economist,* Feb. 4, 1989, p. 68. Ott, op. cit., p. 89. One commentator summarized the structural changes in the industry which occurred in the first decade of deregulation: "The 11 major airlines have shrunk to eight; the eight former local service carriers are now two and they are trying to merge; the eight original low-cost charter airlines have been reduced to one, through bankruptcy and abandonment; 14 former regional airlines have shrunk to only four; over 100 new upstart airlines were certificated by the CAB and about 32 got off the ground and most of those crashed, leaving only a handful still operating; of the 50 top commuters in existence in 1978, 29 have disappeared....

"Today, the top 50 commuter carriers who constitute 90 percent of that industry are captives of the major carriers, in part or in total owned, controlled, and financed by the giant airlines and relegated to serving the big airlines at their hubs." "Airline Deregulation: Hearings before the Subcommittee on Antitrust, Monopolies and Business Rights of the Senate Judiciary Committee," 100th Congress, 1st Sess., 1987, pp. 61–62 (testimony of Morten S. Beyer).

Chapter 11 (*Cont.*)

56 49 U.S.C. §1302(a)(7). Paul Dempsey, "The Rise and Fall of the Civil Aeronautics Board: Opening Wide the Floodgates of Entry," 11 *Transportation Law Journal* 91 (1979).

57 See Paul Dempsey, "Antitrust Law and Policy in Transportation: Monopoly I\$ the Name of the Game," 21 *Georgia Law Review* 538 (1987).

58 Concentration levels in the passenger industry are even more pronounced when one recognizes that before deregulation, America had a healthy charter airline industry, enjoying significant market share. Under deregulation, it nearly vanished. See Melvin Brenner, "Airline Deregulation—A Case Study in Public Policy Failure," 16 *Transportation Law Journal* 184 (1988). In 1977, nonscheduled airlines had 43,000 domestic departures, compared with 18,000 in 1986. Federal Aviation Administration, "Airport Activity Statistics of Certificated Route Carriers," 1977, pp. 747–749; Federal Aviation Administration, "Airport Activity Statistics of Certificated Route Carriers," 1986, pp. 798–800.

59 Alfred Kahn, "Despite Waves of Airline Mergers, Deregulation Has Not Been a Failure," *The Denver Post*, Aug. 31, 1986, p. 3G.

60 Alfred Kahn, "Airline Deregulation—A Mixed Bag, but a Clear Success Nonetheless," 16 *Transportation Law Journal* 234 (1988) [emphasis in original].

61 U.S. General Accounting Office, "Airline Competition," 1990, p. 15.

62 Brian Harris, "Airlines: 1998 Hub Factbook," Lehman Bros., 1998, p. 31.

63 Julius Maldutis, "Airline Competition at the 50 Largest U.S. Airports—Update," Salomon Bros., July 21, 1997; *Consumer Reports,* June 1988, pp. 362–367; General Accounting Office, "Airline Competition," 1990, p. 33; *Aviation Daily,* June 29, 1990, pp. 628–630; *Aviation Daily,* June 29, 1990, pp. 628–630; U.S. General Accounting Office, "Airline Competition," 1990, p. 33; James Ott, "Congress, Airlines Reassessing Deregulation's Impact," *Aviation Week and Space Technology,* Nov. 9, 1987, p. 163; Martha Hamilton, "The Hubbing of America: Good or Bad?" *The Washington Post,* Feb. 5, 1989, pp. H1, H2. Judith Valente and Robert Rose, "Concern Heightens about the Airline Industry's March toward Near Domination by Only a Few Major Carriers," *The Wall Street Journal,* March 10, 1989, p. A8, col. 1. Since Frontier was absorbed, first by People Express and then by Continental (Texas Air), no airport has enjoyed the three-hub major carrier competition that theretofore existed at Denver. Paul Dempsey, "Antitrust Law and Policy in Transportation: Monopoly I\$ the Name of the Game," 21 *Georgia Law Review* 505, 592–593 (1987).

64 Julius Maldutis, "Airline Competition at the 50 Largest U.S. Airports—Update," Salomon Bros., July 21, 1997, p. 4.

65 Ibid., p. 5; hearings before the U.S. Senate Committee on Commerce, Science and Transportation on Airlines, 1987, pp. 170–195 (statement of Dr. Julius Maldutis). This corresponds to a reduction in the number of "effective" competitors in the average of the 50 airports from 4.51 in 1977 to 2.85 in 1987. Paul Stephen Dempsey, *Flying Blind: The Failure of Airline Deregulation,* Economic Policy Institute, 1990, pp. 17, 18.

66 Maldutis, op. cit., p. 2.

67 Ibid., p. 14.

68 Data provided by Denver International Airport, 1998.

69 Stockton, op. cit.

70 U.S. General Accounting Office, "Airline Competition," 1988, pp. 2, 3.

71 Thomas Hamburger, "Fares Rose with NWA's Dominance," *Minneapolis Star-Tribune,* Dec. 1988, p. 1A.

72 Martha Hamilton, op cit., p. H2. "Happiness Is a Cheap Seat," *The Economist,* Feb. 2, 1989, p. 68.

73 "Airlines: Flying Into 'Pockets of Pain'," *USA Today,* Feb. 23, 1998.

74 Address of DOT Assistant Secretary Patrick Murphy before the ABA Forum on Air and Space Law, San Francisco, July 8, 1998.

Chapter 11 (*Cont.*)

75 Concentrated airports were those defined as having more than 60% of enplanements handled by a single airline.

76 U.S. General Accounting Office, "Airline Competition," 1989, pp. 2, 3. The report was subsequently updated and expanded; see U.S. General Accounting Office, "Airline Competition," 1990. The higher fares at concentrated airports do not reflect a premium for nonstop service, since the average number of coupons per traveler at concentrated airports was virtually identical to that at the comparison, unconcentrated airports (2.26 versus 2.28 coupons). And the difference persisted when average trip length was controlled by excluding from the comparison group of airports those where average trip length was significantly longer than for concentrated airports. Thus neither a higher proportion of nonstops nor a higher proportion of short-haul (and thus more costly) flights can explain the fare premium at concentrated airports. The study also found that the increase in fares was generally greater at concentrated airports, and that the increase in fares was especially dramatic when a carrier established dominance during the period. Finally, the study found that in 13 of the 14 concentrated airports, the dominant carrier had higher fares, in some cases very much higher than other carriers at the same airport. Paul Stephen Dempsey, *Flying Blind: The Failure of Airline Deregulation,* Economic Policy Institute, 1990, pp. 18, 19.

77 U.S. General Accounting Office, "Airline Competition," 1990, p. 3.

78 U.S. Department of Transportation, Secretary's Task Force on Competition in the U.S. Domestic Airline Industry, "Executive Summary," 1990, p. 8.

79 U.S. Department of Transportation, Request for Comments in Docket OST-98-3717, April 6, 1998.

80 "Airline Concentration, Competition Concern Senate Subcommittee," *Aviation Daily,* April 10, 1990, p. 67.

81 "Factors Linked with Higher Fares," *Aviation Daily,* April 10, 1990, p. 67. See statement of Kenneth M. Mead before the Subcommittee on Aviation of the Senate Commerce Committee, April 5, 1990.

82 Deregulation architect Alfred Kahn referred to aircraft as "marginal costs with wings."

83 Address of DOT Assistant Secretary Patrick Murphy before the ABA Forum on Air and Space Law, San Francisco, July 8, 1998.

84 Testimony of DOT Assistant Secretary Patrick Murphy before the U.S. Senate Appropriations Subcommittee, May 5, 1998.

85 Market share data are for 1996, the latest year for which data are available, except for ValuJet, which is 1995. Julius Maldutis, "Airline Competition at the 50 Largest U.S. Airports—Update," Salomon Bros., July 23, 1997. 1995 data were used for ValuJet because ValuJet's fleet was grounded for a significant portion of 1996 after its tragic crash in the Everglades.

86 Ibid.

87 Mike Meyers, "Minnesotans Indeed Pay More for Air Fare, DOT Says," *Minneapolis Star-Tribune,* April 25, 1996, p. 1A.

88 Edwin Dolan, *Economics,* 4th ed., 1986, p. 602.

89 William O'Connor, *An Introduction to Airline Economics,* 5th ed., 1995, p. 7.

90 Brenner, Leet, and Schott, op. cit., p. 50.

91 See Robert Hardaway, "The FAA 'Buy-Sell' Slot Rule: Airline Deregulation at the Crossroads," *Journal of Air Law and Commerce,* 1, 49 (1980).

92 Ibid.

93 Often an airport will expand its capacity in order to accommodate a carrier that decides to set up a hub there. The carrier and the airport will typically enter into a long-term lease agreement for space at the facility. The revenues from the lease payments will be used to underwrite the airport bonds sold to pay for the capacity expansion and thereby

Chapter 11 (*Cont.*)

lower the costs of borrowing. As a quid pro quo, the airline may require the airport to include a majority-in-interest clause in the lease agreement giving the airline a large say in any future airport construction activities that would affect its lease payments. U.S. General Accounting Office, "Airline Competition," 1990, p. 26.

94 Bridget O'Brian, "Delta, AMR's American Airlines Plan to Merge Computer Reservations Systems," *The Wall Street Journal,* Feb. 6, 1989, p. B10, col. 1. Janice Castro, "Eastern Goes Bust," *Time,* March 20, 1989, p. 52.

95 See Derek Saunders, "The Antitrust Implications of Computer Reservations Systems," 51 *Journal of Air Law and Commerce* 157 (1985).

96 Thomas Hamburger, "Fighting Back Begins as Costs Go Up, Up, and Away," *Minneapolis Star-Tribune,* Dec. 24, 1988, p. 6A.

97 Ibid. These statistics were quoted by Michael Levine.

98 "Dereg's Falling Stars," *OAG Frequent Flyer,* Aug. 1988, p. 28.

99 Ibid.

100 U.S. General Accounting Office, "Airline Competition: Impact of Computerized Reservations Systems," 1986.

101 Let us say that we could find a major airport with sufficient capacity to allow us to establish a hub. How could, say, an Air Omaha lure passengers away from its rivals' frequent flier programs with their free trips to Hawaii, when it could only offer a free weekend in Cedar Rapids?

102 Domestic commission overrides range from 1% to 5% above the standard 9% to 10% commission. International bonuses can be several times the standard 8% commission. Robert Rose, "Travel Agents' Games Raise Ethical Issue," *The Wall Street Journal,* Nov. 23, 1988, p. B1.

103 See Michael Levine, "Airline Competition in Deregulated Markets: Theory, Firm Strategy, and Public Policy," 4 *Yale Journal of Regulation* 393, 417 (1987).

104 Severin Borenstein, "The Dominant-Firm Advantage in Multiproduct Industries: Evidence from the U.S. Airlines," *Quarterly Journal of Economics* 1237, 1248 (1991).

105 *Cargill, Inc. v. Monfort of Colorado, Inc.,* 479 U.S. 104 (1986).

106 See, e.g., Robert Bork, *The Antitrust Paradox,* 1978, pp. 149–155; Areeda and Turner, "Predatory Pricing and Related Practices under Section 2 of the Sherman Act," 88 *Harvard Law Review* 697, 699 (1975); Easterbrook, "Predatory Strategies and Counterstrategies," 48 *University of Chicago Law Review,* 263, 268 (1981). This view was essentially embraced by the U.S. Supreme Court in *Matsushita Electric Industrial Co. v. Zenith Radio Corp.,* 475 U.S. 574 (1986).

107 Michael Levine, "Airline Deregulation: A Perspective," 60 *Antitrust Law Journal* 687, 689 (1991).

108 Alfred Kahn, "The Macroeconomic Consequences of Sensible Microeconomic Policies," *Society of Government Economists Newsletter,* May 1985, p. 6.

109 "When Free Markets Throttle Competition," *The Economist,* Aug. 24, 1985, p. 19.

110 It has been argued that a rational firm will not engage in predation because the cost of predation will almost always exceed the expected payback. See sources cited at 62 *New York University Law Review* 968, n. 2 (1987).

111 *Continental Airlines v. American Airlines,* 824 F. Supp. 689 (S.D. Tex. 1993). Michael Conway, CEO of America West said, "It's my strong opinion that this action led by American is clearly a predatory action. I think it is clearly designed to further consolidate the industry, which would result in the remaining carriers, particularly American, being able to capture a greater market share and being able to raise prices at a later date." Martha Hamilton, "Fare Wares: For Some a Duel to the Death," *The Washington Post,* June 4, 1992, p. D9.

112 James Ott, "UltrAir Suspends Service, Reviews Charter Option," *Aviation Week and Space Technology,* Aug. 2, 1993, p. 34.

Chapter 11 (*Cont.*)

113 Testimony of Kevin Stamper before the Subcommittee on Antitrust, Business Practices and Competition of the U.S. Senate Committee on the Judiciary, April 1, 1998.

114 Testimony of Mark Kahan before the Subcommittee on Aviation of the U.S. House Committee on Transportation and Infrastructure, April 23, 1998.

115 Tony Kennedy and Greg Gordon, "Government Investigating NWA's Fares," *Minneapolis Star-Tribune,* Feb. 27, 1998, p. 1D.

116 Bill Poling, "Letter to DOT," *Travel Weekly,* Dec. 10, 1993, p. 19.

117 Fred Allvine and John Lindsley, "Increasing Monopolization of the United States Commercial Airline Industry through the Development and Defense of Fortress Hubs," April 1997 [emphasis in original].

118 Testimony of Lewis Jordan before the Subcommittee on Aviation of the U.S. House Transportation and Infrastructure Committee, March 23, 1995.

119 "United Denies Engaging in Predatory Pricing, Despite Study Claiming Otherwise," *Airline Financial News,* Feb. 10, 1997.

120 "Trial Date Set in Pacific Express vs. United Airlines Suit," *Business Wire,* March 23, 1990.

121 "International News," Reuters, Nov. 24, 1984.

122 Martha Hamilton, "Three Airlines Settle in Laker Case," *The Washington Post,* March 18, 1986, p. B3.

123 "Gulf Air Sues Lorenzo," PR Newswire, March 9, 1987.

124 "Trial Date Set in Pacific Express vs. United Airlines Suit," *Business Wire,* March 23, 1990.

125 "United Charged with Engaging in Predatory Pricing," *Aviation Daily,* May 19, 1992, p. 302; "American, Continental, Pan Am Accused of Predatory Pricing," *Aviation Daily,* Oct. 10, 1990, p. 66.

126 These cases are summarized in Paul Dempsey, *Air Transportation: Foundations for the 21st Century,* Coast Aire, 1997, pp. 283–286.

127 U.S. Department of Transportation, "DOT Releases Airline Competition Policy Statement," press release, April 6, 1998.

128 Peter Cartensen, "Evaluating 'Deregulation' of Commercial Air Travel: False Dichotomization, Untenable Theories, and Unimplemented Premises," 46 *Washington and Lee Law Review* 109, 126–126 (1989); Russell Klingaman, "Predatory Pricing and Other Exclusionary Conduct in the Airline Industry," 4 *DePaul Business Law Journal* 281 (1992).

129 Levine, op. cit., pp. 393, 451.

130 Ibid., pp. 393. 452.

131 For a survey of these conclusions drawn from the economics literature see Dempsey and Gesell, op. cit., pp. 71–84.

132 The first article to cast doubt on the applicability of contestability theory to the airline industry was D. Graham, D. Kaplan, and D. Sibley, "Efficiency and Competition in the Airline Industry," *Bell Journal of Economics* 118 (1983).

133 See sources cited in James Brander, "Dynamic Oligopoly Behavior in the Airline Industry," 11 *International Journal of Industrial Organization* 407, 409 (1993).

134 Severin Borenstein, "The Evolution of U.S. Airline Competition," 6 *Journal of Economic Perspectives* 45, 53 (1992).

135 Ibid., p. 54.

136 See sources cited in William Evans and Loannis Kessides, "Structure, Conduct, and Performance in the Deregulated Airline Industry," *Southern Economics Journal,* 1991.

137 Steven Morrison and Clifford Winston, "Empirical Implications and Tests of the Contestability Hypothesis," 30 *Journal of Law and Economics* 53 (1987).

Chapter 11 (*Cont.*)

138 Michael Levine, "Airline Deregulation: A Perspective," 60 *Antitrust Law Journal* 687, 693 (1991).

139 Alfred Kahn, "Surprises from Airline Deregulation," 78 *AEA Papers and Proceedings* 316, 318 (1988).

140 See Ross, "The Economic Theory of Agency: The Principal's Problem," 63 *American Economic Review* 134 (1973).

141 Levine, "Airline Deregulation: A Perspective," op. cit.; see Levine, "Airline Competition in Deregulated Markets: Theory, Firm Strategy, and Public Policy," op. cit.

142 Levine, "Airline Competition in Deregulated Markets: Theory, Firm Strategy, and Public Policy," op. cit.

143 Alfred Kahn, "Market Power Issues in Deregulated Industries," 60 *Antitrust Law Journal* 857 (1991).

144 Elizabeth Bailey, David Graham, and Daniel Kaplan, *Deregulating the Airlines,* MIT Press, 1985, p. 153.

145 Levine, "Airline Deregulation: A Perspective," op. cit.

146 Levine, "Airline Competition in Deregulated Markets: Theory, Firm Strategy, and Public Policy," op. cit.

147 Ibid., pp. 472, 473.

148 Ibid., pp. 393, 418.

149 Charles Rule, "Antitrust and Airline Mergers: A New Era," speech before the International Aviation Club, Washington, D.C., March 7, 1989, pp. 15, 18.

150 Address by DOT Assistant Secretary Patrick Murphy before the ABA Section on Air and Space Law, San Francisco, July 10, 1998.

151 Joan Robinson, *The Economics of Imperfect Competition,* 1933.

152 Edward Chamberlin, *The Theory of Monopolistic Competition,* 1933.

153 See George Russell, "Flying among the Merger Clouds," *Time,* Sept. 29, 1986, pp. 56, 57.

154 See Paul Dempsey, "Antitrust Law and Policy in Transportation: Monopoly I$ the Name of the Game," 21 *Georgia Law Review* 505, 535 n. 182 (1987).

155 Hamburger, "Fares Rose with NWA's Dominance," op. cit.

156 As one commentator noted: "[E]ntry into the industry by new carriers seems remote, and entry onto new routes is far more difficult than many envisioned it would be with deregulation. Many airline observers thought that the 1978 deregulation of pricing and entry would make airline markets 'contestable.' That is, airlines could engage in 'hit-and-run' entry into each other's markets in response to profit opportunities—simply by shifting a plane from one route to another. Instead the evidence compiled in the USAir-Piedmont record, as well as a large body of solid research by economic and legal scholars in the past three years, demonstrates that incumbent airlines are frequently able to charge higher prices on routes where other carriers face barriers to entry." Margaret Guerin-Calvert, "Hubs Can Hurt on Shorter Flights, at Crowded Airports," *The Wall Street Journal,* Oct. 7, 1987, p. 28.

157 Interview with Alfred E. Kahn, *Antitrust,* Fall 1988, pp. 4, 6. In response to the question of whether there was "too much emphasis given to the absence of entry barriers and to the theoretical possibility of entry, as opposed to actual entry," Kahn recounted in the interview his support for pricing regulation in markets having but one or two carriers, as 85% of America's city-pairs today do: "Unquestionably. Certainly one of the assumptions behind airline deregulation was that entry would be relatively easy....The original deregulation bill retained a rate regulatory ceiling on any routes in which a single carrier accounted for 90 percent or more of business. As Chairman of the Civil Aeronautics Board, I testified onbehalf of a unanimous board which had adopted the posture of favoring deregulation, that the ceiling trigger should be changed to 70 percent. *We believed* that while *entry* should

Chapter 11 *(Cont.)*

be legally free and *would be relatively easy, we never thought that would provide adequate protection in markets* that are naturally monopolistic or oligopolistic—*that just won't support more than one or two carriers.* But what happened was that the ideologues began simplistically to parrot the word 'contestability' as though it were a substitute for looking at the realities, even if the realities were manifestly changing, even if survival of the new entrants was becoming more and more questionable, as more and more of them were going out of business, and even as it became clear that domination of hubs was increasingly unchallengeable by new entrants." Kahn also observed, "There is no question that increased concentration is associated with increased fares." Thomas Hamburger, "Fighting Back Begins as Costs Go Up, Up and Away," op. cit. Kahn has acknowledged that the time has come to consider price ceilings in markets dominated by a single carrier. "Ex-Official Suggests Lid on Air Fares," *Rocky Mountain News,* Nov. 5, 1987, p. 100. Said Kahn, "I don't reject the idea as a matter of principle. If price gouging gets bad enough, consumers will demand and deserve protection" (article by Thomas Hamburger cited above). See also "Safety and Re-Regulation of the Airline Industry: Hearings before the Senate Committee on Commerce, Science and Technology," 100th Congress, 1st Sess. (1987), pp. 159, 160. Kahn further said, "[T]he imperfections of competition I have identified suggest the possible desirability of maximum fares on inadequately competitive routes (which I advocated at the time not be abandoned)...." Alfred Kahn, "Airline Deregulation: A Mixed Bag, but a Clear Success Nonetheless," 16 *Transportation Law Journal* 239 (1988).

158 Power, op. cit., p. 73.

159 Melvin Brenner, "Airline Deregulation—A Case Study in Public Policy Failure," 16 *Transportation Law Journal* 189 (1988).

160 Don Phillips, "$3.1 Billion Airport at Denver Preparing for Rough Takeoff," *The Washington Post,* Feb. 13, 1994, p. A10.

161 "Woman of the Year Ginger S. Evans," *Engineering News-Record,* Feb. 11, 1994, p. 34.

162 See Paul Stephen Dempsey, Andrew R. Goetz, and Joseph S. Szyliowicz, *Denver International Airport: Lessons Learned,* McGraw-Hill, 1997.

163 "Greiner Says SEC Staff Case against Denver Consultants Not Legitimate," *Airports,* Dec. 19, 1993.

164 George Doughty, testimony before the U.S. House Transportation Subcommittee on Aviation, May 11, 1995.

165 "United Airlines, the dominant carrier in the Denver market, has been increasing its service to pressure Continental." Bill Mintz, "Denver a Victim of Continental's Fare Strategy," *Houston Chronicle,* July 8, 1994, p. 2. See also Paul Dempsey, "Rip United Airline's Hold from DIA," *Denver Business Journal,* August 25–31, 1995.

166 UAL Corp., 1992 Annual Report, p. 7 [emphasis supplied].

167 UAL Corp., 1993 Annual Report, p. 6 [emphasis supplied].

168 Jeffrey Leib, "GAO Study Encouraging for DIA's Financing," *The Denver Post,* Feb. 14, 1995, p. C3.

169 Continental Airlines, 1994 Annual Report, p. 19.

170 Michelle Mahoney, "Denver Exit Costly One for Continental Airlines," *The Denver Post,* July 10, 1994, p. 2D; "Continental to Close Denver Crew Bases This Fall," *Aviation Daily,* July 8, 1994, p. 38.

171 Mahoney, op. cit., p. A23.

172 Mintz, op. cit.; Michelle Mahoney, "Airline's Memo Jolts Morale," *The Denver Post,* March 31, 1994, p. C1.

173 City and County of Denver, Airport System Revenue Bonds, Series 1994 (Sept. 1, 1994); Michelle Mahoney, "Airline Changes Buffet Denver," *The Denver Post,* Feb. 13, 1994.

Chapter 11 (*Cont.*)

174 Leigh Fisher Associates analysis prepared for the City and County of Denver, 1994.

175 Jeffrey Leib, "United Joins Fare War," *The Denver Post,* Sept. 19, 1996, p. A1.

176 Michelle Mahoney and Jeffrey Leib, "UAL's Success Key to Denver," *The Denver Post,* July 13, 1994, p. C1.

177 See "ESG Aviation Services," *The Airline Monitor,* Nov. 1994, p. 4, and "ESG Aviation Services," *The Airline Monitor,* Sept. 1997, p. 9. Given the circuity of travel mandated by a hub system (the dominant megatrend on the deregulation landscape), a yield measure probably overstates the postderegulation decline in prices, for yields are based on revenue passenger miles (the revenue derived from passengers per mile flown, measured in cents per mile). Nonetheless, as a proxy for prices, it provides a rough approximation of long-term trends.

178 Charles Stein, "United-Delta Alliance Off for Now," *The Boston Globe,* April 25, 1998, p. E1.

179 Testimony of Alfred Kahn before the U.S. Senate Committee on Commerce, Science and Transportation, April 23, 1998.

180 Ibid.

181 Data calculated by Dr. Andrew Goetz of the University of Denver Intermodal Transportation Institute.

182 Ron Chernow, *Titan,* Random House, 1998.

183 "Unfair Airline Practices," *The New York Times,* April 7, 1998.

Index

About the Author

Dr. Paul Stephen Dempsey is Professor of Law and Director of the Transportation Law Program at the University of Denver, Vice Chairman and Director of Frontier Airlines, and the author of several books on aviation law. He co-authored *Denver International Airport: Lessons Learned*, also from McGraw-Hill.